PRINCIPLES OF

FINITE MATHEMATICS

PRINCIPLES OF

FINITE

MATHE

WILLIAM C. SWIFT
DAVID E. WILSON

Professors, Department of Mathematics
Wabash College
Crawfordsville, Indiana

PRENTICE-HALL, INC., Englewood Cliffs, New Jersey 07632

Library of Congress Cataloging in Publication Data

SWIFT, WILLIAM CLEMENT, (Date)
Principles of finite mathematics.

Includes index.
1. Mathematics—1961- I. Wilson, David E.,
(Date) joint author. II. Title.
QA39.2.S87 510 76-23399
ISBN 0-13-701359-0

PRINCIPLES OF
FINITE MATHEMATICS

WILLIAM C. SWIFT / DAVID E. WILSON

10 9 8 7 6 5 4 3 2 1

Printed in the United States of America

Prentice-Hall International, Inc., *London*
Prentice-Hall of Australia Pty. Limited, *Sydney*
Prentice-Hall of Canada, Ltd., *Toronto*
Prentice-Hall of India Private Limited, *New Delhi*
Prentice-Hall of Japan, Inc., *Tokyo*
Prentice-Hall of Southeast Asia Pte. Ltd., *Singapore*
Whitehall Books Limited, *Wellington, New Zealand*

CONTENTS

2 INTUITIVE PROBABILITY 36

3 MATHEMATICAL PROBABILITY 61

4 REPEATED TRIALS 108

5 STATISTICS 157

PART TWO MATRICES

6 GAMES AND MATRICES 181

7 VECTORS 242

PREFACE

Notwithstanding the heightened emphasis and fresh approaches that the subject has enjoyed for more than a generation of students, there are many serious thinkers, in and beyond college, who remain regrettably naive about the power and the dignity, the beauty and excitement, of mathematics. The profound involvement of mathematics in the world, few would question; but that it serves its role so well by dealing in abstract structures, aloof from the realm of applications, calls for a perspective that takes more than a little thought to appreciate. A chain of ideas has to be thoughtfully considered, and it is of little value for us to try to summarize in a few words what we think the links in this chain might be. What we want to say is best said through an example. The evolution of ideas in this book is our example.

Perhaps more than for any other discipline, the learning of mathematics ought to be a process of discovery; its truths lie within our own minds. This premise underlies our approach. We begin with gentle ideas that might occur to anyone—given a little inspiration—and let elaborations and variations develop naturally. The logic is down-to-earth; we are little distracted by subtle difficulties which intrude at a stage beyond the scope of the book.

Arguments for the results are usually within reach, so there are only a few places where the reader is asked to accept assertions purely on faith.

This book evolved from notes developed for a one–semester course. But the topics that naturally found a place would suffice for a full year course proceeding at a comfortable pace. Indeed, the division into Parts One and Two suggests a course over two terms devoted respectively to Probability and Matrices. The design of a one–semester course demands some judgments as to choice of topics and relative levels of emphasis. A simple formula would be to go as far as you can in Part One in the first half-semester; and then as far as you can go in Part Two. The treatment in Part Two depends only on the material through Sec. 3-4; and there are a number of natural stopping points in both parts. The main line of development in the book culminates with the duality theorem in Sec. 8-5. This result is offered as a worthy final objective of a one–semester course.

Except for the passage into Part Two, it is generally inadvisable to skip over portions of the material. But some acceleration may be gained by selective skimping, and effectively just reading some of the topics to preserve continuity, with little emphasis on the exercises. If the exposition has not left too great a burden on the teacher—and we've tried to reduce this burden as much as possible—then relatively little classroom time should be required for clarification. (Indeed, it is our hope that some will benefit from the book without access to an expert consultant at all.)

For their help, encouragement, and constructive criticism during the development of this material, we express gratitude to our colleagues at Wabash College, particularly to Professors Richard Bieberich, Stephen Schmutte, and Richard Strawn. Our thanks extend as well to the host of students who have sustained us by their good response to preliminary versions of the work. We are also indebted to the editors at Prentice-Hall and their consultants who in reviewing the manuscript have offered many suggestions for improvement, a fair number of which are reflected in the book.

William C. Swift
David E. Wilson

PART ONE

PROBABILITY

1

POSSIBILITIES

The seeds of a mathematical development can be found in the most commonplace situations.

> At the freshman get-together marking the beginning of the school year, three students are talking about a mutual acquaintance named Bob. Says one: "This semester Bob is taking Math, German, Econ, and English." Another reports: "He is pledged to a fraternity and is going out for football." And the third: "Bob intends to major in Economics and hopes to go to law school after graduation."

Apparently Bob's friends are interested in different aspects of his affairs. Looking at the nature of the comments, we might say that the one friend is reporting information from the Registrar's Office, and the others are relaying bits of data from the Student Activities Office and the Office of Career Services. The different offices have different purposes, and they collect, organize, and analyze different kinds of information.

Whatever the situation, the first step in a careful analysis is to decide just what the relevant features are that need to be taken into account.

1-1. Basic Outcome Analysis

There is some validity in the accusation that mathematicians, particularly textbook writers, often behave like the character in the joke who loses his dime in one spot but searches for it in another where the light is better. In keeping with this tradition, we shall exchange the review of Freshman Bob's affairs above for a process that is better defined, contrived though it is.

> Adam amuses himself with an ordinary deck of 52 playing cards while three observers look on. Repeatedly, Adam draws a card, shows it to the observers, replaces the card, shuffles, and draws again. An outsider appears and asks: "What was the last draw?" Says one observer: "It was a Club." Another: "It was a Five." And the third: "It was the Five of Clubs."

Evidently the three observers, whom we'll call O_1, O_2, and O_3, have different ideas about the important features of the card drawn. With an obvious bit of coding, their description of the outcomes in an imagined sequence of repetitions of the process is tabulated below:

Occurrence	O_1	O_2	O_3
1	C	5	5C
2	S	A	AS
3	C	Q	QC
4	H	3	3H

On each occurrence the perspective of observer O_1 is expressed by one of the four statements

C: A Club is drawn.
D: A Diamond is drawn.
H: A Heart is drawn.
S: A Spade is drawn.

The crucial property of this set of four statements is that one and only one is true on each occurrence of the process. Such a set is called a *basic outcome set.*

With respect to a given process, a *basic outcome set* is a set of statements such that on each occurrence of the process exactly one of the statements is true.

From the point of view of observer O_2, the basic outcome set contains thirteen statements, expressed as follows:

$$O_2: \{A, 2, 3, 4, 5, 6, 7, 8, 9, 10, J, Q, K\}.$$

The official position is that "A", for instance, is a convenient symbol for the statement "An Ace is drawn," but no harm will come from thinking of the "A" in more concrete terms: "an Ace."

In the interest of economy, the basic outcome set of observer O_2 is further compressed in the fashion

$$O_2: \{A, 2, 3, \ldots, Q, K\}.$$

In general, the "..." device is used to represent the other elements in a set or sequence in instances where the stated entries indicate a clear pattern.

In this manner the basic outcome set of observer O_3, containing fifty-two statements, is expressed as follows:

$$O_3: \{AC, AD, AH, AS, 2C, 2D, \ldots, KH, KS\}.$$

Relevant Statements and Truth Sets

Like a poker player trying to fill out a flush, observer O_1 in our example is interested only in the suit of the card drawn, and his basic outcome set {C, D, H, S} takes into account only the features of the outcome that he regards as relevant.

There are many statements that one can make about the outcome of Adam's process. For example, consider the two statements:

f: A face card is drawn.
r: The card drawn is red.

With respect to the basic outcome analysis of observer O_1,

$$O_1: \{C, D, H, S\},$$

the face-card statement f is not regarded as *relevant* because the basic outcomes do not contain the information to decide whether f is true or false. On the other hand, the basic outcomes do suffice to determine whether or not the card is red. So the red-card statement r is relevant with respect to the analysis of O_1.

With respect to a basic outcome set, a statement p about the outcome of the process is said to be *relevant* if each basic outcome contains enough information to decide whether p is true or false.

In the analysis of observer O_2,

$$O_2: \{A, 2, 3, \ldots, Q, K\},$$

the face-card statement f is relevant, whereas the red-card statement r is not. Of course, both statements are relevant with respect to the analysis of observer O_3, whose basic outcomes take into account both the rank and the suit of the card drawn.

Given a basic outcome analysis of a process, a relevant statement p separates the basic outcomes into those identifying p as true and those identifying p as false.

The subset P of a basic outcome set consisting of precisely those basic outcomes identifying a given relevant statement p as true is called the *truth set* of p.

Thus in the analysis of observer O_1,

$$O_1: \quad \{C, D, H, S\},$$

the truth set of the red-card statement r is

$$R = \{D, H\}.$$

On the other hand, with respect to the analysis of observer O_3,

$$O_3: \quad \{AC, AD, AH, AS, 2C, 2D, \ldots, KH, KS\},$$

we have

$$R = \{AD, AH, 2D, 2H, 3D, \ldots, KD, KH\}.$$

Statement r hasn't changed, but we have different truth sets R for the different basic outcome sets. Of course, in the analysis of observer O_2 the truth set of statement r is not defined simply because r is not relevant.

Although we shall continue to think in terms of statements, the things with which we formally deal are sets and subsets. From this point of view the two statements

A face card is drawn

and

The card drawn is a Jack, Queen, or King

are indistinguishable because they have the same truth set in any basic outcome analysis in which they are relevant.

Two statements are said to be *equivalent* if they must be either both true or both false on each occurrence of the process.

Thus with respect to any given basic outcome analysis, statements with identical truth sets are equivalent, and vice versa.

Example: Carol has three marbles, red, white, and blue, respectively, and she has two urns, Urn I and Urn II. The process is to put each marble into one or the other of the urns.

Not regarding the disposition of the blue marble as relevant, Carol analyzes the process by constructing the following basic outcome set:

$$S = \{\text{RI and WI, RI and WII, RII and WI, RII and WII}\}.$$

Problem: With the natural interpretation of the basic outcomes in the set S, determine which of the following statements are relevant, and if the statement is relevant, exhibit its truth set.

p_1: The red and white marbles go into the same urn.
p_2: Urn I receives at least one marble.
p_3: The white marble goes into Urn II.
p_4: All three marbles are deposited in Urn I.

Solution: Statements p_1 and p_3 are relevant, with truth sets as follows:

$$P_1 = \{\text{RI and WI, RII and WII}\},$$

$$P_3 = \{\text{RI and WII, RII and WII}\}.$$

Statement p_2 is not relevant because the basic outcome "RII and WII" does not tell us whether Urn I receives a marble or not, since the blue marble may or may not go into Urn I. Similarly, p_4 is not relevant because the basic outcome "RI and WI" doesn't contain enough information to decide whether all three marbles go into Urn I.

Strong and Weak Analyses

In Adam's process some recogintion would seem to be due observer O_3 for making a more comprehensive analysis than did the other two observers. The basic outcomes of O_3 contain all the information that either of the others record, and more. Which is to say, the basic outcomes of O_1 and O_2 are all relevant statements with respect to the basic outcome set of O_3. Here we say that the analysis of O_3 is *stronger* than the analyses of O_1 and O_2.

With respect to a given process, a basic outcome set A is said to be *stronger* than a basic outcome set B, and B is *weaker* than A, if each basic outcome in B is a relevant statement with respect to A.

By this stronger-weaker criterion the analyses of O_1 and O_2 are not comparable, one way or the other.

Since observer O_3 takes into account all of the reasonable properties of the outcome of Adam's one-card process, we identify his analysis as the natural strong basic outcome set, or, for short, *the strong analysis* (with emphasis on "the"). Of course, such a characterization presupposes implicit agreement on what is reasonable. For instance, in accepting the basic outcome set of O_3 as the strong analysis we deny the natural relevance of such statements as

The Ace of Spades is drawn upside down.

or

Adam draws the card with his left hand.

In the opening account about the program and activities of Freshman Bob, the idea of the strong analysis is not too meaningful since the reasonably relevant features are virtually inexhaustible. However, an advisor trying to suggest an appropriate mathematics course might imagine a weak but useful basic outcome set in which a typical statement is

The student had three years of high school math
and intends to major in economics.

In practical problems, the challenge is generally to construct the weakest analysis that is amenable to treatment and encompasses the features of the process relevant to the problem.

1-1 EXERCISES

1. Henry has five coins in his pocket, two nickels, two dimes, and a quarter. He takes out two coins. Interpret the following sets, and decide whether they satisfy the criterion to qualify as basic outcome sets:
   (a) {10¢, 15¢, 20¢, 30¢, 35¢}.
   (b) {At least one nickel, At least one dime}.
   (c) {Less than 25¢, More than 25¢}.

2. The process is to draw two marbles in turn from an urn containing two red and two white marbles. Which of the following sets of statements qualify as basic outcome sets?
   (a) {First draw is red, Second draw is red, Neither draw is red}.
   (b) {First draw is red, First draw is white}.
   (c) {Both draws are red, Neither draw is red, Exactly one of the draws is red}.

3. Paul and Steven each have a raisin cookie. The process is to count the raisins. Explain why the following set of four statements is not a basic outcome set, and create a basic outcome set by adjoining one or more statements.

   Each cookie has more than three raisins.

   Each cookie has less than three raisins.

   Each cookie has exactly three raisins.

   One of the cookies has more than three raisins and the other has less than three raisins.

4. John is to paint his car two-tone, the top one color and the bottom another. He chooses from three colors: red, black, and yellow. Indicate a basic outcome set for the choice of colors.

5. The process is to roll a die once. The basic outcome set is taken to be

   $D =$ {An odd number shows, An even number shows}.

   Explain why the statement

   The number 5 shows

   is not relevant in this analysis.

6. Clare's room is lighted by a floor lamp with a 3-way bulb of 100-, 200-, and 300-watt capacity and a table lamp with a 3-way bulb of 50-, 100-, and 150-watt capacity. The process is to determine how the room is to be lighted. Indicate basic outcome sets S_a, S_b, and S_c in which the only features of relevance are, respectively,

(a) The total wattage of illumination.
(b) Which of the lamps are turned on.
(c) The amount of illumination from the table lamp.

7. A coin is flipped three times. The basic outcome set is taken to be

$$W = \{\text{No heads, One head, Two heads, Three heads}\}.$$

(a) Which of the following statements are relevant in this analysis:

(i) At least one head occurs.
(ii) The second flip is a head.
(iii) At most two tails occur.

(b) Exhibit the truth sets for those statements in part (a) which are relevant.

8. A coin is flipped three times. The strong basic outcome set is represented as follows:

$$S = \{\text{HHH, HHT, HTH, HTT, THH, THT, TTH, TTT}\}.$$

Exhibit the truth sets for the following statements:
(a) A head comes up on the second flip.
(b) At least two heads occur.
(c) At most one tail occurs.

9. The process is to flip a coin twice.
(a) Construct the natural strong basic outcome set.
(b) Exhibit a weaker analysis of the process.

10. Two cards are drawn from an ordinary deck. Construct a weak basic outcome set—the fewer the number of basic outcomes the better—such that the following two statements are relevant.

Both draws are Clubs.

The second draw is not a Diamond.

1-2. Trees

Returning to the urn example of the previous section, we pose the following problem:

Construct the strong basic outcome set for Carol's process of depositing the three marbles, R, W, and B, in the two urns, I and II.

Let's start with a weak analysis and successively strengthen it until the reasonably relevant properties of the outcome have all been taken into account.

In response to the question "Into which urn does marble R go?" there is the set of two answers

R goes into I, R goes into II.

This set of two statements is, of course, a basic outcome set for Carol's process. As a pictorial code for this weak analysis, we introduce the graphical device shown in Fig. 1-1.

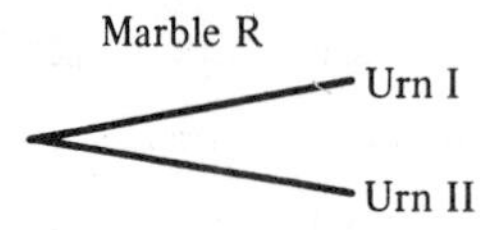

Figure 1-1.

Here we have a very simple example of what we call a *tree*. The tree will have to develop some if it is to represent the strong basic outcome set desired.

The second stage in the tree corresponds to the set of answers to the question "Now that we have taken marble R into account, into which urn does marble W go?" The possibilities are conveyed by adding second-stage components to the initial stage of the tree to obtain the diagram in Fig. 1-2.

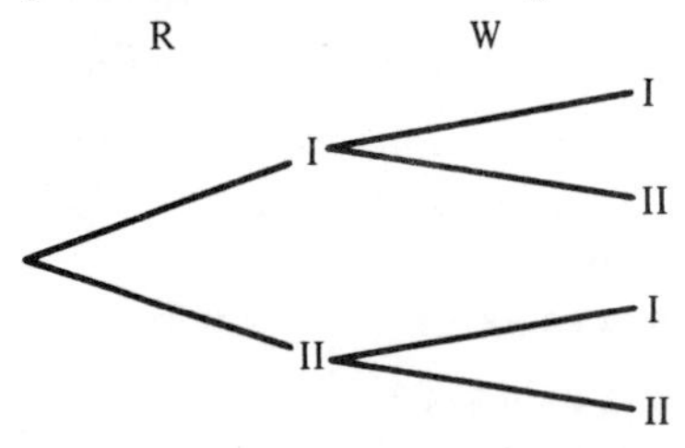

Figure 1-2.

The paths from left to right in this two-stage tree represent the elements of a basic outcome set which is stronger than the earlier two-element set but still weaker than the strong analysis we are seeking. For example, the path in Fig. 1-3 should be viewed as a convenient way of representing the statement "Marble R goes in Urn I and marble W goes in Urn II."

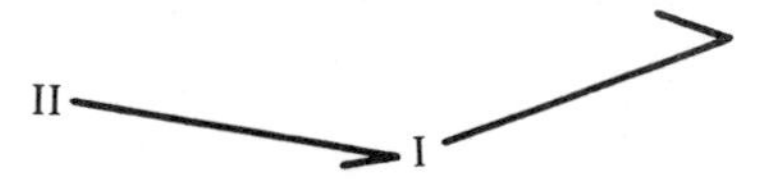

Figure 1-3.

Finally, the tree is completed by adding third-stage components to account for the disposition of marble B. The strong basic outcome set for Carol's process consists of the eight statements represented by the eight paths from left to right in the tree shown in Fig. 1-4.

Most of the terms pertaining to trees should be self-explanatory: *initial point*, *branch points*, *terminal points*, *branches*, *paths*, and *stages*. A single stage of branches and associated points emanating from the initial point or one of the branch points is called a *component*.

Trees furnish a handy device for analyzing many processes. In every case the initial component by itself must represent a basic outcome set for the

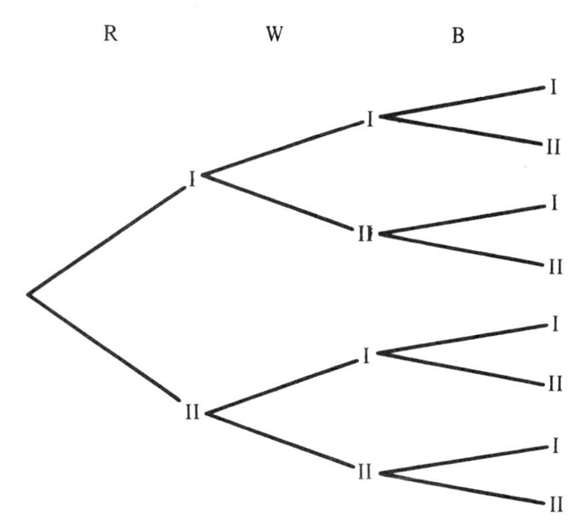

Figure 1-4 Tree for Carol's Process.

process, and the basic outcome analysis is strengthened with the addition of each component.

Example: Diane has three urns. Urn I contains a red marble and a green marble; Urn II contains a red marble and two green marbles; and Urn III contains two red marbles, a green marble, and a yellow marble. See Fig. 1-5.

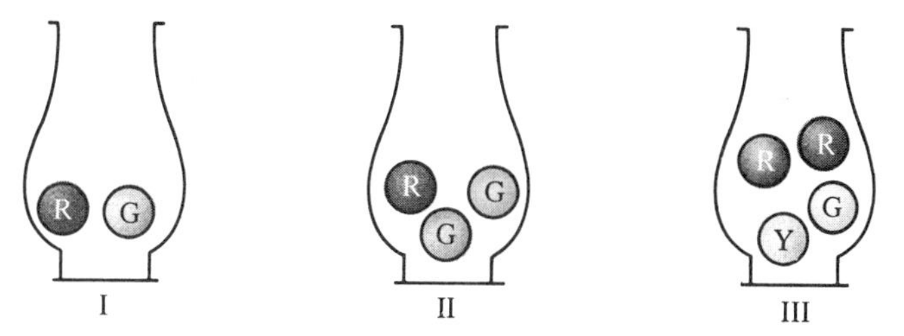

Figure 1-5.

The process is to select one of the urns and withdraw marbles from it one by one until a red marble is drawn.

Problem: Construct a tree to represent an analysis of Diane's process.

Solution: The tree is shown in Fig. 1-6. The initial stage represents the possibilities for the urn selected; the second stage takes into account the color of the first marble drawn, making no relevant distinction between marbles of the same color; and so on. Each path is terminated once a red marble is drawn. The terminal points have been encircled for easier recognition.

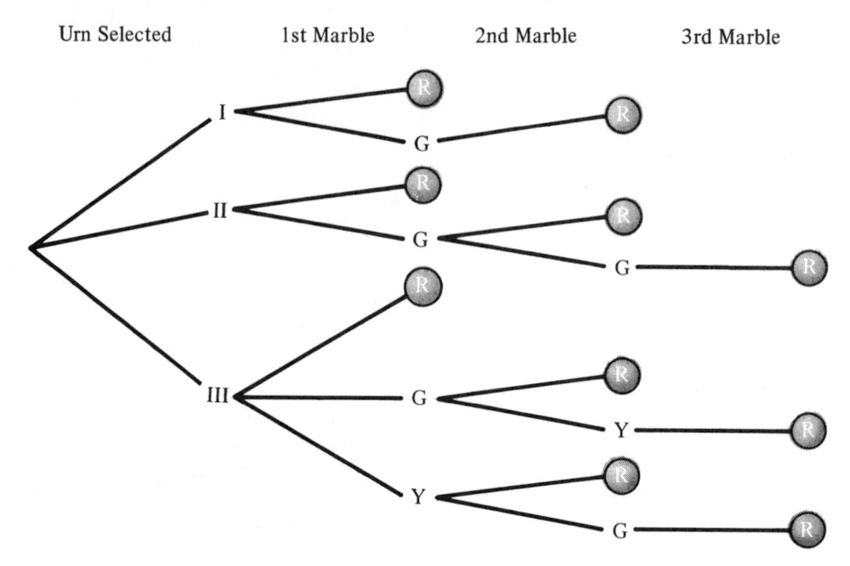

Figure 1-6 Diane's Tree.

Models

To judge from the illustrations thus far, one might guess that the exposition will be overly biased toward serving silly problems concerned with such things as drawing cards or putting marbles in urns or taking them out. Actually, the applications of such examples are more far-reaching than one might suspect. To illustrate the point, we consider the following situation:

> In an economics class of twenty-five students, the instructor questions each student to determine three things: whether the student is also enrolled in a math course, whether he intends to major in economics, and whether he has had prior courses in economics.

Taking the process to be the questioning of a single student, the basic outcome set is represented by the tree shown in Fig. 1-7.

A comparison of this tree with the earlier tree of Fig. 1-4 reveals a close abstract relationship between the instructor's questioning of a student and Carol's process of depositing the three marbles, R, W, and B, into the two urns, I and II. We can think of the three categories in the questioning as corresponding, respectively, to the three marbles in Carol's process (read "$\sim$" as "corresponds to"):

Categories		*Marbles*
Math course	$\sim$	Red marble
Econ major	$\sim$	White marble
Prior econ	$\sim$	Blue marble

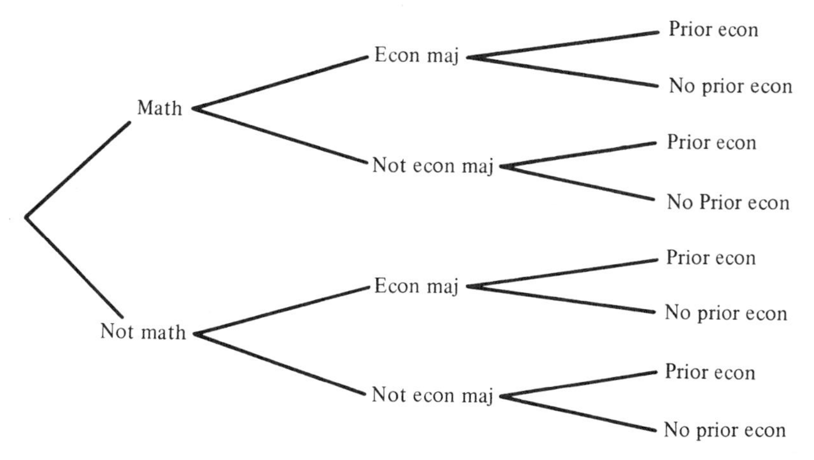

Figure 1-7.

Similarly, the student's possible responses to the respective questions correspond to the two urns, I and II:

Answers		*Urns*
Yes (fits the category)	~	Urn I
No (doesn't fit the category)	~	Urn II

With this association of ingredients in the two processes, each relevant statement in the "student-questioning" process corresponds to a relevant statement in Carol's process. For example,

The student questioned intends to major in economics.

corresponds to

The white marble goes into Urn I.

In particular, the basic outcomes in the one process correspond to basic outcomes in the other. With this association, corresponding relevant statements have truth sets composed of corresponding basic outcomes. For instance, consider the statement

p : The student answers "yes" to exactly two of the three questions.

The truth set of p contains the three basic outcomes

Math, Econ maj, and No prior econ,
Math, Not econ maj, and Prior econ,
Not math, Econ maj, and Prior econ.

The corresponding statement about Carol's process, which we label as p' (read "little p prime"), is as follows:

p' : Urn I receives exactly two marbles.

The three elements in the truth set of p',

RI, WI, and BII,
RI, WII, and BI,
RII, WI, and BI,

correspond, respectively, with the three elements in the truth set of p.

The overall experience of the instructor in questioning the twenty-five students corresponds to a certain sequence of twenty-five occurrences of Carol's process. Insofar as the relevant features of the student-questioning process are concerned, we may set up the associations as given, and turn our attention to Carol's urn process. Whatever we can say about Carol's process can be translated back into a statement about the student-questioning process. We say that Carol's process is an *urn model* for the student-questioning process.

There are many practical problems which are well served by urn models, or playing-card models, or dice models, and other such uncomplicated processes. Thus the background gained in analyzing problems of this sort furnishes a foundation for the analysis of a host of honestly encountered "real-world" problems.

1-2 EXERCISES

1. The process is to draw two marbles in turn from an urn containing two red and two white marbles. The basic outcome analysis is given by the tree in Fig. 1-8.
   (a) Translate the uppermost path as shown in Fig. 1-9 into an ordinary statement describing the outcome.
   (b) By translating the respective paths into statements, present the basic outcome set represented by the tree as a set of statements.

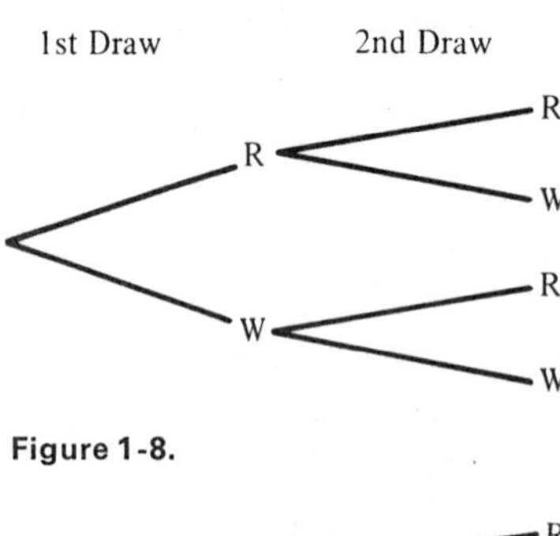

Figure 1-8.

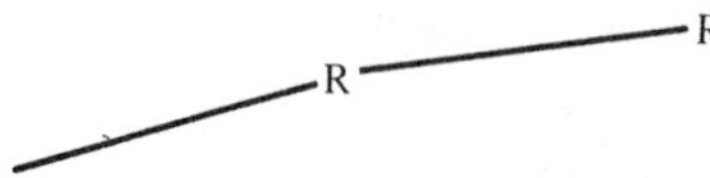

Figure 1-9.

2. Helen has two urns, Urn I containing one red marble and two white marbles and Urn II containing two red marbles and one white marble. The process is to choose an urn and withdraw two marbles in succession. Assuming no relevant distinction between marbles of the same color, construct a tree to represent the basic outcome analysis.

3. Two cards are drawn in order from an ordinary deck. The only feature of relevance is whether the cards drawn are red or black.
   (a) Construct a tree to represent the basic outcome set.
   (b) Explain why the tree in Fig. 1-10 is not a correct response to part (a).

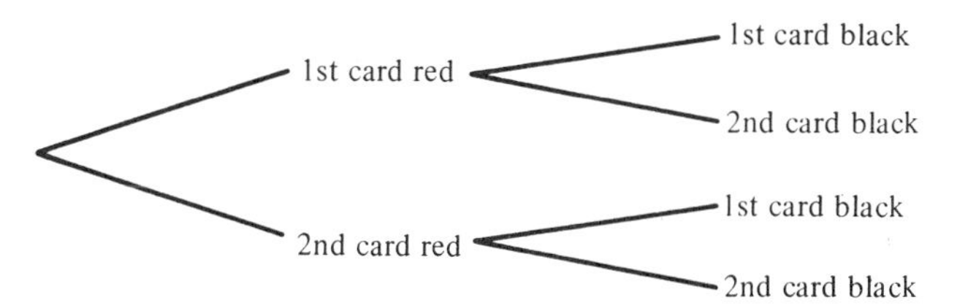

Figure 1-10 Incorrect Tree.

4. Two cards are drawn in order from an ordinary deck.
   (a) Construct a tree to represent a weak basic outcome analysis with respect to which the following two statements are relevant:

   The first card is not a Heart.
   The second card is a Spade.

   (b) Express the elements of the basic outcome set conveyed in part (a) as statements.

5. A coin is flipped three times. Construct a tree that represents the strong basic outcome analysis of the process. Then indicate the paths comprising the truth sets of the following statements by placing an *a*, *b*, or *c* alongside the terminal points.
   (a) The second flip is a Head.
   (b) The third flip is the same as the first.
   (c) At most one Tail occurs.

6. Mike flips a coin until either he obtains two heads in a row or he has obtained a total of three tails. Construct a tree to represent the strong basic outcome analysis.

7. A professor gives oral examinations and follows the practice of asking questions until the student correctly answers two in a row or until a total of three unsatisfactory answers are given. Construct a tree to represent the strong basic outcome analysis.

8. (Dependent on Probs. 6 and 7) Label the terminal points of the paths in the tree of Prob. 7 by placing 1, 2, 3, etc., alongside. Then indicate an interpretation of the process in Prob. 6 as a "coin-flipping" model

of the "oral-examination" process by labeling corresponding terminal points in the two trees with the same number.

As you have set up the correspondence, state the coin-flipping outcome that corresponds to the following oral-examination outcome:

The student gives a satisfactory answer
to only one question.

9. Larry and Matt go duck hunting. The legal limit—which of course they will respect—is five ducks per hunter. In the analysis of their adventure, a typical basic outcome is

Larry bags two ducks and Matt bags three.

(a) Interpret the process of rolling a die twice as a "dice" model of the duck-hunting trip. Indicate your interpretation by exhibiting the "Dice" statements corresponding to the following "Duck-hunting" outcomes:
i : Both boys come home empty-handed.
ii : Larry bags two ducks and Matt bags three.
iii : Together they bag a total of five ducks.
iv : Larry bags more ducks than Matt does.

(b) Describe a process involving marbles and urns which furnishes an urn model of the duck-hunting trip.

1-3. Functions

We now carry the concept of *model building* a crucial step further. What we need for a respectable mathematical treatment of a problem is a *mathematical model.* To this end we must abstract the real-world interpretations out of the problem and confine ourselves to the intellectual structural concepts involved.

A good vehicle for discussion is Carol's process of assigning the three marbles to the two urns. She has a set M of three marbles and a set U of two urns, and the process is to deposit each of the marbles in one or the other of the urns. If we weed out the real-world interpretations, we are left with a purely mathematical process: Given a set M of three elements and a set U of two elements, the process is to associate each of the elements of set M with one of the elements of set U. Such an association of the elements of M with elements of U is said to be a *function* mapping set M into set U. In the finite setting at least, all functions can be thought of as abstractions of the process of putting marbles into urns.

Definition: Given two sets X and Y, a method of associating each element of the set X with exactly one element of the set Y is said to constitute a *function* mapping X into Y.

The set X is called the *domain* of the function, and the set Y is called the *image space*. If the element x_0 in X is associated with the element y_0 in Y, we say that x_0 is *mapped* into y_0 and that y_0 is the *image* of x_0.

Suppose the economics instructor of the previous section compiles a record of the number of credit hours for which each of his twenty-five students is enrolled. For simplicity we assume that the number of credit hours is a whole number between 0 and 30, inclusive. If we identify the students as $E_1, E_2, \ldots, E_{25}$, then the instructor's record determines a function mapping the set

$$X = \{E_1, E_2, \ldots, E_{25}\}$$

into the set

$$Y = \{0, 1, 2, \ldots, 30\}.$$

An urn model of the function is given by the corresponding assigmnent of a set of twenty-five marbles, identified as $E_1, E_2, \ldots, E_{25}$, to a set of thirty-one urns, numbered 0, 1, 2, . . . , 30.

Comment: The reader's past experience with the concept of functions may have been in the context of equations such as

$$y = x^2$$

in which the domain and image space are not finite sets. What this equation does is associate each number x with a corresponding number y. For example, 2 is mapped into 4, -2 is also mapped into 4, $\frac{1}{3}$ is mapped into $\frac{1}{9}$, and $\sqrt{5}$ is mapped into 5. The validity of an urn model for this function is questionable, but if one thinks he can envision a different marble and a different urn for each number, then $y = x^2$ dictates to which urn each marble is assigned, e.g., marble number $-\frac{1}{3}$ goes into urn number $\frac{1}{9}$.

A function can often be efficiently communicated as a certain set of ordered pairs. For example, one of the basic outcomes in Carol's process is as follows:

The red marble goes into Urn II, the white marble into Urn I, and the blue marble into Urn II.

In the mathematical interpretation we have two sets,

$$M = \{\text{R, W, B}\} \quad \text{and} \quad U = \{\text{I, II}\}.$$

Corresponding to the basic outcome above we have the function F mapping M into U:

F maps R into II, W into I, and B into II.

This function is represented as a set of ordered pairs in the following manner:

$$F \sim \{(\text{R, II}), (\text{W, I}), (\text{B, II})\}.$$

In general a function mapping a set X into a set Y corresponds to a set of ordered pairs of the form (x, y), where x is an element of X and y is an element of Y. To qualify as a function it is required that each element in the set X appear exactly once as the first entry in one of the pairs.

Example: Given the sets

$$X = \{1, 2, 3\}, \qquad Y = \{1, 3, 5, 7\},$$

which of the following determine functions mapping X into Y:

a. F_a maps 1 into 7, 2 into 5, and 3 into 3.
b. $y = 2x - 1$.
c. $F_c \sim \{(1, 1), (1, 3), (2, 7), (3, 5)\}$.
d. $F_d \sim \{(1, 1), (2, 2), (3, 3)\}$.
e. $F_e \sim \{(1, 7), (2, 7), (3, 7)\}$.
f. $F_f \sim \{(1, 1), (3, 3)\}$.

Solution: F_a and F_e are acceptable as functions since each element in X has been associated with a unique element in Y, and if we interpret part b in the natural manner, i.e., each number x in X is associated with the number $y = 2x - 1$, then part b also determines a function mapping X into Y. F_c is not a function because the element "1" in X is associated with two different elements in Y. F_d does not associate the element "2" in X with an element in Y, so F_d does not qualify as a function mapping X into Y, and F_f fails for the same reason.

The Set of Functions Mapping *X* into *Y*

In the example above, with

$$X = \{1, 2, 3\} \quad \text{and} \quad Y = \{1, 3, 5, 7\},$$

it would be tedious to list all of the functions mapping X into Y. But it's not difficult to imagine a tree whose paths represent the functions, and the simple form of the tree makes it easy to compute the total number of such functions.

We are concerned with the set of all functions mapping a set X of three elements into a set Y of four elements. The nature of the elements in X and Y is of no consequence in the discussion; it's less distracting to consider the general case in which domain and image space are given as

$$X = \{x_1, x_2, x_3\} \quad \text{and} \quad Y = \{y_1, y_2, y_3, y_4\}.$$

In the tree of Fig. 1-11 the functions mapping X into Y are represented by the paths in a tree of three stages with four branches in each component. The stages correspond to the respective elements of X, and the branches in the components correspond to the possible images from the set Y. The tree

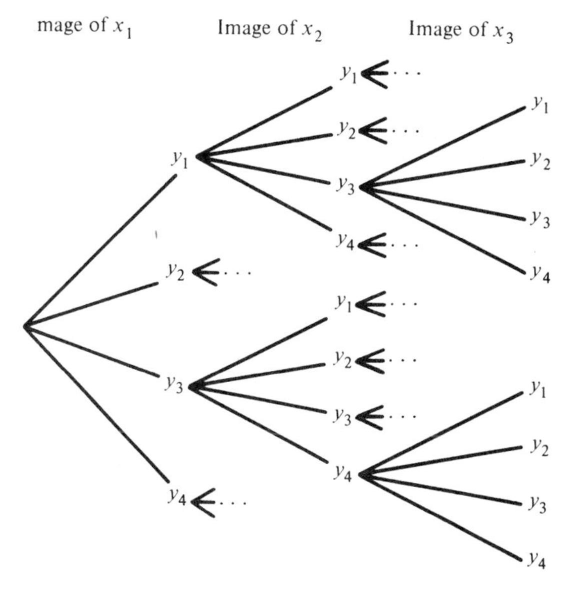

Figure 1-11.

is a little large to draw in detail, but its overall form should be apparent from the paths that are shown.

Each path in the tree represents a particular function mapping the set $X = \{x_1, x_2, x_3\}$ into the set $Y = \{y_1, y_2, y_3, y_4\}$, and vice versa. For instance, consider the path in Fig. 1-12. This path represents the function that associates x_1 with y_3, x_2 with y_4, and x_3 with y_2—or as a set of ordered pairs:

$$\{(x_1, y_3), (x_2, y_4), (x_3, y_2)\}.$$

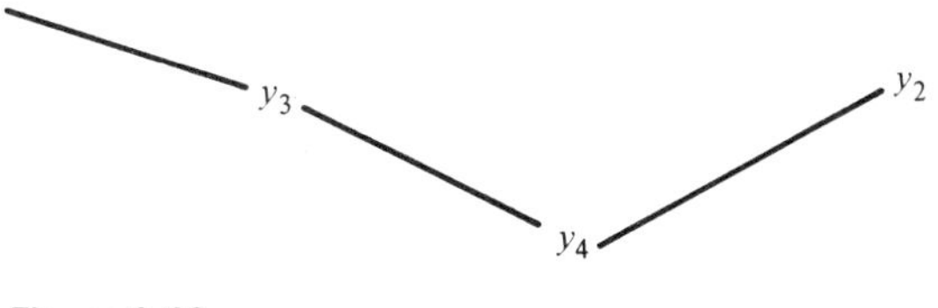

y_3 y_4 y_2

Figure 1-12.

The total number of paths from left to right in the tree of Fig. 1-11 is easy to compute.. Since there are four branches in the initial component and four branches in each second-stage component, the number of distinct paths through the first two stages is $4 \times 4 = 16$, and since there are four branches along which each of the two-stage paths may be completed, the total number of paths in the tree is

$$4 \times 4 \times 4 = 4^3 = 64.$$

We conclude that there are exactly sixty-four distinct functions mapping a given set of three elements into a given set of four elements.

The Multiplicative Principle

The argument for counting the number of paths in a tree in which the components at any given stage have the same number of branches can be generalized. If there are n_1 branches in the initial component and n_2 branches in each second-stage component, then the number of distinct paths through the first two stages is $n_1 \times n_2$, and if there are then n_3 branches in each third-stage component, the number of paths through the first three stages is $n_1 \times n_2 \times n_3$, and so on. The general result is as follows:

> **The Multiplicative Principle:** Given a tree of k stages in which the initial component consists of n_1 branches, each second-stage component consists of n_2 branches, each third-stage component consists of n_3 branches, and so on, with the final kth-stage components each consisting of n_k branches. Then the total number of paths in the tree is
>
> $$n_1 \times n_2 \times n_3 \times \cdots \times n_k.$$

There are many processes which can be analyzed by such well-disciplined trees, often characterized by an analysis of the following form: Part one of the process can occur in n_1 ways, and then part two can occur in n_2 ways, and so on, with part k occurring in n_k ways. The total number of basic outcomes is $n_1 \times n_2 \times \cdots \times n_k$. In the function-counting problem above, the image of x_1 can be any one of the four elements of Y, and then the image of x_2 can be any one of the four elements of Y, and the image of x_3 can then be any one of the four elements of Y. Thus the total number of functions is $4 \times 4 \times 4 = 64$.

The Number of Functions, in General

To consider the general case, let the domain X and image space Y be arbitrary finite sets, of k and n elements, respectively:

$$X = \{x_1, x_2, \ldots, x_k\} \quad \text{and} \quad Y = \{y_1, y_2, \ldots, y_n\}.$$

In the manner of Fig. 1-11 the set of functions mapping X into Y is represented by a tree of k stages, corresponding to the k elements in X, with n branches in each component, corresponding to the n elements of Y into which the respective element in X may be mapped. As in Fig. 1-11, each path in the tree is interpreted as representing a particular function mapping X into Y and vice versa. By the multiplicative principle the total number of paths in the tree is $n \times n \times \cdots \times n$ (k factors), or n^k.

It follows that there are exactly n^k distinct functions mapping a given set of k elements into a given set of n elements.

Example: Cathy has five marbles and three urns. In how many ways can she put the marbles in the urns?

Solution: The basic outcomes in the strong analysis of Cathy's process correspond to the functions mapping a set $M = \{m_1, m_2, m_3, m_4, m_5\}$ of five elements into a set $U = \{\text{I, II, III}\}$ of three elements. Thus the total number of possibilities is 3^5, or 243. The same conclusion is reached by arguing that there are three possible urns to which marble m_1 can be assigned, and then three possible urns to which marble m_2 can be assigned, and similarly for m_3, m_4, m_5.

Example: Suppose we define a 5-letter "word" to be any sequence of five letters, such as "happy" or "xtqsq." How many 5-letter words can be formed from the twenty-six letters of the alphabet?

Solution: The set of basic outcomes for the process of forming a 5-letter word can be visualized as a tree in which the branches in the initial component correspond to the first letter in the word, the branches in the second-stage components correspond to the second letter, and so on. Thus there are five stages with twenty-six branches in each component. It follows from the multiplicative principle that the total number of paths, and the total number of possible words, is 26^5 (which we won't bother to compute).

To analyze the problem in terms of functions, we identify the domain as consisting of the five positions in the word, __ __ __ __ __, with the image space consisting of the twenty-six letters in the alphabet. A function mapping this domain into this image space determines a unique word and vice versa. Thus the number of words is equal to the number of functions mapping a set of five elements into a set of twenty-six elements—again, 26^5.

1-3 EXERCISES

1. The paths in the tree of Fig. 1-13 represent the four functions mapping the set $X = \{a, b\}$ into the set $Y = \{0, 1\}$. For each path, express the corresponding function as a set of ordered pairs.

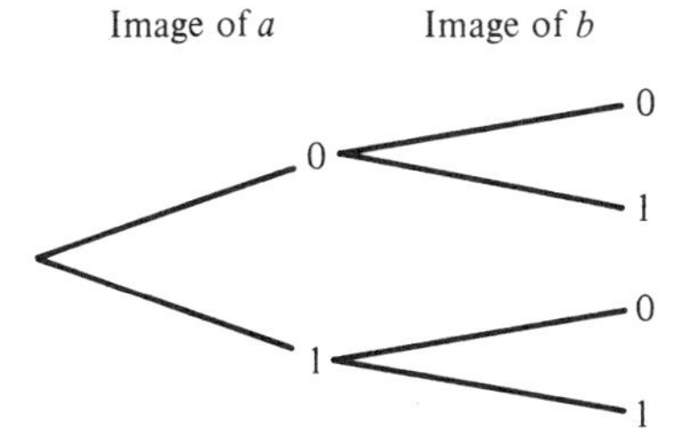

Figure 1-13.

2. (a) Construct a tree to represent the functions mapping $S = \{a, b\}$ into $T = \{1, 2, 3\}$.
   (b) Express the functions of part (a) as sets of ordered pairs.

3. List all of the functions mapping the set $A = \{1, 2, 3\}$ into the set $B = \{1, 2\}$ in their respective "set of ordered pairs" representation.

4. Observers O_1, O_2, and O_3 take a break and are offered coffee, tea, or milk. Compute the number of possibilities if each accepts one and only one drink.

5. Let $X = \{1, 2, 3, 4\}$ and $Y = \{1, 2, 3, \ldots, 10\}$. Which of the following qualify as functions mapping X into Y?
   (a) F_a maps 1 into 3, 3 into 5, 2 into 4, and 4 into 7.
   (b) $y = 3x - 2$.
   (c) $y = 2x - 3$.
   (d) $F_d \sim \{(1, 2), (3, 4)\}$.
   (e) $F_e \sim \{(1, 3), (2, 6), (3, 9), (4, 12)\}$.

6. In how many ways can four marbles of different colors be put into three urns, I, II, and III?

7. How many functions are there mapping the set $\{1, 2, 3, 4, 5\}$ into the set $\{0, 1, 2, \ldots, 9\}$?

8. A lock on a bicycle chain works on the principle that each of 4 dials are set at a particular digit from 0 to 9, inclusive. How many locks can the company manufacture before the same combination must open more than one lock?

9. Let X be the set of positive integers between 100 and 1000 exclusive, i.e.,

$$X = \{101, 102, 103, \ldots, 999\}.$$

   (a) How many numbers in the set X do not contain the digit zero?
   (b) How many numbers in the set X have the property that no digit occurs more than once?

10. Construct a tree to represent those functions mapping $A = \{1, 3, 4, 7\}$ into $B = \{2, 4, 6\}$ which satisfy the following two requirements:

    i: The image of 1 is 2.

    ii: Each element of the image space B is the image of at least one element in the domain A.

    Then count the paths in your tree to determine the total number of such functions.

1-4. Permutations and Combinations

A variation on the function-counting problem is posed by the following process:

Mark has a set of five marbles, $M = \{m_1, m_2, m_3, m_4, m_5\}$, and a set of three urns, $U = \{\text{I, II, III}\}$. The process is to put exactly one of the marbles into each of the urns.

In terms of functions, Mark's process interchanges the prior roles of marbles and urns. Here the basic outcomes may be interpreted as functions mapping the set of three urns into the set of five marbles, rather than the other way around. For example, the function

$$f \sim \{(\text{I}, m_3), (\text{II}, m_2), (\text{III}, m_5)\}$$

corresponds to the outcome "Marble m_3 goes into Urn I, m_2 into II, and m_5 into III."

The novel feature presented by Mark's process is due to the fact that two urns cannot receive the same marble. Thus if Urn I receives marble m_3, then Urn II cannot also receive m_3. In the function model we are limited to functions with the property that the images of the elements in the domain, $\{\text{I, II, III}\}$, are distinct elements of the image space. Such functions are said to be *one-to-one*.

Definition: A function mapping a set X into a set Y is said to be *one-to-one* if distinct elements in X are mapped into distinct elements in Y. Which is to say, no element in the image space Y is the image of more than one element in the domain X.

The number of basic outcomes in the strong analysis of Mark's process is the number of one-to-one functions with domain $U = \{\text{I, II, III}\}$ and image space $M = \{m_1, m_2, m_3, m_4, m_5\}$. The tree representing this set of functions is indicated in Fig. 1-14.

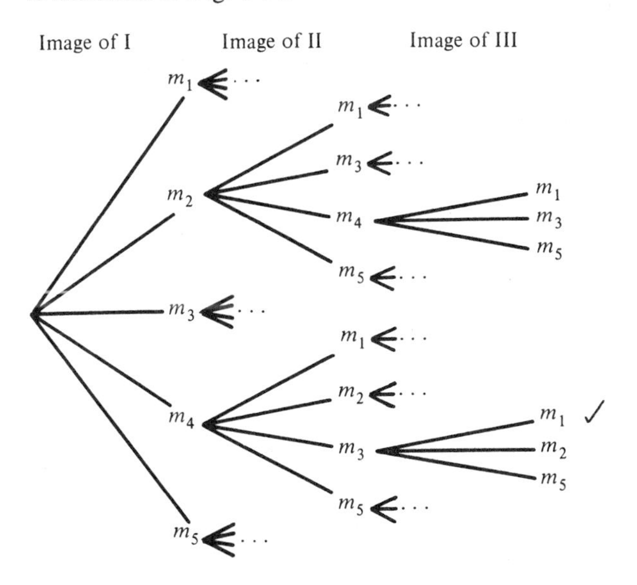

Figure 1-14.

The particular path checked in the tree represents the function

$$g \sim \{(\text{I}, m_4), (\text{II}, m_3), (\text{III}, m_1)\}.$$

In terms of Mark's process, this path corresponds to the outcome "Urn I receives marble m_4, II receives m_3, and III receives m_1."

The tree of Fig. 1-14, representing the one-to-one functions mapping the set U of three elements into the set M of five elements, has three stages, corresponding to the three elements of the domain U. The initial component has five branches corresponding to the five elements of M into which the element I in the set U may be mapped. But each second-stage component, representing the possible images of the element II, has only four branches because the one-to-one requirement forbids us to map II into the element that has served as the image of I in the first stage. Similarly, the third stage components have just three branches. By the multiplicative principle the number of paths in the tree is

$$5 \times 4 \times 3 = 60.$$

Thus there are sixty one-to-one functions mapping a set of three elements into a set of five elements. And there are sixty ways for Mark to assign exactly one of his five marbles to each of the three urns.

If we want to count the one-to-one functions mapping a set $X = \{x_1, x_2, \ldots, x_k\}$ of k elements into a set $Y = \{y_1, y_2, \ldots, y_n\}$ of n elements, we imagine a tree of k stages, corresponding to the elements in the domain X. The initial component has n branches, corresponding to the n elements of Y into which x_1 may be mapped. The second-stage components have $n - 1$ branches, corresponding to the $n - 1$ candidates remaining as possible images of x_2. Similarly, the third-stage components have $n - 2$ branches, because in each such component two of the elements of the image space Y have already been employed in the first two stages—and so on, with the final kth-stage components having $n - (k - 1)$ branches.

According to the multiplicative principle, the number of distinct functions mapping a set of k elements in a one-to-one fashion into a set of n elements is

$$n \times (n - 1) \times (n - 2) \times \cdots \times [n - (k - 1)].$$

We observe that there are k factors in the above product. Of course, k must be less than or equal to n in order to have any one-to-one functions at all. For instance, there are no one-to-one functions mapping a set of five elements into a set of three elements. Happily, the formula produces the value 0 when k is greater than n.

Example: How many 5-letter words can be formed from the twenty-six letters of the alphabet if no letter is allowed to appear more than once in any given word?

Solution: As in the earlier example, we are concerned with functions mapping the set of five positions, _ _ _ _ _, into the set of twenty-six letters. But the nonrepetition requirement confines us to one-to-one functions. From the foregoing argument the number of one-to-one functions mapping a set of $k = 5$ elements into a set of $n = 26$ elements is

$$26 \times 25 \times 24 \times 23 \times 22.$$

And this is the number of admissible words.

Permutations

A one-to-one function mapping the set

$$X = \{1, 2, 3, 4, 5\}$$

into a set Y containing five or more elements may be interpreted as an arrangement in order of some five of the elements of Y. For example let

$$Y = \{p, q, r, s, t, u, v\}.$$

Now consider the one-to-one function mapping X into Y given by the set of ordered pairs

$$\{(1, r), (2, p), (3, t), (4, u), (5, q)\}.$$

This function identifies with the arrangement

$$\text{“}r, p, t, u, q.\text{”}$$

Arrangements of elements from a set are called *permutations*. Conventionally, the symbol $P(n, k)$ represents the total number of permutations of n elements, k at a time. In other words,

$P(n, k)$ is the number of ways one can select k elements from a set S of n elements and arrange them in order.

An arrangement in order of k elements from a set S identifies with a one-to-one mapping of the set

$$K = \{1, 2, 3, \ldots, k\}$$

into the set S. Thus $P(n, k)$ is the number of one-to-one mappings of the set K of k elements into a set S of n elements. By the previous argument,

$$\begin{aligned} P(n, k) &= n \times (n - 1) \times (n - 2) \times \cdots \times (n - (k - 1)) \\ &= n \times (n - 1) \times (n - 2) \times \cdots \times (n - k + 1). \end{aligned}$$

The symbol $n!$ (read "n-factorial") is a universal shorthand for the number

$$n! = 1 \times 2 \times 3 \times \cdots \times n.$$

Thus

$$P(5, 5) = 5 \times 4 \times 3 \times 2 \times 1 = 5!,$$

and, in general,

$$P(n, n) = n!.$$

The factorial notation gives us a symbolic economy in expressing $P(n, k)$. For example, consider

$$P(26, 5) = 26 \times 25 \times 24 \times 23 \times 22.$$

If we want to compute $P(26, 5)$, we simply carry out the multiplication. However, if the aim is to express $P(26, 5)$ as an indicated operation, we may complicate the arithmetic as follows:

$$P(26, 5) = 26 \times 25 \times 24 \times 23 \times 22$$
$$= \frac{(26 \times 25 \times 24 \times 23 \times 22) \times (21 \times 20 \times \cdots \times 3 \times 2 \times 1)}{(21 \times 20 \times \cdots \times 3 \times 2 \times 1)}.$$

Thus in the factorial notation,

$$P(26, 5) = \frac{26!}{21!}.$$

In the general case,

$$P(n, k) = n \times (n-1) \times (n-2) \times \cdots \times (n-k+1)$$
$$= \frac{[n \times (n-1) \times \cdots \times (n-k+1)] \times [(n-k) \times \cdots \times 3 \times 2 \times 1]}{(n-k) \times \cdots \times 3 \times 2 \times 1}.$$

Thus the number of permutations on n elements, k at a time, is given by

$$P(n, k) = n \times (n-1) \times \cdots \times (n-k+1)$$
$$= \frac{n!}{(n-k)!}.$$

It is conventional to set

$$0! = 1.$$

With this convention the factorial expression for $P(n, k)$ gives the correct result for the case in which $k = n$:

$$P(n, n) = \frac{n!}{(n-n)!} = \frac{n!}{0!} = \frac{n!}{1} = n!.$$

Combinations

With the permutations formula in hand we are ready to derive its inseparable companion: the combinations formula. It is this formula that will have the most important role in the applications to follow. Again, there is a simple urn model:

Tony has a set of five marbles $M = \{m_1, m_2, m_3, m_4, m_5\}$ and a single urn. The process is to put exactly three of the marbles into the urn.

It's too bad Tony isn't selecting three of the five marbles and arranging them in order, because then the basic outcomes can be identified as the permutations of five things, three at a time. The number of such permutations is

$$P(5, 3) = 5 \times 4 \times 3 = 60.$$

Among the sixty permutations of the five marbles, three at a time, are

"m_4, m_1, m_3" and "m_1, m_3, m_4."

But whereas these two arrangements count as different permutations, they do not qualify as different basic outcomes in Tony's process. The wording does not suggest any relevant distinction between putting marbles m_4, m_1, and m_3 in the urn and putting marbles m_1, m_3, and m_4 in the urn; at least we don't want to make a distinction between these outcomes.

The basic outcomes in Tony's process correspond to the subsets that can be formed by selecting 3 elements from the 5-element set

$$M = \{m_1, m_2, m_3, m_4, m_5\}.$$

Conventionally, the number of k-element subsets that can be formed from a set of n elements is represented by the symbol $C(n, k)$. In this context, the subsets are commonly called *combinations*; we say that $C(n, k)$ is the number of combinations of n things, k at a time. The number of possibilities in Tony's process is given by $C(5, 3)$.

To find the value of $C(5, 3)$, we make a fresh attack on the problem of finding $P(5, 3)$. The key to the argument is in the characterization of $P(5, 3)$ as the number of ways of choosing three elements from a set of five elements and then arranging them in order. In particular, for the 5-element set

$$M = \{m_1, m_2, m_3, m_4, m_5\}$$

we arrive at the permutation "m_4, m_1, m_3" by first choosing the subset $\{m_1, m_3, m_4\}$ and then selecting the arrangement "m_4, m_1, m_3."

In Fig. 1-15 the permutations of five things, three at a time, are indicated

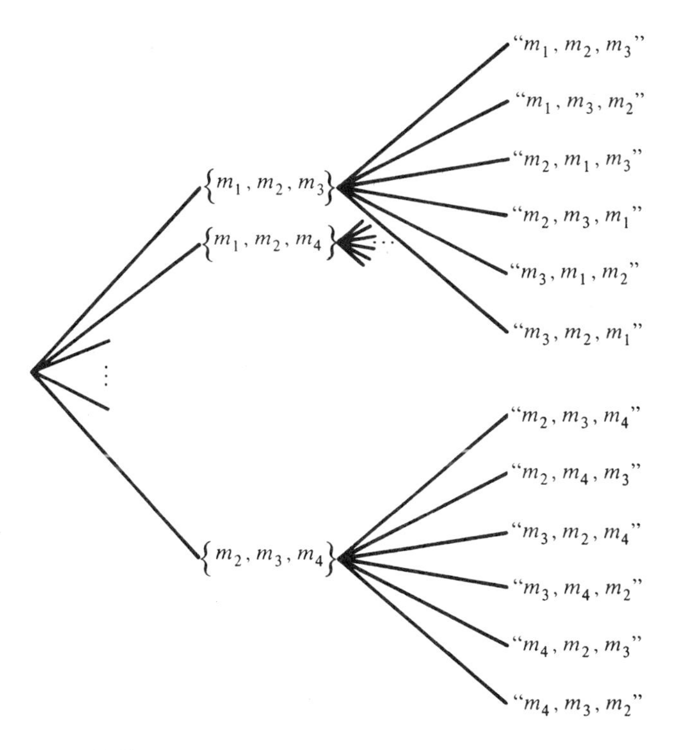

Figure 1-15.

as the paths in a tree of two stages, in which the branches in the initial stage correspond to the possible selections of the 3-element subset from the set of five elements, and the branches in the second stage correspond to the ways in which the three elements can be arranged in order.

There are $C(5, 3)$ branches in the initial component and $P(3, 3) = 6$ branches in each second-stage component. Therefore, by the multiplicative principle the total number of paths, $P(5, 3)$, is given by

$$P(5, 3) = C(5, 3) \times P(3, 3),$$

from which we conclude that

$$C(5, 3) = \frac{P(5, 3)}{P(3, 3)} = \frac{5 \times 4 \times 3}{3 \times 2 \times 1} = \frac{60}{6} = 10.$$

Thus there are ten ways in which Tony can put three of his five marbles in the urn (as can be verified by simply listing them).

To find $C(n, k)$ in general, we think of the permutations of n things, k at a time, as corresponding to the paths in a tree of two stages, with the first stage corresponding to the selection of the subset of k elements to be arranged and the second stage corresponding to their arrangement. There are $C(n, k)$ branches in the initial component and $P(k, k)$ branches in each second-stage component, and $P(n, k)$ paths in all. Therefore, by the multiplicative principle

$$P(n, k) = C(n, k) \times P(k, k).$$

We conclude that the number of k-element subsets that can be formed from a set of n elements is

$$C(n, k) = \frac{P(n, k)}{P(k, k)} = \frac{n \times (n-1) \times (n-2) \times \cdots \times (n-k+1)}{1 \times 2 \times 3 \times \cdots \times k}.$$

In the factorial notation,

$$C(n, k) = \frac{P(n, k)}{P(k, k)} = \frac{n!/(n-k)!}{k!} = \frac{n!}{(n-k)! \times k!}.$$

Comment: From the above formula,

$$\begin{aligned} C(n, n-k) &= \frac{n!}{[n-(n-k)]! \times (n-k)!} \\ &= \frac{n!}{k! \times (n-k)!} \\ &= C(n, k). \end{aligned}$$

In particular, $C(5, 3) = C(5, 2) = 10$. This should hardly be surprising since for every 3-element subset formed from a set of five elements there is a corresponding leftover subset of two elements and vice versa.

Example: There are sixteen members in a certain club. In how many ways can a committee of four be chosen?

Solution: The problem is to find the number of 4-element subsets that can be formed from a set of sixteen elements. The answer is

$$C(16, 4) = \frac{P(16, 4)}{P(4, 4)} = \frac{16 \times 15 \times 14 \times 13}{4 \times 3 \times 2 \times 1} = 1820.$$

Example: (a) How many functions are there mapping the set $X = \{1, 2, 3, 4, 5, 6\}$ into the set $Y = \{0, 1\}$?
(b) How many of the functions in part (a) have the property that the element "1" in Y is the image of exactly four of the elements of X?

Solution: (a) The problem is to find the number of functions mapping a set of six elements into a set of two elements. We recall from the previous section that there are n^k functions mapping a set of k elements into a set of n elements. Setting $k = 6$ and $n = 2$, the answer to part (a) is given by

$$2^6 = 64.$$

(b) Each function of the form required identifies naturally with the 4-element subset of X consisting of those elements mapped into "1". For instance, the function

$$f \sim \{(1, 1), (2, 0), (3, 1), (4, 1), (5, 0), (6, 1)\}$$

identifies with the subset $\{1, 3, 4, 6\}$.

Thus the answer to part (b) is given by $C(6, 4)$, the number 4-element subsets that can be formed from a set of six elements. From the formula derived above,

$$C(6, 4) = \frac{6 \times 5 \times 4 \times 3}{1 \times 2 \times 3 \times 4} = 15.$$

1-4 EXERCISES

1. (a) How may functions are there mapping the set $A = \{1, 3, 5\}$ into the set $B = \{0, 2, 4, 6, 8\}$?
(b) How many of the functions of part (a) are one-to-one?

2. Determine the number of one-to-one functions mapping a set of the form $X = \{x_1, x_2, x_3\}$ into a set of the form $Y = \{y_1, y_2, \ldots, y_{10}\}$.

3. Construct a tree to represent the set of all one-to-one functions mapping the set $S = \{0, 1\}$ into the set $T = \{1, 2, 3, 4\}$.

4. There are twelve horses in a race. Assuming no dead heats (ties), and taking into account only the first three horses across the finish line, how many possibilities are there for the order of finish?

5. Compute: (a) $P(5, 2)$, (b) $P(25, 1)$, (c) $P(10, 3)$, (d) $P(5, 5)$, (e) $P(7, 4)$.

6. Compute: (a) $6!$, (b) $\frac{8!}{3! \times 5!}$, (c) $4! \times 0!$, (d) $\frac{25!}{2! \times 23!}$.

7. Compute: (a) $C(6, 3)$, (b) $C(25, 23)$, (c) $C(10, 10)$, (d) $C(10, 4)$, (e) $C(12, 4)$.

8. In how many ways can six different books be arranged side-by-side on a bookshelf?

9. (a) There are ten 3-element subsets of the set $\{1, 2, 3, 4, 5\}$. List them.
(b) One of the subsets in part (a) is $\{1, 3, 4\}$. List the six permutations of this particular subset.

10. A coin is flippped seven times. A typical basic outcome in the strong analysis is

HTHHTTH.

In this analysis how many basic outcomes are there in the truth sets of the following statements:
(a) Tails occurs exactly three times.
(b) The first two flips are Tails, and exactly three Tails occur in the remaining five flips.

11. (a) In how many ways can a slate of three officers, President, Secretary, and Treasurer, be chosen from a club with twenty members?
(b) In how many ways can a committee of three be chosen from the club with twenty members.

12. Given seven distinguishable marbles and four distinguishable urns, how many possibilities are there for
(a) Putting the marbles in the urns?
(b) Placing exactly one marble in each urn?
(c) Depositing three marbles in a preselected urn?

13. (a) How many functions are there mapping the set $S = \{1, 2, 3, \ldots, 10\}$ into the set $T = \{0, 1\}$?
(b) In how many of the functions in part (a) does the element "1" occur as the image of exactly seven elements of the domain S?

1-5. Frequency Trees

The set of basic outcomes tells us what may happen in advance of a process occurring. After a sequence of occurrences it may be of interest to join each outcome with the number of times it occurred.

To illustrate, we recall Adam's simple one-card process of Sec. 1-1. Let's suppose he draws a card from the shuffled deck one hundred times. Observer O_1, who was interested only in the suit of the card drawn, reports the experience as follows:

> Clubs were drawn 19 times, Diamonds 27 times,
> Hearts 29 times, and Spades 25 times.

His report determines a function F, expressed by the set of ordered pairs

$$F \sim \{(C, 19), (D, 27), (H, 29), (S, 25)\}.$$

Such a function is called a *frequency function.*

Any mapping of a set of basic outcomes into nonnegative integers may be viewed as a frequency function. In a natural way, the frequency function defined over the basic outcomes generates a larger frequency function defined over the set of relevant statements. For instance, in the example at hand, the frequency of the statement

r : The card drawn is red

is computed by adding the frequencies corresponding to the basic outcomes in the truth set R of the statement r. Since $R = \{D, H\}$, and since F maps D into 27 and H into 29, the larger frequency function maps the statement r, or, to be more formal, maps the subset R, into the value $56 = 27 + 29$. The interpretation is immediate: If Adam drew a Diamond 27 times and a Heart 29 times, then he drew a red card 56 times.

For a more substantial example, let's return to the economics instructor of Sec. 1-2, who questions each of his 25 students to determine whether the student is also enrolled in a mathematics course, whether he intends to major in economics, and whether he's had prior courses in economics. Earlier we displayed a tree to represent the basic outcomes; now let's fit a frequency function to the tree. Hypothesizing an overall experience, we record numbers next to the terminal points in this tree to represent the number of students in each basic category. This gives us the tree of Fig. 1-16.

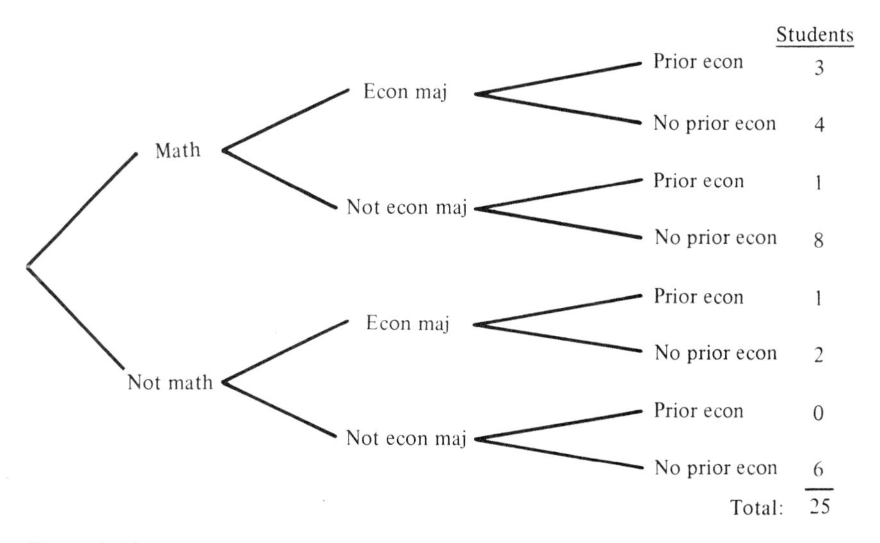

Figure 1-16.

The diagram tells us, for instance, that eight of the twenty-five students are taking a math course, do not intend to major in econ, and have had no prior econ courses.

To have more information readily visible we move the numbers corresponding to the terminal points in the tree to the branches leading to these points, and we add the numbers in each component to arrive at values for the other points in the tree. This gives us Fig. 1-17, which we call the *frequency tree* representing the economics instructor's experience.

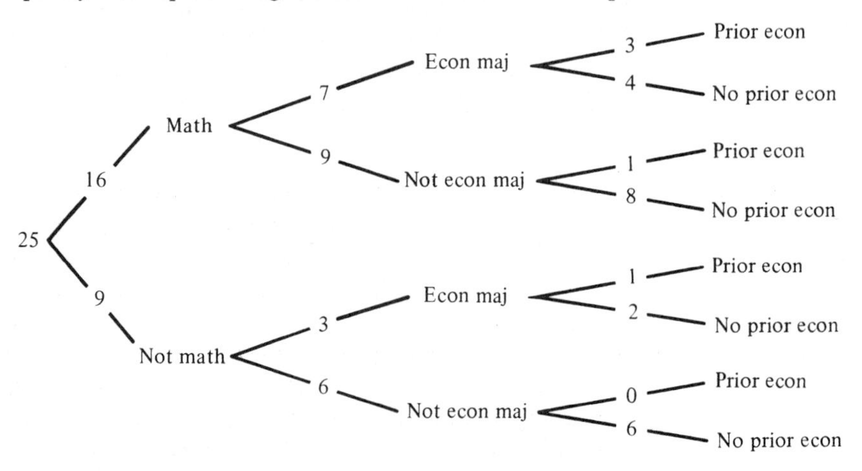

Figure 1-17 Frequency Tree.

From Fig. 1-17 we see, for instance, that three of the twenty-five students are not taking a mathematics course but do intend to major in economics.

In general, a *frequency tree* is realized by assigning nonnegative integers to the points in a tree, recorded on the branches leading to the points, in such a way that the total for the branches in each component is equal to the value assigned the point initiating that component.

Completing a Frequency Tree

The frequency tree for the economics instructor's process was easily completed from the values assigned the terminal points in the tree. In other situations the data given may not transfer to the tree in such a straightforward fashion. The first example below is easy; the second is more challenging.

Example: We have a set of three urns, I, II, III, each containing red and blue marbles. The process is to select one of the urns and withdraw a single marble from it.

In 100 trials, Urn I is selected 30 times and Urn II is selected 28 times. A red marble is drawn from Urn I 9 times, from Urn II 15 times, and from Urn III 24 times.

(a) Construct a frequency tree from the given data.

(b) On how many of the 100 trials is a blue marble drawn from Urn III?

Solution: (a) The tree of Fig. 1-18 represents the basic outcome analysis of the process. The values given in the problem have been entered on the branches leading to the corresponding points. The total number of trials, 100, has been assigned to the initial point.

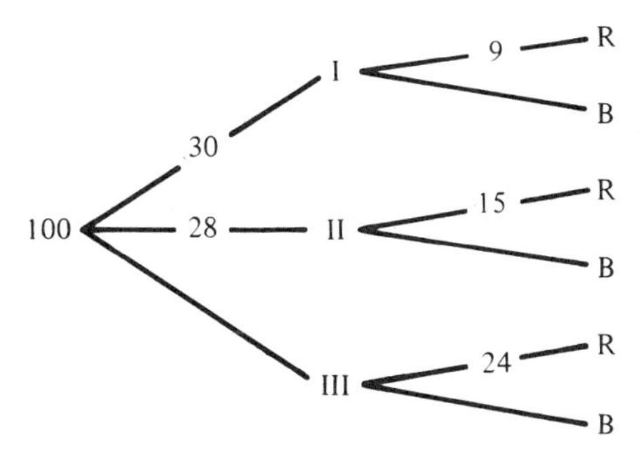

Figure 1-18.

For the branches on which no number has been entered, the appropriate value is determined by the requirement that the sum in each component must equal the value assigned the point initiating that component. A little arithmetic produces the completed frequency tree of Fig. 1-19.

(b) From the frequency tree of Fig. 1-19 we see that a blue marble is drawn from Urn III in 18 of the 100 trials.

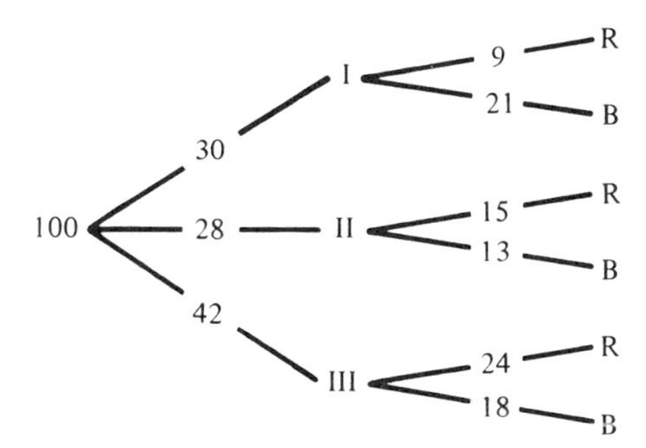

Figure 1-19.

Example: In a group of 50 students it is found that 28 study Math, 19 study Biology, 18 study German, 16 study Math and Biology, 12 study all three subjects, and 18 study none of the three.

(a) How many of the 50 students study only German?
(b) How many study exactly two of the three subjects?
(c) How many study only Math?

Solution: The information given is recorded on the tree in Fig. 1-20. Here, some of the given numbers are assigned to particular branches. In other cases we label the several branches that contribute to a given sum, and record this sum in the legend on the right.

The missing values in the initial component and uppermost second- and third-stage components are easily supplied by the "sum in the component equals the initiating value" requirement. Also, there are two "code-k" branches having a sum of 19, and since the one branch has value 16, the other branch must have value 3. The best we can do for the "code-n" branches is to lift the "n" off the one branch whose value is 12 and reduce the code-n sum to $18 - 12 = 6$.

These first-step adjustments give us the tree of Fig. 1-21, somewhat more complete.

Following the same line of reasoning, the tree of Fig. 1-21 is transformed into the tree of Fig. 1-22. The terminal points comprising the truth sets of the categories in questions (a), (b), and (c) have been so labeled.

It doesn't seem that we can improve on the tree of Fig. 1-22. Let's see if we can answer the questions posed.

(a) There is only one terminal point in the "studies only German" category, and it has value 1. Therefore just one of the fifty students studies German but neither of the other subjects.

(b) There are three terminal points in the "studies exactly two subjects" category. One has value 4, and the other two comprise the pair of code-n branches with sum 5. Therefore, nine of the fifty students study exactly two of the three subjects.

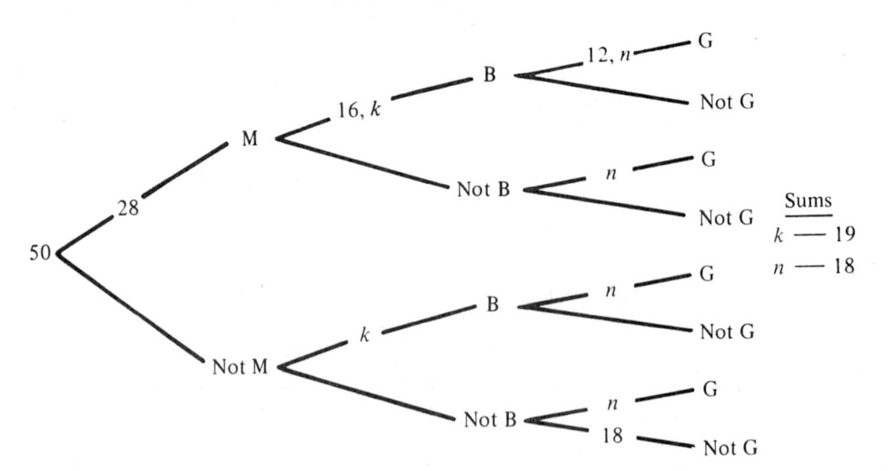

Figure 1-20.

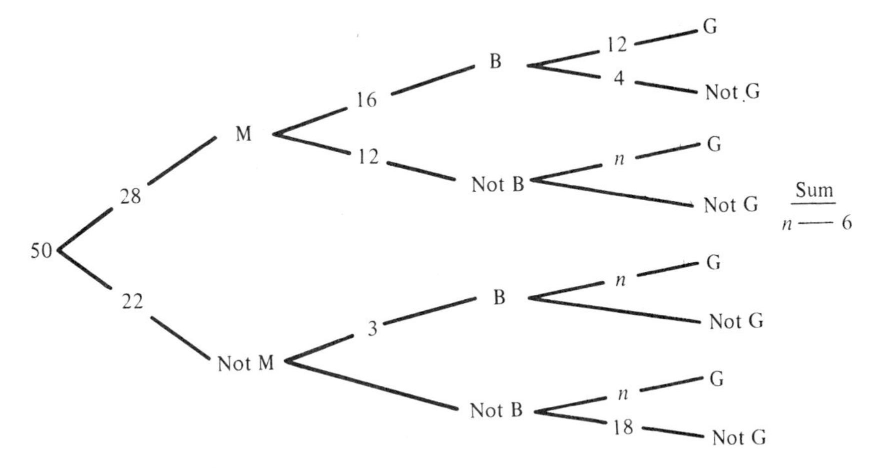

Figure 1-21.

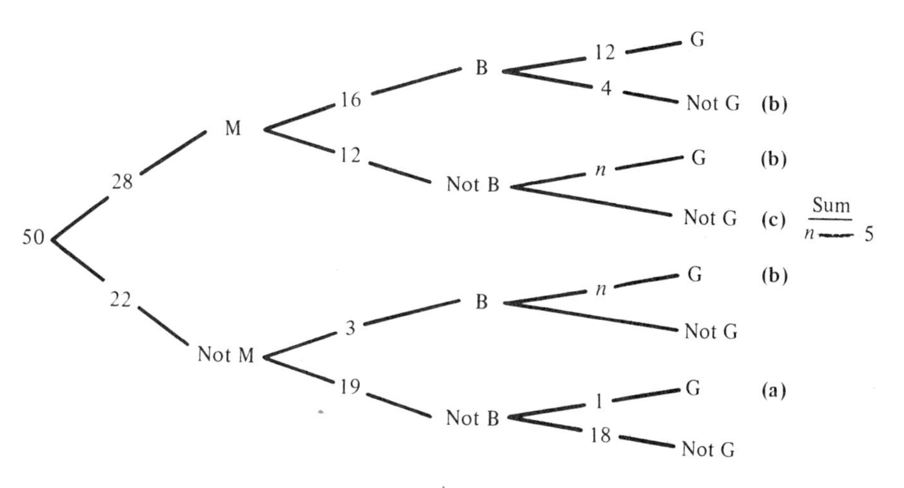

Figure 1-22.

(c) For the one terminal point in the "studies only Math" category, our efforts have not produced a frequency value. We were not given enough information to answer question (c) precisely. But the information in the tree does enable us to place limits on the possible answers.

The frequency tree of Fig. 1-22 may be easily completed if we are given the value for either of the code-n branches. Looking at the lower of the code-n branches, we see that the sum in its component is 3, so the value assigned this branch must be 0, 1, 2, or 3, and with any one of these four values it is possible to complete the tree. We conclude that the number of students studying Math, but neither of the other subjects, is 7, 8, 9, or 10.

1-5 EXERCISES

1. The process is to withdraw two marbles in turn from an urn containing red and white marbles. The experience in 17 trials of the process is given by the frequency tree in Fig. 1-23.

(a) In how many of the 17 trials were two marbles of different colors drawn?

(b) How many times was a red marble drawn in the 34 draws of the 17 trials?

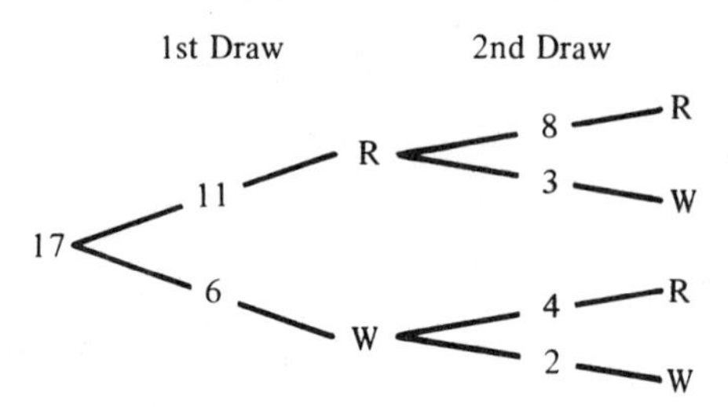

Figure 1-23.

2. The frequency tree shown in Fig. 1-24 reports the results of 100 trials of a process of drawing two marbles from an urn containing red, white,

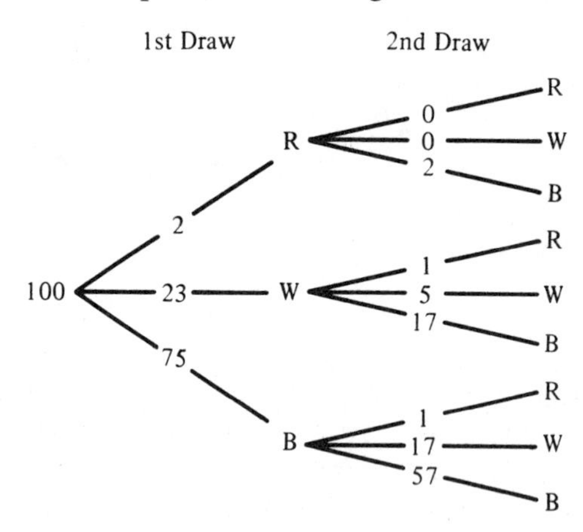

Figure 1-24.

and blue marbles. In how many of the 100 trials did the following outcomes occur?

(a) The second draw was white.

(b) At least one of the draws was white.

(c) Two marbles of the same color were drawn.

3. Complete the frequency tree in Fig. 1-25 for a hypothetical 100 trials of the process of flipping three coins.

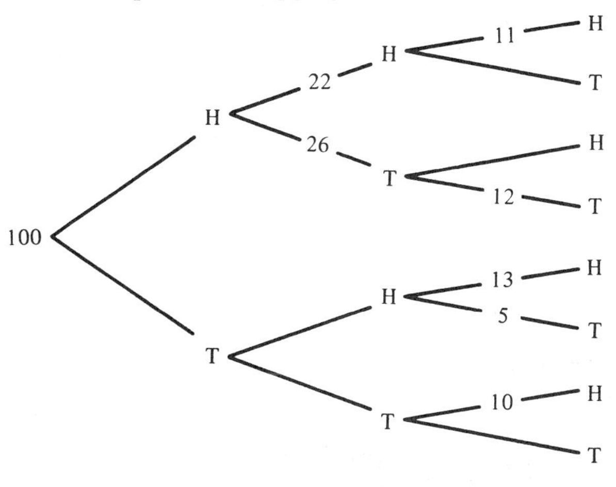

Figure 1-25.

4. The process is to flip a coin three times. Construct a frequency tree for 32 occurrences, given the following information: Heads occurs on the first flip 18 times, on the second flip 15 times, and on the third flip 16 times. HHH occurs 4 times, and TTT occurs 5 times. The first and second flips are Tails 8 times, and the second and third flips are Tails 9 times.

5. Among 50 members in a fraternity, 21 study History, 33 study English, and 28 study Math; 3 study none of the three subjects, and 6 study all three; 10 study both History and Math, and 14 study History and English.

(a) Construct a frequency tree.
(b) How many study exactly one of the three subjects?

6. There are 100 people at a fraternity party, 55 men and 45 women. Of these, 40 are members of the fraternity and 68 are students. Only three of the male students present are not members of the fraternity. If all of the fraternity members are male students, how many of the women present are not students?

7. The sponsor of three television shows, A, B, and C, conducts a survey of 100 households. It is found that 25 watched show A, 30 watched B, and 31 watched C; 10 watched both A and B, 50 watched none of the three shows, and 7 watched all three shows.

(a) Construct as complete a frequency tree as possible.
(b) How many watched exactly two of the three shows?
(c) How many watched only one of the three shows?
(d) At least how many watched both B and C?

2

INTUITIVE PROBABILITY

Had we continued with the parable about Freshman Bob that introduced Chapter 1, we might have found one of his critical friends observing that "Bob may be going out for football, but he probably won't make the team." And in the case of Adam's one-card exercise, we should not be surprised to find observer O_1 ready to lay odds that at least one of the next three cards will be a Spade.

With the developments in Chapter 1, the stage is set for elaborating on the structure of the basic outcome set to take into account the feature of chance that is inherent in many processes. This carries us to the study of *probability*.

2-1. Ideal Frequency Trees

We introduce the notion of probability with a simple urn example. (In fact, it is hard to think of a problem in probability which does not have an urn model.)

Karen has two urns. Urn I contains a red marble and a white marble, and Urn II contains a red marble and two white marbles.

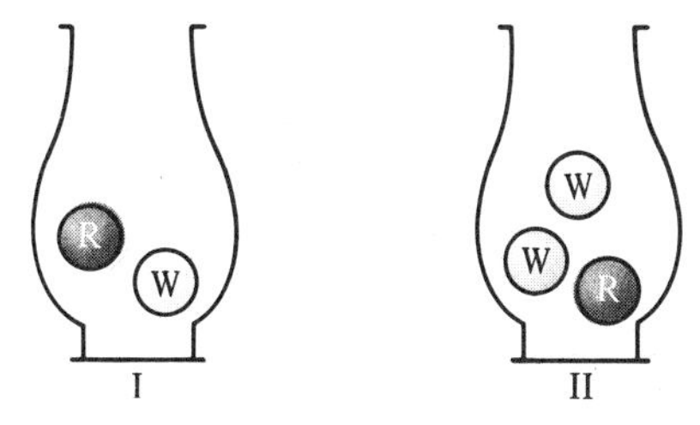

Figure 2-1.

The process is to select one of the urns, withdraw a single marble from it, and observe its color.

Suppose Karen engages in two separate runs of sixty trials each. She reports her results via the frequency trees of Fig. 2-2.

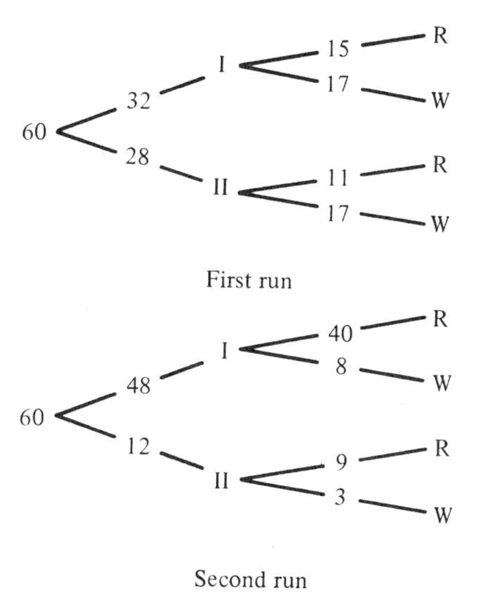

Figure 2-2.

In the second run it looks as if Karen is deliberately making her selection and for some reason favors Urn I and the red marbles. On the other hand, in the first run the results are in line with what one would reasonably expect if she is selecting the urn without bias and picking the marble without peeking. If it is in the nature of the process that the two urns are "equally likely," and within each urn the marbles are all "equally likely," then this is an important property of the process, and our analysis should take it into account.

The notion of "likelihood" of outcomes is most easily attached to processes which can be repeated in arbitrarily long runs. A simple example is the process of flipping a coin. The basic outcomes are, of course, "Head" and "Tail." But we wish to express the fact that the two outcomes are "equally likely." Which is to say, we expect that in a long string of repetitions, each will occur about one-half of the time. To convey this impression, we say that the *probability* of "Head," as of "Tail," is one-half.

In Karen's two-urn process we express the view that she is selecting the urn without bias by assigning a probability of one-half to the outcomes "Urn I" and "Urn II." As we observed, her first run of 60 trials, in which Urn I is selected 32 times and Urn II 28 times, is pretty well in line with this assumption of "equal likelihood." But we go one step further and invent the notion of the *ideal run* in which the fraction of times each outcome occurs is precisely equal to the probability that we feel should be assigned to that outcome. Thus for an ideal run of 60 trials of Karen's process the initial stage of the frequency tree is as shown in Fig. 2-3.

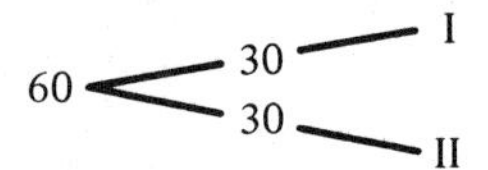

Figure 2-3.

To complete the frequency tree for the ideal run we note that since Urn I is selected 30 times, and since it contains two marbles which we view as equally likely, each of the marbles is selected 15 times. Similarly, each of the three marbles in Urn II is selected 10 times, and since there are one red and two white marbles, the red marble is drawn 10 times and a white marble 20 times. Thus we complete the frequency tree as shown in Fig. 2-4.

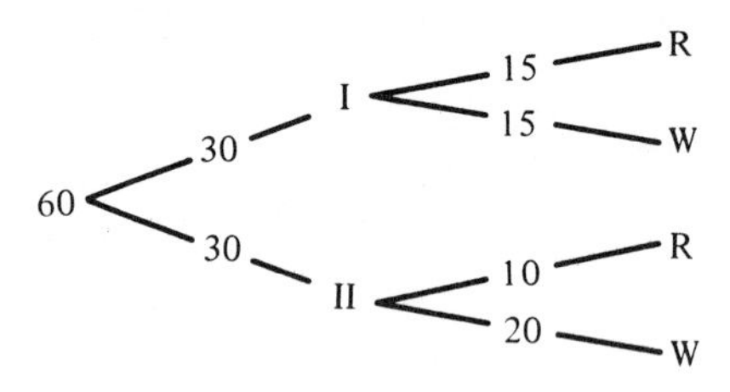

Figure 2-4 Frequency Tree for Ideal Run of Karen's Process.

Generally the frequency trees we shall be considering will be based on ideal runs. We may call them ideal frequency trees, but it is more economical to leave the "ideal" implicit and just call them frequency trees.

In effect, the construction of a frequency tree is a way of assigning probabilities to the relevant outcomes. For example, we see from the tree of Fig. 2-4 that a red marble is drawn in 25 of the 60 trials of the ideal run. Thus the construction assigns a probability of $\frac{25}{60}$, or $\frac{5}{12}$, to the outcome "A red marble is drawn."

In Karen's process it is not necessary to have as many as 60 trials in order to have an ideal run in whole numbers. As the tree in Fig. 2-5 shows, we could get by with only 12 trials. On the other hand, we could have 3000 trials, as shown in Fig. 2-6.

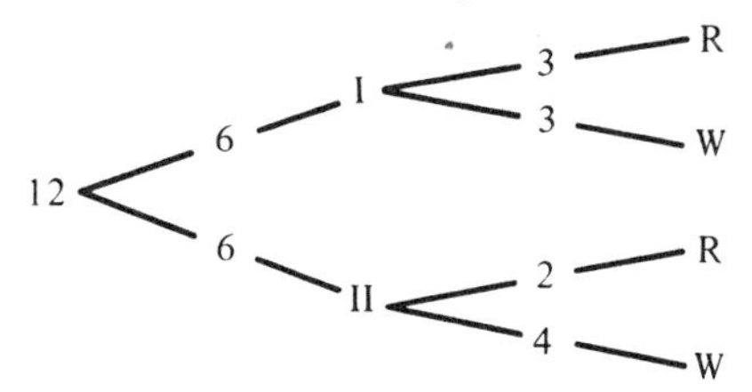

Figure 2-5 12-Trial Frequency Tree.

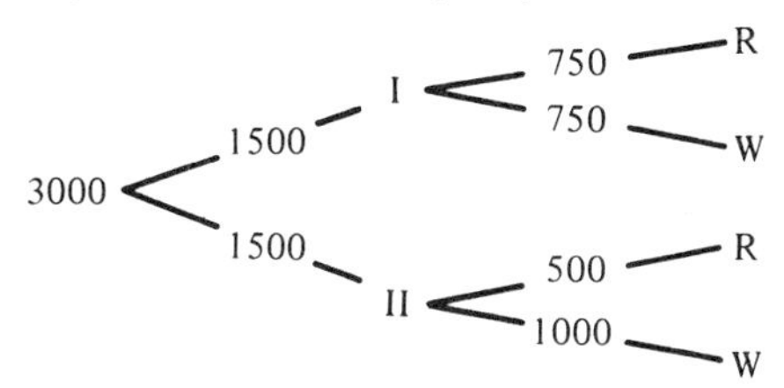

Figure 2-6 3000-Trial Frequency Tree.

It should be clear that we get equivalent frequency trees, in the sense that the probabilities assigned thereby are unchanged, if we divide or multiply each numerical entry in the tree by a constant factor.

Construction of (Ideal) Frequency Trees

As the reader may have detected, there was a little unannounced finagling in the selection of 60 as the number of trials in Karen's process. We could have started with 12 trials, or 3000 trials, but if we start with, say, 100 trials, it is impossible to construct an ideal run of the process in whole numbers. And the advantage of the frequency tree lies in the comfort of working with whole numbers. Let's see how frequency trees may be constructed without advance knowledge of a prudent choice for the number of trials.

In general, the tree that represents the basic outcomes of a process is assembled one component at a time. In the same step-by-step fashion we can assign appropriate frequencies. As an illustration, consider the following process:

> At random, two cards are drawn in order, without replacement, from an ordinary deck of fifty-two cards. For each draw the only feature deemed relevant is whether the card is red or black.

Here, the expression "at random" means that there is no bias in the process, so the fifty-two cards are equally likely candidates for the first draw. "Without replacement," as opposed to "with replacement," means that the first card is not replaced before the second draw. Thus after the first draw, there are fifty-one equally likely candidates for the second draw.

The single-stage frequency tree of Fig. 2-7(a) represents an analysis of the process in which only the color of the first card has been taken into account. In Fig. 2-7(b) the tree is simplified by dividing out the common factor of 26.

The upper second-stage component takes into account the color of the second card for those trials in which the first card is red. This component, with frequencies unassigned, has been joined to the tree of Fig. 2-7(b) to give the tree of Fig. 2-8.

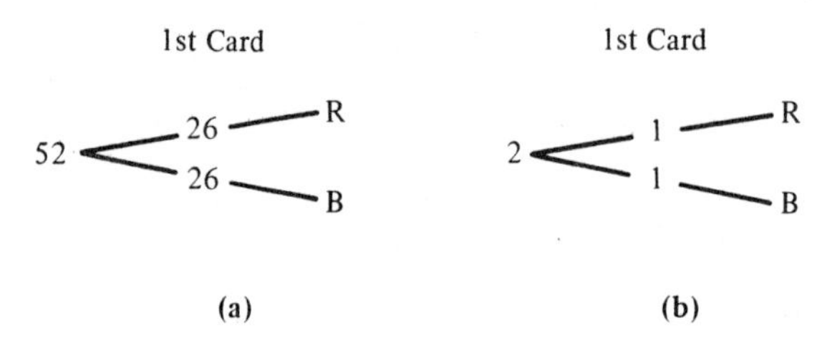

Figure 2-7.

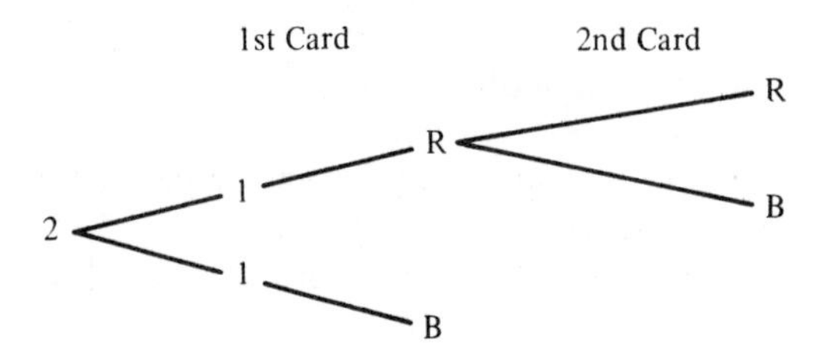

Figure 2-8.

After a red card has been selected on the first draw there are fifty-one equally likely candidates for the second draw, twenty-five of which are red and twenty-six black. We interpret the second-stage component shown in Fig. 2-8 as the analysis of the process of drawing a card from a set of twenty-five red and twenty-six black cards. Thus for this isolated component we assign frequencies as shown in Fig. 2-9. The "— —" device indicates a tem-

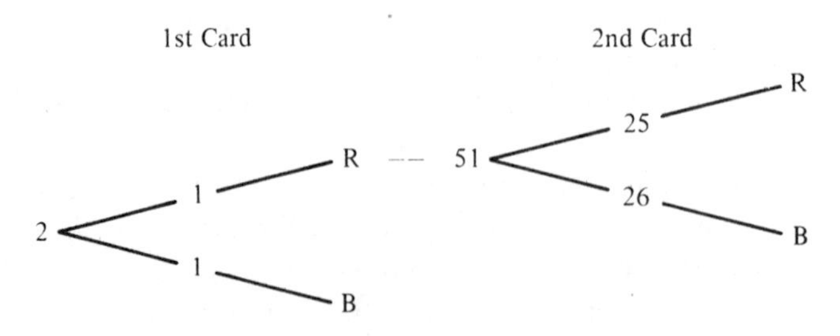

Figure 2-9.

porary disconnection with respect to frequency assignment. The disconnection is removed by multiplying the entries in the initial stage by a factor of 51. This gives us the frequency tree of Fig. 2-10.

The same treatment is accorded the lower second-stage component. For the disconnected tree of Fig. 2-11 we observe that the value 51 assigned the point initiating the component is the same as the sum for the branches in the component. Therefore we remove the disconnection, as shown in Fig. 2-12.

The frequency tree of Fig. 2-12 represents an ideal run of 102 trials for the process of drawing the two cards. We see, for instance, that both cards are red in 25 of the 102 trials; thus the probability of getting two red cards is $\frac{25}{102}$. We also see that the two cards drawn are of different colors in 52 of

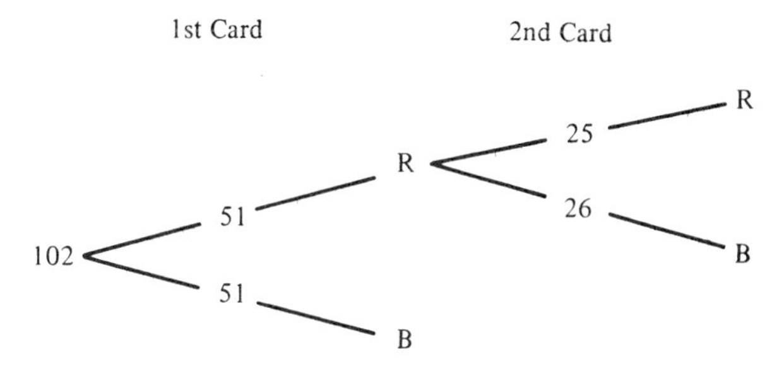

Figure 2-10.

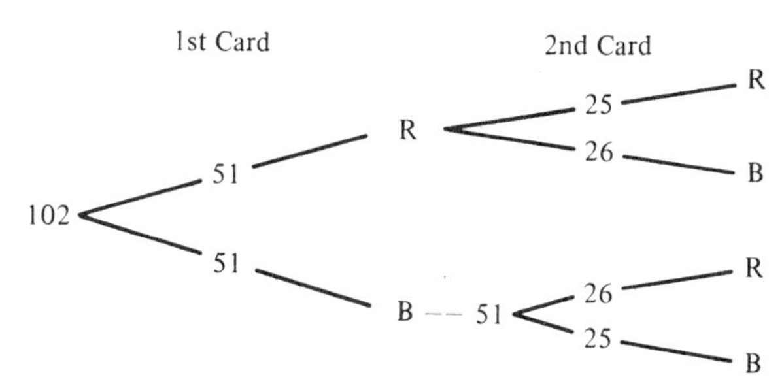

Figure 2-11.

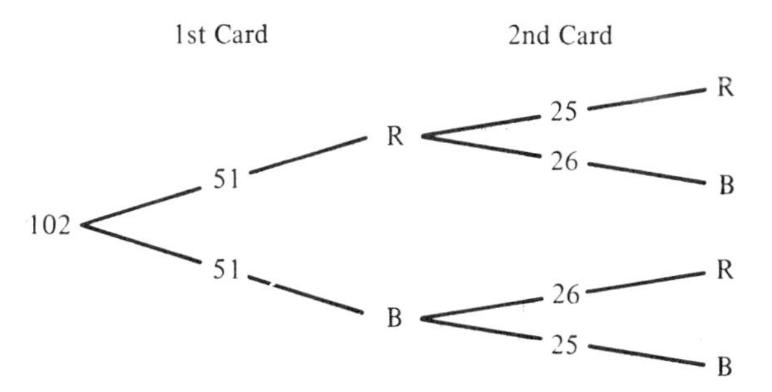

Figure 2-12.

the 102 trials; thus the probability of getting two cards of different colors is $\frac{52}{102}$, or $\frac{26}{51}$.

Example: On rainy days the Mudlarks of the Mudville softball league win two times out of three, but on clear days they win only one time in three. And it rains six days out of ten in Mudville.

(a) Construct a frequency tree.
(b) Find the probability that the Mudlarks win.

Solution: (a) The information given is translated into the frequencies assigned for the isolated components in the tree of Fig. 2-13.

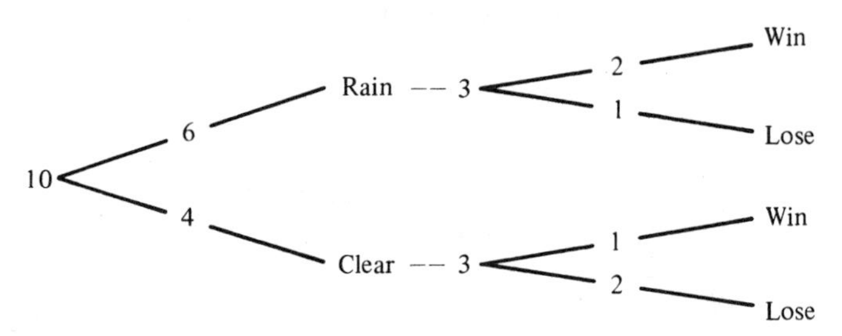

Figure 2-13.

Compatibility is achieved for the upper second-stage component by dividing out the common factor of 2 in the initial component. The component is attached in Fig. 2-14.

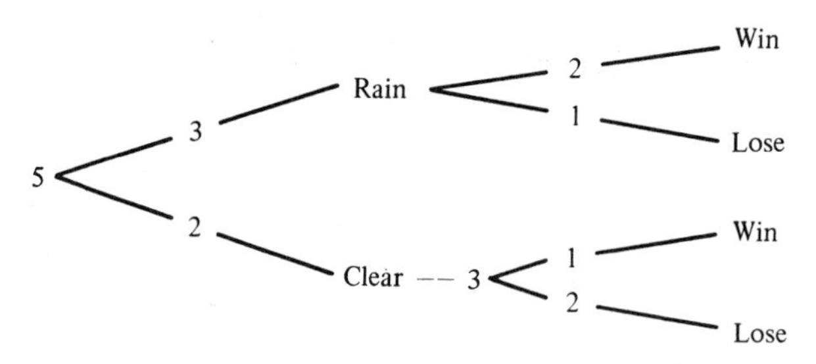

Figure 2-14.

To achieve compatibility for the lower second-stage component the entries in the developing tree are multiplied by 3, and the entries in the component are multiplied by 2, giving a common value of 6 for the point initiating the component and for the sum of the values in the component. The completed frequency tree is shown in Fig. 2-15.

(b) From the frequency tree of Fig. 2-15 we see that the Mudlarks win eight times in an ideal run of fifteen games. Therefore the probability of victory is $\frac{8}{15}$.

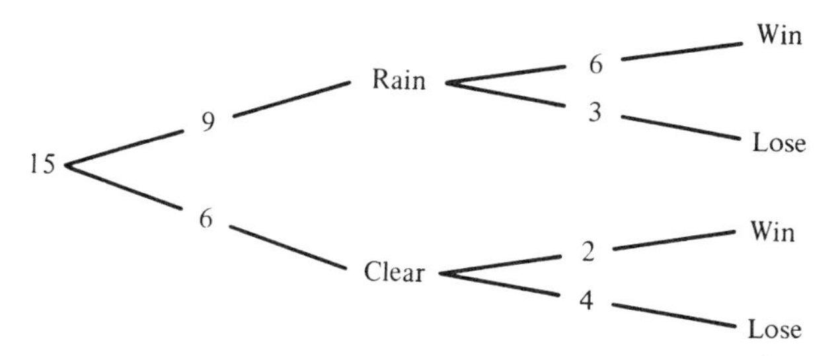

Figure 2-15.

2-1 EXERCISES

1. Kate has two urns. Urn I contains two red and three white marbles, and Urn II contains one red and one white marble. At random an urn is selected and a single marble withdrawn.
   (a) Construct a frequency tree for an ideal run of 100 trials.
   (b) Find the probability that the marble drawn is red.

2. Fred has an urn containing two red and three white marbles. At random he draws two marbles in turn.
   (a) Suppose the process is performed "with replacement"; i.e., the first marble is returned before the second draw. Construct a frequency tree for an ideal run of 100 trials.
   (b) Construct a 100-trial (ideal) frequency tree, assuming that the process is performed "without replacement."
   (c) For both the "with replacement" and "without replacement" processes, find the probability that the two marbles drawn are of the same color.

3. Construct an 8-trial frequency tree for the process of flipping a coin three times. What is the probability that exactly two of the flips are Heads?

4. A thumbtack is so constructed that when it is flipped it lands point down two times out of three, in the long run. The process is to flip the thumbtack three times.
   (a) Construct a frequency tree for an ideal run of 27 trials of the "three-flip" process.
   (b) Find the probability that the thumbtack lands point down exactly twice.

5. Urn I contains one red and two white marbles, and Urn II contains two red and three white marbles. At random, an urn is selected and a single marble withdrawn.
   (a) Construct a frequency tree from the disconnected tree in Fig. 2-16.
   (b) Find the probability that the marble drawn is red.

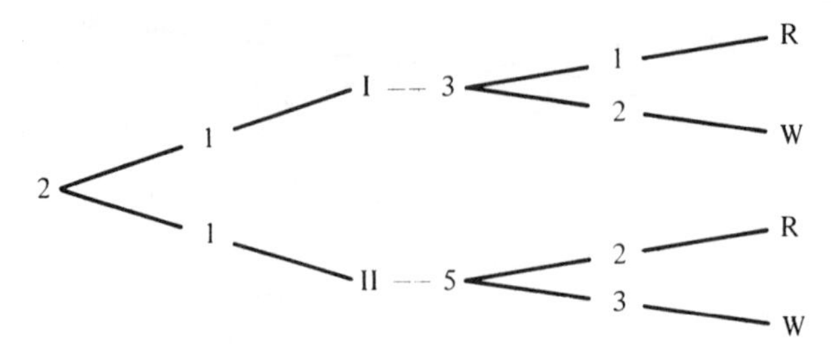

Figure 2-16.

6. Urn I contains six red and two white marbles, and Urn II contains two red and four white marbles. At random, an urn is selected and a single marble is withdrawn.
   (a) Construct a frequency tree.
   (b) Find the probability that the marble drawn is red.

7. Joe deals two cards from an 8-card deck consisting of the Kings and Queens from an ordinary deck. Construct an appropriate frequency tree and find
   (a) The probability that both cards are Kings.
   (b) The probability that one is a King and one is a Queen.

8. Two cards are drawn in order, without replacement, from an 8-card deck consisting of the four Kings and four Queens. Construct a frequency tree, and find the probability that
   (a) Neither card is a Club.
   (b) The second card is a Spade.
   (c) At least one of the cards is black.

9. In a certain town, 50% of the voters are Democrats, 40% are Republicans, and 10% are Independents. A proposed ordinance is supported by 60% of the Democrats, 30% of the Republicans, and 70% of the Independents. Does it look like the ordinance is destined to pass?

10. The process is to flip a coin until either two Heads occur in succession or a total of three Tails have occurred. Find the probability that the process terminates with the occurrence of two successive Heads.

2-2. Probability Trees

From a frequency tree we compute the probability of a relevant outcome by dividing the number of trials in which the outcome occurs by the total number of trials. Thus the probabilities of the outcomes represented by the points in the tree can be made immediately apparent by dividing every entry in the frequency tree by the total number of trials.

As an illustration we recall in Fig. 2-17(a) the frequency tree of the previous section for the process of drawing two cards from a standard deck. In Fig. 2-17(b) the entries have been divided by 102, the total number of trials.

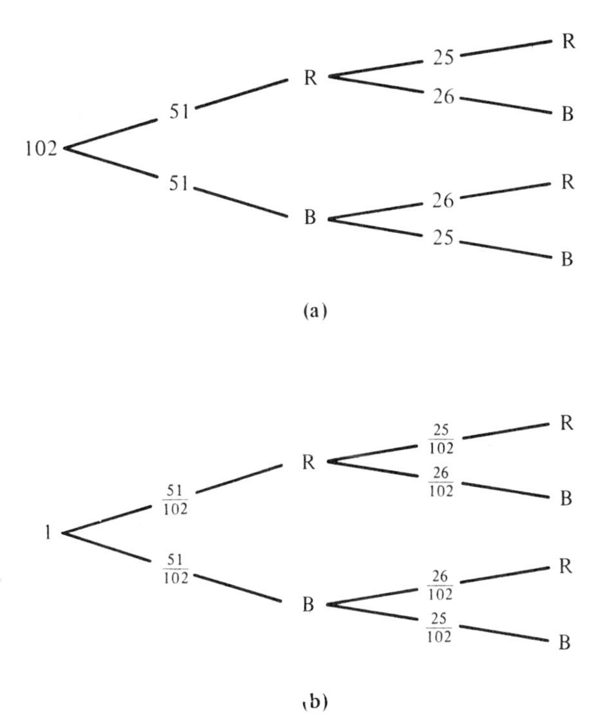

Figure 2-17.

The outcome "Both cards are red" is represented by the path leading to the uppermost terminal point R in both trees. From Fig. 2-17(a) we deduce the probability of this outcome by dividing the value 25 assigned to this point by the total number of trials, 102; this probability, $\frac{25}{102}$, is read directly from the tree of Fig. 2-17(b) as the value associated with the point. The probability that "The cards are of different colors" is found from Fig. 2-17(a) by adding the values assigned the middle two terminal points, which represent the basic outcomes in the truth set of the statement, and dividing by the total number of trials. With respect to Fig. 2-17(b), the same computation is performed by adding the values associated with the middle two terminal points. Thus the probability of the "different colors" outcome is

$$\frac{26}{102} + \frac{26}{102} = \frac{52}{102} = \frac{26}{51}.$$

Trees such as in Fig. 2-17(b), realized by dividing every value in a frequency tree by the total number of trials, will be referred to as *relative frequency trees*. Like a frequency tree, a relative frequency tree is a tree with nonnegative numbers assigned on the branches, with the sum in each component being equal to the value assigned the initiating point. The novel feature of the relative frequency tree is that the value assigned the initial point of the tree must be one.

By the technique of the previous section the construction of a frequency tree is preceded by a preliminary step in which frequency assignments are made for the isolated components in the tree. A relative frequency tree can be constructed in the same manner.

In the "two cards from a deck" example at hand, the "disconnected components" step in the construction of the frequency tree is shown in Fig. 2-18(a). The analogous step for the relative frequency tree is shown in Fig. 2-18(b).

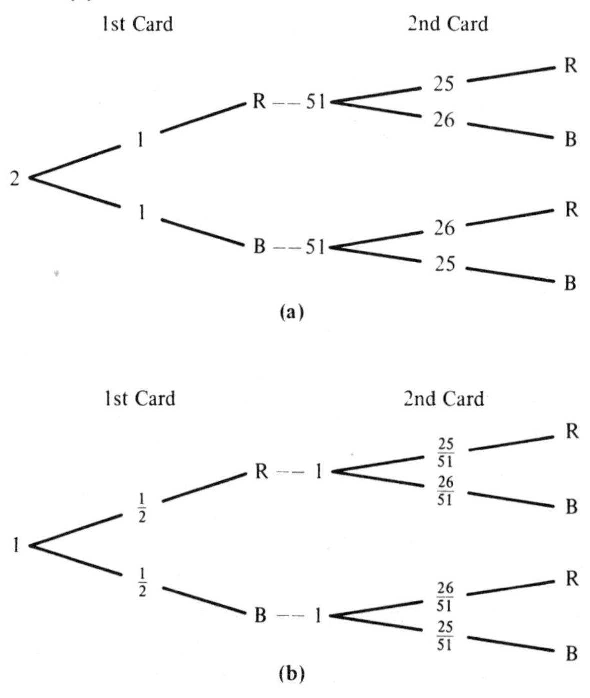

Figure 2-18.

The entries on the branches of the initial component in Fig. 2-18(b) are precisely the probabilities associated with the respective outcomes "first card is red" and "first card is black." In the second-stage components the entries are the probabilities of the respective outcomes in the process as it has been qualified by the experience in the initial stage. For example, in the upper second-stage component of Fig. 2-18(b) the value $\frac{25}{51}$ on the upper branch is the probability of drawing a red card from a deck in which one red card has been removed. It is our intuition for probabilities of this sort that enables us to make the probabilistic analysis; we refer to them as *branch probabilities*.

Actually, the thing that will be of greatest interest is not the frequency tree, or the relative frequency tree, but rather the tree realized by simply recording on each branch the appropriate branch probability. Such a diagram is called a *probability tree*. For the "two cards from the deck" example the probability tree is obtained by assigning the branch probabilities shown in Fig. 2-18(b) to the paths in the original tree. This gives Fig. 2-19.

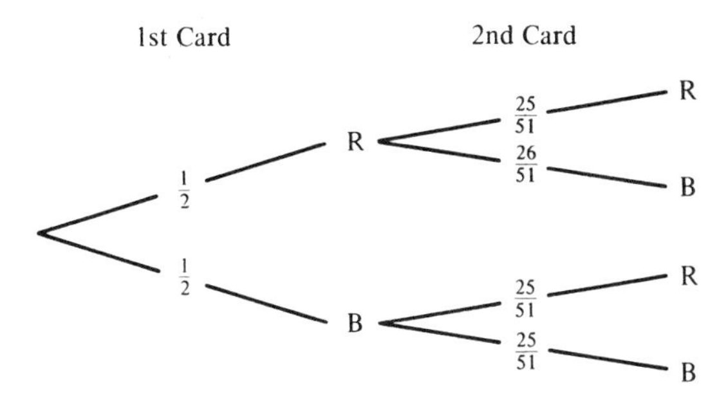

Figure 2-19 Probability Tree for the "Two Cards from a Deck" Process.

In general, a probability tree is a tree in which nonnegative numbers have been assigned to the branches in such a manner that the sum of the values in each component is equal to one.

Comparing Figs. 2-18(a), 2-18(b), and 2-19, we see that the information in the probability tree corresponds to the "disconnected components" step in the construction of the frequency tree. The transformation of the probability tree into the relative frequency tree is easy. The entries in the initial component are correct as they stand. Since the relative frequency tree requires that the sum for the branches in each component equal the value assigned the point initiating the component, compatibility between the first and second stages is achieved by multiplying the entries in each second-stage component by the value assigned the initiating point. In this manner the probability tree of Fig. 2-19 transforms into the relative frequency tree of Fig. 2-20, which is the same tree as in Fig. 2-17(b).

As already noted, the entries in the relative frequency tree are precisely the probabilities associated with the corresponding points. The important fact is that we compute these probabilities from the entries in the probability tree by multiplying the branch probabilities leading to the respective points.

In general, the probability corresponding to a point in a probability tree is the product of the branch probabilities leading to that point.

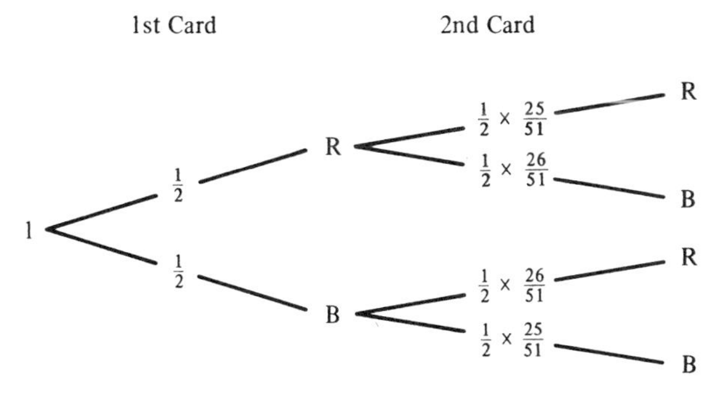

Figure 2-20 Relative Frequency Tree for the "Two Cards from a Deck" Process.

Another Example

To better appreciate the general applicability of the probability tree, let's look at another process.

From an ordinary 52-card deck, all of the cards except the Jacks, Queens, and Kings are discarded. This 12-card Royal Deck, as it will be called, is shuffled and three cards are drawn.

We are interested in the probability that

(0) There are no Jacks among the three;
(1) There is exactly one Jack;
(2) There are exactly two Jacks;
(3) There are exactly three Jacks.

In the interest of a tractable analysis, we treat the problem as though the three cards are drawn in order, without replacement, although the wording attaches no relevance to the order in which the cards are drawn. With respect to each card the only relevant property is whether or not it is a Jack. We know how many Jacks and how many cards remain at each stage. This information is translated into the probability tree of Fig. 2-21.

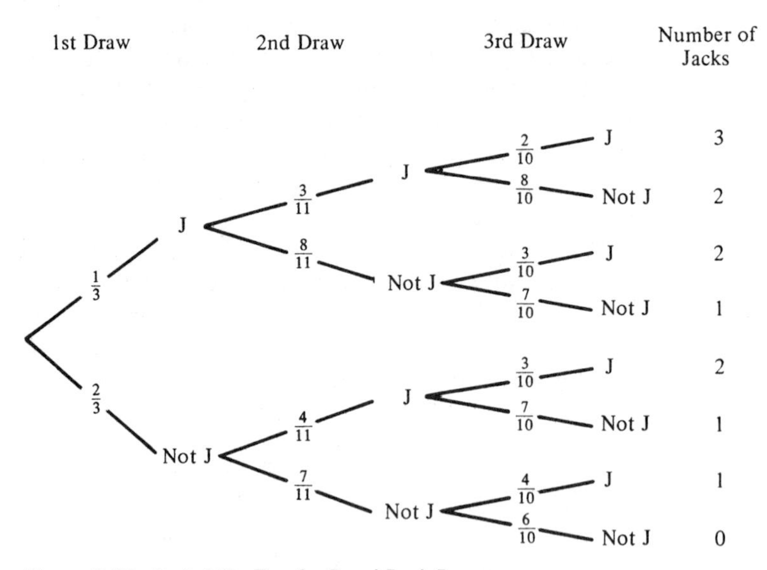

Figure 2-21 Probability Tree for Royal Deck Process.

Alongside each terminal point in the figure is recorded the number of Jacks in the corresponding outcome. Thus, for instance, the truth set of statement (2), the "two Jacks" outcome, consists of the paths with terminal points labeled "2."

As shown in Fig. 2-22, the probability tree is converted into a relative frequency tree by assigning to each point the product of the branch probabilities along the path leading to the point.

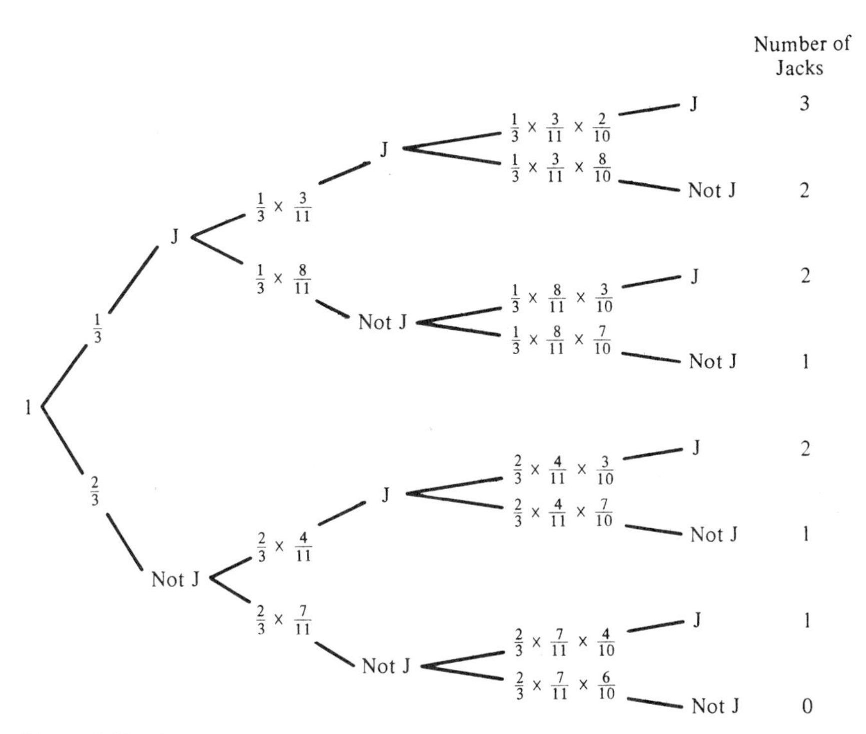

Figure 2-22 Relative Frequency Tree for Royal Deck Process.

The branch point J in the initial component represents the statement "The first draw is a Jack." The value $\frac{1}{3}$ has been assigned on its branch in both trees because "Jack" is one of three equally likely outcomes for the first draw.

In the "Not J" branch of the upper second-stage component a branch probability of $\frac{8}{11}$ is assigned in the probability tree of Fig. 2-21 because there are 11 equally likely cards remaining after the first draw, 8 of which are not Jacks. This second stage point, "Not J," corresponds to the statement "The first draw is a Jack and the second draw is not a Jack." As indicated in the probability tree, the first draw is a Jack one-third of the time, and of these instances the second draw is not a Jack eight-elevenths of the time, in the long run. It follows that both statements should be true in eight-elevenths of one-third of the trials. Thus the probability that both statements are true is $\frac{1}{3} \times \frac{8}{11}$, which is precisely the value in the relative frequency tree of Fig. 2-22.

Carrying the argument one step further produces the product $\frac{1}{3} \times \frac{8}{11} \times \frac{3}{10}$ as the appropriate probability for the completion of this path to the terminal point "J" in the third stage.

The relative frequency tree of Fig. 2-22 is transformed into a frequency tree by multiplying each entry by the common denominator of 330. This tree, shown in Fig. 2-23, represents an ideal run of 330 trials of the Royal Deck process.

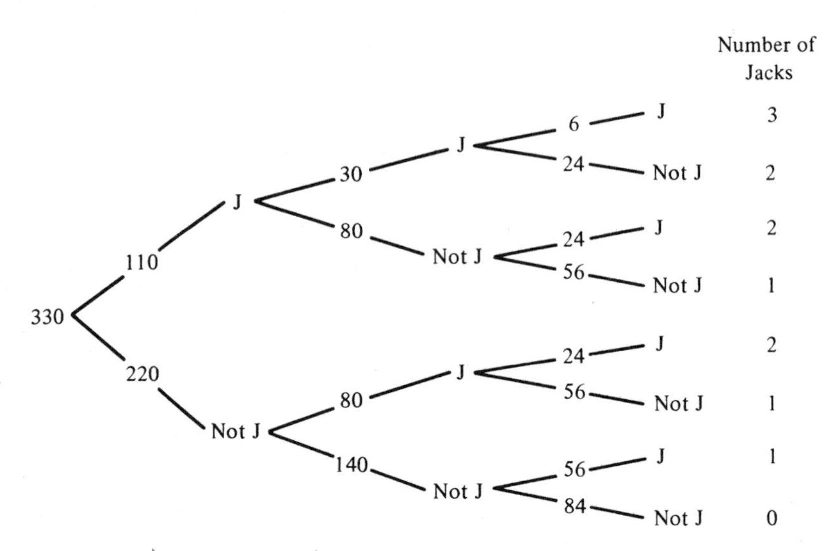

Figure 2-23 330-Trial Frequency Tree for the Royal Deck Process.

Now it is easy to find the probabilities of interest. We see that in the ideal run of 330 trials there are no Jacks in 84 of the cases. Thus the probability of statement (0), "no Jacks," is $\frac{84}{330}$, or $\frac{14}{55}$. Similarly, "exactly one Jack" occurs in $56 + 56 + 56 = 168$ of the 330 trials; thus the probability of statement (1) is $\frac{168}{330}$, or $\frac{28}{55}$. The probability of statement (2), "exactly two Jacks," is $(24 + 24 + 24) \div 330$, or $\frac{12}{55}$. And the probability of statement (3) is $\frac{6}{330}$, or $\frac{1}{55}$.

The same computations may be made directly from the relative frequency tree since the entries on the branches are the same as in the frequency tree except for a common factor of 330, which does not affect the ratios calculated.

More to the point, we could have computed the probabilities directly from the probability tree. Let's confine ourselves to the probability tree of Fig. 2-21 and see how statement (2) may be handled. We see that there are three paths in the truth set of the "two Jacks" outcome. Their respective probabilities are computed by multiplying the branch probabilities along the path, and the probability of the "two Jacks" statement is then calculated as the sum of the probabilities for the three paths in its truth set. Therefore the probability that there are exactly two Jacks is

$$\left(\frac{1}{3} \times \frac{3}{11} \times \frac{8}{10}\right) + \left(\frac{1}{3} \times \frac{8}{11} \times \frac{3}{10}\right) + \left(\frac{2}{3} \times \frac{4}{11} \times \frac{3}{10}\right),$$

or

$$3 \times \frac{24}{330} = \frac{72}{330} = \frac{12}{55}.$$

This is essentially a repetition of the arithmetic that gave the same answer from the frequency tree.

2-2 EXERCISES

1. The probability tree shown in Fig. 2-24 is for the process of drawing twice, without replacement, from an urn containing two red and four white marbles.

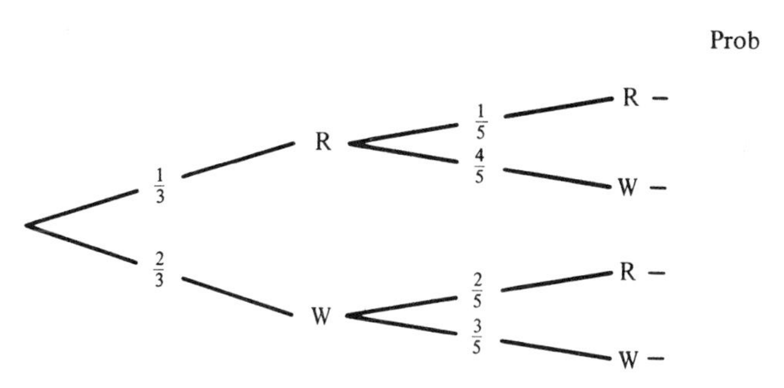

Figure 2-24.

(a) In the "Prob" column on the right, enter the probabilities for the respective paths, and verify that their sum is one.
(b) Find the probability that the second draw is red.
(c) Convert the probability tree to a frequency tree for an ideal run of 15 trials. Then refer to the frequency tree to find the probability that the second draw is red.

2. The process is to draw three marbles, without replacement, from an urn containing three red and two white marbles.
(a) Construct a probability tree. Display the probabilities for the respective paths alongside the terminal points, and verify that their sum is one.
(b) Find the probability that the second draw is white.
(c) Find the probability that exactly one of the draws is white.

3. Urn I contains two red and four white marbles, and Urn II contains three red and two white marbles. At random, an urn is selected and a single marble is drawn from it.
(a) Construct a probability tree.
(b) From the probability tree find the probability that a red marble is drawn.
(c) Convert the probability tree to a frequency tree. Now refer to the frequency tree to find the probability that a red marble is drawn.

4. Two cards are drawn in order, without replacement, from a 12-card Royal Deck consisting of the Jacks, Queens, and Kings from an ordinary deck. Construct a probability tree, and find the probability that
(a) The first card is a King and the second card is a Queen.

(b) At least one of the cards is a King.
(c) Neither of the cards is a Jack.

5. Two dice are rolled. The only feature of relevance for each die is whether or not it is a "six." Construct an appropriate probability tree, and find the probability that
(a) Both dice show a "six."
(b) Neither die shows a "six."
(c) Exactly one of the dice shows a "six."

6. An exhibitor has a large number of firecrackers; 80% are brand A, and 20% are brand B. The probability that a brand A firecracker will fire is .95; for a brand B firecracker, the probability is .9. Suppose a firecracker is chosen at random and tested. What is the probability that it fires?

7. At random three cards are drawn in order, without replacement, from an ordinary 52-card deck. Only the suit of the respective cards is regarded as relevant.
(a) Imagine a probability tree, and exhibit the single path corresponding to the basic outcome "The first card is a Heart, the second is a Diamond, and the third is a Heart."
(b) Find the probability of the outcome in part (a).

8. A coin is flipped ten times. What is the probability that all ten flips are Heads? (Imagine the corresponding path in the probability tree.)

9. Suppose three cards are drawn in order, without replacement, from an ordinary 52-card deck.
(a) Find the probability that all three cards are Hearts by imagining a probability tree and considering the single path corresponding to this coutcome.
(b) Find the probability of the "Three Hearts" outcome of part (a) by counting permutations: There are $P(52, 3)$ equally likely permutations of three cards from a deck of fifty-two, and satisfying the outcome there are $P(13, 3)$ permutations of three cards from among the thirteen Hearts in the deck.
(c) Similarly, find the probability called for in parts (a) and (b) by counting combinations: All subsets of three cards from the deck of fifty-two are equally likely, and some of these are 3-element subsets of the set of thirteen Hearts.

10. There are 18 men and 18 women in a certain club. Suppose a "clean-up-after-the-party" committee of four is selected by drawing names out of a hat. What is the probability that all four are men?

11. Diane has two urns. Urn I contains one red and two white marbles, and Urn II contains two red and two white marbles. At random, Diane chooses an urn and withdraws marbles one-by-one, without replacement, until a red marble is drawn, and then she stops. Construct a probability tree, and find the probability that the number of draws required is (a) 1, (b) 2, (c) 3.

12. The process is to roll a pair of dice and then to roll those dice not showing a "six" a second time. Find the probability that
 (a) Both dice end up "six."
 (b) Neither die ends up "six."
 (c) Exactly one die ends up "six."
 Also,
 (d) Verify that your answers to parts (a), (b), and (c) total one.

2-3. Conditional Probability

The key to the transformation of a tree into a probability tree is an intuitive sense of the branch probabilities associated with the branches in the individual components of the tree. Consider, for example, the simple process of drawing two cards from a deck. Recognizing the equally likely possibilities, we know that if the first card is red, then the probability that the second card is also red is $\frac{25}{51}$. Thus a branch probability of $\frac{25}{51}$ is assigned, as shown in Fig. 2-25.

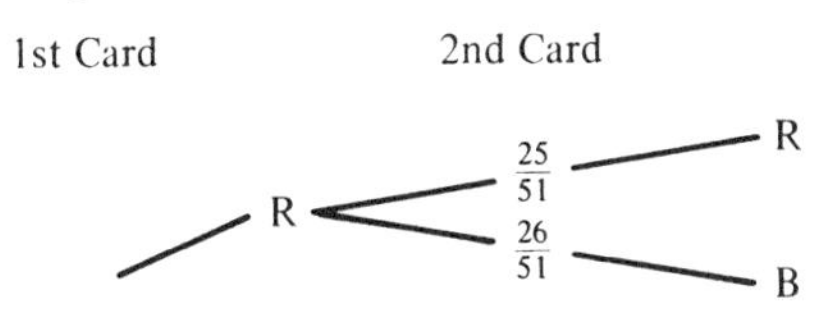

Figure 2-25 Portion of the Probability Tree for the Two-Card Process.

We may interpret the single second-stage component shown in Fig. 2-25 as representing the analysis of the process of drawing a card from a deck composed of twenty-five red and twenty-six black cards. We can equally well identify the process as qualified by the experience of the first stage—the first card is red—in the following language:

> The process is the same as before—two cards are drawn from a deck—but with the understanding that all occurrences are ignored except those in which the first card is red. In other words, we void the trial and treat it as though it had never occurred if the first card is not red.

Here we identify the requirement that the first card is red as a *condition* on the process. The branch probability of $\frac{25}{51}$ that the second card is red, subject to this condition, is identified as a *conditional probability*. We say

> The (conditional) probability that the second card is red, given that the first card is red, is $\frac{25}{51}$.

The modifier "conditional," shown in parentheses, is generally left understood since the "given that" phrase signals that the probability is a conditional probability.

The concept of conditional probability is not confined to intuitively evident cases of the branch probability variety. Staying with the two-card process at hand, let's consider the following:

Problem: Find the probability that the second card is red, given that two cards of different colors are drawn.

The underlying interpretation is that the two-card process is performed a great many times, but only those trials are recorded in which two cards of different colors are drawn. For the trials recorded we are interested in the fraction of times that the second card is red.

Since the experience to be expected in a long string of trials is embodied in an ideal run, we can compute conditional probabilities from a frequency tree for the unqualified process. We recall in Fig. 2-26 the frequency tree for the two-card process. The outcome and condition of interest are shown with the tree. The numbers recorded to the right of the terminal points in the tree take into account those trials in which the condition is satisfied and also those trials in which both condition and outcome are true.

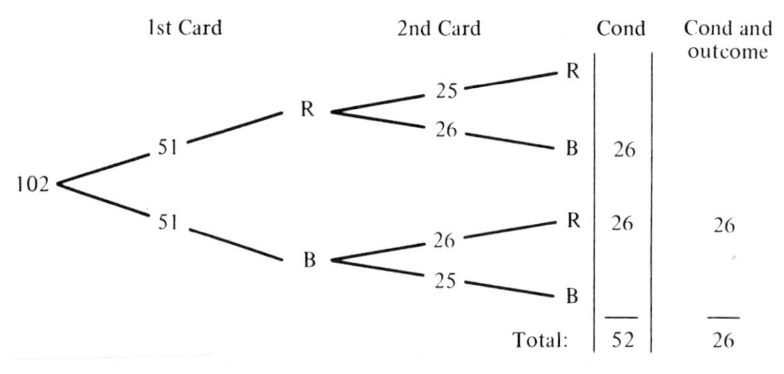

Figure 2-26.

We see that in the ideal run of 102 trials of the unqualified process, the condition—two cards of different colors are drawn—occurs 52 times. In these 52 trials, which are the only ones of interest, the outcome "the second card is red" is true 26 times. Therefore, counting only those trials in which the condition is satisfied, the fraction of times that the given outcome occurs in the long run is $\frac{26}{52}$, or $\frac{1}{2}$, which gives us the solution of the problem posed:

Solution: In the two-card process the probability that the second card is red, given that two cards of different colors are drawn, is one-half.

Essentially the same computation can be made directly from the probability tree. By multiplying branch probabilities we compute the probabilities corresponding to the terminal points in the tree; these probabilities are in a fixed ratio to the frequencies of the corresponding terminal points in a frequency tree. The probability tree corresponding to Fig. 2-26 is shown in Fig. 2-27.

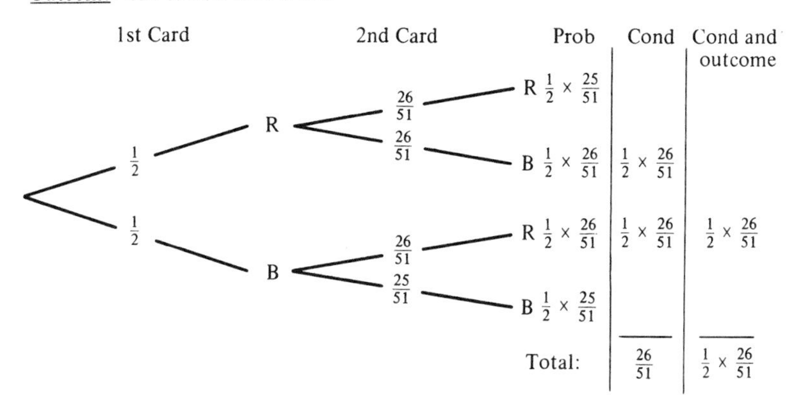

Figure 2-27.

By the reasoning that motivated the computation of the conditional probability from the frequency tree, we compute the conditional probability from the probability tree as the total from the "Cond and Outcome" column on the right divided by the total from the "Cond" column. We obtain the same answer as before:

$$\left(\frac{1}{2} \times \frac{26}{51}\right) \div \frac{26}{51} = \frac{1}{2}.$$

We note that the total from the "Cond and Outcome" column—$\frac{1}{2} \times \frac{26}{51}$—is the probability that both condition and outcome are true, and the total from the "Cond" column—$\frac{26}{51}$—is the probability that the condition is true. And the conditional probability is the one divided by the other.

For another example we return to Karen's two-urn process of Sec. 2-1.

Example: Karen has two urns, I and II. Urn I contains two marbles, one red and one white, and Urn II contains three marbles, one red and two white. At random, Karen selects an urn and withdraws a single marble.

(a) Find the probability that a red marble is drawn, given that Urn I is selected.

(b) Find the probability that Urn I is selected, given that a red marble is drawn.

Solution: (a) Here the conditional probability is reasoned intuitively.

Since Urn I contains two equally likely marbles, one of which is red, the probability that a red marble is drawn, given that Urn I is selected, is one-half.

(b) In Fig. 2-28 we recall the frequency tree for an ideal run of 12 trials in Karen's process. The information bearing on the problem is included.

Condition: A red marble is drawn.
Outcome: Urn I is selected.

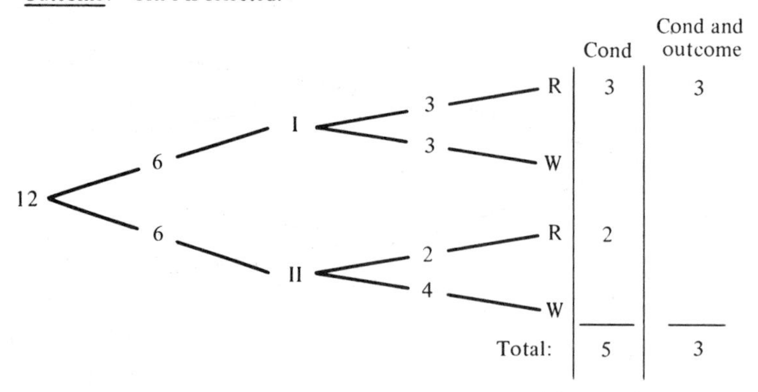

Figure 2-28.

Voiding all trials in which the condition is not satisfied yields an ideal run of 5 trials for the qualified process. In these 5 trials the given outcome occurs 3 times. Therefore the probability that Urn I is selected, given that a red marble is drawn, is $\frac{3}{5}$.

Attacking part (b) on the basis of a probability tree we simplify in Fig. 2-29 by displaying the probabilities of the terminal points multiplied by a common factor of 12. This adjustment does not

Condition: A red marble is drawn.
Outcome: Urn I is selected.

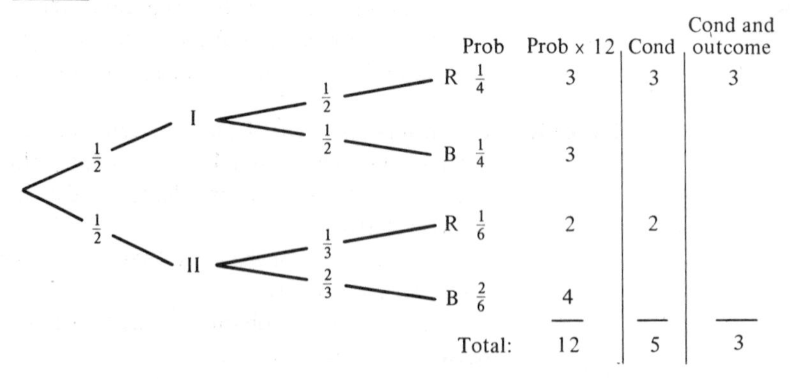

Figure 2-29.

affect the final ratio and does render unmistakable the equivalence to the frequency tree approach.

We compute the desired conditional probability by dividing the probability that condition and outcome are both true by the probability that the condition is true. Looking at the totals in the columns of Fig. 2-29, and correcting for the common factor of 12 that has been introduced, the desired ratio is

$$\frac{3}{\not{1}\not{2}} \div \frac{5}{\not{1}\not{2}} = \frac{3}{5}.$$

Of course, the common factor of 12 divides out in the operation, so the answer of $\frac{3}{5}$ is more easily obtained by simply dividing the total in the "Cond and Outcome" column of Fig. 2-29 by the total in the "Cond" column.

In general, the notion of conditional probability can be applied to any pair of relevant statements in a chance process. Supported by the reasoning underlying the foregoing illustrations, the conclusion is as follows:

Given two relevant statements, identified as the "condition" and the "outcome," in some chance process, the (conditional) probability of the outcome, given the condition, is equal to the probability that both condition and outcome are true divided by the probability that the condition is true.

Example: At random, two cards are drawn, without replacement, from a 12-card Royal Deck consisting of the Jacks, Queens, and Kings from an ordinary deck. Find the probability that

(a) The second card is a Jack, given that the first card is a Queen.

(b) The second card is a Jack, given that the first card is not a Queen.

(c) The second card is a Jack, given that both cards are Jacks.

(d) The second card is a Jack, given that neither card is a Jack.

Solution: (a) In effect, this is a branch probability that is intuitively assigned. After drawing a Queen there are 11 equally likely candidates for the second drawn, 4 of which are Jacks. Thus the answer is $\frac{4}{11}$.

(b) A probability tree, with pertinent information included, is shown in Fig. 2-30. The probabilities associated with the terminal points have been multiplied by a common factor of 33; it should be clear that the entries in the "Prob $\times$ 33" column represent an ideal run of 33 trials.

Dividing the total in the "Cond and Outcome" column by

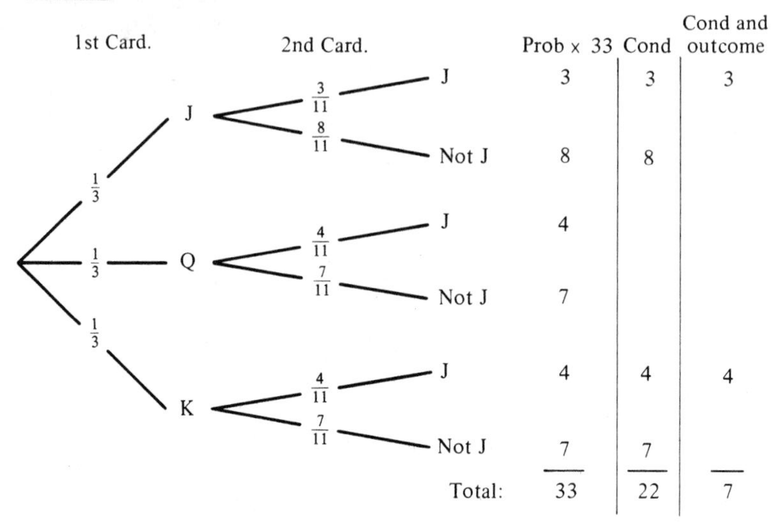

Figure 2-30.

the total in the "Cond" column, the answer to part (b) is seen to be $\frac{7}{22}$.

(c) Clearly the answer is one—the second card is always a Jack if both cards are Jacks.

(d) Clearly the answer is zero—the second card is never a Jack if neither is a Jack.

2-3 EXERCISES

1. Suppose two cards are drawn in order from an ordinary 52-card deck, at random and without replacement.

(a) What is the probability that the second card is a Spade, given that the first card is a Spade?

(b) What is the probability that the second card is a King, given that the first card is not a King?

2. Among the 50 members of a fraternity, 30 study Math and 23 study English, and 12 of the members study both subjects. Suppose a fraternity member is chosen at random and questioned.

(a) What is the probability that the student questioned studies English, given that he studies Math?

(b) What is the probability that he does not study English, given that he does not study Math?

3. As in Prob. 1 of Exercises 2-1, Kate has two urns: Urn I containing two red and three white marbles and Urn II containing one red and one white marble. At random an urn is selected and a single marble withdrawn.
   (a) Find the probability that a red marble is drawn, given that Urn I is selected.
   (b) Construct a 100-trial frequency tree; then find the probability that Urn I is selected, given that a red marble is drawn.
   (c) Construct a probability tree, and use it to find the conditional probability of part (b).

4. At random, two marbles are drawn without replacement from an urn containing three red and four white marbles. Construct a probability tree; then find the probability that
   (a) The second marble is red.
   (b) At least one of the marbles is red.
   (c) The second marble is red, given that at least one of the marbles is red.
   (d) At least one of the marbles is red, given that the second marble is red.

5. In a class of 40 students, 25 take test T_1 and the other 15 take test T_2. Of those taking T_1, 5 fail, and of those taking T_2, 4 fail. The process is to record the test experience of a student chosen at random.
   (a) Find the probability that the student fails, given that he takes test T_1.
   (b) Find the probability that the student takes test T_1, given that he fails.

6. A coin is flipped three times. Construct a probability tree; then find the probability that the second flip is a Head, given that exactly two of the flips are Heads.

7. At random, two cards are drawn in order and without replacement from a 12-card Royal Deck consisting of the Jacks, Queens, and Kings from an ordinary deck.
   (a) Construct a probability tree that takes into account the ranks—Jack, Queen, or King—of the cards drawn.
   (b) Find the probability that the second card is not a Jack, given that the first card is not a Jack.
   (c) Find the probability that the second card is not a Jack, given that the first card is not a Queen.

8. In a variation of Prob. 7, let the two cards be drawn from the Royal Deck *with replacement*; i.e., the first card is returned before the second draw.
   (a) Construct a probability tree that takes into account the ranks of the cards drawn.

(b) Find the probability that the second card is a Jack, given that the first card is not a Jack.

(c) Find the probability that the second card is a Jack, given that at least one of the cards is a Jack.

9. Recall Prob. 9 of Exercises 2-1: Of the voters in a certain town, 50% are Democrats, 40% are Republicans, and 10% are Independents. A proposed ordinance is supported by 60% of the Democrats, 30% of the Republicans, and 70% of the Independents. Find the probability that a randomly chosen voter is an Independent, given that he favors the ordinance.

10. Recall from Sec. 2-1: The Mudlarks of the Mudville softball league win two times out of three on rainy days and one time in three on clear days. And it rains six days out of ten in Mudville.

(a) Find the probability that it is a rainy day, given that the Mudlarks win.

(b) Suppose that the Mudville weather prognosticator predicts rain 60% of the time and that it actually rains on 80% of those days when it has been predicted. Find the probability that rain has not been predicted, given that it is a clear day.

11. A distributor receives his supply of a certain product from three sources: 50% comes from Source A, 30% from Source B, and 20% from Source C. Of those items from Source A, 2% are defective; from Source B, 3% are defective; and from Source C, 5% are defective. Given that a randomly chosen item is defective, what is the probability that it comes from (a) Source A, (b) Source B, (c) Source C?

12. The food-handlers in a community are subjected to a test to detect the presence of a certain infection. If a person has the infection, there is a probability of .9 that the test is positive. If a person does not have the infection, there is a probability of .02 that the test is positive. Suppose that 5% of those tested actually have the infection. What is the probability that a person has the infection, given that the test is positive?

3

MATHEMATICAL PROBABILITY

Common sense, not mathematics, tells us that if we flip a coin a million times it will fall Heads about half the time. And, with a little added thought, ordinary reason tells us that if we flip a coin twice, the probability of Two Heads is one-fourth. It is such primitive agreement about the nature of probability that motivates the mathematical model. Then we may look to the model to tell us, for instance, the probability that in a million flips the coin will fall Heads at least 49.9% of the time.

3-1. The Mathematical Perspective

With two observers following the action, Ben engages in the process of flipping a coin twice. The first observer analyzes Ben's process as follows:

Basic outcome	*Probability*
No Head	$\frac{1}{3}$
One Head	$\frac{1}{3}$
Two Heads	$\frac{1}{3}$

"Clearly in error" announces the second observer. To set his friend straight, he presents his analysis via the probability tree of Fig. 3-1.

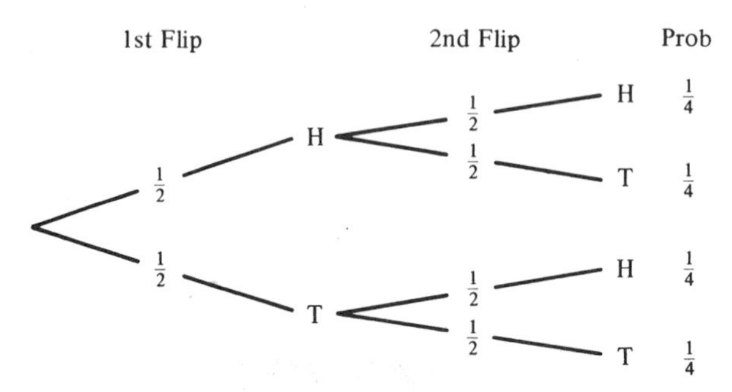

Figure 3-1.

"Merely a difference of opinion" rejoins the first. In the hope of settling the matter they take the dispute to the Mathematician for arbitration. They ask, "Which of these two analyses is mathematically correct?" And the Mathematician answers, "They both look correct to me."

> **Comment:** Let it be made clear that the Mathematician here introduced is to be seen as a caricature in the extreme. The only things of interest to him are abstract structural concepts, such as set and function, which form the building blocks of mathematics. Of things not purely mathematical, such as marbles in urns and decks of cards, our Mathematician is sublimely oblivious (a lofty perspective worthy of appreciation, but not to the exclusion of a firm grip on reality.)

Let's look now at the process of drawing a card from a small deck, and insert the Mathematician into the picture.

At random Bob draws a single card from a 12-card Royal Deck consisting of the Jacks, Queens, and Kings from an ordinary deck. Observing both suit and rank of the card drawn, Bob makes the strong analysis and constructs a basic outcome set consisting of twelve elements corresponding to the twelve cards. Since the cards are regarded as equally likely, each basic outcome is assigned a probability of $\frac{1}{12}$.

How does Bob report his analysis to the Mathematician? Statements such as "The Jack of Clubs is drawn" are meaningless to the Mathematician, but he understands the purely mathematical aspects of the situation completely.

What Bob transmits to the Mathematician is a system consisting of a set of 12 elements, designated

$$S = \{\text{JC, JD, JH, JS, QC, QD, QH, QS, KC, KD, KH, KS}\},$$

with a *probability weight* of $\frac{1}{12}$ assigned to each of the elements. Whereas Bob sees "JC" as a code for a certain basic outcome, the Mathematician sees "JC" simply as a symbol for a particular element of the set S. The Mathematician verifies that the given weights are nonnegative numbers adding up to one and accepts the system.

From the system of elements and weights, the Mathematician constructs a function that associates each subset of the set S with a number. This function is called the *probability function*, and is designated "Pr." For each subset X of S, the value assigned by the function Pr is obtained by adding the probability weights for the elements in X. This value, *the probability of* X, is designated $\Pr(X)$. The set S together with the function Pr is said to be a *probability space*.

Suppose Bob wants the Mathematician's help in finding the probability of the statement "A Jack is drawn." From the arguments of the previous chapter, the probability of a statement is the sum of the probabilities of the basic outcomes in its truth set, and these basic probabilities are precisely the weights which Bob gave the Mathematician. And the value assigned by the probability function Pr to a subset of S is the sum of the weights of its member elements. Thus the probability of a relevant statement is the value assigned by the function Pr to the truth set of the statement. So Bob phrases his request as follows:

$$\text{Let } J = \{\text{JC, JD, JH, JS}\}, \text{ find } \Pr(J), \text{ please.}$$

The Mathematician sums the weights of $\frac{1}{12}$ each for the 4 elements of the subset J and responds that

$$\Pr(J) = \frac{4}{12} = \frac{1}{3}.$$

Thus if Bob wishes to avail himself of the Mathematician's talents, he must translate his statements into subsets, i.e., represent the relevant statements by their truth sets. This switching from statement to subset and back may get confusing, so we introduce the word *event* to serve the dual purpose. To the Mathematician the event corresponding to the statement "A Jack is drawn" is the subset

$$J = \{\text{JC, JD, JH, JS}\}.$$

Bob may prefer to think of the event as the statement itself or as the actual happening.

Grammar and Set Theory

A digression is in order here to bring the operations of set theory into the exposition.

In the illustration above we had Bob approaching the Mathematician for $\Pr(J)$, where

$$J = \{\text{JC, JD, JH, JS}\}$$

is the event corresponding to the statement "The card drawn is a Jack." Now suppose Bob is interested in the event "The card drawn is *not* a Jack." This is the *negation* of the former statement. The truth set of the negation is the subset

$$\{QC, QD, QH, QS, KC, KD, KH, KS\},$$

which consists of those elements of the set S that do not belong to J.

In the present setting the set S is the source of all the elements from which events are assembled. For this reason we say that S is the *universal set* in the context of Bob's analysis.

In the language of set theory: Given a universal set S and a subset E of S, the subset consisting of precisely those elements of S which do not belong to E is called the *complement* of E and is denoted by $\bar{E}$.

It should be clear that if E is the truth set of a relevant statement with respect to a basic outcome set S, then $\bar{E}$ is the truth set of the negation of that statement.

With event J as before, let C be the event "The card drawn is a Club," i.e.,

$$C = \{JC, QC, KC\}.$$

Now consider the statement "The card drawn is a Jack *and* it is a Club." The truth set of this statement,

$$\{JC\},$$

is the subset consisting of those elements of S which belong to both J and C.

In general, if E_1 and E_2 are subsets of a universal set S, then the subset consisting of the elements belonging to both E_1 and E_2 is called the *intersection* of E_1 and E_2 and is denoted by $E_1 \cap E_2$.

Finally, the statement "The card drawn is a Jack *or* it is a Club" has as its truth set the subset

$$\{JC, JD, JH, JS, QC, KC\},$$

consisting of those elements which belong to either J or C or both. This subset is called the *union* of J and C and is denoted by $J \cup C$.

In general, $E_1 \cup E_2$, the union of E_1 and E_2, consists of the elements belonging to either or both of E_1 and E_2.

Comment: In the present context it is appropriate to read "$\bar{J}$" as "not J," "$J \cap C$" as "J and C," and "$J \cup C$" as "J or C."

We have been treating, and shall continue to treat, the notion of "set" as a primitive intuitive concept. Complementation, union, and intersection are viewed in the same perspective. But a couple of universally accepted conventions deserve attention.

A subset of a set S is, of course, a set whose elements are drawn from S. Contrary to what the prefix "sub" may suggest, the entire set S is regarded as a subset of itself. At the other extreme, we admit a subset which contains no elements at all. This *empty* subset is denoted by the symbol "$\varnothing$." Thus if A and B have no elements in common, then $A \cap B = \varnothing$. In the probability space, where the function Pr is required to map each subset of the universal set S into a number, we define $\Pr(\varnothing)$ to be zero.

Analyses Without Processes

The Mathematician sees the mathematical structure as existing apart from the real-world process it is designed to serve. To better appreciate this perspective, let us view a simple probability space from a purely mathematical point of view.

We are offered a system consisting of a set S containing 4 elements,

$$S = \{h_1, h_2, h_3, h_4\},$$

with probability weights assigned as follows:

Element	h_1	h_2	h_3	h_4
Weight	.1	.2	.3	.4

Since the weights are nonnegative numbers whose sum is one, we accept the system as a basis for generating a probability space.

In the present context, the universal set S is called the *sample set*. As before, the subsets of the sample set are called *events*. The probability function Pr is defined by associating each event with the sum of the weights of its member elements. Since the sample set S is so small, it is feasible to display the entire probability space. We give it the name "$\mathcal{H}$," for "hypothetical."

The Probability Space $\mathcal{H}$
The Sample Set, $S = \{h_1, h_2, h_3, h_4\}$

Event	Pr	Event	Pr	Event	Pr
$\varnothing$	0	$\{h_1, h_2\}$	.3	$\{h_1, h_2, h_3\}$	.6
$\{h_1\}$	.1	$\{h_1, h_3\}$	.4	$\{h_1, h_2, h_4\}$	.7
$\{h_2\}$	.2	$\{h_1, h_4\}$	.5	$\{h_1, h_3, h_4\}$	.8
$\{h_3\}$	.3	$\{h_2, h_3\}$	.5	$\{h_2, h_3, h_4\}$	.9
$\{h_4\}$	.4	$\{h_2, h_4\}$	.6	S	1
		$\{h_3, h_4\}$	.7		

Simple as it is, the space $\mathcal{H}$ is representative of all probability spaces of the sort that we shall be considering. For the space with sample set S of twelve elements corresponding to the analysis of Bob's one-card process it would be tedious to display all of the events, but it doesn't take much imagination to conceive of the listing.

Example: The probability space $\mathcal{G}$ has sample set

$$S = \{g_1, g_2, g_3, g_4, g_5, g_6\}.$$

The probability function is generated by the following assignment of weights:

Element of S	g_1	g_2	g_3	g_4	g_5	g_6
Weight	.05	.10	.25	.35	.15	.10

Consider the events

$$A = \{g_1, g_2, g_3\} \quad \text{and} \quad B = \{g_2, g_4, g_6\}.$$

Find

(a) $\Pr(A)$, $\Pr(B)$.
(b) $\Pr(\bar{A})$, $\Pr(\bar{B})$.
(c) $\Pr(A \cup B)$, $\Pr(A \cap B)$, $\Pr(A \cup \bar{A})$, $\Pr(A \cap \bar{A})$.

Solution: (a) Adding the weights of the member elements, we have

$$\Pr(A) = \Pr(\{g_1, g_2, g_3\}) = .05 + .10 + .25 = .40$$
$$\Pr(B) = \Pr(\{g_2, g_4, g_6\}) = .10 + .25 + .10 = .55.$$

(b) Since $A = \{g_1, g_2, g_3\}$, and since the complement $\bar{A}$ consists of those elements of the sample set S that do not belong to A, it follows that

$$\bar{A} = \{g_4, g_5, g_6\}.$$

Therefore,

$$\Pr(\bar{A}) = .35 + .15 + .10 = .60.$$

Similarly,

$$\Pr(\bar{B}) = \Pr(\{g_1, g_3, g_5\}) = .05 + .25 + .15 = .45.$$

(c) Since the union of two sets consists of those elements belonging to either, or both, of the sets, we have

$$A \cup B = \{g_1, g_2, g_3, g_4, g_6\}.$$

Thus

$$\Pr(A \cup B) = .05 + .10 + .25 + .35 + .10 = .85.$$

Recalling that the intersection of two sets consists of those elements belonging to both of the sets, we have

$$A \cap B = \{g_2\}.$$

Consequently,

$$\Pr(A \cap B) = .10.$$

From the meaning of union, intersection, and complement,

$$A \cup \bar{A}$$

is the universal set S and

$$A \cap \bar{A}$$

is the empty set $\varnothing$. Therefore,

$$\Pr(A \cup \bar{A}) = \Pr(S) = 1,$$

and

$$\Pr(A \cap \bar{A}) = \Pr(\varnothing) = 0.$$

Since the structure of probability spaces is motivated by the interpretation of probability for actual processes, it may seem that the consideration of a space without reference to a process is putting the cart before the horse. Let's look at the space $\mathcal{G}$ and see if we can find a horse to fit the cart.

Example: Invent a process for which the probability space $\mathcal{G}$ of the previous example is an appropriate mathematical model.

Solution: (This solution is indicative of a universal solution to problems of this type—but everyone to his own taste.)

George has an urn containing twenty marbles: one red, two white, five blue, seven yellow, three green, and two black. The process is to withdraw one marble at random and observe its color.

Regarding the marbles as equally likely, George analyzes the process as follows:

	Basic outcome	*Probability*
g_1:	The red marble is drawn	.05
g_2:	One of the two white marbles is drawn	.10
g_3:	One of the five blue marbles is drawn	.25
g_4:	One of the seven yellow marbles is drawn	.35
g_5:	One of the three green marbles is drawn	.15
g_6:	One of the two black marbles is drawn	.10

The basic outcomes have been coded to establish a correspondence with the elements of the sample set S in the probability space $\mathcal{G}$. With this correspondence the truth set of each relevant statement about the process translates into an event in the space, and the probability associated with the statement agrees with the value assigned by the probability function Pr to the event. Recognizing this correspondence, we see the probability space $\mathcal{G}$ as a mathematical model for George's process.

3-1 EXERCISES

1. The universal set S consists of 8 elements, as follows:

$$S = \{s_1, s_2, s_3, s_4, s_5, s_6, s_7, s_8\}.$$

For $A = \{s_1, s_2, s_5, s_6\}$, $B = \{s_1, s_3, s_5, s_7\}$, and $C = \{s_2, s_4, s_6, s_8\}$, exhibit the following subsets:

$$\bar{A}, \quad \bar{B}, \quad A \cup B, \quad A \cap B, \quad \bar{A} \cup C, \quad A \cap \bar{C}.$$

2. In the mathematical model of the process of flipping a coin three times the sample set is represented as follows:

$$S = \{\text{HHH, HHT, HTH, HTT, THH, THT, TTH, TTT}\}.$$

A weight of $\frac{1}{8}$ is assigned to each of the elements of S. Consider the events

$$A = \{\text{HHT, HTH, THH}\} \quad \text{and} \quad B = \{\text{HHH, HHT, THH, THT}\}.$$

(a) With the obvious interpretation of the elements of the sample set S, express the events A and B as statements about the actual process.

(b) Express the following events as statements:

$$\bar{A}, \quad A \cup B, \quad A \cap \bar{B}.$$

(c) Exhibit the events in part (b) as subsets of S, and find the probability of each.

3. The probability space $\mathcal{K}$ has the sample set

$$S = \{k_1, k_2, k_3, k_4, k_5\}.$$

The probability function is generated by the following assignment of weights:

Element of S	k_1	k_2	k_3	k_4	k_5
Weight	.25	.15	.2	.3	.1

Consider the events

$$A = \{k_1, k_2, k_3\} \quad \text{and} \quad B = \{k_1, k_3, k_5\}.$$

Find (a) $\Pr(A)$, (b) $\Pr(\bar{A})$, (c) $\Pr(A \cap B)$, (d) $\Pr(A \cup B)$.

4. In the probability space $\mathcal{K}$ of Prob. 3, let X and Y be the events

$$X = \{k_1, k_2\}, \qquad Y = \{k_4, k_5\}$$

Find (a) $\Pr(X)$, (b) $\Pr(\bar{Y})$, (c) $\Pr(X \cap \bar{Y})$, (d) $\Pr(X \cup \bar{Y})$.

5. In the probability space $\mathcal{K}$ of Prob. 3, let the events A and B be as follows. Find $\Pr(A \cap B)$ and $\Pr(A \cup B)$.

(a) $A = \{k_1, k_3, k_4\}$, $B = \{k_2, k_5\}$.

(b) $A = \varnothing$, $B = \{k_1, k_3, k_5\}$.

(c) $A = \{k_1, k_2, k_3\}$, $B = S$.

6. Invent a process for which the probability space $\mathcal{K}$ of Prob. 3 is an appropriate mathematical model. (You might draw a marble from an urn containing marbles of several colors. How many colors, and how many marbles of the respective colors should there be?) List the basic outcomes of your process, and for each basic outcome, indicate the corresponding element of the sample set S.

7. For the probability space $\mathfrak{M}$ the elements of the sample set S and the weights generating the probability function are as follows:

Element of S	m_1	m_2	m_3
Weight	.2	.5	.3

(a) Display the entire probability space $\mathfrak{M}$, exhibiting each event X together with $\Pr(X)$.

(b) Invent a process for which the space $\mathfrak{M}$ is an appropriate mathematical model.

8. Suppose the process is to choose at random a marble from an urn containing three red, four white, and five black marbles. As a mathematical model we construct a probability space with the sample set

$$S = \{R, W, B\}.$$

Display all of the events in the space, along with the values assigned by the probability function Pr.

9. For the probability space $\mathcal{A}$ the elements of the sample set S and the weights generating the probability function are as follows:

Element of S	a_1	a_2	a_3	a_4
Weight	.2	.1	.4	.3

Display the entire probability space $\mathcal{A}$.

10. In the probability space $\mathcal{E}$ with the sample set $S = \{e_1, e_2, e_3\}$, events $A = \{e_1, e_2\}$ and $B = \{e_2, e_3\}$ have the following probabilities:

$$\Pr(A) = \frac{7}{12} \quad \text{and} \quad \Pr(B) = \frac{8}{12}.$$

What weights are assigned to the elements of S?

3-2. The Formal System

The examples of probability spaces in the previous section have been intended to motivate the formal definition, not to substitute for it. But before moving to the definition it is well that we clarify the raw materials that we presume to have on hand for the construction and development of the system.

There are three fundamental concepts which we take as the primitive ingredients from which mathematical systems are constructed. These concepts are *set*, *function*, and *number*.

A grasp of the elementary notion of "set" is presumably part of everyone's mental equipment. The notion carries with it the ideas of elements, subsets, complements, unions, and intersections. The terminology may have been troublesome on first encounter, but it is assumed that this difficulty has been largely resolved by usage. We view the concept of set and its subsidiary concepts as premathematical—common logical notions in terms of which we are free to build mathematical systems.

Similar comments apply to the "function" concept. The definition given for "function" in Chapter 1 was intended only to clarify the use of the word by tying it to the essentially synonymous expression "method of associating." It is presupposed that everyone knows what "method of associating," and therefore "function," means.

We take the same position with respect to the concept "number," but this demands some apology. If by "number" we mean only the *natural numbers*, 1, 2, 3, 4, etc., then there is little uneasiness in premising "number" as a fundamental logical concept. Indeed one may be inclined to accept natural numbers as a part of the "set" family of ideas; for example, it should seem quite natural to speak of a set of *three* elements. In fact, the natural numbers have a definite role in the exposition as logical entities of this character.

However, in general we use the word "number" to refer to the elements of a much larger set, the *real numbers*. For a working criterion, we imagine a line, unbounded in either direction, with a number scale superimposed. This is the *number line* indicated in Fig. 3-2. Each point of the number line is to be seen as corresponding to a real number and vice versa. Thus the following are real numbers:

$$-\frac{5}{2}, \quad 3, \quad \sqrt{2}, \quad 1.234, \quad \pi = 3.14159\ldots.$$

But ∞ (infinity) and $\sqrt{-1}$ (where $\sqrt{-1} \times \sqrt{-1} = -1$) are not real numbers, because they do not correspond to points on the number line.

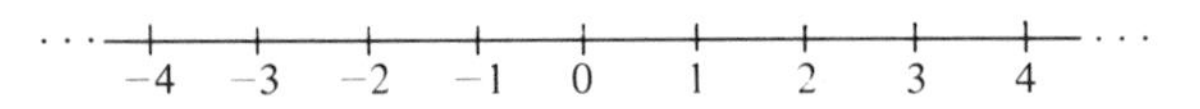

Figure 3-2 The Number Line.

Presumably everyone knows that real numbers are things that can be added, subtracted, multiplied, and divided. Also, real numbers can be compared, "less than" or "greater than," and they can be raised to powers. It is, moreover, assumed that there is sufficient agreement about the symbolism for representing numbers and about the rules governing their manipulation to permit a good deal of progress before subtle difficulties begin to arise.

It takes a lot for granted, but, in the interest of efficiency, we premise the real number system as an intellectual given, available for use in the construction of more advanced mathematical systems.

Comment: From the point of view of pure mathematics, a more satisfying course would be to evolve the real numbers from the primitive concepts of set, function, and natural number. This can be done, but it would take much too long.

Accompanying the fundamental notions of set, number, and function, it goes without saying that we must assume a certain familiarity with the terms of ordinary language. Finally, although debate about valid lines of argument has continued from antiquity to the present, it is sufficient for us to presuppose the ability to follow a line of reasoning based on "common sense." And so we do.

With this attempt to establish the ground on which we stand, we return to the subject.

Definition of a Probability Space

Earlier we displayed examples of probability spaces. We saw that one needed a sample set S and a function Pr mapping the subsets of the sample set into real numbers. The requirement was that the value of Pr for an event is the sum of nonnegative weights associated with the member elements, with the sum of the weights for all the elements of S being one. The formal definition is stated in more sophisticated terms.

Definition: A *probability space* consists of two things:

1. A set S, called the *sample set*.
2. A function Pr, called the *probability function*, mapping the subsets of S, called *events*, into real numbers.

The following *axioms* are imposed:

Axiom 1. For each event X,

$$0 \leq \Pr(X) \leq 1;$$

i.e., the value assigned to X by the function Pr is a real number between 0 and 1, inclusive.

Axiom 2. $\Pr(\varnothing) = 0$ and $\Pr(S) = 1$.

Axiom 3. If X and Y are events such that $X \cap Y = \varnothing$, then

$$\Pr(X \cup Y) = \Pr(X) + \Pr(Y).$$

The language is fancier, but a probability space as determined by the formal definition can still be viewed as having been generated by an assignment of weights to the elements of the sample set S. For each element s in S the corresponding weight is $\Pr(\{s\})$—for symbolic efficiency $\Pr\{s\}$—the value assigned by the function Pr to the subset consisting of the single element s. Such a singleton subset $\{s\}$ is called an *elementary event*; its probability $\Pr\{s\}$ is referred to as a weight, or as an *elementary probability*.

The reconciliation of the formal definition with the earlier construction of probability spaces is set forth in the following *theorem*. (In general, any result deduced by logical reasoning from the axioms of a given mathematical system is called a theorem.)

Theorem 1: Given a nonempty event A in a probability space, let

$$A = \{a_1, a_2, \ldots, a_k\};$$

i.e., we label the distinct elements of A as "a_1", "a_2", . . . , "a_k". Then

$$\Pr(A) = \Pr\{a_1\} + \Pr\{a_2\} + \cdots + \Pr\{a_k\}.$$

Proof: If A consists of a single element,

$$A = \{a_1\},$$

the conclusion of the theorem is immediate.

If A consists of two elements,

$$A = \{a_1, a_2\},$$

we let

$$X = \{a_1\} \quad \text{and} \quad Y = \{a_2\}.$$

Thus

$$X \cup Y = A \quad \text{and} \quad X \cap Y = \varnothing.$$

Applying Axiom 3,

$$\begin{aligned} \Pr(A) &= \Pr(X \cup Y) = \Pr(X) + \Pr(Y) \\ &= \Pr\{a_1\} + \Pr\{a_2\}. \end{aligned}$$

Therefore, the theorem is valid for all events consisting of two elements.

If A contains three elements,

$$A = \{a_1, a_2, a_3\},$$

we let

$$X = \{a_1, a_2\} \quad \text{and} \quad Y = \{a_3\}.$$

Since X contains two elements, it follows from the prior argument that

$$\Pr(X) = \Pr\{a_1\} + \Pr\{a_2\}.$$

And since $X \cap Y = \varnothing$, we have, by Axiom 3,

$$\begin{aligned} \Pr(A) &= \Pr(X \cup Y) = \Pr(X) + \Pr(Y) \\ &= \Pr\{a_1\} + \Pr\{a_2\} + \Pr\{a_3\}. \end{aligned}$$

Thus for every event of three elements, the probability is the sum of the elementary probabilities corresponding to the member elements.

For an event of four elements,

$$A = \{a_1, a_2, a_3, a_4\},$$

we let $X = \{a_1, a_2, a_3\}$ and $Y = \{a_4\}$, and proceed as before. And so on. For every value of k, the argument repeats itself until it reaches events consisting of k elements. Always the conclusion is that the probability of A is the sum of the elementary probabilities of the elementary events corresponding to the member elements.

Comment: The argument above, like the ones to follow throughout the book, is simplified by the relaxed position we've assumed with respect to the primitive concepts taken for granted. If the foundation were on a deeper level, the crucial phrase "and so on" in the above proof would have to be replaced by a careful argument employing the "Axiom of Mathematical Induction" for the natural numbers.

Empirical Versus Mathematical

The probability space as a formal mathematical system exists on a level apart, but it has been motivated by, and is intended to serve, real-world processes. To emphasize the distinction, the actual processes are identified as *empirical.*

Some fundamental empirical truths are explicitly reflected in the axioms of the mathematical model. Other empirical truths suggest theorems in the system.

For example, consider the assertion

$$\Pr(\bar{A}) = 1 - \Pr(A),$$

where A is any event. If we interpret the assertion empirically, that is, with respect to a real-world process, it should be clear that this is a true statement. We think of $\Pr(A)$ as the fraction of the time that A occurs in the long run, and since not–A occurs precisely when A does not, it follows that $\Pr(\bar{A}) = 1 - \Pr(A)$.

Of course, there is no mention of the concept "long-run fraction" in the definition of a probability space. The argument above is founded not on the formal system but on the empirical interpretation underlying the mathematical structure. We say that it is a *heuristic* argument. As important as heuristic arguments are for supporting the plausibility of results, mathematical proofs must be of an abstract character, confined to the structure and axioms as set forth in the definition of the system.

If the formal probability space is a full and faithful model of the empirical situations that it has been designed to serve, then it should be possible to prove, in the mathematical setting, statements which are true in the empirical interpretation. So if we can't reach the result $\Pr(\bar{A}) = 1 - \Pr(A)$, we should consider adding it as a fourth axiom. But we can reach it.

Theorem 2: Let A be an event in a probability space. Then

$$\Pr(\bar{A}) = 1 - \Pr(A).$$

Proof: From the meaning of complement, no element of the sample set S can belong to both A and also to its complement $\bar{A}$, and every element of S must belong to either A or to $\bar{A}$. Therefore,

$$A \cap \bar{A} = \varnothing \quad \text{and} \quad A \cup \bar{A} = S \tag{1}$$

Since $A \cap \bar{A} = \varnothing$, we may apply Axiom 3 with $X = A$ and $Y = \bar{A}$, giving

$$\Pr(A \cup \bar{A}) = \Pr(A) + \Pr(\bar{A}). \tag{2}$$

From Eq. (1), $A \cup \bar{A} = S$. Making this substitution in Eq. (2), we have

$$\Pr(S) = \Pr(A) + \Pr(\bar{A}). \tag{3}$$

From Axiom 2,

$$\Pr(S) = 1. \tag{4}$$

Inserting this value in Eq. (3),

$$1 = \Pr(A) + \Pr(\bar{A}). \tag{5}$$

Subtracting $\Pr(A)$ from both sides and interchanging sides, we have the desired result:

$$\Pr(\bar{A}) = 1 - \Pr(A).$$

General Remarks

We have lifted the concept of probability from its empirical interpretation as a "long-run fraction" and installed it in a mathematical model constructed from the abstract notions of set, number, and function.

There are distinct advantages to the elevation of perspective offered by the construction of a mathematical model. Given the model, we can bring a vast reservoir of mathematical results and techniques to bear on the problem. Also, there is the fact that similar, and sometimes diverse, types of empirical problems yield to the same mathematical treatment.

A more subtle reason for constructing a mathematical model is that empirical situations are often loosely defined and subjective by their nature, so agreement on a model is essential to clarifying the problem. The subject at hand furnishes a good illustration.

Anyone with a practical understanding of the terms involved should agree that the probability space as defined mathematically is consistent with the underlying motivation—"long-run fraction" does possess the properties postulated for the theoretical function Pr. But, whereas there is no ambiguity in the mathematical system, it is virtually impossible to give a satisfying empirical definition of even so simple an expression as the following:

"In flipping a coin the probability of Heads is one-half."

In every effort at a precise definition the fundamental difficulty manifests itself in the use of terms such as "expect" or "probably" or "very likely," and the attempt to define these terms leads one around in circles. In effect, what we do is agree that the corresponding probability space is a good model for the process and go on from there.

In the applications that concern us, probability can always be thought of as a "long-run fraction." But there are serious difficulties in a less confined use of the term. For example, suppose a scientist declares

> "The probability that there are other forms of intelligent life in the universe is greater than .99."

It's not easy to make much sense out of this statement on the basis of "long-run fractions." Indeed, reasonable men may differ as to whether the statement makes any sense at all. One resolution is to have the scientist enlarge on the framework of his statement and express it in the context of a probability space. Then the reasonable men can debate whether or not his probability space is a good model of the situation.

3-2 EXERCISES

1. Suppose the sample set S consists of eight elements, as follows:

$$S = \{s_1, s_2, s_3, s_4, s_5, s_6, s_7, s_8\}.$$

For each of the following pairs of events A and B, exhibit the events

$$A \cap B, \quad A \cap \bar{B}, \quad \bar{A} \cap B, \quad \bar{A} \cap \bar{B}.$$

(a) $A = \{s_1, s_2, s_3, s_4\}$ and $B = \{s_2, s_4, s_6, s_8\}$.
(b) $A = \{s_1, s_2, s_3, s_4\}$ and $B = \{s_1, s_2\}$.
(c) $A = \{s_1, s_2, s_3, s_4\}$ and $B = \{s_7, s_8\}$.
(d) $A = \{s_1, s_2, s_3, s_4\}$ and $B = S$.

2. Let A and B be events in a probability space. Give an argument to support the fact that each element of the sample set S must belong to exactly one of the following four events:

$$A \cap B, \quad A \cap \bar{B}, \quad \bar{A} \cap B, \quad \bar{A} \cap \bar{B}.$$

3. For events A and B in a probability space it is given that

$$\Pr(A \cap B) = .25 \quad \text{and} \quad \Pr(A \cap \bar{B}) = .15.$$

From this information, deduce that

$$\Pr(A) = .4.$$

(Use the result of Prob. 2.)

4. For events A and B in a probability space it is given that

$$\Pr(A \cap B) = .25, \qquad \Pr(A \cap \bar{B}) = .4,$$
$$\Pr(\bar{A} \cap B) = .2, \qquad \Pr(\bar{A} \cap \bar{B}) = .15.$$

Find

$$\Pr(A), \quad \Pr(B), \quad \Pr(A \cup B), \quad \Pr(\bar{A} \cup B).$$

5. Let A and B be subsets of a universal set S. Give arguments to support the following set-theoretic identities:

$$A \cup B = A \cup (\bar{A} \cap B), \tag{1}$$

$$B = (A \cap B) \cap (\bar{A} \cap B), \tag{2}$$

$$A \cap (\bar{A} \cap B) = \varnothing, \tag{3}$$

$$(A \cap B) \cap (\bar{A} \cap B) = \varnothing. \tag{4}$$

Then use these identities in establishing the following important result:

Theorem: Let A and B be events in a probability space. Then

$$\Pr(A \cup B) = \Pr(A) + \Pr(B) - \Pr(A \cap B).$$

6. For events A and B in a probability space it is given that

$$\Pr(A) = .50, \quad \Pr(B) = .70, \quad \Pr(A \cap B) = .25.$$

Find
(a) $\Pr(A \cup B)$.
(b) $\Pr(\bar{A} \cup B)$.

7. (This exercise shows that Axiom 2 for a probability space says more than necessary.) Axiom 2 for a probability space states

$$\Pr(\varnothing) = 0 \quad \text{and} \quad \Pr(S) = 1.$$

Suppose we change the axiom by deleting $\Pr(\varnothing) = 0$. In other words, replace Axiom 2 by

Axiom 2′: $\Pr(S) = 1$.
From Axioms 1, 2′, and 3, deduce as a theorem the assertion

$$\Pr(\varnothing) = 0.$$

(*Hint:* $\varnothing \cap X = \varnothing$ and $\varnothing \cup X = X$ for all events X.)

3-3. Conditional Probability and Independent Events

Only the most basic of the empirical features of probability have been translated into the formal definition of a probability space. Other notions are brought to life in the model by well motivated definitions within the system. Recalling the intuitive interpretation, let's lift the concept of conditional probability into the mathematical system.

In the discussion of Sec. 2-3 we addressed ourselves to two relevant statements in a chance process, identified as the "outcome" and "condition," respectively. The (conditional) probability of the outcome, given the condition, was interpreted empirically as the fraction of the time that the outcome occurs in the long run if we count only those trials in which the condition is true. In terms of the probabilities for the unqualified process, the conclusion was that the probability of the outcome, given the condition, is equal to the probability that both the outcome and condition occur divided by the probability that the condition occurs. The definition of conditional probability in the mathematical system is motivated by this observation.

In the mathematical model, let A and B be events interpreted as "outcome" and "condition," respectively. The probability that both outcome and condition are true is represented by $\Pr(A \cap B)$, and the probability that the condition is true is represented by $\Pr(B)$. From the empirical interpretation it follows that the (conditional) probability of A, given B, is expressed by the ratio

$$\frac{\Pr(A \cap B)}{\Pr(B)}.$$

With this background, we make the following definition:

Definition: Given two events A and B in a probability space, the (conditional) *probability of A, given B*, represented symbolically as $\Pr(A \mid B)$, is defined as follows:

$$\Pr(A \mid B) = \frac{\Pr(A \cap B)}{\Pr(B)}$$

provided $\Pr(B) \neq 0$. [$\Pr(A \mid B)$ is not defined when $\Pr(B) = 0$.]

Comment: In effect, the condition B restricts each event A to $A \cap B$, i.e., reduces the subset A to those elements that are also contained in B. The division by $\Pr(B)$ restores the requirement that the elementary probabilities for all surviving elements add up to one.

Example: A probability space is generated by weights assigned to the elements of the sample set S as shown in the following tabulation:

Element of S	e_1	e_2	e_3	e_4	e_5	e_6	e_7
Weight	.1	.2	.15	.1	.15	.2	.1

For the events

$$A = \{e_2, e_3, e_4, e_5\} \quad \text{and} \quad B = \{e_1, e_3, e_5, e_7\},$$

find

$$\Pr(A \mid B) \quad \text{and} \quad \Pr(B \mid A).$$

Solution: Adding the weights assigned the respective elements, we have

$$\Pr(A) = .2 + .15 + .1 + .15 = .6,$$

$$\Pr(B) = .1 + .15 + .15 + .1 = .5.$$

Also

$$\Pr(A \cap B) = \Pr(B \cap A) = \Pr\{e_3, e_5\} = .15 + .15 = .3.$$

Therefore,

$$\Pr(A \mid B) = \frac{\Pr(A \cap B)}{\Pr(B)} = \frac{.3}{.5} = .6,$$

and

$$\Pr(B \mid A) = \frac{\Pr(B \cap A)}{\Pr(A)} = \frac{.3}{.6} = .5.$$

Example: Mary has two urns. Urn I contains two red marbles and three black marbles, and Urn II contains three red marbles and seven black marbles. At random she selects an urn and withdraws a single marble.

Construct a probability space to serve as a model for this process. Then, by interpretation within the model, find

(a) The probability that a red marble is drawn, given that Urn I is selected.
(b) The probability that Urn I is selected, given that a red marble is drawn.

Solution: The empirical process is represented by the probability tree of Fig. 3-3. The paths in the tree, which represent the elements of the sample set S, have been labeled e_1, e_2, e_3, and e_4. The probability weight for each path has been computed as the product of the branch probabilities.

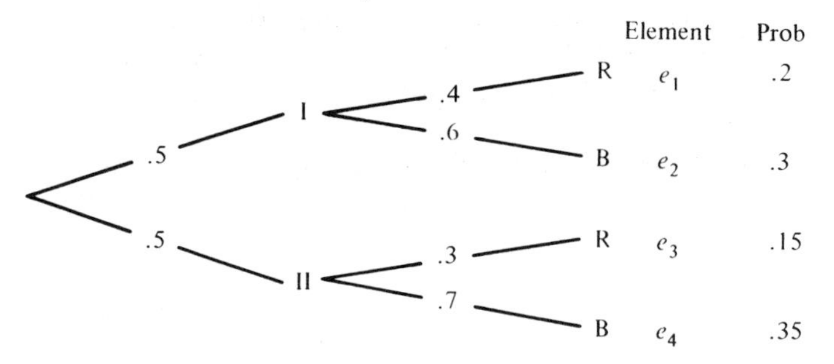

Figure 3-3.

The probability space representing the process is generated by the following assignment of weights to the elements of the sample set S:

Element of S	e_1	e_2	e_3	e_4
Weight	.2	.3	.15	.35

The statement "A red marble is drawn" corresponds to the event

$$R = \{e_1, e_3\},$$

and the statement "Urn I is selected" corresponds to the event

$$I = \{e_1, e_2\}.$$

Thus the solution to part (a) is provided by $\Pr(R \mid I)$, and the solution to part (b) is $\Pr(I \mid R)$.

We have

$$\Pr(R) = \Pr\{e_1, e_3\} = .2 + .15 = .35,$$
$$\Pr(I) = \Pr\{e_1, e_2\} = .2 + .3 = .5,$$
$$\Pr(R \cap I) = \Pr(I \cap R) = \Pr\{e_1\} = .2.$$

Therefore, for part (a)

$$\Pr(R \mid I) = \frac{\Pr(R \cap I)}{\Pr(I)} = \frac{.2}{.5} = .4,$$

and for part (b)

$$\Pr(I \mid R) = \frac{\Pr(I \cap R)}{\Pr(R)} = \frac{.2}{.35} = \frac{4}{7}.$$

Independent Events

The notion of independent events arises naturally in the discussion of conditional probability. To introduce the concept in an empirical setting we return to a Royal Deck process, considered earlier.

At random two cards are drawn in order, without replacement, from a 12-card Royal Deck consisting of the Jacks, Queens, and Kings from a conventional deck. We are interested in

1. The probability that the second card is a Jack.
2. The probability that the second card is a Jack, given that the first card is a Club.

The two events of relevance are "The second card is a Jack," which we represent as "$J2$," and "The first card is a Club," labeled "$C1$." In the analysis given by the probability tree of Fig. 3-4, both statements are relevant, and the branch probabilities are easy to assign.

The elementary probabilities, computed by multiplying the branch probabilities along the respective paths, are displayed in the figure, and the paths comprising the events of interest have been checked.

Adding the weights for the basic outcomes in the event $J2$, we conclude

$$\Pr(J2) = \frac{1}{132} \times (3 + 9 + 8 + 24) = \frac{44}{132} = \frac{1}{3}.$$

Comment: In retrospect, it should be obvious that we could have assigned the appropriate value for $\Pr(J2)$ immediately. Given no information about the first card, there are three equally likely possibilities for the second card, Jack, Queen, and King. Since $J2$ is one of three equally likely outcomes, it follows that $\Pr(J2) = \frac{1}{3}$. (Of course, it's not

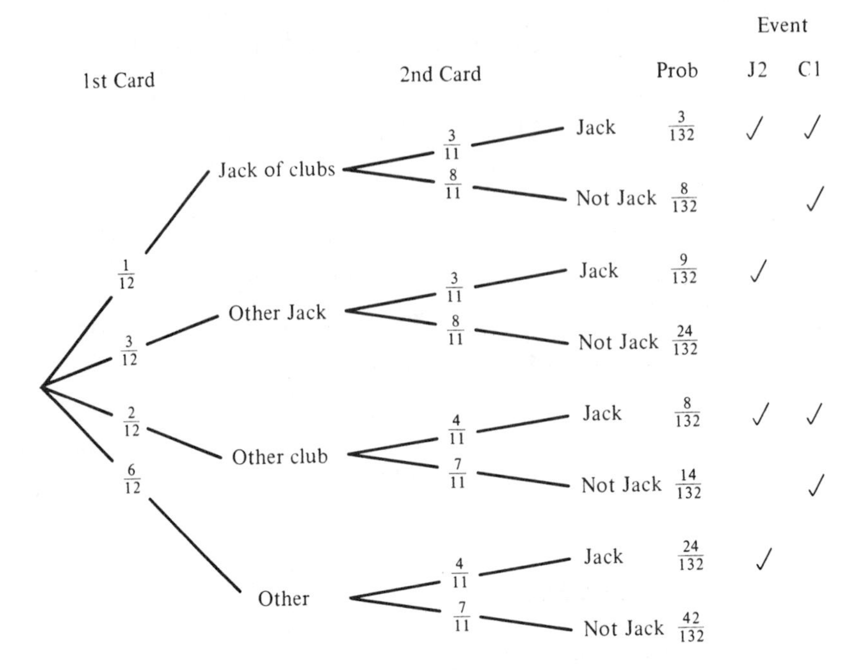

Figure 3-4 Probability Tree for Royal Deck Process.

unusual that something doesn't become "obvious" until some energy has been expended in a more complicated attack.)

To find $\Pr(J2 \mid C1)$, the probability that "The second card is a Jack, given that the first card is a Club," we appeal to the mathematical definition of conditional probability:

$$\Pr(J2 \mid C1) = \frac{\Pr(J2 \cap C1)}{\Pr(C1)}.$$

The elements comprising the event $J2 \cap C1$ are represented by the two paths checked in Fig. 3-4 as satisfying both events. Thus

$$\Pr(J2 \cap C1) = \frac{3}{132} + \frac{8}{132} = \frac{11}{132} = \frac{1}{12}.$$

Similarly from Fig. 3-4,

$$\Pr(C1) = \frac{1}{132} \times (3 + 8 + 8 + 14) = \frac{33}{132} = \frac{1}{4}$$

(of course).

Therefore,

$$\Pr(J2 \mid C1) = \frac{\Pr(J2 \cap C1)}{\Pr(C1)} = \frac{\frac{1}{12}}{\frac{1}{4}} = \frac{4}{12} = \frac{1}{3}.$$

The last result should hardly come as a surprise. The knowledge that the first card is a Club gives no information with respect to the likelihood that the first card is also a Jack or that the second card is a Jack. Hence the probability that "The second card is a Jack" is the same whether or not "The first card is a Club." What has been observed is simply that

$$\Pr(J2 \mid C1) = \Pr(J2).$$

Here we say that the event $J2$ is *independent* of the event $C1$.

In general, the relation

$$\Pr(A \mid B) = \Pr(A)$$

for events A and B in some process expresses the fact that the probability of A is the same whether or not B is true. Motivated by this interpretation, we make the formal definition:

Definition: Given two events A and B in a probability space, with $\Pr(B) \neq 0$, then A is said to be *independent of* B provided

$$\Pr(A) = \Pr(A \mid B).$$

It is to be hoped that, given some feel for the mathematical content of the situation, one will be led to mathematical results that can be translated back into interesting empirical implications. In fact, this is the whole point of applied mathematics. A modest example of this method of argument is at hand.

It can be argued empirically that if event A is independent of B, then it ought also to be the case that B is independent of A. But the mathematical argument is easier.

Theorem 3: Let A and B be events in a probability space, with $\Pr(A) \neq 0$ and $\Pr(B) \neq 0$. Suppose A is independent of B. Then B is independent of A.

Proof: Since A is independent of B, we have, by definition,

$$\Pr(A) = \Pr(A \mid B) = \frac{\Pr(A \cap B)}{\Pr(B)}. \tag{1}$$

Multiplying both sides by $\Pr(B)$ and dividing both sides by $\Pr(A)$ [remembering that $\Pr(A)$ and $\Pr(B)$ are nonzero], we have

$$\Pr(B) = \frac{\Pr(A \cap B)}{\Pr(A)}. \tag{2}$$

From the meaning of set intersection we know that $A \cap B = B \cap A$. Thus from the definition of conditional probability, Eq. (2) may be written

$$\Pr(B) = \frac{\Pr(A \cap B)}{\Pr(A)} = \frac{\Pr(B \cap A)}{\Pr(A)} = \Pr(B \mid A).$$

Which is to say, B is independent of A.

Inspection of the above argument suggests a more symmetrical definition embracing the notion of independence of events.

Definition: Events A and B in a probability space are said to be *independent* provided

$$\Pr(A \cap B) = \Pr(A) \times \Pr(B).$$

Comment: Here we do not require $\Pr(A) \neq 0$ or $\Pr(B) \neq 0$. Suppression of these requirements presents no difficulties and is of advantage in theoretical developments.

The next theorem follows immediately from the definitions.

Theorem 4: Given events A and B in a probability space, with $\Pr(B) \neq 0$, then A is independent of B precisely when A and B are independent events.

Example: A coin is flipped three times. Events A, B, and C correspond to the following outcomes:

A : The first flip is a Head,

B : At least two of the flips are Heads,

C : The three flips are all the same.

Consider the pairs of events A and B, A and C, B and C, determining in each case whether the events in the pair are independent.

Solution: From the nature of the empirical process it is immediate that A and B are not independent events. Clearly the knowledge that A is true—the first flip is a Head—increases the probability that B is true—at least two of the flips are Heads. Which is to say, $\Pr(B \mid A)$ is greater than $\Pr(B)$, so B is not independent of A. Empirical arguments can also be given for the other pairs of events, but these arguments are not so straightforward; it's more comforting to seek security in a mathematical analysis.

In the table of Fig. 3-5 the elements of the sample set are listed as HHH, HHT, etc., corresponding to the equally likely basic outcomes in the process. For the events of interest the weights of the elements contained are repeated in the respective columns and totaled to obtain the probabilities.

We see that

$$\Pr(A) \times \Pr(B) = \frac{1}{2} \times \frac{1}{2} = \frac{1}{4},$$

and

$$\Pr(A \cap B) = \frac{3}{8}.$$

Since

$$\Pr(A) \times \Pr(B) \neq \Pr(A \cap B),$$

the events A and B are not independent.

Element	Wgt	A	B	C	$A \cap B$	$A \cap C$	$B \cap C$
HHH	$\frac{1}{8}$	$\frac{1}{8}$	$\frac{1}{8}$	$\frac{1}{8}$	$\frac{1}{8}$	$\frac{1}{8}$	$\frac{1}{8}$
HHT	$\frac{1}{8}$	$\frac{1}{8}$	$\frac{1}{8}$		$\frac{1}{8}$		
HTH	$\frac{1}{8}$	$\frac{1}{8}$	$\frac{1}{8}$		$\frac{1}{8}$		
HTT	$\frac{1}{8}$	$\frac{1}{8}$					
THH	$\frac{1}{8}$		$\frac{1}{8}$				
THT	$\frac{1}{8}$						
TTH	$\frac{1}{8}$						
TTT	$\frac{1}{8}$			$\frac{1}{8}$			
Total:	1	$\frac{1}{2}$	$\frac{1}{2}$	$\frac{1}{4}$	$\frac{3}{8}$	$\frac{1}{8}$	$\frac{1}{8}$

Figure 3-5.

Moving to the next pair of events,

$$\Pr(A) \times \Pr(C) = \frac{1}{2} \times \frac{1}{4} = \frac{1}{8},$$

and

$$\Pr(A \cap C) = \frac{1}{8}.$$

Therefore, A and C are independent events.

Finally,

$$\Pr(B) \times \Pr(C) = \frac{1}{2} \times \frac{1}{4} = \frac{1}{8},$$

and

$$\Pr(B \cap C) = \frac{1}{8}.$$

Thus B and C are independent events.

3-3 EXERCISES

1. In the hypothetical probability space $\mathcal{H}$ of Sec. 3-1, the sample set is $S = \{h_1, h_2, h_3, h_4\}$, and the probability function is generated by the following assignment of weights:

Element	h_1	h_2	h_3	h_4
Weight	.1	.2	.3	.4

Find $\Pr(A \mid B)$ for the following pairs of events:

(a) $A = \{h_1, h_2\}$, $B = \{h_1, h_4\}$.
(b) $A = \{h_1, h_3\}$, $B = \{h_1, h_2, h_3\}$.
(c) $A = \{h_1, h_2, h_4\}$, $B = \{h_1, h_3, h_4\}$.

2. Consider the probability space generated by the following table:

Element of S	e_1	e_2	e_3	e_4	e_5	e_6
Weight	$\frac{2}{18}$	$\frac{1}{18}$	$\frac{2}{18}$	$\frac{4}{18}$	$\frac{7}{18}$	$\frac{2}{18}$

Compute $\Pr(A)$ and $\Pr(A \mid B)$ to test the following pairs of events for independence; then check by computing $\Pr(A) \times \Pr(B)$ and $\Pr(A \cap B)$.

(a) $A = \{e_3, e_5\}$, $B = \{e_3, e_4\}$.
(b) $A = \{e_1, e_2, e_4, e_5\}$, $B = \{e_3, e_5\}$.
(c) $A = \{e_2, e_3, e_4, e_6\}$, $B = \{e_1, e_2, e_5, e_6\}$.

3. Meg has six urns: one red, two green, and three yellow. Each urn contains two marbles, one of the same color as the urn and the other white. At random, an urn is selected and a marble withdrawn. Test the following events for independence:

A : The marble drawn is not red.

B : The urn selected is not green.

4. Observer B_1, whom we encountered early in the chapter, analyzes the process of flipping a coin twice with the following model:

Outcome	HH	HT	TH	TT
Probability Weight	$\frac{1}{3}$	$\frac{1}{6}$	$\frac{1}{6}$	$\frac{1}{3}$

Are the events "First flip is a Head" and "Second flip is a Tail" independent in his analysis?

5. A coin is flipped three times. Find all independent pairs of events from among the following:

A : The first flip is a Head.

B : The second and third flips are Tails.

C : The second flip is the same as the first.

D : Exactly two of the flips are Heads.

6. A poll of 100 students shows that 25 live off campus and that 15 of these 25 own bicycles. In all, 64 of the 100 students own bicycles. Suppose a student is chosen at random from among the hundred and questioned. Are the outcomes "Owns a bicycle" and "Lives off campus" independent?

7. The probability tree of Fig. 3-6 represents the weak analysis of a process in which only the events A and B are required to be relevant. Find (a) $\Pr(B)$, (b) $\Pr(A \cup B)$, (c) $\Pr(A \cup \bar{B})$, (d) $\Pr(A \mid B)$, (e) $\Pr(B \mid A)$.

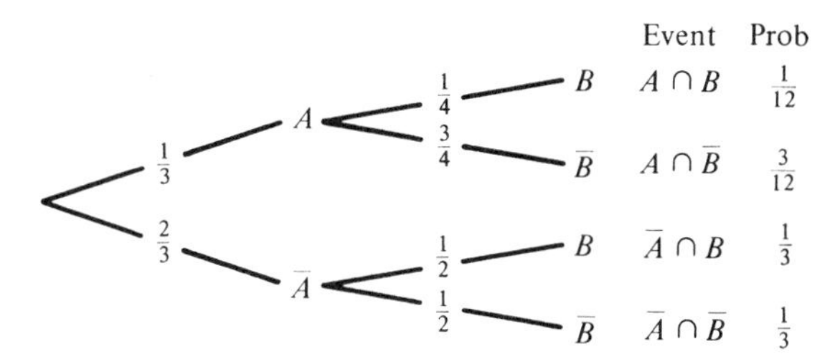

Figure 3-6.

8. Two cards are drawn in order, at random and without replacement, from an ordinary 52-card deck. Without drawing a tree but using the fact that

$$\Pr(A \cap B) = \Pr(A) \times \Pr(B \mid A),$$

find the probability that
(a) Both cards are Spades. (*Hint:* Let A be the event "First card is a Spade" and B the event "Second card is a Spade.")
(b) Both cards are Kings.
(c) The first card is a Spade and the second card is a Club.

9. As in Prob. 8, two cards are drawn in order, at random and without replacement, from an ordinary 52-card deck. Use the result from the previous section,

$$\Pr(\bar{A}) = 1 - \Pr(A),$$

to assist in finding the probability that
(a) At least one of the cards is not a Spade.
(b) At least one of the cards is not a King.

10. Two cards are drawn in order, at random and without replacement, from a 12-card Royal Deck consisting of the Jacks, Queens, and Kings. Use the result from Prob. 5 of Exercises 3-2,

$$\Pr(A \cup B) = \Pr(A) + \Pr(B) - \Pr(A \cap B),$$

together with the fact that

$$\Pr(A \cap B) = \Pr(A) \times \Pr(B \mid A)$$

to find the probability that
(a) At least one of the cards is a King. (*Hint:* A: "First card is a King," and B: "Second card is a King.")
(b) The first card is a King or the second card is a Queen.
(c) At least one of the cards is a Spade.

11. (a) Suppose A and B are independent, $\Pr(A) = .7$ and $\Pr(B) =$

$\Pr(B \mid A) = .4$. Find $\Pr(B \mid \bar{A})$. (*Hint:* Construct the weak probability tree for events A and B as in Prob. 7)

(b) In general, show that if A and B are independent events in a probability space, then $\bar{A}$ and B are also independent. (What light does this result shed on Prob. 3 above?)

12. Suppose A, B, and C are events in a probability space such that A is independent of B and A is also independent of C. Does it necessarily follow that A is independent of $B \cap C$?

3-4. Random Variables

For many applications of probability it is necessary to enlarge on the structure of a probability space by adjoining functions mapping the sample set into real numbers. As a preliminary example we consider the following speculative venture.

In the "Triple-Toss" booth at the local Charity Bazaar, the pain of giving is eased by the excitement of gambling. The player bets a dollar on one of the six numbers 1, 2, 3, 4, 5, 6, and the House rolls three dice. If the player's number comes up once, he gets back \$2; if twice, he gets back \$3; and if all three dice show his number, the player gets back \$5 for his \$1 investment. Otherwise, of course, he loses his dollar.

We pretend that the player engages in a large number of plays of the game. After each play, he records his return as one of the four numbers 0, 2, 3, 5, and when he is through he computes his average return by adding up the recorded numbers and dividing by the number of plays. Theoretically this average may be anywhere between 0 and 5, inclusive, but that's academic. To be more realistic, let's see what the "laws of probability" suggest.

Naturally it is assumed that the six numbers are equally likely on each of the three dice. For each of the dice the only outcomes of relevance are "success," meaning the player's number comes up, and "failure." Thus for a given die, the probability assigned S—"success"—is $\frac{1}{6}$. We analyze with the probability tree of Fig. 3-7.

The only feature of relevance to the player is the amount of his return, R. Thus we weaken the analysis, extracting the appropriate probabilities from the tree. The probabilities suggest an ideal run of 216 plays, and we include this information in the following analysis:

Basic outcome	*Probability*	*Frequency in 216-trial ideal run*
$R = 0$	$\frac{125}{216}$	125
$R = 2$	$\frac{75}{216}$	75
$R = 3$	$\frac{15}{216}$	15
$R = 5$	$\frac{1}{216}$	1

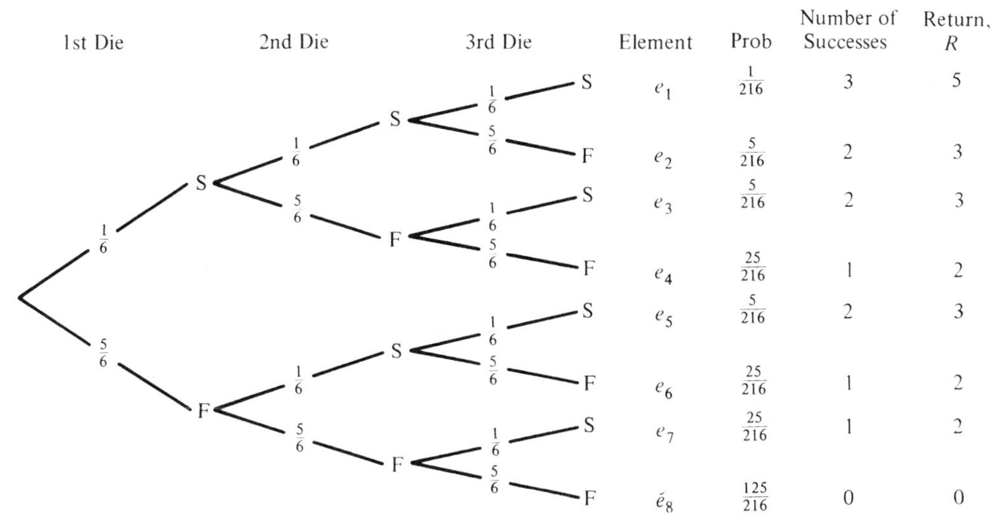

Figure 3-7 Probability Tree for "Triple-Toss."

We see that in an ideal run of 216 plays the player's return totals

$$0 \times 125 + 2 \times 75 + 3 \times 15 + 5 \times 1 = 200.$$

The ideal average return is $\frac{200}{216}$, or $\frac{25}{27}$, which is about .926. The arithmetic is essentially the same if the ideal average is computed by simply multiplying each value of R by its associated probability and summing the products:

$$0 \times \frac{125}{216} + 2 \times \frac{75}{216} + 3 \times \frac{15}{216} + 5 \times \frac{1}{216} = \frac{200}{216} = .926.$$

The conclusion is that in the long run the player can expect to get back an average of about 92.6 cents per game for his one-dollar investment; which is to say, he can expect to lose about 7.4 cents per game, on the average in the long run.

Comment: The player now translates this result into a decision about a course of action and decides that Triple-Toss does merit his participation. Of course if this had been a business proposition, he would have interpreted the result otherwise. As is often the case, the theoretical result has to be weighed against subjective factors which are not built into the mathematical model, such as "fun of the game" and "it's all for charity anyhow."

Also there is the fact that the mathematical model assumes that the player will engage in a large number of plays, perhaps more than time will allow. So he must use some judgment about how seriously he regards the theoretical result of minus 7.4 cents per play.

We are accustomed to associating probability weights with the elements of a basic outcome set, thereby generating a probability space. In the analysis of the Triple-Toss game we have associated another number, the value of R, with each basic outcome. Formally, R is a function mapping the sample set into the set of real numbers. Such a function is said to be a *random variable*.

> **Definition:** Let $\mathcal{P}$ be a probability space with sample set S. Let X be a function that associates each element of S with a real number. Then X is said to be a *random variable* defined on $\mathcal{P}$.

As a symbolic convention, we let $X = x_0$, where x_0 is a real number, represent the event consisting of those elements of S for which the value assigned by X is x_0.

In the Triple-Toss example the sample set consists of the elements, e_1, $e_2, \ldots, e_8$, corresponding to the 8 paths in the tree of Fig. 3-7. For each element the return is as tabulated in the figure; this is the number assigned to the element by the random variable R. In addition to R, we have tabulated another random variable in Fig. 3-7. This variable, which we call K, represents the "number of successes." For instance, if the path is such that the player's number comes up twice, then the value assigned by K is 2, and that path belongs to the event $K = 2$.

Another random variable of interest in the example is the one corresponding to the player's net gain. If the path is such that R is 2, then we know that the return is two dollars, but since there is a one-dollar investment before the play, the net gain is $2 - 1$, or one dollar. And if R is 0, the net gain is -1. The value of the net gain is always one less than the value of R, so we represent the "net gain" or "winnings" random variable as $R - 1$. In the natural way we think of $R - 1$ as the difference between two random variables, where "1" represents the *constant* random variable which associates each element of the sample set with the number 1. (Constant random variable and number are represented by the same symbol, but sometimes the symbol will be punctuated with quotation marks to emphasize that it represents the random variable.)

Expected Value

Focusing on the random variable R in the example at hand, we note that this function maps each element in the sample set S into one of four numbers.

As with all functions, the set of all values which the function R assigns,

$$\{0, 2, 3, 5\},$$

is called the *range* of R.

Every value in the range of R determines a nonempty event, as follows:

$$R = 0 : \{e_8\},$$

$$R = 2 : \{e_4, e_6, e_7\},$$

$$R = 3 : \{e_2, e_3, e_5\},$$

$$R = 5 : \{e_1\}.$$

The probability associated with each value in the range of R is the probability of the corresponding event. Thus for the event $R = 2$ we have

$$\Pr(R = 2) = \Pr\{e_4, e_6, e_7\} = \frac{25}{216} + \frac{25}{216} + \frac{25}{216}$$

$$= \frac{75}{216}.$$

The ideal average which we computed earlier is found by multiplying each value in the range of R by its associated probability and adding. This sum of products is called the *expected value*, or *mean*, of the random variable R and is symbolically represented as $E(R)$:

$$E(R) = 0 \times \Pr(R = 0) + 2 \times \Pr(R = 2)$$
$$+ 3 \times \Pr(R = 3) + 5 \times \Pr(R = 5)$$
$$= 0 \times \frac{125}{216} + 2 \times \frac{75}{216} + 3 \times \frac{15}{216} + 5 \times \frac{1}{216}$$
$$= \frac{200}{216} = \frac{25}{27} = .926.$$

In line with the earlier discussion we interpret the *expected value* of R, $E(R)$, as the average value which one would expect to be closely approximated in a long string of repetitions of the process.

In general, the concept of expected value applies to any random variable defined over a probability space.

Definition: Let X be a random variable defined over a probability space $\mathcal{P}$. Let the range of X, i.e., the set of values assigned to the elements of the sample set by the function X, be the set

$$\{x_1, x_2, x_3, \ldots, x_n\}.$$

Then the *expected value*, or *mean*, of X, denoted $E(X)$, is defined as follows:

$$E(X) = x_1 \times \Pr(X = x_1) + x_2 \times \Pr(X = x_2)$$
$$+ x_3 \times \Pr(X = x_3) + \cdots + x_n \times \Pr(X = x_n).$$

For the random variable K—"number of successes"—in the Triple-Toss system tabulated in Fig. 3-7, we have

$$E(K) = 0 \times \Pr(K = 0) + 1 \times \Pr(K = 1)$$
$$+ 2 \times \Pr(K = 2) + 3 \times \Pr(K = 3)$$
$$= 0 \times \frac{125}{216} + 1 \times \frac{75}{216} + 2 \times \frac{15}{216} + 3 \times \frac{1}{216}$$
$$= \frac{108}{216} = \frac{1}{2}.$$

Comment: Remembering that K represents the number of times the player's number comes up in the roll of three dice, the result $E(K) = \frac{1}{2}$

could have been anticipated. Clearly, on any given die the chosen number comes up one time in six, in the long run, so for three dice the chosen number should come up three times in six. Thus the ideal average value of K ought to be $\frac{3}{6}$, or $\frac{1}{2}$.

Example: An absentminded professor has four keys, exactly two of which will unlock the door to the Computer Center. Of course, he can't remember which two. So he tries them at random, one after the other, until he gets the door open.

Find the expected number of tries required.

Solution: Letting K_1 represent the basic outcome "the first key opens the door," and analogously for K_2 and K_3, we analyze the process via the probability tree of Fig. 3-8. The random variable X in the display corresponds to the number of tries required.

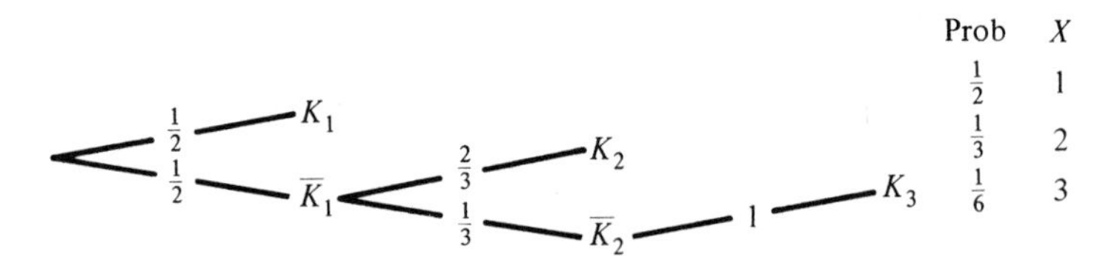

Figure 3-8.

The range of X is the set $\{1, 2, 3\}$, with

$$\Pr(X = 1) = \frac{1}{2}, \qquad \Pr(X = 2) = \frac{1}{3}, \qquad \Pr(X = 3) = \frac{1}{6}.$$

The computation of the expected value of X is straightforward:

$$\begin{aligned} E(X) &= 1 \times \Pr(X = 1) + 2 \times \Pr(X = 2) + 3 \times \Pr(X = 3) \\ &= 1 \times \frac{1}{2} + 2 \times \frac{1}{3} + 3 \times \frac{1}{6} = \frac{3 + 4 + 3}{6} = \frac{10}{6} \\ &= \frac{5}{3}. \end{aligned}$$

We conclude that the expected number of tries required for the professor to open the door is one and two-thirds.

3-4 EXERCISES

1. A probability space with random variable X is generated by the following table:

Element of S	e_1	e_2	e_3	e_4	e_5
Weight	.1	.1	.2	.2	.4
Random Variable X	-1	1	2	2	3

(a) Find $\Pr(X = -1)$, $\Pr(X = 1)$, $\Pr(X = 2)$, and $\Pr(X = 3)$.
(b) Find $E(X)$, the expected value of X.

2. Suppose the random variable X has the range $\{-2, 0, 1, 3\}$, with $\Pr(X = -2) = .35$, $\Pr(X = 0) = .15$, $\Pr(X = 1) = .3$, and $\Pr(X = 3) = .2$. Find $E(X)$.

3. A single marble is drawn at random from an urn containing ten marbles marked, respectively, with the following numbers: 0, 1, 1, 2, 2, 2, 3, 3, 3, 3. Find the expected value for the random variable corresponding to the number on the marble drawn.

4. In betting on "Black" in Roulette, there is a probability of $\frac{18}{38}$ of winning a dollar, and a probability of $\frac{20}{38}$ of losing a dollar.
(a) What is the expected amount lost per play?
(b) About how much should one expect to lose in 100 plays?

5. An urn contains four marbles marked, respectively, with the numbers 1, 2, 2, 3. The process is to withdraw two marbles, at random and without replacement, taking note of the sum of the numbers on the marbles drawn.
(a) Construct a probability tree to represent the process. Tabulate the probability weights alongside the terminal points; also record the value assigned by the random variable X corresponding to the sum of the numbers on the marbles drawn. Compute $E(X)$.
(b) Suppose the two marbles are drawn "with replacement," the first marble being returned before the second draw. Follow the direction of part (a) for the altered process.

6. The process is to flip a coin three times. The random variable K associates each outcome with the number of Heads that occur. Find $E(K)$.

7. A thumbtack is so constructed that when it is flipped in the air it lands with its point down 60% of the time. The process consists of flipping the thumbtack three times. Let K be the random variable corresponding to the number of times the tack lands point down in the "three-flip" process.
(a) Construct a probability tree, recording the probability weights for the paths and the values assigned by K.
(b) Find $\Pr(K = 0)$, $\Pr(K = 1)$, $\Pr(K = 2)$, and $\Pr(K = 3)$.
(c) Find $E(K)$.

8. (a) For a friendlier variation, let's change the rules in the "Triple-Toss" example and have the House return the player \$10, rather than \$5, when all three dice show the player's number. In the other cases the return is \$0, \$2, or \$3, as before. Now find the expected value of the amount lost per play.
(b) For a perfectly friendly Triple-Toss game the expected loss per play should be zero. If the other returns are as before, what should the jackpot payoff in the "three-of-your-kind" case be for a perfectly friendly game.

9. Suppose our absentminded professor nightly stores his keys in one of six drawers. And every morning he randomly searches the drawers, one after the other, until he finds them. Find the expected number of drawers in his search.

10. A coin is flipped until either two Heads occur in succession or until a total of three Tails have been observed. Find the expected number of flips required.

11. The process is to roll a die twice.
    (a) Find the expected value for the sum of the numbers that appear.
    (b) Find the expected value for the product of the numbers that appear.

12. An urn contains three red marbles and three white marbles. The process consists of withdrawing marbles, at random and without replacement, until a red marble is drawn. Find the expected number of draws required.

13. On each play of a simple game (of the type to be discussed in Chapter Six) Rob and Carl each flip a coin, simultaneously. Rob pays Carl two cents if both coins land Heads and four cents if both land Tails, and Carl pays Rob three cents if one coin lands Heads and the other lands Tails.
    (a) Find the expected value of Rob's "winnings" per play if both players are giving their coins honest flips.
    (b) Find the expected value of Rob's "winnings" per play if Rob cleverly biases his flip so that it lands Heads with probability $\frac{7}{12}$ and Tails with probability $\frac{5}{12}$.

14. In a "best two-out-of-three" playoff, teams A and B play until one or the other has won two games. Suppose the probability is .6 that team A wins in any one game.
    (a) Find the probability that team A wins the playoff.
    (b) Find the expected number of games in the playoff.
    (c) Find the expected number of games in the playoff won by (i) team A, (ii) team B, (iii) the team losing the playoff.

3-5. Combination of Random Variables

Since the values assigned by a random variable are real numbers, the arithmetical operations on real numbers furnish a natural device for constructing new random variables from those already defined. Suppose, for example, that X and Y are given random variables defined over some probability space. If for each element of the sample set we add the values assigned by X and Y, respectively, we determine a new random variable which, naturally enough, we represent as $X + Y$. In an analogous manner we can construct random variables XY, $2X - 4Y$, X^2, and what have you.

We also have the constant random variables which map each element of the sample set into a fixed real number. For instance, given a probability space, the random variable "5" associates each element of the sample set with the real number 5.

Let's recall the hypothetical probability space $\mathcal{H}$ of Sec. 3-1 and define thereon random variables X and Y. The tabulation in Fig. 3-9 conveys the system, together with several additional random variables constructed according to the general rules indicated above.

	Element of S	e_1	e_2	e_3	e_4
	Weight	.1	.2	.3	.4
Random variables	X	1	2	−1	2
	Y	3	0	3	−1.5
	$X+Y$	4	2	2	0.5
	XY	3	0	−3	−3
	$2X-4Y$	−10	4	−14	10
	X^2	1	4	1	4
	5	5	5	5	5
	$Y+5$	8	5	8	3.5

Figure 3-9 System $\mathcal{H}$ with Random Variables.

The computation of the expected values for the various random variables is straightforward:

$$\begin{aligned} E(X) &= 1 \times \Pr(X=1) + 2 \times \Pr(X=2) \\ &\quad + (-1) \times \Pr(X=-1) \\ &= 1 \times .1 + 2 \times .6 + (-1) \times .3 = 1, \\ E(Y) &= 3 \times \Pr(Y=3) + 0 \times \Pr(Y=0) \\ &\quad + (-1.5) \times \Pr(Y=-1.5) \\ &= 3 \times .4 + 0 \times .2 + (-1.5) \times .4 = .6 \end{aligned}$$

Similarly,

$$\begin{aligned} E(X+Y) &= 4 \times .1 + 2 \times .5 + .5 \times .4 = 1.6, \\ E(XY) &= 3 \times .1 + 0 \times .2 + (-3) \times .7 = -1.8, \\ E(2X-4Y) &= (-10) \times .1 + 4 \times .2 + (-14) \times .3 + 10 \times .4 \\ &= -.4, \\ E(X^2) &= 1 \times .4 + 4 \times .6 = 2.8, \\ E(5) &= 5 \times 1 = 5, \\ E(Y+5) &= 8 \times .4 + 5 \times .2 + 3.5 \times .4 = 5.6. \end{aligned}$$

The point in carrying out the above computations is to call attention to some important properties of expected value. First we note that

$$E(X + Y) = 1.6 = 1 + .6 = E(X) + E(Y).$$

We see that the expected value of the sum of the random variables X and Y is the sum of the expected values. In the same vein, we observe that

$$E(2X - 4Y) = -.4 = 2 \times 1 - 4 \times .6 = 2E(X) - 4E(Y).$$

Also,

$$E(5) = 5 \quad \text{and} \quad E(Y + 5) = E(Y) + E(5).$$

It begins to look as if the expected value of a combination of random variables is just the corresponding combination of the expected values. However, before getting carried away, it is well to note that in our example

$$E(XY) = -1.8,$$

whereas

$$E(X) \times E(Y) = 1 \times .6 = .6.$$

Thus it is not necessarily the case that the expected value of the product equals the product of the expected values.

Rather than attempt long-winded proofs for the general results that do apply, we shall be content to examine the spirit of the arguments as revealed in the setting of the system $\mathcal{H}$ tabulated in Fig. 3-9. To begin, let's take another look at $E(X)$. Since the range of X is the set $\{1, 2, -1\}$, we have, by definition,

$$E(X) = 1 \times \Pr(X = 1) + 2 \times \Pr(X = 2) + (-1) \times \Pr(X = -1).$$

The event $X = 1$ is, of course, the subset $\{e_1\}$, since e_1 is the only element of the sample set for which the value assigned by X is 1.

In line with the universal convention for expressing the value assigned by a function we write

$$X(e_1) = 1.$$

Similarly, the event $X = 2$ is the subset $\{e_2, e_4\}$, since $X(e_2) = X(e_4) = 2$, and these are the only elements mapped into the value 2. Finally, $X = -1$ is the subset $\{e_3\}$.

With these observations and notations in mind, and remembering that the probability of an event is the sum of the probability weights of the member elements, we rewrite the three terms appearing in the expression for $E(X)$ as follows:

$$\begin{aligned} 1 \times \Pr(X = 1) &= 1 \times \Pr\{e_1\} = X(e_1) \times \Pr\{e_1\}, \\ 2 \times \Pr(X = 2) &= 2 \times \Pr\{e_2, e_4\} \\ &= 2 \times \Pr\{e_2\} + 2 \times \Pr\{e_4\} \\ &= X(e_2) \times \Pr\{e_2\} + X(e_4) \times \Pr\{e_4\}, \\ (-1) \times \Pr(X = -1) &= -1 \times \Pr\{e_3\} = X(e_3) \times \Pr\{e_3\}. \end{aligned}$$

Since $E(X)$ is the sum of these three terms, we have

$$E(X) = X(e_1) \times \Pr\{e_1\} + X(e_2) \times \Pr\{e_2\} + X(e_3) \times \Pr\{e_3\} + X(e_4) \times \Pr\{e_4\}.$$

The specific values in the range of the random variable X as well as the specific probability weights have disappeared in this expression for $E(X)$. Indeed the nature of the argument is such that we would be led to this expression regardless of what values X assigns, or what the probability weights may be. Furthermore, the argument extends without complication to sample sets with an arbitrary number of elements.

Content with this suggestion of a general proof, we have the following result—in effect, an alternative definition of expected value.

Theorem 5: Let X be a random variable defined on a probability space $\mathcal{P}$. Let the sample set of $\mathcal{P}$ be given by

$$S = \{e_1, e_2, \ldots, e_n\}.$$

Then the expected value of X is

$$E(X) = X(e_1) \times \Pr\{e_1\} + X(e_2) \times \Pr\{e_2\} + \cdots + X(e_n) \times \Pr\{e_n\}.$$

We observed earlier in the space $\mathcal{H}$ with random variables X and Y that

$$E(X + Y) = E(X) + E(Y).$$

With the new characterization of the expected value we could have anticipated this result. First,

$$\begin{aligned} E(X + Y) = {} & (X + Y)(e_1) \times \Pr\{e_1\} \\ & + (X + Y)(e_2) \times \Pr\{e_2\} \\ & + (X + Y)(e_3) \times \Pr\{e_3\} \\ & + (X + Y)(e_4) \times \Pr\{e_4\} \end{aligned} \tag{1}$$

From the interpretation of the random variable $X + Y$, we know that

$$(X + Y)(e_1) = X(e_1) + Y(e_1); \tag{2}$$

i.e., the value assigned by $X + Y$ to e_1 is the sum of the values assigned by X and Y, respectively, and similarly for the elements e_2, e_3, e_4. Substituting Eq. (2) and its obvious analogs into Eq. (1), we have

$$\begin{aligned} E(X + Y) &= [X(e_1) + Y(e_1)] \times \Pr\{e_1\} + \cdots \\ &\quad + [X(e_4) + Y(e_4)] \times \Pr\{e_4\} \\ &= [X(e_1) \times \Pr\{e_1\} + \cdots + X(e_4) \times \Pr\{e_4\}] \\ &\quad + [Y(e_1) \times \Pr\{e_1\} + \cdots + Y(e_4) \times \Pr\{e_4\}] \\ &= E(X) + E(Y). \end{aligned}$$

The argument extends to situations in which the sample set has an arbitrary number of elements and also to sums of an arbitrary number of random variables. The general result is as follows:

Theorem 6: Let $X_1, X_2, \ldots, X_n$ be random variables defined on a probability space. Then

$$E(X_1 + X_2 + \cdots + X_n) = E(X_1) + E(X_2) + \cdots + E(X_n).$$

Comment: A heuristic argument for this result is provided by the empirical interpretation of expected value as a "long-run average." Imagine that we have a chance process which produces numbers X and Y on each occurrence. Suppose that after many occurrences the average value of X is 3 and the average value of Y is 4. This is the empirical analogue of $E(X) = 3$ and $E(Y) = 4$. It should be clear that if we had added the X and Y values on each occurrence, then the average of the sums recorded would be $3 + 4 = 7$, corresponding to the fact that $E(X + Y) = E(X) + E(Y)$.

Similarly, suppose we multiply the X value by 2 on each occurrence. Since the X average is 3, the $2X$ average is 6, which suggests the result $E(2X) = 2E(X)$. Finally, if X corresponds to the constant random variable "3," i.e., X always produces the value 3, then certainly the average value of X is 3, which corresponds to the formal result $E(3) = 3$. Thus two more theorems suggest themselves.

A variation of the argument given in support of Theorem 6 leads to the following result.

Theorem 7: Let X be a random variable defined on a probability space, and let c be a real number. Then

$$E(cX) = cE(X).$$

Earlier we had observed that for the system $\mathcal{H}$ tabulated in Fig. 3-9

$$E(2X - 4Y) = 2E(X) - 4E(Y).$$

This conclusion is easily reached using Theorems 6 and 7:

$$\begin{aligned} E(2X - 4Y) &= E[2X + (-4)Y] \\ &= E(2X) + E[(-4)Y]. \qquad \text{(Theorem 6)} \\ &= 2E(X) + (-4)E(Y) \qquad \text{(Theorem 7)} \\ &= 2E(X) - 4E(Y). \end{aligned}$$

We conclude with the most obvious theorem of all. In this case it is easy to give a general proof.

Theorem 8: Given a probability space and a real number c, then for the constant random variable "c," which assigns the value c to each element of the sample set, we have

$$E(\text{"}c\text{"}) = c.$$

Proof: (Here it seems advisable consistently to punctuate with quotation marks to distinguish the constant random variable "c" from the number c.) Since the range of "c" is the set $\{c\}$, we have

$$E(\text{“}c\text{”}) = c \times \Pr(\text{“}c\text{”} = c).$$

Here, the event "c" $= c$ is the entire sample set S. Therefore,

$$E(\text{“}c\text{”}) = c \times \Pr(S) = c \times 1 = c.$$

3-5 EXERCISES

1. A probability space with random variables X and Y is generated by the following table:

Element of S	e_1	e_2	e_3	e_4	e_5
Weight	.1	.1	.2	.2	.4
X	-1	1	2	2	3
Y	-2	-1	0	1	2

Compute $E(X)$, $E(Y)$, and $E(X + Y)$.

2. A probability space with random variables X and Y is generated by the following table:

Element of S	e_1	e_2	e_3	e_4
Weight	$\frac{1}{6}$	$\frac{1}{3}$	$\frac{1}{6}$	$\frac{1}{3}$
X	-2	.5	0	3
Y	3	0	-3	0

(a) Extend the table to include the random variables $X + Y$, XY, and $2X + 1$.

(b) Find $E(X)$, $E(Y)$, $E(X + Y)$, $E(XY)$, and $E(2X + 1)$.

3. A probability space with random variables X and Y is generated by the following table:

Element of S	e_1	e_2	e_3	e_4	e_5	e_6
Weight	$\frac{2}{18}$	$\frac{1}{18}$	$\frac{2}{18}$	$\frac{4}{18}$	$\frac{7}{18}$	$\frac{2}{18}$
X	1	-1	2	2	-2	0
Y	-2	1	2	0	-1	-4

(a) Extend the table to include the random variables $2X - Y$, X^2, and XY.
(b) Find $E(2X - Y)$ from the definition of expected value, and verify that

$$E(2X - Y) = 2E(X) - E(Y).$$

(c) Find $E(X^2)$ and $E(XY)$.

4. (a) An urn contains three marbles marked, respectively, with the numbers 0, 1, 5. At random two marbles are drawn without replacement. Let X_1 and X_2 be the random variables corresponding to the numbers on the first and second marbles drawn, respectively. Construct a probability tree and find $E(X_1)$, $E(X_2)$, $E(X_1 + X_2)$, and $E(X_1X_2)$.
(b) Alter the process by assuming that the two marbles are drawn with replacement, and follow the directions in part (a).

5. Henry has five coins in his pocket, two nickels, two dimes, and a quarter. At random he takes out three coins, in order and without replacement. Let X_1, X_2, and X_3 be the random variables corresponding to the number of cents represented by the first, second, and third draws, respectively.
(a) Find $E(X_1)$, $E(X_2)$, $E(X_3)$, and $E(X_1 + X_2 + X_3)$.
(b) What is the interpretation of $E(X_1 + X_2 + X_3)$?

6. In each play of a certain game there is a probability of .4 of winning a dollar and a probability of .6 of losing a dollar. Find the expected value of the total amount lost if the game is played (a) once, (b) twice, (c) five times.

7. An assortment of shims (small strips of metal of precise thickness) contains equal numbers of shims of thickness 5, 10, 15, and 20 mils (thousandths of an inch). Suppose four shims are chosen at random and stacked together. Find the expected value of the thickness of the stack.

8. Box A contains three nickels, two dimes, and one quarter, and box B contains five pennies, two nickels, and one quarter. At random, three coins are taken from box A and two coins from box B. Find the expected value of the total amount of money drawn.

9. Let X and Y be random variables on a probability space. Suppose that the range of X is the set $\{0, 1\}$ and that the range of Y is the set $\{1, 2\}$. Suppose further that the events $X = 1$ and $Y = 1$ are independent. Then it follows by the result of Prob. 10(b) in Exercises 3-3 that the events $X = 1$ and $Y = 2$ are also independent. Prove

$$E(XY) = E(X) \times E(Y).$$

[In general, random variables X and Y are said to be *independent* if, for any pair of values x_i and y_j in the ranges of X and Y, respectively, the events $X = x_i$ and $Y = y_j$ are independent. It can be shown that for independent random variables it is always the case that

$$E(XY) = E(X) \times E(Y).$$

As an example, note the random variables X_1 and X_2 in Prob. 4(b) above.]

10. Use the alternative definition of expected value given in Theorem 5, and prove Theorem 7.

3-6. The Standard Deviation

There are two important numbers to be associated with a random variable. One of these is the expected value, which one "expects" to be closely approximated by the average value of the random variable in a large number of trials. The other number measures the extent to which the values of the random variable are separated from the mean; it is called the *standard deviation*.

To evolve the standard deviation we appeal to the probabilistic interpretation of the data in the following report:

In a certain university the number of courses for which full-time students are enrolled varies from three to seven. The Registrar reports that 10% of the students take three courses, 30% take four, 40% take five, 10% take six, and 10% take seven.

Corresponding to the process of recording the number of courses taken by a randomly chosen student, we have a random variable X with range $\{3, 4, 5, 6, 7\}$. The Registrar's data translate into the following assignment of probabilities:

Value of X	3	4	5	6	7
Probability	.1	.3	.4	.1	.1

For a sense of the relative weights of the values in the range of a random variable it is helpful to view them in the ideal-run setting. In an ideal run of 10 trials the values assigned by the random variable X above are as follows:

Values of X: 3, 4, 4, 4, 5, 5, 5, 5, 6, 7.

The average of the ideal-run values is, of course, the expected value of X. Adding the numbers and dividing by 10, we find that

$$E(X) = 4.8.$$

Thus the average number of courses for which students are enrolled is 4.8.

When the random variable is clear from context it is conventional to represent the expected value or mean by the single symbol μ (mu). If we subtract the mean $\mu = 4.8$ from the respective values of X for the 10-trial ideal run above, the differences are called the *deviations from the mean*, or just the *deviations*. These deviations may be positive or negative, and from the way they are formed their sum must be zero. If we drop the minus signs, i.e., replace each deviation by its absolute value, we obtain the *absolute deviations:*

Values of X: ($\mu = 4.8$)	3,	4,	4,	4,	5,	5,	5,	5,	6,	7
Deviations:	-1.8,	$-.8$,	$-.8$,	$-.8$,	.2,	.2,	.2,	.2,	1.2,	2.2
Absolute Deviations:	1.8,	.8,	.8,	.8,	.2,	.2,	.2,	.2,	1.2,	2.2

One way of characterizing how the values are spread about the mean is to compute the mean of the absolute deviations. As one might guess, this value is known as the *mean absolute deviation.* In the example at hand the sum of the absolute deviations is 8.4, and since there are 10 values, the mean absolute deviation is .84.

The mean absolute deviation gives us an idea of the extent to which the values are separated from the mean, but it is not very useful for answering more penetrating questions. For example, suppose we are interested in the probability that for 100 students chosen at random the total number of student courses is between 460 and 500. As we shall see in the next chapter, questions of this type are served by the *standard deviation*, which is similar in spirit to the mean absolute deviation but obtained by a more complicated procedure.

First, square the deviations. In our example the *square deviations* are as follows:

Square deviations: 3.24, .64, .64, .64, .04, .04, .04, .04, 1.44, 4.84.

Next, find the mean of the square deviations. This value is called the *mean square deviation.* In anticipation of the next step it is denoted σ^2 (sigma square). For the ten values above we have $\sigma^2 = 1.160$.

Finally, the *standard deviation* is the nonnegative square root of the mean square deviation σ^2. Needless to say, it is denoted by the symbol σ. In our example the standard deviation is

$$\sigma = \sqrt{1.16} = 1.077$$

(We'll use the "=" symbol in cases like this, but it should be noted that 1.077 is an approximation to three decimal places.)

The Theoretical Setting

We have tied the standard deviation to a sequence of values of the random variable corresponding to an ideal run; in effect, we have set up a string of trials in which the values occur in the proportions to be expected in a long sequence of trials. The ideal-run interpretation of deviations, absolute deviations, and square deviations translate into corresponding random variables. For the random variable X the deviations correspond to the random variable $X - \mu$, where $\mu = E(X)$; the absolute deviations correspond to $|X - \mu|$; and the square deviations correspond to $(X - \mu)^2$. (Recall that for any real number x the expression $|x|$ represents the absolute value of x; i.e., $|x| = x$ if x is zero or positive, and for negative x the absolute value $|x|$ is the positive number obtained by dropping the minus sign.)

The following multipart definition brings the constructions into the formal system.

Definition: Let X be a random variable defined on a probability space $\mathcal{P}$. Let $\mu = E(X)$ be the mean, or expected value, of X. With respect to the random variable X we define the following:

1. The *deviation from the mean*, or just the *deviation*, is the random variable
$$X - \mu.$$
2. The *absolute deviation* is the random variable
$$|X - \mu|.$$
3. The *mean absolute deviation* is $E(|X - \mu|)$, the expected value of the absolute deviation.
4. The *square deviation* is the random variable
$$(X - \mu)^2.$$
5. The *mean square deviation* (also called the *variance*), denoted as σ^2, is the value
$$\sigma^2 = E[(X - \mu)^2].$$
6. The standard deviation, denoted as σ, is the nonnegative square root of σ^2,
$$\sigma = \sqrt{\sigma^2}.$$

To illustrate an important result about the mean square deviation σ^2, we return to the hypothetical space $\mathcal{H}$ with random variable X considered in Sec. 3-5. There we computed $\mu = E(X) = 1$.

The tabulation of Fig. 3-10 displays the system, together with several random variables derived from X.

The random variable $X^2 - 2\mu X + \mu^2$ is included in the tabulation to illustrate the fact that combinations of random variables may be manipulated as if the random variables were real numbers. In particular, $(X - \mu)^2$ is the function that assigns to each element e_i of the sample set the value $[X(e_i) - \mu]^2$. Since $X(e_i)$ and μ are real numbers, we have

$$[X(e_i) - \mu]^2 = [X(e_i)]^2 - 2\mu X(e_i) + \mu^2.$$

The value on the right is, of course, the value assigned the element e_i by the random variable $X^2 - 2\mu X + \mu^2$. Thus $(X - \mu)^2$ and $X^2 - 2\mu X + \mu^2$ are one and the same function, as the table displays.

Element	e_1	e_2	e_3	e_4
Weight	.1	.2	.3	.4
Random variable: X $[\mu = E(X) = 1]$	1	2	−1	2
Deviation: $X - \mu$	0	1	−2	1
Square deviation: $(X - \mu)^2$	0	1	4	1
$X^2 - 2\mu X + \mu^2$	0	1	4	1
X^2	1	4	1	4

Figure 3-10.

The computation of σ^2 for the random variable X is straightforward:

$$\sigma^2 = E[(X - \mu)^2] = 0 \times .1 + 1 \times .6 + 4 \times .3 = 1.8.$$

From the identity, $(X - \mu)^2 = X^2 - 2\mu X + \mu^2$, we derive an expression for σ^2 which is often useful, both for computations and theoretical development:

$$\sigma^2 = E[(X - \mu)^2] = E(X^2 - 2\mu X + \mu^2).$$

From Theorems 6 and 7 of the previous section it follows that

$$\sigma^2 = E(X^2) - 2\mu E(X) + E(\mu^2).$$

But $E(X) = \mu$, and by Theorem 8, $E(\mu^2)$, the expected value of the constant random variable "μ^2," is just μ^2. Thus

$$\sigma^2 = E(X^2) - 2\mu \times \mu + \mu^2 = E(X^2) - \mu^2,$$

which gives the following result:

Theorem 9: Let X be a random variable defined on a probability space. Then

$$\sigma^2 = E[(X - \mu)^2] = E(X^2) - \mu^2.$$

In the illustration considered above $E(X^2) = 1 \times .4 + 4 \times .6 = 2.8$. Since $\mu^2 = 1$, we have, as before,

$$\sigma^2 = E(X^2) - \mu^2 = 2.8 - 1 = 1.8.$$

The table below tabulates the random variables X, $(X - \mu)^2$, and X^2 for the earlier process of recording the number of courses taken by a randomly chosen student. (The mathematical structure may appear incomplete in that the sample set is not identified; in cases like this we interpret the range of the random variable X as the sample set.)

Probability	.1	.3	.4	.1	.1
Random Variable: X ($\mu = 4.8$)	3	4	5	6	7
Square Deviation: $(X - \mu)^2$	3.24	.64	.04	1.44	4.84
X^2	9	16	25	36	49

Appealing to the definition of expected value, we find the mean square deviation σ^2 as follows:

$$\begin{aligned}\sigma^2 = E[(X - \mu)^2] &= 3.24 \times .1 + .64 \times .3 + .04 \times .4 \\ &\quad + 1.44 \times .1 + 4.84 \times .1 \\ &= .324 + .192 + .016 + .144 + .484 \\ &= 1.160.\end{aligned}$$

This, of course, is just a repetition of the arithmetic that earlier led us to the value $\sigma^2 = 1.16$ in the ideal-run setting.

To apply Theorem 9, we first compute

$$\begin{aligned}E(X^2) &= 9 \times .1 + 16 \times .3 + 25 \times .4 + 36 \times .1 + 49 \times .1 \\ &= .9 + 4.8 + 10.0 + 3.6 + 4.9 \\ &= 24.2\end{aligned}$$

and

$$\mu^2 = (4.8)^2 = 23.04.$$

And we obtain, as before,

$$\sigma^2 = E(X^2) - \mu^2 = 24.2 - 23.04 = 1.16.$$

The following example sets the stage for some concluding remarks.

Example: We have three urns. Urn I contains four marbles, marked, respectively, with the numbers 1, 3, 3, 5; Urn II contains four marbles, marked with the number 2, 2, 4, 4; and Urn III contains five marbles, marked with the numbers 1, 2, 3, 4, 5.

Process I is to draw at random a single marble from Urn I and record the number on the marble drawn. Processes II and III are defined in the same manner, the marble being drawn from Urns II and III, respectively.

For each process find the mean, mean absolute deviation, and standard deviation for the random variable corresponding to the number recorded.

Solution: For Process I we have a random variable X_1 with range $\{1, 3, 5\}$. The tabulation below gives the probabilities and includes the derived random variables of interest:

Probability	.25	.5	.25
Random Variable: X_1 Mean: $\mu = 3$	1	3	5
Absolute Deviation: $\|X_1 - \mu\|$ Square Deviation: $(X_1 - \mu)^2$	2 4	0 0	2 4

From the table, the mean absolute deviation is

$$E(|X_1 - \mu|) = 2 \times .25 + 0 \times .5 + 2 \times .25$$
$$= 1.$$

The mean square deviation is

$$\sigma_1^2 = E[(X_1 - \mu)^2] = 4 \times .25 + 0 \times .5 + 4 \times .25$$
$$= 2,$$

and the standard deviation is

$$\sigma_1 = \sqrt{2} = 1.414.$$

For Process II the tabulation is as follows:

Probability	.5	.5
Random Variable: X_2 Mean: $\mu = 3$	2	4
Absolute Deviation: $\|X_2 - \mu\|$ Square Deviation: $(X_2 - \mu)^2$	1 1	1 1

The mean absolute deviation is

$$E(|X_2 - \mu|) = 1 \times .5 + 1 \times .5$$
$$= 1.$$

The mean square deviation is

$$\sigma_2^2 = E[(X_2 - \mu)^2] = 1 \times .5 + 1 \times .5$$
$$= 1,$$

and the standard deviation is

$$\sigma_2 = \sqrt{1} = 1.$$

Similarly, for Process III we have

Probability	.2	.2	.2	.2	.2
Random Variable: X_3 Mean: $\mu = 3$	1	2	3	4	5
Absolute Deviation: $\|X_3 - \mu\|$ Square Deviation: $(X_3 - \mu)^2$	2 4	1 1	0 0	1 1	2 4

The mean absolute deviation is

$$E(|X_3 - \mu|) = 2 \times .2 + 1 \times .2 + 0 \times .2 + 1 \times .2 + 2 \times .2$$
$$= 1.2.$$

The mean square deviation is

$$\sigma_3^2 = E[(X_3 - \mu)^2] = 4 \times .2 + 1 \times .2 + 0 \times .2$$
$$+ 1 \times .2 + 4 \times .2$$
$$= 2.0,$$

and the standard deviation is

$$\sigma_3 = \sqrt{2} = 1.414.$$

We find that the mean is 3 for all three processes. The mean absolute deviation is 1 for Processes I and II and 1.2 for Process III. The standard deviation for Process I is $\sqrt{2}$; for Process II, it is 1; and for Process III, it is $\sqrt{2}$, the same as for Process I.

Since the mean is 3 for all three processes in the above example, one would expect that in many trials of any of the processes the average of the recorded numbers would be close to 3. But here's an interesting question:

In which of the three processes is the probability highest that in 100 trials the average of the recorded numbers is between 2.9 and 3.1?

The answer to this question is provided by the developments to be reached in Sec. 4-6. For Process I, the probability that the average of the recorded numbers in 100 trials is between 2.9 and 3.1 is about .52. For Process II, the probability is about .68. And for Process III, the probability is almost exactly the same as for Process I, about .52. Thus Process II provides the highest probability.

The important fact is that the probabilities cited above were based only on the values of the mean μ and standard deviation σ for the respective processes and the fact that $n = 100$ qualifies as a sufficiently large number of trials to justify a relatively simple attack on the problems. Since Processes I and III produce the same mean and standard deviation, we obtain the same approximate probabilities for the question posed. The fact that Processes I and II have the same mean absolute deviation is of little value.

3-6 EXERCISES

1. The heights in inches of the starting lineup of a basketball team are reported as follows:

72, 78, 79, 80, 76.

Find the mean and standard deviation for this set of data; that is, find the mean and standard deviation of the random variable "height in inches" in the process of considering a randomly chosen player from among the five (for which process the given data express an ideal-run experience).

2. The enrollments in eight sections of the course Mathematics 101 are as follows:

$$36, \quad 44, \quad 28, \quad 32, \quad 41, \quad 36, \quad 39, \quad 32.$$

Find the mean and standard deviation for this set of data.

3. At random, Tom chooses two coins from the four nickels and one dime in his pocket. Find the mean and standard deviation for the amount of money in Tom's choice.

4. A probability space with random variable X is generated by the following table:

Element of S	e_1	e_2	e_3	e_4	e_5
Weight	.1	.2	.3	.3	.1
X	3	2	1	0	2

(a) Find the mean μ of X.
(b) Extend the table to include the random variables X^2, $X - \mu$, and $(X - \mu)^2$.
(c) Compute the mean square deviation σ^2 by the formula of Theorem 9, $\sigma^2 = E(X^2) - \mu^2$.
(d) Verify the answer to part (c) by applying the definition $\sigma^2 = E[(X - \mu)^2]$.

5. A probability space with random variable X is generated by the following table:

Element of S	e_1	e_2	e_3	e_4
Weight	.3	.2	.15	.35
X	-2.5	1.5	2	-1

Find
(a) The mean, $\mu = E(X)$.
(b) The mean absolute deviation, $E(|X - \mu|)$.
(c) The mean square deviation, $\sigma^2 = E[(X - \mu)^2] = E(X^2) - \mu^2$.
(d) The standard deviation σ.

6. Suppose the random variable Y has range $\{-2, 0, 1\}$, and that for the values in the range we have

$$\Pr(Y = -2) = .3, \qquad \Pr(Y = 0) = .3, \qquad \Pr(Y = 1) = .4.$$

Find the mean μ and mean square deviation σ^2 of Y.

7. The random variable V has range $\{-2, -1, 0, 1, 2\}$; the probabilities corresponding to the values in the range of V are as follows:

Value of V	-2	-1	0	1	2
Prob.	$\frac{4}{24}$	$\frac{8}{24}$	$\frac{7}{24}$	$\frac{4}{24}$	$\frac{1}{24}$

Find the values of μ and σ^2 for V.

8. In a batch of small raisin cookies, 10% of the cookies contain one raisin, 20% contain two, 40% contain three, 20% contain four, and 10% contain five. Find the mean and standard deviation for the number of raisins per cookie.

9. On each play of a certain game a player wins one dollar, with probability .2; or he wins nine dollars, with probability .05; or he loses one dollar, with probability .75. Find the mean and standard deviation for the player's "winnings" per play.

10. Find the mean μ and mean square deviation σ^2 for the random variable K corresponding to "number of Heads" in the process of flipping a coin n times where (a) $n = 1$, (b) $n = 2$, (c) $n = 3$.

11. An urn contains two red and two white marbles.

(a) The process is to draw two marbles, at random and with replacement. Find the mean and standard deviation of the random variable R corresponding to the number of red marbles drawn.

(b) Same as part (a), except that the process is performed "without replacement."

12. An urn contains two red and two white marbles. The process is to withdraw marbles one by one, at random and without replacement, until a red marble is drawn. Let X be the random variable corresponding to the number of draws required. For X, find μ, σ^2, and σ.

13. Let X_1, X_2, and X_3 be random variables on a probability space such that

$$E(X_1{}^2) = E(X_2{}^2) = E(X_3{}^2) = .5$$

and

$$E(X_1X_2) = E(X_2X_3) = E(X_3X_1) = .25.$$

Let X be the "sum" random variable

$$X = X_1 + X_2 + X_3.$$

Find $E(X^2)$.

4

REPEATED TRIALS

Underlying the concept of probability is the notion of repeating the same process many times in succession. To this point the mathematical structure takes into account only the outcomes for a single occurrence. To handle the many occurrences, we consider larger processes which themselves consist of a number of trials of a given process. In this way we progress from talking about a single flip of a coin to considering at once 100 flips, or 10,000 flips, and finding the probability that, say, more than 52% of the trials produce Heads.

4-1. Preliminary Examples

For an illustration of a larger process consisting of repeated trials of a simple process, we return to the "Triple-Toss" game of Sec. 3-4. Adopting an insignificant change in perspective, we assume that one die is rolled three times, rather than three dice all at once. Thus the process consists of three successive trials of the *component process* of rolling the die once.

A certain number between 1 and 6, inclusive, is preselected. If the die shows this number, the trial is labeled S, for success, and otherwise F, for failure. The analysis of the component process is simple indeed:

Element	S	F
Weight	$\frac{1}{6}$	$\frac{5}{6}$

The overall process consists of three trials of the component process. A typical basic outcome for the three-trial process would be the following: Success on the first trial, failure on the second, and success on the third. The above outcome is represented as SFS. Labeling the outcome in this manner, we see at a glance what happens on each of the three trials. To work this information into the mathematical system, we create three random variables, X_1, X_2, and X_3. For a given element of the sample set the value assigned by X_1 is either 1 or 0, depending on whether the corresponding outcome represents success or failure on the first trial, and analogously for the random variables X_2 and X_3. In particular, for the element represented as SFS,

$$X_1(\text{SFS}) = 1, \qquad X_2(\text{SFS}) = 0, \qquad X_3(\text{SFS}) = 1.$$

Displayed in Fig. 4-1 is the probability tree for the "Three Rolls of the

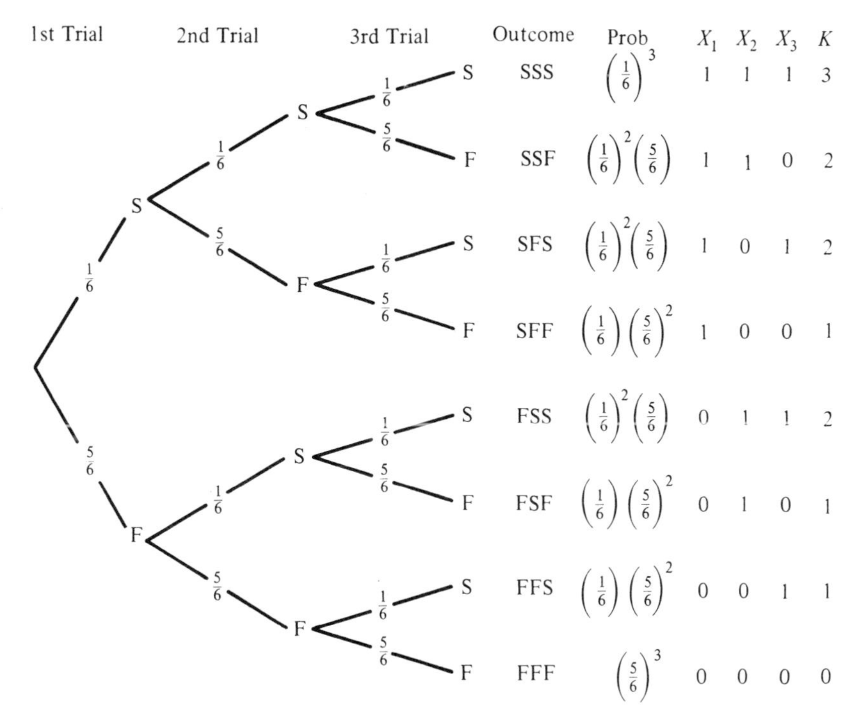

Figure 4-1 "Three Rolls of the Die" Analysis.

Die" process, with the branch probabilities as assigned by the simple analysis of the component process. The outcomes are coded in the manner suggested, and the random variables, X_1, X_2, and X_3, tabulated. Also tabulated is the random variable K corresponding to *the total number of successes.*

From a mathematical point of view, the model which we have constructed corresponds to a probability space with four random variables, X_1, X_2, X_3, and K. We note the following properties.

1. The range of each of the random variables X_1, X_2, and X_3 is the set $\{0, 1\}$. Put succinctly, for $i = 1, 2, 3$, the range of X_i is $\{0, 1\}$.
2. A simultaneous assignment of values, 0 or 1, to the respective X_i determines a unique element of the sample set. Which is to say, if (x_1, x_2, x_3) is a triple of numbers, each of which is either 0 or 1, then there is exactly one element e in the sample set such that
$$X_1(e) = x_1, \qquad X_2(e) = x_2, \qquad X_3(e) = x_3.$$
3. The random variable K is the sum of the random variables X_1, X_2, and X_3; i.e.,
$$K = X_1 + X_2 + X_3.$$
4. For a given element of the sample set, the probability weight is
$$\left(\frac{1}{6}\right)^k \left(\frac{5}{6}\right)^{3-k},$$
where k is the value assigned that element by the random variable K.

Let's examine these four properties in terms of the empirical motivation. Underlying property 1 is the fact that X_1 represents the number of successes on the first trial; the value assigned by X_1 is either 1 or 0, depending on whether we have success or failure on the first trial. And similarly for X_2 and X_3.

With respect to property 2, we recall the basic outcome SFS considered earlier. From the interpretation of the random variables X_i, we have
$$X_1(\text{SFS}) = 1, \qquad X_2(\text{SFS}) = 0, \qquad X_3(\text{SFS}) = 1.$$
Furthermore, SFS is the only element in the basic outcome set which is mapped into the values 1, 0, and 1, respectively, by the random variables X_1, X_2, and X_3. In general, for any triple of ones and zeros there is exactly one basic outcome mapped into these respective values by the X_i. For instance, the triple (0, 0, 1) corresponds to the outcome FFS, and only to this outcome.

Concerning property 3, we need only remember that the random variable K corresponds to the total number of successes. Thus for the outcome SFS, the value assigned by K is 2 because there are two successes in the three trials. Since X_1, X_2, and X_3 convey the number of successes, 1 or 0, on the respective trials, it is immediate that the value assigned by K is the sum of the values value assigned by the X_i. In short, $K = X_1 + X_2 + X_3$.

Property 4 recalls the fact that the probability of a path in a probability tree is the product of its branch probabilities. From the analysis of the component process, each of these branch probabilities is either $\frac{1}{6}$ or $\frac{5}{6}$ depending on whether the branch leads to an S or an F. In particular, the path corresponding to the outcome SFS is as shown in Fig. 4-2. The probability weight assigned this path is $(\frac{1}{6})^2(\frac{5}{6})$ because there are two points S and one

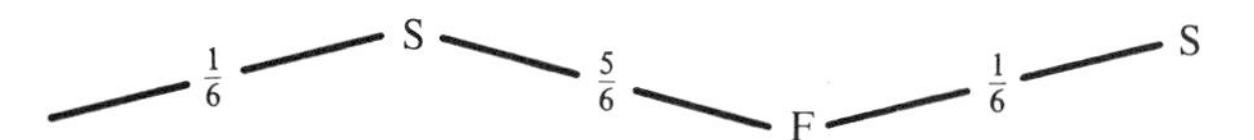

Figure 4-2.

point F along the path. The same interpretation tells us that $K(\text{SFS}) = k$, where $k = 2$, the exponent in the factor $(\frac{1}{6})^2$. Furthermore, $3 - k = 1$, the number of points F along the path, and also the implicit exponent in the factor $\frac{5}{6}$. In short, the weight assigned the path SFS is

$$\left(\frac{1}{6}\right)^2\left(\frac{5}{6}\right) = \left(\frac{1}{6}\right)^k\left(\frac{5}{6}\right)^{3-k}, \qquad \text{where } k = 2 = K(\text{SFS}).$$

The same argument applies to all of the paths in the tree. For those paths for which $k = 0$ or $3 - k = 0$ it must be remembered that any number raised to power zero is 1. (Conventionally, 0^0 is not defined, but for theoretical completeness we interpret 0^0 to be 1 in the present context.) Thus for the path SSS, we have $K(\text{SSS}) = k = 3$, and the probability weight is

$$\left(\frac{1}{6}\right)^k\left(\frac{5}{6}\right)^{3-k} = \left(\frac{1}{6}\right)^3\left(\frac{5}{6}\right)^0 = \left(\frac{1}{6}\right)^3.$$

Another Example

For a variation on the previous example, consider the following problem:

> An urn contains five marbles, two white and three black. The process is to withdraw a marble at random, five times in succession, always replacing the marble drawn before the next trial.

Problem: Find the probability that a white marble is drawn on exactly three of the five trials.

Here, the process consists of five trials of a component process in which the outcomes of relevance are identified as S, success, meaning a white marble is drawn, and F, failure, if a black marble is drawn. With the random variable X_i representing the number of successes on the ith trial, the analysis of the component process is as follows:

Element	S	F
Weight	.4	.6
Random Variable, X_i	1	0

The five-trial process can be represented by a probability tree of five stages, with every component of the same form. Figure 4-3 displays a few of the paths.

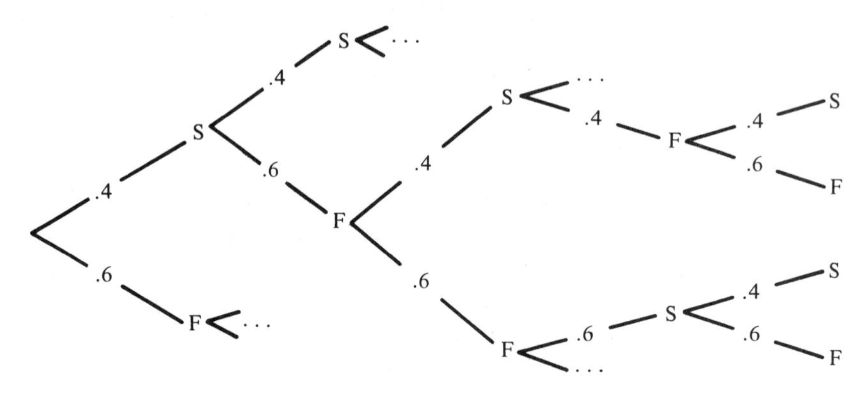

Figure 4-3.

Consistent with the previous example, the paths in the tree, representing the basic outcomes in the process, are labeled in the form

$$\text{SSSSS, SSSSF}, \ldots, \text{SFSFF, SFFSS}, \ldots, \text{FFFFF}.$$

The branch probabilities are either .4 or .6, depending on whether the branch leads to an S or an F. Thus the probability weight assigned a basic outcome is $(.4)^k(.6)^{5-k}$, where k is the number of points S along the path and $5 - k$ is the number of points F.

As before, we interpret the random variable X_i as the number of successes, 1 or 0, on the ith trial, for $i = 1, 2, 3, 4, 5$. Again, the random variable K corresponds to the total number of successes: $K = X_1 + X_2 + X_3 + X_4 + X_5$.

In Fig. 4-4 we indicate the analysis of the "Five Draws from the Urn" process, depending on a few of the elements of the sample set to indicate the general pattern.

Element	Prob wgt	X_1	X_2	X_3	X_4	X_5	K
SSSSS	$(.4)^5$	1	1	1	1	1	5
SSSSF	$(.4)^4(.6)$	1	1	1	1	0	4
⋮	⋮			⋮			
SFSFF	$(.4)^2(.6)^3$	1	0	1	0	0	2
SFFSS	$(.4)^3(.6)^2$	1	0	0	1	1	3
⋮	⋮			⋮			
FFFFF	$(.6)^5$	0	0	0	0	0	0

Figure 4-4 "Five Draws from the Urn" Analysis.

The mathematical model which we extract from the tabulation of Fig. 4-4 is almost word for word the same as in the "Three Rolls of the Die" example. Two adjustments are needed: Previously we were dealing with a three-trial process, now we have five trials; and the probability of success on each trial is .4, whereas it was $\frac{1}{6}$ in the earlier example.

Mathematical Model for the "Five Draws from the Urn" Process: We have a probability space with six random variables, X_1, X_2, X_3, X_4, X_5, and K. The following four properties are satisfied:

1. For $i = 1, 2, 3, 4, 5$, the range of X_i is the set $\{0, 1\}$.
2. Each simultaneous assignment of values, 0 or 1, to the respective X_i determines a unique element of the sample set.
3. $K = X_1 + X_2 + X_3 + X_4 + X_5$.
4. If e is an element of the sample set, then $\Pr\{e\} = (.4)^k(.6)^{5-k}$, where $K(e) = k$.

The problem asked for the probability of getting a white marble on exactly three of the five draws. Since we have identified the drawing of a white marble as "success," we are looking for the probability of exactly three successes. And since the random variable K corresponds to the total number of successes in the five trials, we interpret the problem as requesting the value of $\Pr(K = 3)$, the probability of the event $K = 3$.

One of the elements in the event $K = 3$ appears among those explicitly listed in the tabulation of Fig. 4-4. This is the element SFFSS, which corresponds to the basic outcome

"A white marble is drawn on the first, fourth, and fifth trials, and a black marble on the other two trials."

From the probability tree of Fig. 4-2, and as required by property 4, the probability weight assigned the element SFFSS is

$$(.4)^3(.6)^2 = (.4)^k(.6)^{5-k}, \qquad \text{where } k = 3 = K(\text{SFFSS}).$$

It's not difficult to list all of the elements of the sample set comprising the event $K = 3$:

SSSFF, SSFSF, SSFFS, SFSSF, SFSFS,
SFFSS, FSSSF, FSSFS, FSFSS, FFSSS.

We see that there are exactly ten elements in the event $K = 3$. From the reasoning that assigns the weight $(.4)^3(.6)^2$ to the path SFFSS, and which generally underlies property 4 in the mathematical model, we know that the probability weight of each of these ten elements is $(.4)^3(.6)^2$. Therefore, adding the weights of the elements in the event, we have

$$\Pr(K = 3) = 10 \times (.4)^3(.6)^2 = .2304.$$

To answer the question posed,

The probability of getting exactly three white marbles in five draws from the urn is .2304.

Actually, it wasn't necessary to list the elements in the event $K = 3$ to deduce that there are ten of them. The desired elements are precisely those that are represented by a sequence of three S's and two F's in some order. The total number of such elements is the same as the number of ways of assigning the three S's to the five locations in the sequence—F's automatically go into the vacancies remaining. The problem reduces to the number of ways of selecting three positions from the set of five possible positions, or just of selecting a subset of three elements from a set of five elements. This is precisely the "combinations of five things, three at a time" problem considered in Sec. 1-4. There we saw that the number of subsets of three elements that can be formed from a set of five elements is given by $C(5, 3)$, where

$$C(5, 3) = \frac{5 \times 4 \times 3}{1 \times 2 \times 3} = 10.$$

The combinations argument can be applied equally well to deduce the probability of the events $K = k$ for $k = 0, 1, 2, 3, 4, 5$. For instance, suppose we want the probability of $K = 2$. The elements in the event $K = 2$ are those which can be represented by a sequence of two S's and three F's, such as FSFFS. The number of such sequences is equal to $C(5, 2)$, the number of ways of selecting the two positions in the sequence to be filled by the symbol S. From property 4, the probability weight of each of these elements is $(.4)^2(.6)^3$. Thus

$$\begin{aligned}\Pr(K = 2) &= C(5, 2) \times (.4)^2(.6)^3 \\ &= \frac{5 \times 4}{1 \times 2} \times (.4)^2(.6)^3 \\ &= .3456.\end{aligned}$$

Extending the argument to the other values of k, it follows that in the "Five Draws from the Urn" example,

$$\Pr(K = k) = C(5, k) \times (.4)^k(.6)^{5-k}$$

for $k = 0, 1, 2, 3, 4, 5$.

Comment: We were not ready to talk about $C(5, 0)$ in Sec. 1-4. However, it should be observed that the general formula $C(n, k) = \dfrac{n!}{k!(n-k)!}$ gives the value $C(n, 0) = 1$ by the convention $0! = 1$. This is consistent with the fact that there is exactly one subset, the empty set $\varnothing$, containing no elements. And, in the present context, there is exactly one sequence of five symbols, S or F, containing no S's, FFFFF. And the probability of the event $K = 0$ is indeed given by

$$\Pr(K = 0) = C(5, 0) \times (.4)^0(.6)^{5-0} = (.6)^5.$$

With the background furnished by the "3-trial" and "5-trial" examples, we are ready to move to the general case, to be considered in the next section.

4-1 EXERCISES

1. The process is to draw twice, with replacement, from an urn containing two white and three black marbles. The probability tree of Fig. 4-5 represents this two-trial process, with "success" identified as "A white marble is drawn." The columns on the right are interpreted as in the examples in the text. Only the entries for the uppermost path are shown. Supply the remaining entries in the columns.

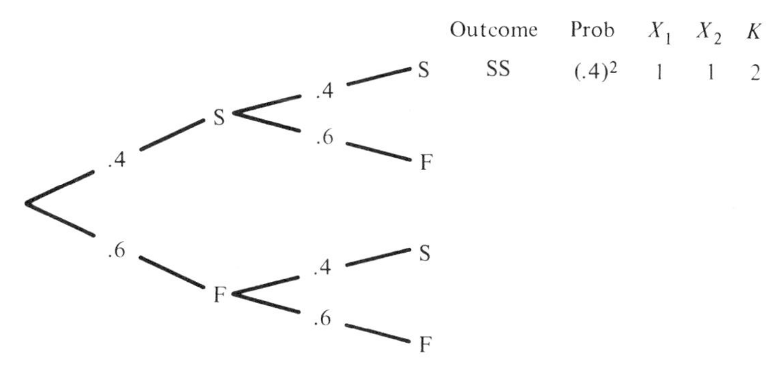

Figure 4-5.

2. A thumbtack is so constructed that when it is flipped it falls point down two-third of the time. The process consists of flipping the tack three times.

(a) In the manner of Prob. 1 above, and Fig. 4-1 of the text, construct a probability tree for the three-trial process, identifying "success" as "The tack falls point down." Alongside the terminal points, supply the appropriate entries in columns labeled Outcome, Prob, X_1, X_2, X_3, K.

(b) From the tree and legend of part (a), find $\Pr(K = 2)$.

3. (a) Complete the tabulation begun below for the process of drawing three times, with replacement, from an urn containing one red marble and three white marbles. Success is interpreted as "The red marble is drawn."

Element	*Prob*	X_1	X_2	X_3	K
SSS	$(\frac{1}{4})^3$	1	1	1	3
SSF	$(\frac{1}{4})^2(\frac{3}{4})$	1	1	0	2
⋮	⋮	⋮	⋮	⋮	⋮

(b) From your tabulation in part (a), find $\Pr(K = k)$ for $k = 0, 1, 2, 3$.

4. The probability of a certain type of chain breaking when subjected to a 10,000-pound load is .4. Suppose three of the chains are chosen at random and subjected to this test. What is the probability that at least one of the three breaks?

5. A coin is flipped four times.
   (a) Identifying "Heads" as "success," display all of the basic outcomes in the form SSSS, SSSF, SSFS, etc.
   (b) What probability weight should be assigned each basic outcome in part (a)?
   (c) With the random variable K representing the total number of successes in the four trials, find $\Pr(K = k)$ for $k = 0, 1, 2, 3, 4$.

6. An airplane has four engines of marginal quality. For each engine the probability of failure during a planned trip is .1. The situation is critical if more than one engine fails. Find the probability of a critical situation arising during the trip.

7. Suppose the probability that Team A defeats Team B in a certain game is .4. If they play a four-game series, what is the probability that Team A wins
   (a) Exactly one game?
   (b) Exactly two games?
   (c) At least two games?

8. A multiple-choice test has five questions, each with four possible answers, only one of which is correct. If the student is just guessing, what is the probability that he gets more than two correct answers?

9. A coin is flipped six times. For $k = 0, 1, 2, 3, 4, 5, 6$, find the probability that it lands Heads exactly k times.

10. Compute the mean μ and mean square deviation σ^2 for the random variable K corresponding to the number of times Heads occurs in
   (a) Two flips of a coin.
   (b) Three flips of a coin.

4-2. Binomial Processes

In the "Three Rolls of a Die" example of the previous section we have $n = 3$ trials with the probability of success on each trial given by $p = \frac{1}{6}$. And in the "Five Draws from the Urn" process we have $n = 5$ trials with a probability $p = .4$ of success on each trial. Except for the different values of n and p, the same model serves both processes. There are many important processes—called *binomial processes*—which fit the general pattern suggested by these two examples.

Empirical Description of a Binomial Process: Let n be a positive integer, $n = 1, 2, 3, \ldots$; let p be a real number between 0 and 1; and let $q = 1 - p$. The corresponding binomial process consists of n trials of some component process, where the relevant outcomes in each trial are "success" and "failure." Furthermore, regardless of what happens on the other trials, the probability of success on each trial is p, and thus the probability of failure is q.

The mathematical model which we associate with a binomial process is a generalization of the models constructed in the examples of the previous section.

Mathematical Model for a Binomial Process for $n = 1, 2, 3, \ldots$ and p, where $0 \leq p \leq 1$: We have a probability space with $n + 1$ random variables, $X_1, X_2, \ldots, X_n$ and K. The following properties are satisfied:

1. For $i = 1, 2, \ldots, n$, the range of X_i is the set $\{0, 1\}$.
2. Each simultaneous assignment of values, 0 or 1, to the respective X_i determines a unique element of the sample set.
3. $K = X_1 + X_2 + \ldots + X_n$.
4. If e is an element of the sample set, then $\Pr\{e\} = p^k \times q^{n-k}$, where $K(e) = k$ and $q = 1 - p$.

As in the examples we interpret the random variable X_i as corresponding to the number of successes, 0 or 1, on the ith trial. The elements of the sample set may be represented as sequences of n symbols, where the ith symbol in the sequence is S if X_i assigns the value 1 and is F if X_i assigns the value 0. The event $K = k$, for $k = 0, 1, 2, \ldots, n$, consists of those elements which are represented by sequences of symbols in which k of the symbols are S and $n - k$ are F. Finally, the probability weight assigned each element in the event $K = k$ is $p^k \times q^{n-k}$, as expressed in property 4.

The number of elements in the event $K = k$ is equal to $C(n, k)$, the number of ways of choosing k of the n positions in the sequence to be filled by the symbol S.

Therefore, for $k = 0, 1, 2, \ldots, n$, the probability of exactly k successes in a binomial process of n trials with probability p of success on each trial is

$$\Pr(K = k) = C(n, k) \times p^k \times q^{n-k}.$$

We introduce the symbol $b(k; n, p)$ to represent $\Pr(K = k)$ in a binomial process for n and p.

Definition: For fixed n and p, the function mapping $k = 0, 1, 2, \ldots, n$ into the value

$$b(k; n, p) = C(n, k) \times p^k \times q^{n-k}$$

is called the *binomial distribution* for n and p.

Comment: In algebra, a *binomial* is the sum of two literal terms, such as "$a + b$." The *binomial theorem* expresses the expansion of $(a + b)^n$

for $n = 0, 1, 2, \ldots$. Looking at small values of n, we have

$$(a + b)^0 = 1,$$
$$(a + b)^1 = a + b,$$
$$(a + b)^2 = (a + b)(a + b) = a(a + b) + b(a + b)$$
$$= a^2 + ab + ba + b^2 = a^2 + 2ab + b^2,$$
$$(a + b)^3 = a^3 + 3a^2b + 3ab^2 + b^3.$$

By reasoning not far removed from that involved in finding $\Pr(K = k)$, it can be shown that, in general,

$$(a + b)^n = C(n, n)a^n + C(n, n - 1)a^{n-1}b^1$$
$$+ C(n, n - 2)a^{n-2}b^2 + \ldots + C(n, n - k)a^{n-k}b^k$$
$$+ \ldots + C(n, 0)b^n.$$

If we expand $(p + q)^n$, the successive terms are precisely $C(n, k)p^kq^{n-k}$ for $k = n, n - 1, \ldots, 2, 1, 0$—hence the names *binomial* process and *binomial* distribution.

Example: Suppose a coin is flipped six times. For $k = 0, 1, 2, 3, 4, 5, 6$, find the probability that the coin falls Heads exactly k times.

Solution: Identifying "Head" as "success" on any given flip, we recognize the six flips of the coin as a binomial process with $n = 6$ and $p = \frac{1}{2}$. The probability of exactly k successes—the coin falls Heads k times—is given by

$$\Pr(K = k) = b\left(k; 6, \frac{1}{2}\right) = C(6, k)\left(\frac{1}{2}\right)^k\left(\frac{1}{2}\right)^{6-k}.$$

We note that, for all k,

$$\left(\frac{1}{2}\right)^k\left(\frac{1}{2}\right)^{6-k} = \left(\frac{1}{2}\right)^6 = \frac{1}{64}.$$

Therefore,

$$\Pr(K = k) = C(6, k) \times \frac{1}{64}.$$

For the respective values of k, the solution to the problem is as follows:

$$\Pr(K = 0) = C(6, 0) \times \frac{1}{64} = \frac{1}{64},$$

$$\Pr(K = 1) = C(6, 1) \times \frac{1}{64} = \frac{6}{64},$$

$$\Pr(K = 2) = C(6, 2) \times \frac{1}{64} = \frac{6 \times 5}{1 \times 2} \times \frac{1}{64} = \frac{15}{64},$$

$$\Pr(K = 3) = C(6, 3) \times \frac{1}{64} = \frac{6 \times 5 \times 4}{1 \times 2 \times 3} \times \frac{1}{64} = \frac{20}{64}.$$

For the next two values we recall from Sec. 1-4 that $C(n, k) = C(n, n - k)$ for all n and k:

$$\Pr(K = 4) = C(6, 4) \times \frac{1}{64} = C(6, 2) \times \frac{1}{64} = \frac{15}{64},$$

$$\Pr(K = 5) = C(6, 5) \times \frac{1}{64} = C(6, 1) \times \frac{1}{64} = \frac{6}{64}.$$

Finally,

$$\Pr(K = 6) = C(6, 6) \times \frac{1}{64} = 1 \times \frac{1}{64} = \frac{1}{64}.$$

Example: Four times in succession a marble is drawn, at random and with replacement, from an urn containing one white marble and two black marbles. What is the probability that at least half of the draws produce the white marble?

Solution: We identify the four draws as a binomial process of $n = 4$ trials with probability $p = \frac{1}{3}$ of success—the white marble is drawn—on each trial. The probability of success on at least half of the four trials is obtained by summing the probabilities of the three events

$$K = 2, \quad K = 3, \quad \text{and} \quad K = 4.$$

We have

$$\Pr(K = 2) = b\left(2; 4, \frac{1}{3}\right) = C(4, 2)\left(\frac{1}{3}\right)^2\left(\frac{2}{3}\right)^2$$

$$= \frac{4 \times 3}{1 \times 2} \times \frac{2^2}{3^4} = \frac{24}{3^4},$$

$$\Pr(K = 3) = b\left(3; 4, \frac{1}{3}\right) = C(4, 3)\left(\frac{1}{3}\right)^3\left(\frac{2}{3}\right)$$

$$= 4 \times \frac{2}{3^4} = \frac{8}{3^4},$$

$$\Pr(K = 4) = b\left(4; 4, \frac{1}{3}\right) = C(4, 4)\left(\frac{1}{3}\right)^4$$

$$= \frac{1}{3^4}.$$

Therefore, the probability that at least half of the draws produce the white marble is

$$\frac{24 + 8 + 1}{3^4} = \frac{33}{3^4} = \frac{11}{3^3} = \frac{11}{27}.$$

Example: The probability that a certain type of fuse fails under a prescribed test is .02. What is the probability that of one hundred fuses subjected to this test exactly two fail?

Solution: Violating language a bit, we identify a fuse failing the test as "success." With this understanding, the testing of the lot of 100 fuses is seen as a binomial process with $n = 100$ and $p = .02$. The value of interest is $\Pr(K = 2)$:

$$\Pr(K = 2) = b(2; 100, .02) = C(100, 2)(.02)^2(.98)^{98}$$
$$= \frac{100 \times 99}{2} \times .0004 \times (.98)^{98}$$
$$= 1.98 \times (.98)^{98}.$$

(One may be forgiven for leaving the answer in this form.)

Resorting to a calculator, one finds that $(.98)^{98}$ is approximately .138. Thus

$$\Pr(K = 2) = 1.98 \times .138 = .273.$$

Therefore, the probability that exactly two fuses fail in the lot of 100 fuses tested is about .27.

4-2 EXERCISES

1. Consider a binomial process of $n = 5$ trials with probability $p = \frac{1}{3}$ of success on each trial. Employ the formula

$$b\left(k; 5, \frac{1}{3}\right) = C(5, k)\left(\frac{1}{3}\right)^k\left(\frac{2}{3}\right)^{n-k}$$

to find the probability of exactly $k = 3$ successes.

2. An urn contains one red marble and three white marbles. The process consists of withdrawing a marble four times, at random and with replacement. Employ the formula

$$\Pr(K = k) = b(k; n, p) = C(n, k)p^k q^{n-k}$$

with $n = 4$ and $p = \frac{1}{4}$; and find the probability that the red marble is drawn exactly k times, where $k = 0, 1, 2, 3, 4$.

3. For $n = 5$ and $p = \frac{1}{3}$, compute the value of $b(k; 5, \frac{1}{3})$ for $k = 0, 1, 2, 3, 4, 5$, and verify that the sum of these values is one.

4. Compute (a) $b(3; 6, .5)$, (b) $b(2; 5, \frac{1}{3})$, (c) $b(4; 4, .75)$, (d) $b(9; 10, \frac{1}{2})$.

5. Display the following values of $b(k; n, p)$, but don't carry out the arithmetic operations: (a) $b(16; 20, .8)$, (b) $b(8; 12; .6)$, (c) $b(20; 25, \frac{3}{4})$, (d) $b(96; 100, .95)$.

6. Three times in succession a marble is drawn at random from an urn containing two red and three white marbles. Find the probability that a red marble is drawn on exactly one of the three trials if it is assumed that the draws are performed

(a) With replacement.
(b) Without replacement.

7. Consider the process of flipping three coins. Suppose the process is performed four times in succession.
   (a) Find the probability that the outcome "Two of the three coins fall Heads" occurs on exactly two of the four trials?
   (b) Find the probability that a total of exactly 8 Heads occurs in the overall process of flipping the three coins four times.

8. A multiple-choice test has ten questions, each with five possible answers, only one of which is correct. If the student is just guessing, what is the probability that he gets exactly three of the ten questions correct? (Unless a calculator is handy, don't carry out the multiplications involved.)

9. For a certain type of fuse the probability of failure is .01. If 100 fuses are tested, what is the probability that
   (a) None fail? (b) One fails?
   (c) Two fail? (d) More than two fail?
   (It's sufficient to indicate the arithmetic involved.)

10. On the average, Joe gets a hit once in three times at bat. Assuming that he will appear at the plate five times in the upcoming game, we would like to know the probability that he will get exactly three hits in the game.
    (a) Can you think of reasons why one might be uneasy in identifying Joe's five trips to the plate as a binomial process?
    (b) Treat the five trips to the plate as a binomial process, and find the probability that Joe gets exactly three hits.

11. An urn contains two red and four green marbles. At random three marbles are drawn in succession, with replacement. Find the expected number of times a red marble is drawn.

12. Consider a binomial process of $n = 3$ trials with probability $p = \frac{1}{3}$ of success on each trial. With the usual identification of the random variables X_1, X_2, X_3, K, find the following expected values: (a) $E(X_2)$, (b) $E(X_1{}^2)$, (c) $E(X_1 X_2)$, (d) $E(K)$, (e) $E(K^2)$.

13. In a binomial process with $n = 10$ and $p = \frac{3}{4}$, which is greater,

$$\Pr(K = 7) \quad \text{or} \quad \Pr(K = 8)?$$

4-3. Mean and Standard Deviation for a Binomial Process

For binomial prcesses involving a small number of trials the formula

$$b(k; n, p) = C(n, k)p^k q^{n-k}$$

serves well for finding probabilities associated with the number of successes, but for large values of n the formula presents a prohibitive exercise in arithmetic. As we shall see in the next section, the analysis of binomial processes for large n yields to a beautifully unified treatment. The resulting technique for finding $\Pr(K = k)$ depends on the mean μ and standard deviation σ for the random variable K. Our immediate concern is to deduce these values.

In advance of a mathematical argument it is worth noting that the value of $\mu = E(K)$ is easily reasoned from the empirical character of binomial processes. For example, consider the process of flipping a coin 100 times. Since Head and Tail are equally likely, each should occur about half the time in the long run. Thus in many repetitions of the 100-trial process, the average number of Heads obtained should be very close to 50, which is the empirical interpretation of the fact that $\mu = E(K) = 50$ in a binomial process with $n = 100$ and $p = .5$.

In the general case the probability of success on each trial is p, where $0 \leq p \leq 1$, which means that the fraction of successes in many repetitions of the component process should be close to p. So in many repetitions of blocks of n trials of the component process, the average number of successes should be close to np. Thus the empirical interpretation suggests that for a binomial process of n trials with probability p of success on each trial the mean of the random variable K is

$$\mu = E(K) = np.$$

To establish this result by a more mathematical argument we recall that in a binomial process for given n and p we have random variables $X_1, X_2, \ldots, X_n$, corresponding to the number of successes, 0 or 1, on the respective trials, and we have the random variable

$$K = X_1 + X_2 + \ldots + X_n,$$

corresponding to the total number of successes. By Theorem 6 of Sec. 3-5,

$$\begin{aligned} \mu = E(K) &= E(X_1 + X_2 + \ldots + X_n) \\ &= E(X_1) + E(X_2) + \ldots + E(X_n). \end{aligned}$$

For $i = 1, 2, \ldots, n$, the range of X_i is the set $\{1, 0\}$. Thus

$$E(X_i) = 1 \times \Pr(X_i = 1) + 0 \times \Pr(X_i = 0) = \Pr(X_i = 1).$$

But "$X_i = 1$" is the event "success on the ith trial." The motivation for the model is that the probability of success on each trial is precisely p. Thus if our model is faithful to its motivation, it must be the case that $E(X_i) = \Pr(X_i = 1) = p$ for $i = 1, 2, \ldots, n$. Indeed this result can be deduced in the model, but it is so clearly suggested that we take it for granted (heuristic argument).

Therefore,

$$\begin{aligned} \mu = E(K) &= E(X_1) + E(X_2) + \ldots + E(X_n) \\ &= p + p + \ldots + p \qquad (n \text{ terms}) \\ &= np. \end{aligned}$$

Having established the value of the mean μ, we turn to the question of the standard deviation σ. Unfortunately there seems to be no easy empirical argument for the value of σ, and the mathematical argument requires some attention. But a clue is provided by the two simplest cases, where $n = 1$ and $n = 2$.

A binomial process with $n = 1$ is merely a single trial of the component process. In this case the random variable K is the same as the random variable X_1. As the table below shows, K is also the same as K^2.

Element	S	F
Weight	p	$1 - p = q$
Ran Var: K	1	0
$\mu = p$		
Ran Var: K^2	1	0

We apply Theorem 9 of Sec. 3-6:

$$\sigma^2 = E(K^2) - \mu^2.$$

But here

$$E(K^2) = E(K) = \mu = p.$$

Thus

$$\sigma^2 = E(K^2) - \mu^2 = p - p^2 = p(1 - p) = pq$$

In a binomial process with $n = 2$ the range of the random variable K—total number of successes—is the set $\{0, 1, 2\}$. The probabilities corresponding to the values in the range of K are provided by the formula of the previous section:

$$\Pr(K = 0) = b(0; 2, p) = C(2, 0)p^0q^2 = q^2,$$
$$\Pr(K = 1) = b(1; 2, p) = C(2, 1)p^1q^1 = 2pq,$$
$$\Pr(K = 2) = b(2; 2, p) = C(2, 2)p^2q^0 = p^2.$$

From the earlier argument we know that

$$\mu = E(K) = 2p.$$

To find $\sigma^2 = E[(K - \mu)^2]$ we again appeal to the result

$$\sigma^2 = E(K^2) - \mu^2.$$

Since the range of K is the set $\{0, 1, 2\}$, the range of K^2 is the set $\{0, 1, 4\}$. The probabilities are as before:

$$\Pr(K^2 = 0) = \Pr(K = 0) = q^2,$$
$$\Pr(K^2 = 1) = \Pr(K = 1) = 2pq,$$
$$\Pr(K^2 = 4) = \Pr(K = 2) = p^2.$$

We have

$$\begin{aligned} E(K^2) &= 0 \times \Pr(K^2 = 0) + 1 \times \Pr(K^2 = 1) + 4 \times \Pr(K^2 = 4) \\ &= 0 \times q^2 + 1 \times 2pq + 4 \times p^2 \\ &= 2pq + 4p^2. \end{aligned}$$

Therefore,

$$\begin{aligned}\sigma^2 = E(K^2) - \mu^2 &= (2pq + 4p^2) - (2p)^2 \\ &= 2pq + 4p^2 - 4p^2 \\ &= 2pq.\end{aligned}$$

Earlier we found that in a binomial process with $n = 1$ the mean square deviation of the random variable K is given by

$$\sigma^2 = pq.$$

Now we see that in a binomial process with $n = 2$ the value is

$$\sigma^2 = 2pq.$$

A reasonable guess would be that with $n = 3$ we obtain $\sigma^2 = 3pq$; and with $n = 4$, $\sigma^2 = 4pq$; and so on. As we shall see, it's a good guess.

In advance of the general argument, let's state the result, together with the earlier result about the value of μ.

Theorem 1: For a binomial process of n trials with probability p of success on each trial, the mean of the random variable K is given by

$$\mu = E(K) = np.$$

The mean square deviation and standard deviation of K are, respectively,

$$\sigma^2 = E[(K - \mu)^2] = npq,$$
$$\sigma = \sqrt{\sigma^2} = \sqrt{npq}.$$

Example: Find the mean μ and standard deviation σ of the random variable K for binomial processes determined by the following values of n and p:

(a) $n = 100$ and $p = .5$.
(b) $n = 150$ and $p = .6$.
(c) $n = 1000$ and $p = .9$.

Solution:

(a) $\mu = np = 100 \times .5 = 50,$
$\sigma = \sqrt{npq} = \sqrt{(np)(1 - p)} = \sqrt{50 \times .5} = \sqrt{25} = 5.$
(b) $\mu = np = 150 \times .6 = 90,$
$\sigma = \sqrt{npq} = \sqrt{90 \times .4} = \sqrt{36} = 6.$
(c) $\mu = np = 1000 \times .9 = 900,$
$\sigma = \sqrt{npq} = \sqrt{900 \times .1} = \sqrt{90} = 9.49.$

Derivation of $\sigma^2 = npq$

The result $\sigma^2 = npq$ has already been verified for the cases $n = 1$ and $n = 2$. Let's establish it for the case $n = 3$; but now we employ an argument that is easily adapted to the general case. Again we appeal to Theorem 9 of Sec.

3-6:

$$\sigma^2 = E[(K - \mu)^2] = E(K^2) - \mu^2. \tag{1}$$

Since $n = 3$,

$$K = X_1 + X_2 + X_3.$$

Therefore,

$$E(K^2) = E[(X_1 + X_2 + X_3)^2]. \tag{2}$$

As observed in Sec. 3-6, arithmetical combinations of random variables may be manipulated as if we were dealing with real numbers. In particular,

$$\begin{aligned} K^2 &= (X_1 + X_2 + X_3)^2 = (X_1 + X_2 + X_3)(X_1 + X_2 + X_3) \\ &= X_1X_1 + X_1X_2 + X_1X_3 + X_2X_1 + X_2X_2 + X_2X_3 + X_3X_1 \\ &\qquad + X_3X_2 + X_3X_3 \end{aligned} \tag{3}$$

We see that K^2 is the sum of three random variables of the form X_i^2 plus six random variables of the form X_jX_m, where $j \neq m$.

The symmetry of the roles of the various X_i assures that $E(X_i^2)$ is the same for all $i = 1, 2, 3$, and $E(X_jX_m)$ is the same for all $j, m = 1, 2, 3$ and $j \neq m$. Thus by Theorem 6 of Sec. 3-5, which states that the expected value of a sum of random variables is the sum of the expected values, it follows from Eqs. (2) and (3) that

$$E(K^2) = 3E(X_i^2) + 6E(X_jX_m), \qquad j \neq m. \tag{4}$$

The computation of $E(X_i^2)$ is easy. The random variable X_i assigns either the value 1 or 0; in either case X_i^2 assigns the same value as X_i. Which is to say, X_i and X_i^2 are one and the same random variable, and, as we already know, the expected value of X_i is p. Thus

$$E(X_i^2) = E(X_i) = p. \tag{5}$$

The computation of $E(X_jX_m)$ is more involved. Let's focus on X_1X_2. Since X_1 and X_2 both have range $\{1, 0\}$, the range of X_1X_2 is also $\{1, 0\}$. The only elements for which the random variable X_1X_2 assigns the value 1 are those for which X_1 and X_2 both assign the value 1. In set terminology the event $X_1X_2 = 1$ is precisely the intersection

$$(X_1 = 1) \cap (X_2 = 1).$$

The interpretation of the events $X_1 = 1$ and $X_2 = 1$ are "success" on the first and second trials, respectively. The mathematical structure was motivated on the assumption that the experience on the first trial does not affect the probabilities of success or failure on the second trial. We say they are *independent trials*. In particular, the events $X_1 = 1$ and $X_2 = 1$ are independent events. From the development in Sec. 3-3, this means

$$\begin{aligned} \Pr\{(X_1 = 1) \cap (X_2 = 1)\} &= \Pr(X_1 = 1) \times \Pr(X_2 = 1) \\ &= p \times p = p^2. \end{aligned}$$

Again this is a result that can be proved in the formal system. We take it for granted on the basis of the foregoing heuristic argument.

Since the range of X_1X_2 is the set $\{1, 0\}$, we have

$$\begin{aligned} E(X_1X_2) &= 1 \times \Pr(X_1X_2 = 1) + 0 \times \Pr(X_1X_2 = 0) \\ &= \Pr(X_1X_2 = 1) = \Pr\{(X_1 = 1) \cap (X_2 = 1)\} \\ &= p^2. \end{aligned}$$

And in general

$$E(X_jX_m) = p^2 \qquad \text{for } j \neq m. \tag{6}$$

Substituting Eqs. (5) and (6) in Eq. (4), we have

$$E(K^2) = 3E(X_i^2) + 6E(X_jX_m) = 3p + 6p^2. \tag{7}$$

Finally, we establish the fact that $\sigma^2 = 3pq$ for the binomial process with $n = 3$ by substituting Eq. (7), together with the earlier result $\mu = 3p$, into Eq. (1):

$$\begin{aligned} \sigma^2 = E[(K - \mu)^2] &= E(K^2) - \mu^2 \\ &= 3p + 6p^2 - (3p)^2 \\ &= 3p + 6p^2 - 9p^2 \\ &= 3p - 3p^2 \\ &= 3p(1 - p) \\ &= 3pq. \end{aligned}$$

The argument for the general binomial process follows the same lines. We have

$$\sigma^2 = E[(K - \mu)^2] = E(K^2) - \mu^2,$$

where

$$\mu^2 = (np)^2.$$

The expansion of K^2 follows a simple pattern:

$$\begin{aligned} K^2 &= (X_1 + X_2 + \cdots + X_n)^2 \\ &= (X_1 + X_2 + \cdots + X_n)(X_1 + X_2 + \cdots + X_n) \\ &= (X_1^2 + X_1X_2 + \cdots + X_1X_n) + (X_2X_1 + X_2^2 + \cdots + X_2X_n) \\ &\quad + \cdots + (X_nX_1 + X_nX_2 + \cdots + X_n^2). \end{aligned}$$

In the expansion of K^2 there are n terms of the form X_i^2, and there are $n - 1$ terms of the form X_1X_j, $1 \neq j$; $n - 1$ terms of the form X_2X_j, $2 \neq j$; and so on. Thus K^2 is the sum of n terms of the form X_i^2 plus $n(n - 1)$ terms of the form X_iX_j, $i \neq j$.

The arguments that $E(X_i^2) = p$ and $E(X_jX_m) = p^2$, $j \neq m$, apply to the general case. Therefore,

$$\begin{aligned} E(K^2) &= nE(X_i^2) + n(n - 1)E(X_jX_m), \qquad j \neq m, \\ &= np + n(n - 1)p^2. \end{aligned}$$

Finally,

$$\begin{aligned}\sigma^2 = E(K^2) - \mu^2 &= [np + n(n-1)p^2] - n^2p^2 \\ &= np[1 + (n-1)p - np] \\ &= np(1-p) \\ &= npq,\end{aligned}$$

where $q = 1 - p$. This completes the argument for Theorem 1.

4-3 EXERCISES

1. Consider the process of flipping a coin four times. Let the random variable K correspond to the number of times Heads occurs in the four flips.

(a) Find $\Pr(K = k)$ for $k = 0, 1, 2, 3, 4$.

(b) Using the results of part (a), find the mean μ and mean square deviation σ^2 of K directly from the definitions of Sec. 3-6:

$$\mu = E(K), \qquad \sigma^2 = E[(K - \mu)^2].$$

(c) Verify that the results in part (b) are in agreement with the $n = 4$ and $p = \frac{1}{2}$ case of the general formulas

$$\mu = np \quad \text{and} \quad \sigma^2 = npq.$$

2. A coin is flipped

(a) 100 times.

(b) 10,000 times.

(c) 1,000,000 times.

For each of these "many flips" processes, find μ, σ^2, and σ for the random variable K corresponding to the total number of Heads.

3. Apply Theorem 1 to find μ, σ^2, and σ for the random variable K in binomial processes in which the values of n and p are as follows:

(a) $n = 150$ and $p = .4$.

(b) $n = 150$ and $p = .6$.

(c) $n = 400$ and $p = .8$.

(d) $n = 1000$ and $p = .1$.

4. Compute μ, σ^2, and σ for the random variable K for binomial processes in which n and p are as follows:

(a) $n = 432$ and $p = \frac{1}{4}$.

(b) $n = 768$ and $p = \frac{1}{4}$.

(c) $n = 1200$ and $p = \frac{1}{4}$.

(Note that $1200 = 768 + 432$. Is this fact reflected in the values of μ, σ^2, and σ?)

5. A thumbtack is so constructed that, in the long run, it falls point down two-thirds of the time. Suppose the tack is flipped 450 times. For the random variable K corresponding to the number of times that the tack falls point down, find the mean μ and the standard deviation σ.

6. A die is rolled 180 times, the occurrence of a "six" being identified as "success." Compute μ and σ for the random variable K, "total number of successes."

7. The probability of a certain type of flashbulb firing under prescribed conditions is .9. Suppose 100 of the bulbs are tested under these conditions. Let the random variable K correspond to the number of bulbs firing. Find the mean μ and standard deviation σ for K.

8. The probability that a certain type of chain breaks under a prescribed load is .4. Suppose 24 of the chains are subjected to this test. For the random variable K corresponding to the number of chains breaking, find μ and σ.

9. Long experience indicates that the probability of a certain type of fuse being defective is about .003. Accepting this figure, what is the expected number of defectives in a lot of 12,000 fuses? And what is the standard deviation for the number of defectives in the lot of 12,000?

10. An urn contains four marbles marked, respectively, with the numbers 0, 1, 1, 2.
    (a) Suppose a single marble is drawn at random from the urn. For the random variable X corresponding to the number on the marble drawn, find μ, σ^2, and σ.
    (b) Suppose a marble is drawn from the urn two times in succession, at random and with replacement. For the random variable Y corresponding to the sum of the numbers on the marbles drawn, find μ, σ^2, and σ.

4-4. The Normal Curve

For background to the discussion of the unified treatment of binomial processes involving a large number of trials, we consider a simple example of a practical problem.

A buyer of large numbers of flashbulbs requires that at least 50% of the bulbs must be good by a certain stringent criterion. A potential supplier asserts that about 80% of his brand of flashbulbs will pass the buyer's test. A sample of 25 of these bulbs is subjected to the test, and it is found that 16 of the bulbs are good.

The testing of the 25 flashbulbs is viewed as a binomial process of $n = 25$ trials. For the probability of success on each trial the values of interest are

$p = .8$, as the supplier claims, and $p = .5$, the minimum level of quality tolerated by the buyer. Two questions present themselves:

Question 1: Assuming that the supplier is correct and that the probability of a bulb being good is .8, what is the probability of getting a result as bad as that observed—only 16 good in a box of 25? In technical terms, we are interested in the value of $\Pr(K \leq 16)$ in a binomial process with $n = 25$ and $p = .8$. (Here, $K \leq 16$ is the event consisting of those elements in the sample set for which the value assigned by the random variable K is 16 or less.)

Question 2: Assuming that the probability of a bulb being good is .5, just meeting the standard set by the buyer, what is the probability of getting a result as good as that observed? In other words, find $\Pr(K \geq 16)$ in a binomial process with $n = 25$ and $p = .5$.

With respect to Question 1, the event $K \leq 16$ is the union of the events

$$K = 0,\ K = 1, \ldots, K = 16.$$

Since no two of these events contain an element in common—we say they are *disjoint* events—it follows that

$$\begin{aligned}\Pr(K \leq 16) &= \Pr(K = 0) + \Pr(K = 1) + \cdots + \Pr(K = 16) \\ &= b(0; 25, .8) + b(1; 25, .8) + \cdots + b(16; 25, .8).\end{aligned}$$

The arithmetic is tedious, but the computation is straightforward by the formula developed in Sec. 4-2. For instance,

$$\begin{aligned}b(16; 25, .8) &= C(25, 16) \times (.8)^{16} \times (.2)^{9} \\ &= C(25, 9) \times (.8)^{16} \times (.2)^{9} \\ &= \frac{25 \times 24 \times \cdots \times 17}{1 \times 2 \times \cdots \times 9} \times (.8)^{16} \times (.2)^{9}.\end{aligned}$$

Turning a computer loose on the calculation shows that, to three decimal places, $b(16; 25, .8) = .029$.

Since the computer is programmed and ready to go, we have it compute all values of the form $b(k; 25, .8)$ for $k = 0, 1, 2, \ldots, 25$. Also, with respect to Question 2, we shall have the computer find all values of the form $b(k; 25, .5)$. The results appear in the graphs of Fig. 4-6(a) and (b).

The graphs do not display those values of k for which the associated probability is less than .001. With respect to Fig. 4-6(a) it can be shown that for $n = 25$ and $p = .8$, $\Pr(K < 13)$ is only about .001. Similarly, with respect to Fig. 4-6(b), for $n = 25$ and $p = .5$, the values of $\Pr(K < 5)$ and $\Pr(K > 20)$ are also about .001.

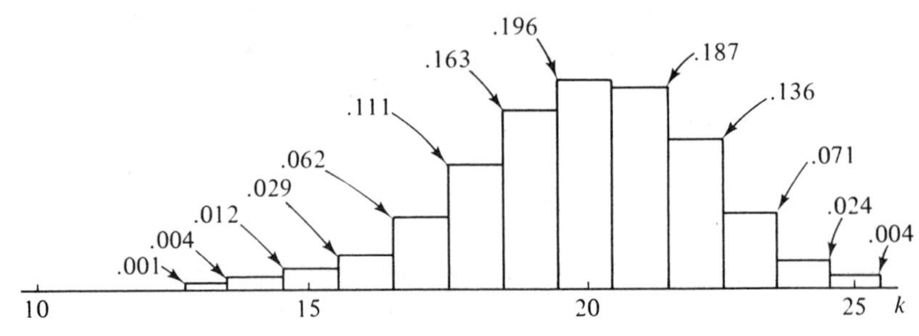

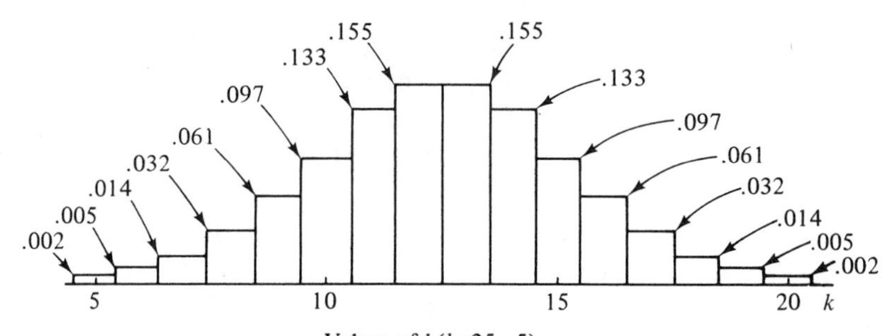

Figure 4-6 (a) Binomial Distribution for $n = 25$ and $p = .8$.
(b) Binomial Distribution for $n = 25$ and $p = .5$.

With nearly all of the work done for us, it is a simple matter to answer the questions posed.

Answer to Question 1: In the binomial process with $n = 25$ and $p = .8$,

$$\begin{aligned}\Pr(K \leq 16) &= \Pr(K < 13) + \Pr(K = 13) + \Pr(K = 14) \\ &\quad + \Pr(K = 15) + \Pr(K = 16) \\ &= .001 + .001 + .004 + .012 + .029 \\ &= .047.\end{aligned}$$

Answer to Question 2: In the binomial process with $n = 25$ and $p = .5$,

$$\begin{aligned}\Pr(K \geq 16) &= \Pr(K = 16) + \cdots + \Pr(K = 20) + \Pr(K > 20) \\ &= .061 + .032 + .014 + .005 + .002 + .001 \\ &= .115.\end{aligned}$$

We see that the probability is relatively small—less than .12—that one would find as many as 16 good bulbs in a sample of 25 if, in general, only 50% of the bulbs are good. Thus the buyer may feel somewhat assured that the bulbs do meet his criterion. On the other hand, it is unlikely—of probability less than .05—that as few as 16 good bulbs would be found in a sample of 25 if, as the supplier asserts, 80% of all the bulbs are good. But the buyer is accustomed to suppliers exaggerating the quality of their products.

The Adjusted Graph of a Binomial Distribution

The foregoing flashbulb problem has been introduced as an excuse for constructing the graphs of Fig. 4-6(a) and (b), on which we now focus. Notice how much the two graphs look alike. True, they don't peak at the same values and one is shorter and fatter than the other, but these differences are largely removed by a couple of simple adjustments.

In the previous section we found the mean μ and the standard deviation σ for the random variable K in a binomial process. These are the values that enter into the adjustments of the graphs.

The adjustment of the horizontal scale in the graph of the binomial distribution is based on assigning to each value k in the range of the random variable K a corresponding "x value," where the x value is the number of standard deviations by which k differs from the mean.

In the $n = 25$ and $p = .8$ case, the mean μ and standard deviation σ are, respectively,

$$\mu = np = 25 \times .8 = 20$$

and

$$\sigma = \sqrt{npq} = \sqrt{25 \times .8 \times .2} = \sqrt{4} = 2.$$

Measuring the number of standard deviations by which several values of k are removed from the mean μ, we see that

$k = 20$ corresponds to $x = 0$,
$k = 21$ corresponds to $x = .5$,
$k = 22$ corresponds to $x = 1$,
$k = 19$ corresponds to $x = -.5$.

The algebraic expression for the relation between k and x is

$$k = \mu + \sigma x = 20 + 2x,$$

or, equivalently,

$$x = \frac{k - \mu}{\sigma} = \frac{k - 20}{2}.$$

In general the adjustment of the horizontal scale for the graph of a binomial process is realized by assigning each value k in the range of K the x value

$$x = \frac{k - \mu}{\sigma},$$

where $\mu = np$ is the mean of K and $\sigma = \sqrt{npq}$ is the standard deviation.

For the binomial process with $n = 25$ and $p = .5$, we have

$$\mu = np = 25 \times .5 = 12.5$$

and

$$\sigma = \sqrt{npq} = \sqrt{25 \times .5 \times .5} = \sqrt{6.25} = 2.5.$$

Thus the adjustment of the horizontal scale is realized by assigning the following x value for each $k = 0, 1, 2, \ldots, 25$:

$$x = \frac{k - \mu}{\sigma} = \frac{k - 12.5}{2.5}.$$

For example, $k = 10$ corresponds to

$$x = \frac{10 - 12.5}{2.5} = -1.$$

The other adjustment that we make of the graph of a binomial distribution is as follows:

The adjustment of the vertical scale is realized by multiplying each value of $b(k; n, p)$ by the standard deviation σ. Since $k = \mu + \sigma x$, the adjusted graph displays the values of

$$\sigma \times b(\mu + \sigma x; n, p).$$

In Fig. 4-7(a) and (b) we have reconstructed the graphs of Fig. 4-6(a)

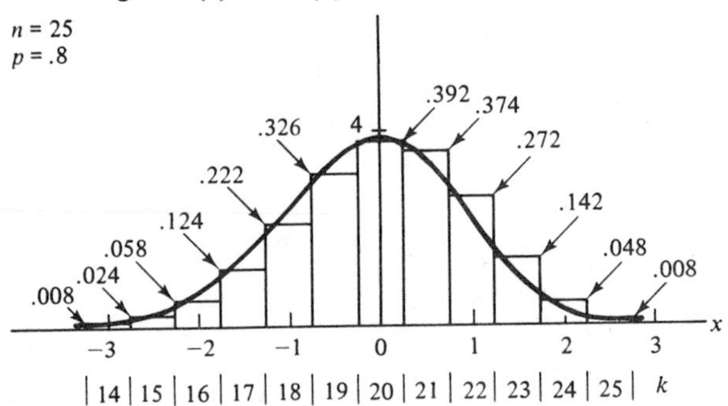

Adjusted graph $b(k; 25, .8)$
Value shown is $\sigma \times b(\mu + \sigma \times x; 25, .8)$,
where $\mu = 20$ and $\sigma = 2$.

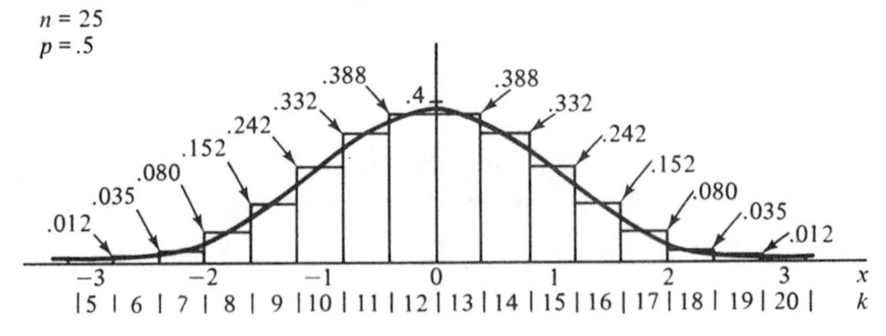

Adjusted graph of $b(k; 25, .5)$
[Value shown is $\sigma \times b(\mu + \sigma \times x; 25, .5)$,
where $\mu = 12.5$ and $\sigma = 2.5$.]

Figure 4-7 (a) Value shown is $\sigma \times b(\mu + \sigma \times x; 25, .8)$, where $\mu = 20.0$ and $\sigma = 2.0$.
(b) Value shown is $\sigma \times b(\mu + \sigma \times x; 25, .5)$, where $\mu = 12.5$ and $\sigma = 2.5$.

and (b), making the two adjustments. As a basis for comparison the same smooth curve is shown overlaid on both graphs. Their similarity is unmistakable.

The Normal Distribution

The fact that the adjusted graphs of Fig. 4-7(a) and (b) are both closely-approximated by the same smooth curve is not a peculiarity of the particular values of n and p considered.

The mathematical arguments are beyond our scope but it can be shown that this same curve approximates the adjusted graphs of all binomial distributions where n is large and p is not too close to 0 or 1. This curve, the famous "bell-shaped curve" of probability and statistics, is called the *normal curve*. The value on the normal curve corresponding to a value of x is designated $n(x)$ (read "little n of x"). The function defined by $n(x)$ is called the *normal distribution* (or *Gaussian* distribution).

Values of $n(x)$ for nonnegative x are shown to two-decimal-place accuracy in Table 4-1. The normal curve is symmetrical about $x = 0$, i.e., $n(x) = n(-x)$ for all x, so for negative x the value of $n(x)$ is the same as $n(|x|)$.

Table 4-1

x	$n(x)$	x	$n(x)$	x	$n(x)$
0	.40	1.0	.24	2.0	.05
.1	.40	1.1	.22	2.1	.04
.2	.39	1.2	.19	2.2	.04
.3	.38	1.3	.17	2.3	.03
.4	.37	1.4	.15	2.4	.02
.5	.35	1.5	.13	2.5	.02
.6	.33	1.6	.11	2.6	.01
.7	.31	1.7	.09	2.7	.01
.8	.29	1.8	.08	2.8	.01
.9	.27	1.9	.07	2.9	.01
1.0	.24	2.0	.05	$x \geq 3.0$	.00

For instance, $n(-2) = n(2) = .05$, to two decimal places. These values are employed in the examples that follow, but serious work with the normal curve generally depends on a different tabulation, to be taken up in the next section.

Example: A coin is flipped 100 times. Use the normal curve approximation to estimate the probability that the number of Heads occurring is (a) 45, (b) 50, (c) 52, (d) 60.

Solution: The process is interpreted as a binomial process with $n = 100$ and $p = .5$. Thus

$$\mu = np = 100 \times .5 = 50,$$
$$\sigma = \sqrt{npq} = \sqrt{100 \times .5 \times .5} = \sqrt{25} = 5.$$

(a) The desired value is $b(45; 100, .5)$. For $k = 45$ the corresponding x value is

$$x = \frac{k - \mu}{\sigma} = \frac{45 - 50}{5} = -1.$$

The normal curve approximates the adjusted graph of the binomial distribution, so in this case we have the following near-equality:

$$n(-1) \doteq \sigma \times b(45; 100, .5)$$

or

$$b(45; 100, .5) \doteq \frac{n(-1)}{\sigma} = \frac{n(1)}{\sigma} = \frac{.24}{5} = .048.$$

(The "$\doteq$" indicates that a theoretical approximation is involved.)

We conclude that the probability of getting exactly 45 Heads in 100 flips of a coin is about .05. [It can be shown that the value of $b(45; 100, .5)$ is .04847, to four significant figures.]

(b) Here, $k = 50$; the corresponding x value is $x = 0$. We have

$$b(50; 100, .5) \doteq \frac{n(0)}{\sigma} = \frac{.40}{\sigma} = .080.$$

Thus the probability of exactly 50 Heads is about .08.

(c) For $k = 52$,

$$x = \frac{k - \mu}{\sigma} = \frac{52 - 50}{5} = .4.$$

We have

$$b(52; 100, .5) \doteq \frac{n(.4)}{\sigma} = \frac{.37}{5} = .074.$$

Thus the desired probability is about .07.

(d) For $k = 60$,

$$x = \frac{k - \mu}{\sigma} = \frac{60 - 50}{5} = 2.$$

We have

$$b(60; 100, .5) \doteq \frac{n(2)}{\sigma} = \frac{.05}{5} = .01,$$

from which we conclude that the probability of getting exactly 60 Heads in 100 flips of a coin is about .01.

The numbers are not quite so convenient in the next example. Here a calculator comes in handy.

Example: Employ the normal curve to find the approximate value of $b(50; 80, .7)$.

Solution: Since $n = 80$ and $p = .7$,

$$\mu = np = 80 \times .7 = 56,$$

$$\sigma = \sqrt{npq} = \sqrt{80 \times .7 \times .3} = \sqrt{16.8} = 4.10.$$

For $k = 50$ the corresponding x value is

$$x = \frac{k - \mu}{\sigma} = \frac{50 - 56}{4.10} = -\frac{6}{4.10} = -1.46.$$

Interpolating in the table of values of $n(x)$,

$$n(-1.46) = n(1.46) = .14.$$

Using the normal curve approximation, we have

$$b(50; 80, .07) \doteq \frac{n(-1.46)}{\sigma} = \frac{.14}{4.10} = .034.$$

Thus we conclude that the value of $b(50; 80, .7)$ is about .03.

4-4 EXERCISES

1. (a) Employ the formulas

$$\mu = np \quad \text{and} \quad \sigma = \sqrt{npq}$$

to determine the mean and standard deviation (for the random variable K) in a binomial process with $n = 100$ and $p = .8$.

(b) With μ and σ as determined in part (a), employ the formula

$$x = \frac{k - \mu}{\sigma}$$

to find the x values corresponding to (i) $k = 86$, (ii) $k = 76$.

2. (a) Find the mean μ and standard deviation σ for a binomial process with $n = 900$ and $p = .9$.

(b) With μ and σ as found in part (a), compute the x values corresponding to (i) $k = 800$, (ii) $k = 828$.

3. For a binomial process with $n = 150$ and $p = .4$, find the values of k corresponding to the following values of x on the adjusted scale: (a) $x = 0$, (b) $x = 1$, (c) $x = -2$.

4. (a) To two-decimal-place accuracy, find the probability of obtaining exactly six Heads in nine flips of a coin. Use the formula

$$b(k; n, p) = C(n, k)p^k q^{n-k},$$

with $n = 9$, $k = 6$, and $p = .5$.

(b) For a binomial process with $n = 9$ and $p = .5$, show that $\mu = 4.5$ and $\sigma = 1.5$. Also show that the x value corresponding to $k = 6$ is $x = 1$.

(c) Refer to Table 4-1 to compute the value

$$\frac{n(x)}{\sigma} = \frac{n(1)}{1.5},$$

which furnishes the normal curve approximation to $b(6; 9, .5)$. Compare with your answer in part (a).

5. Suppose a coin is flipped 64 times. Employ the normal curve approximation,

$$b(k; n, p) \doteq \frac{n(x)}{\sigma},$$

to estimate the probability of obtaining exactly (a) 32 Heads, (b) 39 Heads, (c) 28 Heads.

6. Suppose a card is drawn from an ordinary deck 48 times, at random and with replacement. Use the normal curve approximation to estimate the probability of obtaining exactly (a) 12 Spades, (b) 15 Spades, (c) 8 Spades.

7. Suppose 40% of the voters support a proposed ordinance. Find the approximate probability that among 150 randomly chosen voters the number supporting the ordinance is (a) 57, (b) 60, (c) 72.

8. In general, two-thirds of the shoppers entering a certain store make a purchase. Find the approximate probability that, of the next 72 shoppers, the number making a purchase is (a) 42, (b) 48, (c) 52.

9. The probability of a certain type of flashbulb failing a prescribed test is .1. Suppose 100 of these bulbs are subjected to this test. Find the approximate probability that the number failing is (a) 9, (b) 10, (c) 11, (d) 12.

10. From an urn containing one red marble and four white marbles a marble is drawn 100 times, at random and with replacement. With the random variable K corresponding to the number of times the red marble is drawn, use the normal curve approximation to find the approximate value of $\Pr(18 \leq K \leq 22)$.

4-5. The Area Under the Normal Curve

For most applications of the normal curve the analysis depends not so much on the value $n(x)$ but rather on the area under the curve between given values of x. To illustrate the interpretation of the area under the curve we have extracted in Fig. 4-8(a) and (b) corresponding portions of the graphs of Figs. 4-6(b) and 4-7(b).

In the graph of Fig. 4-8(a) the value $b((15; 25, .5) = .097$ is shown as the height of a rectangle which has base width 1, extending from 14.5 to 15.5 on the k scale. Thus the area of the rectangle is .097, the value assigned $k =$

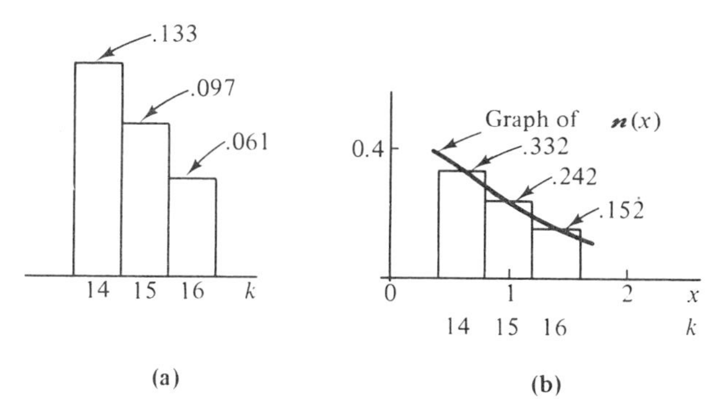

Figure 4-8 (a) Portion of Figure 4-6(b).
(b) Portion of Figure 4-7(b).

15. In the adjustment of the graph in Fig. 4-7(b) this rectangle has been transformed into a rectangle of height

$$\sigma \times b(15; 25, .5) = 2.5 \times .097 = .242.$$

The base of the transformed rectangle extends from the point $x = .8$, the x value corresponding to the midpoint $k = 14.5$, to the point $x = 1.2$, the x value for $k = 15.5$. Thus the base width of the transformed rectangle is $1.2 - .8$, or .4. Since the height of this rectangle is .242, the area is $.242 \times .4$, which is .097 to the three-decimal-place accuracy with which we have been working.

In short, the transformed rectangle has the same area as the original rectangle, and this area is precisely the value of $b(15; 25, .5)$.

In general, the adjustment of the graph of a binomial distribution may be viewed as transforming rectangles over the k scale into rectangles over the x scale. From the manner in which the x values are defined the base widths of 1 on the k scale are transformed into base widths of $1/\sigma$ on the x scale, where σ is the standard deviation. Since the adjustment dictates that the height of the rectangles are multiplied by σ, it follows that the area of the transformed rectangle is the same as the area of the original rectangle, and this common area is precisely $b(k; n, p)$.

Referring to Fig. 4-8(b), we see that the height of the rectangle corresponding to $k = 15$ is closely approximated by the value $n(1)$ on the normal curve. The smoothness of the curve assures that the area of the rectangle is approximately equal to the area under the normal curve between $x = .8$ and $x = 1.2$, the x values for the midpoints $k = 14.5$ and $k = 15.5$, respectively. Thus an approximation to the value of $\Pr(K = 15) = b(15; 25, .5)$ is provided by the area under the normal curve between $x = .8$ and $x = 1.2$. The technique for finding this area is beyond our scope, but its value can be shown to be about .097, which, of course, is in agreement with the fact that $\Pr(K = 15) = .097$.

Now suppose we want to find the value of $\Pr(14 \leq K \leq 16)$ in the binomial process with $n = 25$ and $p = .5$. The desired value is the sum of the areas of the three rectangles, in both Fig. 4-8(a) and Fig. 4-8(b) The sum of the corresponding areas under the normal curve for the three values of k is the area above the segment from x_1 to x_2 on the x scale, where x_1 corresponds to $k_1 = 13.5$ and x_2 corresponds to $k_2 = 16.5$. We have

$$x_1 = \frac{k_1 - \mu}{\sigma} = \frac{13.5 - 12.5}{2.5} = \frac{1}{2.5} = .4$$

and

$$x_2 = \frac{k_2 - \mu}{\sigma} = \frac{16.5 - 12.5}{2.5} = \frac{4}{2.5} = 1.6.$$

It can be shown that the area under the normal curve from $x_1 = .4$ to $x_2 = 1.6$ is about .290. Thus, in the binomial process with $n = 25$ and $p = .5$,

$$\Pr(14 \leq K \leq 16) \doteq .290.$$

The technique of approximating probabilities by areas under the normal curve is valid for all binomial processes in which n is large and p is not too close to 0 or 1. In these cases, the value of $\Pr(k_a \leq K \leq k_b)$ is approximately equal to the area under the normal curve between the values x_1 and x_2, corresponding, respectively, to the midpoints $k_1 = k_a - \frac{1}{2}$ and $k_2 = k_b + \frac{1}{2}$:

$$x_1 = \frac{k_1 - \mu}{\sigma}, \qquad x_2 = \frac{k_2 - \mu}{\sigma},$$

where $\mu = np$ and $\sigma = \sqrt{npq}$.

Also, the value of $\Pr(k_a \leq K)$ is approximately equal to the area under the normal curve for values greater than x_1, and $\Pr(K \leq k_b)$ is approximately equal to the area under the normal curve for values less than x_2.

The validity of the approximation requires that n be "large" and p "not too close" to 0 or 1. For two-decimal-place accuracy, a good criterion is to require that the mean number of successes, $\mu = np$, and also the mean number of failures, $nq = n - \mu$, be at least 10. Three-decimal-place accuracy is justified if μ and $n - \mu$ are both at least 25.

Of course, we need values for the areas under the normal curve to apply the approximation technique. To this end, we let $\mathfrak{N}(x)$ (read "big n of x") represent the area under the normal curve for the interval 0 to x, where $x \geq 0$, as shown in Fig. 4-9. Values of $\mathfrak{N}(x)$ are displayed in Table 4-2.

As observed in the previous section, the normal curve is symmetrical about $x = 0$; for example, $n(-1.2) = n(1.2)$. It follows that the area under

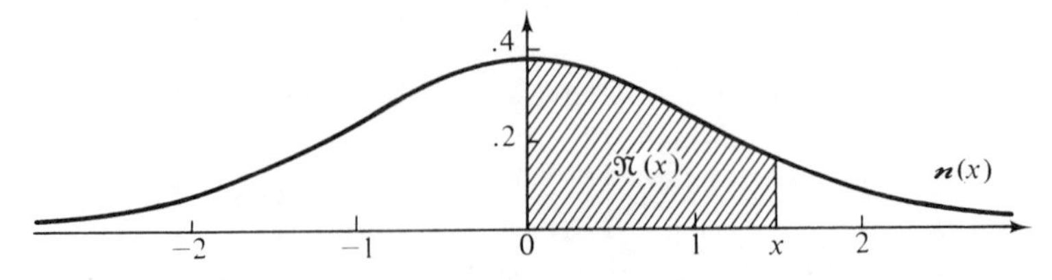

Figure 4-9.

Table 4-2 Table of Values of $\mathfrak{N}(x)$

x	$\mathfrak{N}(x)$	x	$\mathfrak{N}(x)$	x	$\mathfrak{N}(x)$	x	$\mathfrak{N}(x)$
0	0	1.0	.341	2.0	.477	3.0	.4987
.1	.040	1.1	.364	2.1	.482	3.2	.4993
.2	.079	1.2	.385	2.2	.486	3.4	.4997
.3	.118	1.3	.403	2.3	.489	3.6	.4998
.4	.155	1.4	.419	2.4	.4918	3.8	.4999
.5	.191	1.5	.433	2.5	.4938	4.0 +	.5–
.6	.226	1.6	.445	2.6	.4953		
.7	.258	1.7	.455	2.7	.4965		
.8	.288	1.8	.464	2.8	.4974		
.9	.316	1.9	.471	2.9	.4981		
1.0	.341	2.0	.477	3.0	.4987		

the normal curve from -1.2 to 0 is the same as from 0 to 1.2, $\mathfrak{N}(1.2) = .385$, and likewise for all negative values of x.

The area under the positive half of the normal curve is precisely one-half; from the symmetry of $n(x)$ about $x = 0$ it follows that the area under the entire curve is precisely one. The fact that the total area is one corresponds to the fundamental property that, for given n and p, the sum of $b(k; n, p)$ for $k = 0, 1, 2, \ldots, n$ must be precisely one.

Example: In a binomial process with $n = 600$ and $p = .4$, find the approximate value of $\Pr(230 \leq K \leq 260)$.

Solution: The mean μ and standard deviation σ are

$$\mu = np = 600 \times .4 = 240$$

and

$$\sigma = \sqrt{npq} = \sqrt{600 \times .4 \times .6} = \sqrt{240 \times .6} = \sqrt{144} = 12.$$

Since $\mu = 240$ and $n - \mu = 360$ are both quite large, the normal curve technique furnishes a very good approximation. The cut-off midpoints of interest are

$$x_1 = \frac{k_1 - \mu}{\sigma} = \frac{229.5 - 240}{12} = \frac{-10.5}{12} = -.875$$

and

$$x_2 = \frac{k_2 - \mu}{\sigma} = \frac{260.5 - 240}{12} = \frac{20.5}{12} = 1.71.$$

The value of $\Pr(230 \leq K \leq 260)$ is approximately equal to the area under the normal curve between $x_1 = -.875$ and $x_2 = 1.71$, as shaded in the sketch of Fig. 4-10.

Interpolating in Table 4-2, the area under the curve from $x_1 = -.875$ to $x = 0$, which is the same as from 0 to .875, is $\mathfrak{N}(.875) = .309$. And from $x = 0$ to $x_2 = 1.71$ the area is $\mathfrak{N}(1.71) = .456$. Thus the area from $-.875$ to .1.71 is

$$.309 + .456 = .765,$$

which gives us the solution to the problem:

$$\Pr(230 \leq K \leq 260) \doteq .765.$$

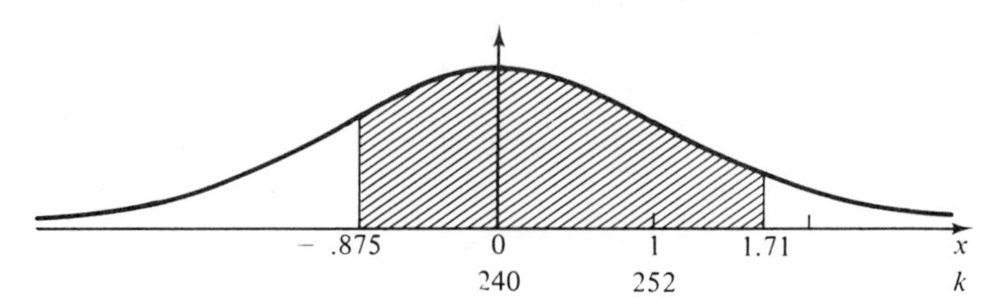

Figure 4-10.

Example: A coin is flipped 10,000 times. Find the approximate probability that Heads occurs on at least 51% of the trials.

Solution: The 10,000 flips is a binomial process with $n = 10{,}000$ and $p = .5$. The problem calls for the value of $\Pr(K \geq 5100)$.

The mean and standard deviation are

$$\mu = np = 10{,}000 \times .5 = 5000,$$
$$\sigma = \sqrt{npq} = \sqrt{2500} = 50.$$

Clearly, the normal curve approximation technique is justified. Also, it should be clear that, in this instance, very little precision is sacrificed if we simplify and work with $k = 5100$ rather than the midpoint value $k = 5100.5$. The corresponding x value is

$$x = \frac{k - \mu}{\sigma} = \frac{5100 - 5000}{50} = 2.$$

Applying the normal curve technique, we approximate $\Pr(K \geq 5100)$ as the area under the normal curve from $x = 2$ on out, as shown in Fig. 4-11: The area to the right of $x = 2$ is computed as the area under the positive half of the curve, which is precisely .5, minus the area from 0 to 2:

$$.5 - \mathfrak{N}(2) = .500 - .477 = .023.$$

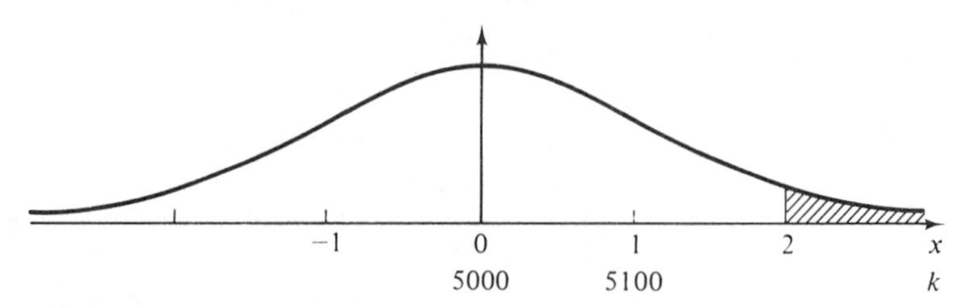

Figure 4-11.

We conclude that the probability of getting at least 51% Heads in 10,000 flips of a coin is about .023.

Example: In a damp shipment of firecrackers the probability of a randomly chosen specimen firing is assumed to be .9.

(a) Suppose a random sample of 100 firecrackers is tested. What is the probability that fewer than 93 fire?

(b) If a sample of 400 is tested, what is the probability that more than 390 fire?

Solution: (a) The test is interpreted as a binomial process with $n = 100$ and $p = .9$. The value desired is $\Pr(K < 93)$.

We have

$$\mu = np = 100 \times .9 = 90,$$

$$\sigma = \sqrt{npq} = \sqrt{9} = 3.$$

Here, the value of $n - \mu$ is 10, barely justifying the normal curve technique for a two-decimal-place approximation.

The midpoint value of interest in finding

$$\Pr(K < 93) = \Pr(K \leq 92)$$

is $k = 92.5$. (Simplifying as in the previous example, we can obtain a rough approximation with $k = 92$ or $k = 93$, but the two-decimal-place accuracy would be lost.) The x value corresponding to $k = 92.5$ is

$$x = \frac{k - \mu}{\sigma} = \frac{92.5 - 90}{3} = \frac{2.5}{3} = .833.$$

By the approximation property the value of $\Pr(K < 93)$ is approximately equal to the area under the normal curve up to $x = .833$, as shown in Fig. 4-12: The area up to $x = .833$ is equal to the area under the negative half of the curve plus the area from 0 to .833. Thus the area of the region shaded is

$$.5 + \mathfrak{N}(.833) = .500 + .298 = .798.$$

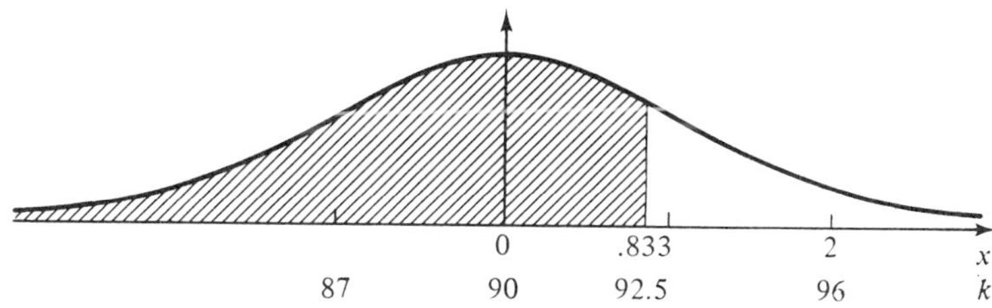

Figure 4-12.

We conclude that the probability that fewer than 93 firecrackers fire in the tested sample of 100 is about .80.

(b) The test of 400 firecrackers is seen as a binomial process with $n = 400$ and $p = .9$. The problem calls for $\Pr(K > 390)$.

Mean and standard deviation are

$$\mu = np = 400 \times .9 = 360,$$

$$\sigma = \sqrt{npq} = \sqrt{36} = 6.$$

Here we ignore the midpoint correction and take $k = 390$, giving

$$x = \frac{k - \mu}{\sigma} = \frac{390 - 360}{6} = \frac{30}{6} = 5.$$

It follows that the value of $\Pr(K > 390)$ is approximately equal to the area under the normal curve from $x = 5$ on out. From Table 4-2 this area is zero, to at least four decimal places.

We conclude that the probability of more than 390 of the 400 firecrackers firing is zero, for all reasonable intents and purposes.

4-5 EXERCISES

1. Consider binomial processes for the following values of n and p. Determine whether the normal curve technique is justified for approximating probabilities of the form $\Pr(k_a \leq K \leq k_b)$ to two-decimal-place accuracy.

(a) $n = 10$ and $p = .4$.
(b) $n = 100$ and $p = .9$.
(c) $n = 25$ and $p = .4$.
(d) $n = 10{,}000$ and $p = .4$.
(e) $n = 1000$ and $p = .005$.

2. (a) Consider a binomial process with $n = 400$ and $p = .5$. Compute the x values, x_1 and x_2, corresponding, respectively, to

$$k_1 = 189.5 \quad \text{and} \quad k_2 = 210.5.$$

(b) Sketch the normal curve, shading the region under the curve between the values x_1 and x_2 computed in part (a). Approximate the value of $\Pr(190 \leq K \leq 210)$ by finding the area under the normal curve between x_1 and x_2. Refer to Table 4-2.

3. Consider a binomial process with $n = 100$ and $p = .2$. Find the corresponding areas under the normal curve to approximate the following:

(a) $\Pr(15 \leq K \leq 25)$.
(b) $\Pr(26 \leq K)$.
(c) $\Pr(K = 20)$.
(d) $\Pr(K \leq 25)$.

4. Consider a binomial process with $n = 400$ and $p = .9$. Apply the normal curve technique to find the approximate value for the following:

(a) $\Pr(350 \leq K \leq 370)$.
(b) $\Pr(345 \leq K)$.
(c) $\Pr(K < 360)$.

5. A die is rolled 180 times. What is the approximate probability that a "six" occurs on more than 35 of the rolls?

6. A thumbtack is so constructed that, when flipped, it lands point down two-thirds of the time. Suppose the tack is flipped 72 times. What is the probability (approximately) that it lands point up on less than 20 of the 72 flips?

7. Suppose the probability of a flashbulb being defective is $p = \frac{1}{8}$. If 700 are tested, what is the probability that less than 80 are found to be defective?

8. A true-false test has 100 questions. Suppose the probability that a certain student knows the answer to any given question is .8. A passing score is 72 or better. What is the probability that the student passes the test?

9. Let it be assumed that 51% of the voters across the country favor a certain proposed constitutional amendment. Suppose 2500 voters are chosen at random and questioned. Find the probability that less than half of those questioned favor the amendment.

10. For each atom of a certain radioactive material the probability of its decaying in any one minute is .00001. Suppose we have a sample in which 1,000,000,000 atoms of this material are present. Find the probability that the number of these atoms decaying in the next minute is between 9800 and 10,200.

11. A telephone exchange serves 5000 customers. Suppose the maximum number of incoming calls that the equipment can accept in the next 15 seconds is 35 and that the probability of a randomly chosen customer placing a call in the next 15 seconds is .005. Find the probability that the number of incoming calls exceeds the capacity of the equipment during the 15-second period.

12. Suppose 500,000 grains of pollen have settled on the calm surface of a swimming pool whose area is 5000 square feet. One square foot of the pool's surface is chosen, and the number of pollen grains thereon is counted. Find the probability that the count reveals more than 115 grains of pollen.

13. Find the probability that the fraction of times Heads occurs is between .49 and .51, inclusive, if a coin is flipped
    (a) 400 times,   (b) 2500 times.

14. Suppose a coin is flipped 10,000 times. Find the smallest value of m for which the following statement is true: The probability is greater than .99 that the number of Heads is between $5000 - m$ and $5000 + m$, inclusive.

15. Joe and Jim are cutting cards at random. If a cut produces a black

card, Joe pays Jim one cent; otherwise Jim pays Joe one cent. Find the probability that after 100 plays one or the other has lost more than ten cents.

16. Experience indicates that of those making reservations for a certain airline flight only 80% actually show for the flight. Suppose the plane can accommodate 100 passengers. How many reservations can the airline make for the flight if they want to be 99% sure that the number of passengers showing does not exceed the capacity of the plane?

4-6. Repeated Trials, in General

Thus far, the analysis of repeated trials has been confined to binomial processes in which the component process has only two relevant outcomes. But many of the results can be extended to repeated trials of more complicated processes.

We return to the Triple-Toss game of Secs. 3-4 and 4-1 in which the House tosses three dice, and the player gets back 0, 2, 3, or 5 dollars for a one-dollar wager, depending on whether 0, 1, 2, or 3 of the dice show his number. Since we want to talk about a large number of plays, we view the game from the House's perspective. From the computation in Sec. 3-4, the House's profit per play has an expected value of .074 dollars. Thus in 10,000 player games of a busy night the House's profit should be in the neighborhood of \$740. But "neighborhood" is a little vague for the sponsors—they would be happier with the answers to questions of the sort,

"In 10,000 player games of Triple-Toss, what is the probability that the House's profit will be at least \$600?"

To analyze the larger process, we treat the problem as though a single player engages in 10,000 plays. Thus we have $n = 10{,}000$ trials of a component process on which is defined a random variable, labeled W_i, corresponding to the House's profit on a single play.

In Sec. 3-4 we considered the random variable $R - 1$, the player's profit. Since the House's profit is the negative of the player's profit, the random variable W_i is the same as $1 - R$. Recalling the earlier analysis, the pertinent data for the component process are as follows:

Values of R	0	2	3	5
Values of $W_i = 1 - R$	1	-1	-2	-4
Probability	$\frac{125}{216}$	$\frac{75}{216}$	$\frac{15}{216}$	$\frac{1}{216}$

We let μ_i and σ_i represent the mean and standard deviation, respectively, of the random variable W_i.

$$\mu_i = E(W_i) = 1 \times \frac{125}{216} + (-1) \times \frac{75}{216} + (-2) \times \frac{15}{216} + (-4) \times \frac{1}{216}$$
$$= \frac{16}{216} = .0741.$$

According to Theorem 9 of Sec. 3-6, the mean square deviation of W_i is

$$\sigma_i^2 = E[(W_i - \mu_i)^2] = E(W_i^2) - \mu_i^2.$$

We have

$$E(W_i^2) = 1 \times \frac{125}{216} + 1 \times \frac{75}{216} + 4 \times \frac{15}{216} + 16 \times \frac{1}{216}$$
$$= \frac{276}{216} = 1.278$$

and

$$\mu_i^2 = (.0741)^2 = .005.$$

Therefore,

$$\sigma_i^2 = 1.278 - .005 = 1.273$$

and

$$\sigma_i = \sqrt{1.273} = 1.13.$$

Having determined the mean μ_i and standard deviation σ_i for the random variable W_i of the component process, we are ready to attack the larger process of $n = 10{,}000$ trials. We have random variables

$$W_1, W_2, W_3, \ldots, W_{10{,}000},$$

where W_i corresponds to the House's profit on the ith play. The total profit in 10,000 plays is given by the random variable

$$H = W_1 + W_2 + W_3 + \cdots + W_{10{,}000}.$$

By Theorem 6 of Sec. 3-5 the expected value of the sum is the sum of expected values; thus the expected value μ of the random variable H is the sum of the expected values of the component random variables W_i. As computed above, W_i has expected value $\mu_i = .0741$, for $i = 1, 2, 3, \ldots,$ 10,000. Therefore,

$$\mu = E(H) = E(W_1) + E(W_2) + \cdots + E(W_{10{,}000})$$
$$= 10{,}000 \times E(W_i) = 741.$$

The procedure thus far is reminiscent of the general analysis of binomial processes in Sec. 4-3. There we saw that for the random variable X_i of the component process the mean is $E(X_i) = p$. And for the random variable $K = X_1 + X_2 + \cdots + X_n$ the mean is np. That is, the mean of K is just n times the mean of X_i. It should be clear from the treatment of $E(H)$ above that this result is equally valid for "repeated-trial" processes in general.

We recall also that in the analysis of binomial processes in Sec. 4-3 the mean square deviation of the component random variable X_i was found to be pq, and the mean square deviation of the "total number of successes" random variable K was found to be npq, n times the mean square deviation of X_i. Although we shall forego the details, the derivation of σ^2 for a binomial process can be extended to produce the analogous conclusion in the general "repeated-trials" situation.

In particular, for the $n = 10{,}000$ plays of the Triple-Toss game the mean square deviation of

$$H = W_1 + W_2 + \cdots + W_{10{,}000}$$

is 10,000 times the common value σ_i^2 for the mean square deviation of the W_i.

Representing the mean square deviation of H as σ^2, we have

$$\sigma^2 = 10{,}000 \times \sigma_i^2 = 10{,}000 \times 1.273 = 12{,}730.$$

Thus the standard deviation of H is

$$\sigma = \sqrt{10{,}000 \times \sigma_i^2} = \sigma_i \times \sqrt{10{,}000}.$$

Earlier we found that $\sigma_i = 1.13$. Thus

$$\sigma = 1.13 \times \sqrt{10{,}000} = 1.13 \times 100 = 113.$$

In short, the House's profit in 10,000 plays has a mean of \$741 and a standard deviation of \$113, both rounded off to the nearest whole number.

In the general case we consider a process consisting of n independent trials of a component process on which is defined a random variable W_i. As in the binomial case, "independent" means that the probabilities in any given trial are unaffected by the experience on other trials. For example, in the Triple-Toss example the event $W_3 = -2$ is independent of the event $W_2 = 1$ and also of the event $(W_1 = -4) \cap (W_2 = 1)$.

We represent the mean and standard deviation of W_i as μ_i and σ_i, respectively. Then for the random variable $H = W_1 + W_2 + \cdots + W_n$ the mean is given by $\mu = n\mu_i$, and the mean square deviation of H is $\sigma^2 = n \times \sigma_i^2$. Thus the standard deviation of H is $\sigma = \sqrt{n} \times \sigma_i$.

The Normal Curve Approximation

Our ability to treat binomial processes has been greatly facilitated by the technique of approximating probabilities by the corresponding areas under the normal curve. Happily this technique applies in the general repeated-trial case much as it does for binomial distributions. The result may be stated as follows:

Let $H = W_1 + W_2 + \cdots + W_n$ be the "sum" random variable in a "repeated-trials" process as described above. For n sufficiently large, probabilities of the form $\Pr(h_1 \leq H \leq h_2)$ are approximated to an arbitrary degree of accuracy, i.e., number of decimal places, by the area under the normal curve between x_1 and x_2, where

$$x_1 = \frac{h_1 - \mu}{\sigma} \quad \text{and} \quad x_2 = \frac{h_2 - \mu}{\sigma}.$$

Similarly, $\Pr(h_1 \leq H)$ is approximated by the area under the curve to the right of x_1, and $\Pr(H \leq h_2)$ by the area up to x_2.

With this result, it's a relatively simple matter to give the Triple-Toss sponsors the answer to the question posed earlier:

Find the probability that the House's profit in 10,000 plays of Triple-Toss will be at least \$600.

In our model the total profit corresponds to the random variable H, which, as we've seen, has mean $\mu = 741$ and standard deviation $\sigma = 113$. To find $\Pr(H \geq 600)$ we transform $h_1 = 600$ into the corresponding x value:

$$x_1 = \frac{h_1 - \mu}{\sigma} = \frac{600 - 741}{113} = \frac{-141}{113} = -1.25.$$

By the general result stated above, the value of $\Pr(H \geq 600)$ is approximately equal to the area under the normal curve to the right of $x_1 = -1.25$. This region is depicted in Fig. 4-13.

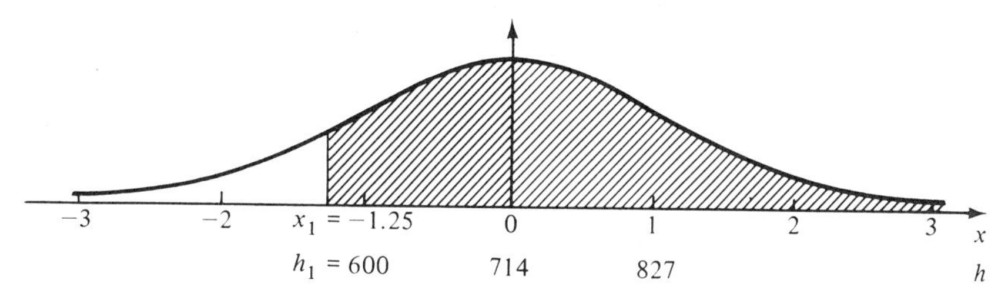

Figure 4-13.

From Table 4-2, $\mathfrak{N}(1.25) = .394$. Thus the area under the curve from $x_1 = -1.25$ to $x_2 = 0$ is .394, and from $x_2 = 0$ on out, the area is .5. It follows that the area to the right of $x_1 = -1.25$ is $.394 + .5 = .894$; i.e., $\Pr(H \geq 600) \doteq .89$.

We conclude that there is a probability of about 89% that in 10,000 player games of Triple-Toss the House will realize a profit of \$600 or more.

Comment: In a binomial process problem we gain greater precision by setting $h_1 = 599.5$, rather than $h_1 = 600$. However, in other problems of this general nature the range of the random variable H may not even consist of whole numbers, so "splitting the difference" requires some discretion. As it has been stated, the approximation result does not call for such a correction. Some accuracy can be gained by a prudent adjustment of the value of h_1, but general directions are not easy to give.

Also, we take a relaxed position with respect to the accuracy of the normal curve technique. Here we'll be content with "rough" approximations and not try to set criteria for two- or three-decimal-place precision.

Example: An urn contains four marbles, two red and two white. A single trial of the process consists of drawing marbles from the urn at random, one-by-one and without replacement, until a red marble is drawn. Find the approximate probability that in 900 trials of the process the total number of draws required is between 1475 and 1525, inclusive.

Solution: For each trial of the process the random variable corresponding to the number of draws required is represented as W_i. The analysis of the component process is conveyed by the probability tree of Fig. 4-14. The mean μ_i of the random variable W_i is computed in the usual fashion:

$$\begin{aligned}\mu_i = E(W_i) &= 1 \times \frac{1}{2} + 2 \times \frac{1}{3} + 3 \times \frac{1}{6} \\ &= \frac{3 + 4 + 3}{6} \\ &= \frac{5}{3}.\end{aligned}$$

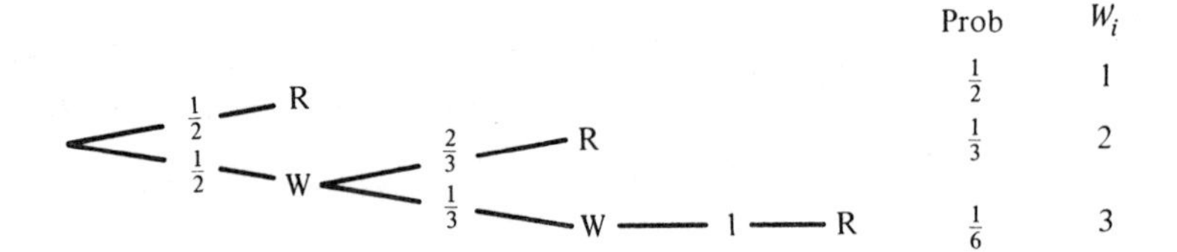

Figure 4-14.

To find the mean square deviation σ_i of the random variable W_i we use Theorem 9 of Sec. 3.6:

$$\sigma_i^2 = E[(W_i - \mu_i)^2] = E(W_i^2) - \mu_i^2.$$

Here,

$$\begin{aligned}E(W_i^2) &= 1 \times \frac{1}{2} + 4 \times \frac{1}{3} + 9 \times \frac{1}{6} \\ &= \frac{3 + 8 + 9}{6} \\ &= \frac{10}{3}\end{aligned}$$

and

$$\mu_i^2 = \left(\frac{5}{3}\right)^2 = \frac{25}{9}.$$

Thus

$$\sigma_i^2 = E(W_i^2) - \mu_i^2 = \frac{10}{3} - \frac{25}{9} = \frac{5}{9}.$$

The total number of draws required in $n = 900$ trials corresponds to the random variable

$$H = W_1 + W_2 + \cdots + W_{900}.$$

From the general result above the mean μ and mean square deviation σ^2 of H are as follows:

$$\mu = n \times \mu_i = 900 \times \frac{5}{3} = 1500,$$

$$\sigma^2 = n \times \sigma_i^2 = 900 \times \frac{5}{9} = 500.$$

Therefore the standard deviation of H is

$$\sigma = \sqrt{\sigma^2} = \sqrt{500} = 22.36.$$

The desired value, $\Pr(1475 \leq H \leq 1525)$, is approximated by the corresponding area under the normal curve. The x values for $h_1 = 1475$ and $h_2 = 1525$ are as follows:

$$x_1 = \frac{h_1 - \mu}{\sigma} = \frac{1475 - 1500}{22.36} = -\frac{25}{22.36} = -1.12,$$

$$x_2 = \frac{h_2 - \mu}{\sigma} = \frac{1525 - 1500}{22.36} = -\frac{25}{22.36} = 1.12.$$

Thus $\Pr(1475 \leq H \leq 1525)$ is approximately equal to the area under the normal curve between -1.12 and 1.12, shown in Fig. 4-15.

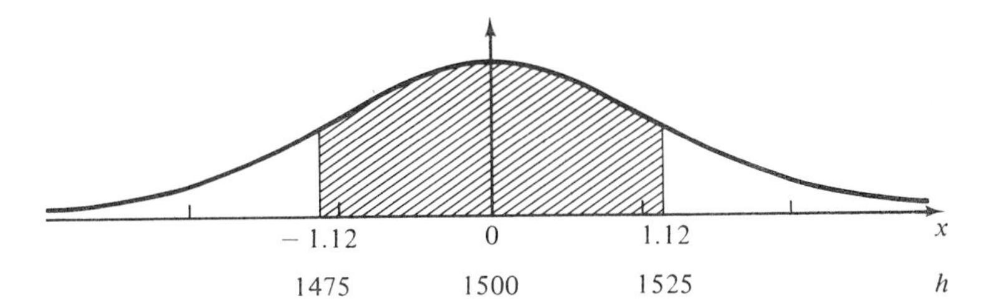

Figure 4-15.

The area from -1.12 to 0 is the same as the area from 0 to 1.12. Interpolating in Table 4-2, it follows that the region shaded has area

$$2\mathfrak{N}(1.12) = 2 \times .368 = .736.$$

We conclude that the probability is about .74, or roughly three-fourths, that between 1475 and 1525 draws are required in 900 trials of the urn process.

The next example illustrates that all we need to know for a repeated-trials analysis is the mean μ_i and standard deviation σ_i of the random variable in the component process.

Example: As the quality-control department at the thumbtack factory well knows, their little boxes of "about 50" thumbtacks may contain anywhere from 48 to 53 tacks. Frequent inspections indicate that the number of tacks per box has a mean of 50.1 and a standard deviation of .8.

Problem: Use the normal curve technique to find the approximate probability that a package of 100 boxes contains at most 5000 thumbtacks.

Solution: The process of counting the tacks in the 100 boxes is interpreted as $n = 100$ repeated trials of counting tacks in individual boxes. For the component process we let the random variable W_i correspond to the number of tacks in a single box. For W_i, the mean and standard deviation are as given:

$$\mu_i = 50.1 \quad \text{and} \quad \sigma_i = .8.$$

For the sum random variable,

$$H = W_1 + W_2 + \cdots + W_{100},$$

the general result stated earlier dictates that the mean and standard deviation are, respectively,

$$\mu = n \times \mu_i = 100 \times 50.1 = 5010$$

and

$$\sigma = \sqrt{n} \times \sigma_i = \sqrt{100} \times .8 = 10 \times .8 = 8.$$

We approximate $\Pr(H \leq 5000)$ as the area under the normal curve up to x, where

$$x = \frac{5000 - \mu}{\sigma} = \frac{5000 - 5010}{8} = -\frac{10}{8} = -1.25.$$

This region is shaded in Fig. 4-16.

From Table 4-2, we see that $\mathfrak{N}(1.25) = .394$. Since the area from -1.25 to 0 is .394, it follows that the area to the left of -1.25 is $.5 - .394 = .106$.

We conclude that there is a probability of about .1 that the package of 100 boxes of "about 50" thumbtacks will contain at most 5000 tacks. (And thus there is a probability of about .9 that the 100 boxes will contain at least 5000 tacks.)

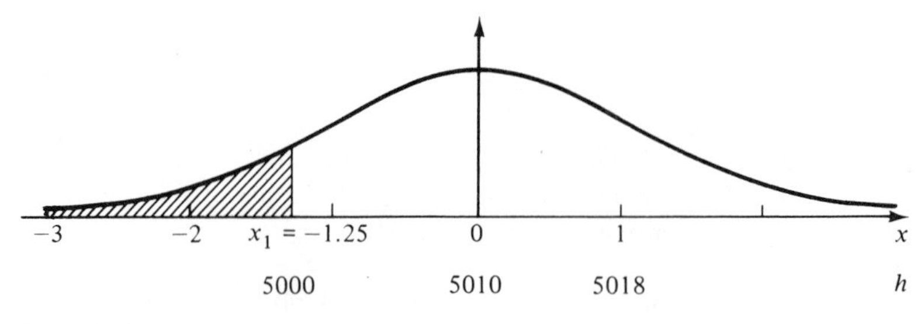

Figure 4-16.

4-6 EXERCISES

1. An urn contains three marbles marked, respectively, with the numbers 1, 2, and 3. The process is to withdraw one marble at random and record the number marked thereon.

(a) Find the mean μ_i and mean square deviation σ_i^2 of the random variable W_i corresponding to the number on the marble drawn.

(b) Suppose the process is performed 150 times. With H representing the random variable corresponding to the sum of the recorded numbers, find the mean μ and standard deviation σ of H.

(c) Use the nomal curve technique to compute the approximate probability that the sum of the recorded numbers in 100 trials is at least 220.

2. An urn contains two red marbles and one green marble. The process is to withdraw marbles, at random and without replacement, until a red marble is drawn and then stop.

(a) Find the mean μ_i and mean square deviation σ_i^2 of the random variable W_i corresponding to the number of draws required on a single trial.

(b) Suppose the process is performed 225 times. Let H represent the random variable corresponding to the total number of draws required. Find the mean and standard deviation σ of H.

(c) Use the normal curve technique to find the approximate probability that the number of draws required in 225 trials is between 290 and 310, inclusive.

3. Consider a process consisting of $n = 100$ independent trials of a component process with random variable W_i as given in the following table:

Values of W_i	1	2	3
Probability	.2	.5	.3

(a) Find the mean μ and standard deviation σ of the "sum" random variable

$$H = W_1 + W_2 + \cdots + W_{100}.$$

(b) Find the approximate value of $\Pr(H \leq 200)$.

4. On each play of a certain game of chance there is a probability of .9 that the player loses a dollar and a probability of .1 that he wins eight dollars. Find the approximate probability that the player comes out ahead in 100 plays of the game.

5. In little boxes of a "dozen" screws packaged by the Casual Count Company there is a probability of

   .06 that a box contains 11 screws,
   .88 that it contains 12 screws, and
   .06 that it contains 13 screws.

   Use the normal curve technique to find the approximate probability that in a random selection of 100 boxes the average number of screws per box is between 11.95 and 12.05.

6. Suppose the number of black jelly beans in little bags of twenty-five has a mean of 4 and a standard deviation of 1.5. Find the approximate probability that a gross of 144 bags contains more than 600 black jelly beans.

7. Taking into account all of the full-time students in a certain large university, it is found that the number of credit hours per student has a mean of 15.8 and a standard deviation of .9. Suppose the average number of credit hours per student is computed for a randomly chosen sample of 100 students. Find the approximate probability that this average lies between 15.7 and 15.9.

8. According to the records of a certain gas station, the amount of gasoline purchased per motorist has a mean of 12.6 gallons, with a standard deviation of 3.3 gallons. Estimate the probability that 100 typical motorists will purchase a total of (a) less than 1200 gallons, (b) more than 1300 gallons.

9. On each play of a certain game the player wins \$1, with probability .2; or he wins \$9, with probability .05; or he loses \$1, with probability .75. What is the probability that the player comes out ahead if he plays the game (a) twice, (b) 80 times, (c) 2000 times?

10. Use the normal curve technique to get a rough estimate of the probability that the player loses more than five dollars in 25 plays of the Triple-Toss game described in the text. (Since the number of trials is so small, one may seriously question the reliability of the method. For purposes of comparison, let it be revealed that the desired probability is .264, to three decimal places.)

11. The number of chocolate chips in a certain kind of cookie has a mean of 5 and a standard deviation of 2. Find the values of k for which the following statement is true: The probability is greater than .9 that a box of 100 cookies contains between $500 - k$ and $500 + k$ chocolate chips.

4-7. The Ubiquitous Normal Curve

In addition to its role in repeated-trials processes, the normal curve arises in the study of many other theoretical problems in probability. On the practical side, we find the normal curve approximating the shape of graphed data in a wide variety of empirical settings.

For example, consider the following situation:

> A test consisting of 100 true-false questions is administered to a group of 900 students. A graph of the results is constructed to display the number of students answering k questions correctly, where k ranges from 0 to 100.

If we assume that every student is just guessing at each question, then we may justifiably anticipate that the graph will resemble that of the binomial distribution with $n = 100$ and $p = .5$.

But the students are not just guessing. Nonetheless, it is reasonable to expect the graphed data to approximate the characteristic bell shape of the normal curve. For example, one should not be surprised if the graph of the results looks like Fig. 4-17.

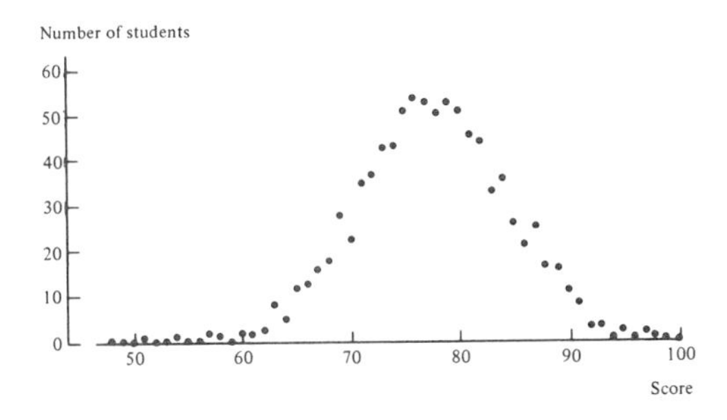

Figure 4-17 Hypothetical Data for Text of 100 T-F Questions.

Just as the Registrar in the illustration of Sec. 3-6 computed the mean and standard deviation for the number of courses carried by the students in the University, so the grader may compute the mean and standard deviation of the grades of the 900 students. Suppose he finds a mean of 77.4 by averaging the scores, and he finds a standard deviation of 6.8 by taking the square root of the average square deviation from the mean. The grader then may report to the administrator as follows:

> "The scores are roughly distributed normally with a mean of 77.4 and a standard deviation of 6.8."

Knowing these two values, the administrator can easily supply approximate answers to other general questions about the distribution of the scores. For instance, suppose he wants to know the approximate percentage of students who scored better than 70 on the test. With a mean of 77.4 and a standard deviation of 6.8, we see that one standard deviation below the mean is 70.6. According to Table 4-2, the probability that the outcome is to the right of one standard deviation below the mean is about .84. Thus the administrator would assume that roughly 84% of the 900 students scored better than 70.

Another example where an approximation by the normal curve can be expected is the following:

Example: The yardstick factory is proud of the quality of its product. In a routine check, the length of each yardstick in a lot of 1000 is measured to the nearest one-thousandth of an inch. Computations with the 1000 measurements produce a mean of 36.013 inches and a standard deviation of .027 inch for the lot of 1000 yardsticks. The measurements are observed to be approximately distributed normally.

Question: About how many of the 1000 yardsticks are less than 36 inches long?

Solution: The value 36.000 is .013 to the left of $\mu = 36.013$. Thus the corresponding x value is

$$x_1 = \frac{-.013}{\sigma} = \frac{-.013}{.027} = -.48.$$

The information is conveyed in Fig. 4-18. The area of the region shaded in the figure is

$$.5 - \mathfrak{N}(.48) \doteq .5 - .18 = .32.$$

Since this region corresponds to yardstick measurements of less than 36 inches, we conclude that about 32% of the 1000 yardsticks, or roughly 320 of them, are less than 36 inches long.

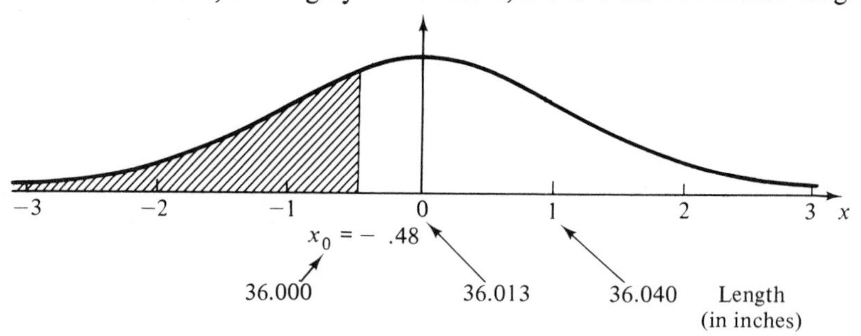

Figure 4-18.

In advance of any measurements we know that the yardsticks are not all exactly 36 inches long. Nor are they all exactly the same length. But why not? The answer is that there are presumably many factors contributing to the variation—small imperfections in the gears, blades getting dull, hard and soft spots in the wood, etc., not to mention inaccuracies in the measuring apparatus.

In cases such as this, where the variation in the measurements represents the accumulation of many small effects operating more or less independently of each other, one may expect the graphed data to closely approximate the characteristic bell shape of the normal curve.

4-7 EXERCISES

1. The grades on a test taken by a large group of students are found to be distributed normally, with a mean of 76.4 and a standard deviation of 8.0. About what percentage of the students received a grade between 70 and 80?

2. Tests at the factory indicate that the lifetimes for a certain type of light-bulb are distributed normally with a mean of 1240 hours and a standard deviation of 160 hours. Suppose a bulb is chosen at random and tested. What is the probability that its lifetime is
(a) Less than 1000 hours?
(b) Between 1000 and 1400 hours?
(c) Greater than 1400 hours?

3. The yearly incomes of apprentice workers in a nationwide labor union are found to be distributed normally (roughly) with a mean of $7420 and a standard deviation of $560. Find the approximate percentages of these workers having yearly incomes in the following ranges:
(a) Less than $6000.
(b) $6000 to $7000.
(c) $7000 to $8000.
(d) Greater than $8000.

4. A standard exercise in a certain engineering course is to measure the length of the football field. Over several years with many students the set of measurements produces a mean of 99.936 yards, with a standard deviation of .048 yard. As should be expected, the measurements are found to be distributed normally. About what percentage of the measurements have the length of the field as more than 100 yards?

5. The quality-control department at the measuring cup factory reports that the liquid volume of the "one-cup" cups has a mean of 7.96 ounces, with a standard deviation of .11 ounces. Assuming a normal distribution, what is the probability that a randomly chosen cup has a volume of less than 8 ounces?

6. For a large shipment of "five-pound" bags of grapefruit it is found that the actual weights of the individual bags are roughly distributed normally with a mean of 5.1 pounds and a standard deviation of .2 pounds.
(a) If one bag is chosen at random, what is the approximate probability that it weighs more than 5 pounds?
(b) Suppose four bags are chosen at random. Find the approximate probability that the sum of their weights is more than 20 pounds. (As one might suspect, the "four-bag" weights will also conform to a normal distribution. For the standard deviation, recall the previous section.)

7. The number of raisins in boxes of Caveman Natural Breakfast Cereal conforms to a normal distribution with a mean of 400 raisins and a standard deviation of 15 raisins per box.
   (a) If one box is selected at random, find the approximate probability that the number of raisins therein is between 390 and 410, inclusive.
   (b) If 100 boxes are selected at random, find the approximate probability that the average number of raisins per box is between 399 and 401.

5

STATISTICS

Many of the practical applications of probability fall into the realm of *statistics*, which may be characterized as the science of analyzing numerical data. Since statistics is indispensable to many aspects of decision making in our world, it behooves us to look into the rationale underlying its procedures. As we shall see, there is a certain subtlety in the logic of a statistical analysis, and the conclusions are easily misinterpreted. (Thus have the honest, but misunderstood, efforts of the statistician sometimes cast him as the butt of the old saw "Figures never lie, but liars figure.")

5-1. Hypothesis Testing

The following situation is representative of a fundamental problem in statistics:

> The better to gauge his prospects for election, Candidate A has an aide question 400 randomly selected voters as to whether they favor him or his opponent.

To clarify the practical purpose of the voter poll it is assumed that Candidate A is considering several possible campaign strategies, his choice to depend on the outcome of the poll. In the interest of simplicity we further assume that the candidate has decided in advance how he will respond to different outcomes. Focusing on a particular response, let's say that the candidate's position is that the favor of 220 or more of the 400 voters will be interpreted as strong evidence that he enjoys the support of a majority of all the voters, in which case he will assume a posture appropriate to a "sure winner."

The poll of the 400 voters is interpreted as a binomial process of $n = 400$ trials, with "success" on each trial corresponding to the event "the voter favors Candidate A." Thus the probability p of success is taken to be equal to the fraction of all voters favoring the candidate. Of course the value of p is unknown. But the candidate's position is that the occurrence of the event $K \geq 220$, i.e., 220 or more successes, is to be interpreted as strong evidence that $p > .5$.

Comment: The critical reader may protest that the voter poll does not qualify as a binomial process in that the trials are not really independent. For instance, if there are 1,000,000 voters in all and 500,000 favor Candidate A, then $p = .5$ on the first trial. But if the first voter supports the candidate, then the probability that the second supports him is only 499,999/999,999 since we've removed the first voter from the pool. Thus the second trial is not quite independent of the first.

One answer to the objection is to interpret that the voters are chosen "with replacement," so we don't disallow the unlikely possibility that the first and second voters are one and the same. The more obvious answer, and the one we adopt, is that in problems of this sort the number of trials is so small relative to the larger population that the experience on the trials does not significantly affect the value of p. Thus p is assumed to be constant from trial to trial.

We identify the event $K \geq 220$ as the *critical outcome*. And, giving priority to the assumption that p falls in the undesirable range, we identify the inequality $p \leq .5$ as the *null hypothesis*. The position of the candidate with respect to the 400-trial binomial process may now be paraphrased in the conventional language of *hypothesis testing*:

If the critical outcome—$K \geq 220$—occurs, then the null hypothesis—$p \leq .5$—*is rejected.*

One may be tempted to defend the candidate's position on the grounds that if the critical outcome occurs, then the probability of the null hypothesis $p \leq .5$ is small. But here's where the subtlety lies—the voters are there, they have their preferences, and either $p \leq .5$ or $p > .5$, with no question of probability about it. In short, p is fixed but unknown. What does underlie the candidate's "sure-winner" position is the fact that the probability of the critical outcome $K \geq 220$ is small if the null hypothesis $p \leq .5$ is true. Let's see how small this probability is.

The Level of Significance

The voter-poll process has been interpreted as a binomial process with $n = 400$ and p unknown. The null hypothesis—which we symbolize as H_0—is the assumption that $p \leq .5$. The critical outcome—symbolized as C—is the event $K \geq 220$. The ingredients in the process, which we identify as a *hypothesis test*, may be displayed on one line:

$$n = 400; \qquad H_0: p \leq .5; \qquad C: K \geq 220.$$

The understanding is that the occurrence of the event C translates into the action "*The null hypothesis is rejected.*" If C does not occur, we regard the test as *inconclusive*.

As has been said, the test derives its validity from the fact that the probability of the event C is small if it is assumed that the null hypothesis H_0 is true. Thus the occurrence of C—unlikely if H_0 is true—is interpreted as evidence that H_0 is false.

We consider first the extreme case $p = .5$. With $n = 400$ and $p = .5$ the mean μ and standard deviation σ are, respectively,

$$\mu = np = 400 \times .5 = 200,$$
$$\sigma = \sqrt{npq} = \sqrt{400 \times .5 \times .5} = \sqrt{100} = 10.$$

The normal curve technique is employed to find the approximate value of $\Pr(K \geq 220)$, the probability of the critical outcome C. The x value corresponding to $k_0 = 220$ is

$$x_0 = \frac{k_0 - \mu}{\sigma} = \frac{220 - 200}{10} = 2.$$

Comment: For arithmetic convenience we shall abandon the finer precision of the "half-interval correction"; thus here we work with $k_0 = 220$, rather than $k_0 = 219.5$. Of course, some accuracy is lost. Indeed, there may be occasions where we are guilty of treating approximations as being more precise than a critical evaluation would justify.

Since $k_0 = 220$ corresponds to $x_0 = 2$, the value of $\Pr(K \geq 220)$ is approximated by the area under the normal curve from $x_0 = 2$ on out, as shaded in Fig. 5-1.

From Table 4-2 in Sec. 4-5, the area under the normal curve from 0 to $x_0 = 2$ is

$$\mathfrak{N}(2) = .477.$$

Thus the area to the right of $x_0 = 2$, and the approximate probability of the

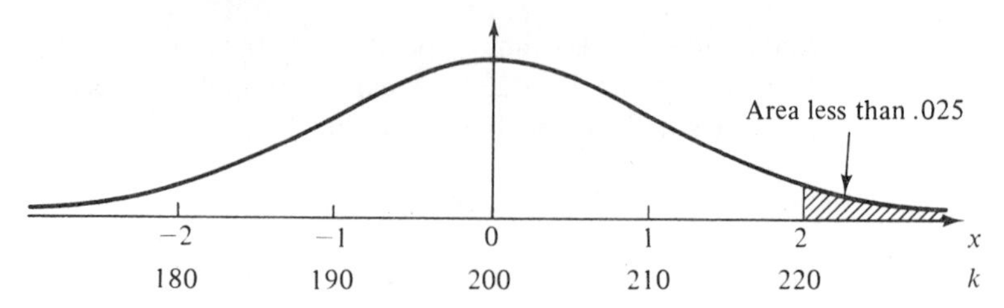

Figure 5-1.

critical outcome, is given by

$$\Pr(K \geq 220) \doteq .5 - .477 = .023.$$

Allowing for some inaccuracy, the conclusion we reach is that in a binomial process with $n = 400$ and $p = .5$

$$\Pr(K \geq 220) < .025.$$

It should be clear that if the value of p is less than .5, then the value of $\Pr(K \geq 220)$ is less than the value in the $p = .5$ case. Therefore, if the null hypothesis—$H_0: p \leq .5$—is true, the probability of the critical outcome—C: $K \geq 220$—is less than .025.

We've established that .025 is an upper bound on $\Pr(K \geq 220)$, the probability of the critical outcome C, subject to the null hypothesis, H_0: $p \leq .5$. This being the case, we say that .025 is the *level of significance* of the hypothesis test. Symbolically, Sig: .025.

Let's look again at the empirical interpretation. Candidate A accepts the favor of 220 or more of the voters polled as strong evidence that he enjoys the support of more than half of all the voters. This position is based on the fact that the probability is less than .025—the level of significance of the test—that the poll would reveal 220 or more supporters if indeed the candidate does not have the support of more than half of the voters.

Variations

The passage from the particular hypothesis test to the general case is exemplified by variations on the voter-poll example.

As a first variation, let's suppose the aide and his team of workers poll 1600 voters, rather than 400. The candidate still desires evidence that $p > .5$, i.e., he wants to reject the null hypothesis $p \leq .5$, and he is still happy with a .025 level of significance.

Question: What is the appropriate value of k_0 such that the favor of k_0 or more of the 1600 voters signals the rejection of the null hypothesis? In other words, for what k_0 is the event "$K \geq k_0$" the critical outcome?

The ingredients in the new hypothesis test involve one unknown, k_0, to be determined:

$$n = 1600; \qquad H_0: p \leq .5; \qquad C: K \geq k_0; \qquad \text{Sig: } .025.$$

In the binomial process with $n = 1600$ and $p = .5$, the mean and standard deviation are as follows:

$$\mu = np = 1600 \times .5 = 800,$$

$$\sigma = \sqrt{npq} = \sqrt{1600 \times .5 \times .5} = \frac{\sqrt{1600}}{2} = \frac{40}{2} = 20.$$

We arrived at the .025 level of significance in the original example from the fact that the area under the normal curve to the right of $x = 2$ is a little less than .025. It follows that in any binomial process for which the normal curve approximation is applicable the value of $\Pr(K \geq k_0)$ is less than .025 if the x value corresponding to k_0 is 2 or greater, which is to say, if k_0 is 2 or more standard deviations greater than the mean. With $n = 1600$ and $p = .5$ the smallest k_0 that satisfies this requirement is

$$k_0 = \mu + 2\sigma = 800 + 2 \times 20 = 840.$$

Thus for $n = 1600$ and $p = .5$

$$\Pr(K \geq 840) < .025.$$

It should be clear that this inequality remains valid if p is strictly less than .5. Accordingly, the critical outcome C is defined to be the event $K \geq 840$.

We see that 840 or more successes in 1600 trials furnishes essentially the same degree of evidence that $p > .5$ as does 220 or more successes in 400 trials.

As a second variation let's make the original test more severe and adjust the critical outcome so that the level of significance is pushed down from .025 to .01. The ingredients in the test again involve an unknown k_0, to be determined:

$$n = 400; \qquad H_0: p \leq .5; \qquad C: K \geq k_0; \qquad \text{Sig: } .01.$$

With $n = 400$ and $p = .5$, the same values as in the original test, we have

$$\mu = 200 \quad \text{and} \quad \sigma = 10.$$

The requirement on k_0 is that

$$\Pr(C) = \Pr(K \geq k_0) < .01.$$

This demands that the area under the normal curve from the corresponding x_0 on out must be less than .01. From Table 4-2 we see that $\mathfrak{N}(2.4)$ is a little greater than .49, so the area to the right of $x = 2.4$ is a little less than .01. Accordingly, we take k_0 to be the smallest integer for which the corresponding x_0 is 2.4 or greater. In this case

$$k_0 = \mu + 2.4\sigma = 200 + 2.4 \times 10 = 224.$$

We conclude that for $n = 400$ and $p = .5$

$$\Pr(K \geq 224) < .01.$$

And the inequality remains valid if $p < .5$. Thus the critical outcome is defined to be the event C: $K \geq 224$.

Empirically, the new test demands the favor of 224 or more of the 400 voters before the candidate rejects the null hypothesis $p \leq .5$, that is, before he accepts the outcome as sufficient evidence that he has the support of more than half of all the voters. But the evidence is somewhat stronger than in the original test in that the probability of obtaining the critical outcome $K \geq 224$, if the null hypothesis is true, is now less than .01.

As a final variation let's suppose that we are dealing with a strange candidate who, being an avid supporter of the two-party system, is anxious not to oversell his case. Let's say that he does not desire the support of as much as 60% of the electorate. Thus the null hypothesis is $H_0 \geq .6$. The hypothesis is rejected if the poll does not reveal too many supporters. Thus the critical outcome is of the form C: $K \leq k_0$. Continuing with 400 voters in the poll and returning to a .025 level of significance, the ingredients in the test are as follows:

$$n = 400; \qquad H_0: p \geq .6; \qquad C: K \leq k_0; \qquad \text{Sig: } .025.$$

With $n = 400$ and $p = .6$, mean and standard deviation are

$$\mu = np = 400 \times .6 = 240,$$

$$\sigma = \sqrt{npq} = \sqrt{400 \times .6 \times .4} = \sqrt{96} = 9.80.$$

The value of k_0 is determined by the requirement

$$\Pr(C) = \Pr(K \leq k_0) < .025.$$

The inequality is satisfied if the area under the normal curve up to the corresponding x_0 is less than .025, as shown in Fig. 5-2.

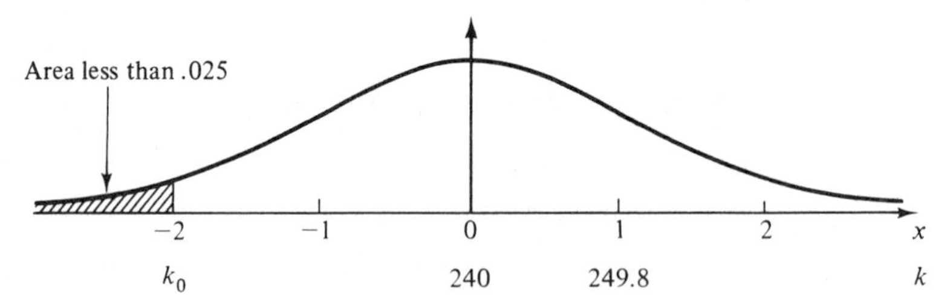

Figure 5-2.

As observed earlier, the area from $x = 2$ on out is a little less than .025; thus the area up to $x = -2$ is also a little less than .025. Thus k_0 is taken to be the greatest integer which is two or more standard deviations below the mean. Two standard deviations below the mean we find

$$\mu - 2\sigma = 240 - 2 \times 9.8 = 220.4.$$

Accordingly, we set

$$k_0 = 220.$$

The argument shows that with $n = 400$ and $p = .6$,

$$\Pr(K \leq 220) < .025.$$

The inequality is preserved if $p > .6$. The critical outcome in this last variation is C: $K \leq 220$.

The General Case

Underlying a hypothesis test is some chance process in which the probability p of success is unknown. The test is a binomial process composed of n trials of the process. Its aim is to give evidence that $p > p_0$ (or $p < p_0$) in that the occurrence of a critcal outcome of the form $K \geq k_0$ (or $K \leq k_0$) signals the rejection of the null hypothesis $p \leq p_0$ (or $p \geq p_0$). The level of significance is an upper bound on the probability of the critical outcome under the assumption that the null hypothesis is true.

In general terms the essential ingredients in a hypothesis test are of two forms:

$$n = n_0; \qquad H_0: p \leq p_0; \qquad C: K \geq k_0; \qquad \text{Sig: } \alpha_0;$$

or

$$n = n_0; \qquad H_0: p \geq p_0; \qquad C: K \leq k_0: \qquad \text{Sig: } \alpha_0.$$

The requirement is that for any value of p satisfying the null hypothesis H_0 we have

$$\Pr(C) < \alpha_0.$$

(α_0 is read "alpha sub zero" or "alpha-nought.")

Example: A buyer contemplates purchasing a large number of flashbulbs. But first he wants some assurance that more than 60% will fire under prescribed marginal conditions.

Problem: Design a hypothesis test such that the occurrence of the critical outcome gives the buyer the desired assurance at a .05 level of significance.

Solution: For arithmetic convenience we take the number of trials to be $n = 150$. The ingredients in the test are as follows:

$$n = 150; \qquad H_0: p \leq .6; \qquad C: K \geq k_0; \qquad \text{Sig: } .05.$$

The problem is to determine the appropriate value of k_0.

With $n = 150$ and $p = .6$, the mean and standard deviation are as follows:

$$\mu = np = 150 \times .6 = 90,$$
$$\sigma = \sqrt{npq} = \sqrt{150 \times .6 \times .4} = \sqrt{36} = 6.$$

Since the level of significance is .05, k_0 is determined by the requirement

$$Pr(C) = Pr(K \geq k_0) < .05.$$

Thus the desired k_0 has the property that the area under the normal curve to the right of the corresponding x_0 is a little less than .05, as shown in Fig. 5-3.

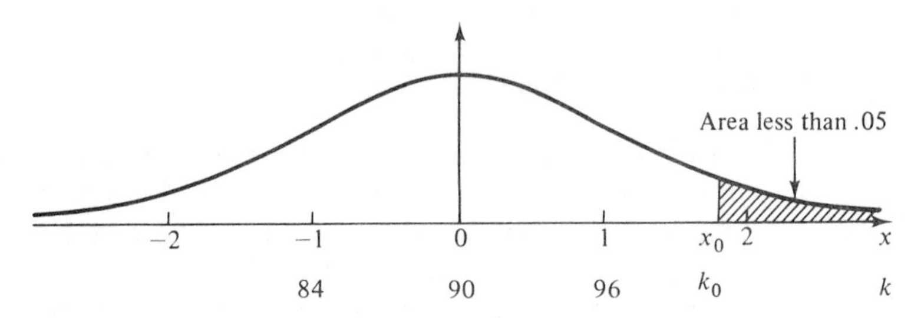

Figure 5-3

From Table 4-2 we see that $\mathfrak{N}(1.7)$ is a little more than .45, so the area from $x = 1.7$ on out is a little less than .05. We associate $x = 1.7$ with a .05 level of significance, just as we earlier associated $x = 2$ with a .025 level and $x = 2.4$ with a .01 level.

The k value corresponding to $x = 1.7$ is

$$k = \mu + \sigma x = 90 + 6 \times 1.7 = 90 + 10.2 = 100.2.$$

It follows that for $n = 150$ and $p \leq .6$,

$$\Pr(K \geq 101) < .05.$$

The critical outcome is the event

$$C\colon K \geq 101.$$

The interpretation is that the flashbulb buyer should test a sample of 150 bulbs. If more than 100 fire, he rejects the null hypothesis—the probability of a bulb firing is $p \leq .6$—at a .05 level of significance. In rejecting the null hypothesis the buyer is left with the desired assurance that more than 60% of the lot of flashbulbs are good.

5-1 EXERCISES

1. An urn contains three marbles, either two reds and one white or one red and two whites—we know not which. At random and with replacement, a marble is drawn 50 times.

(a) Assume there's only one red marble in the urn. Then show that the probability is less than .01 that a red marble would be drawn on more than 25 of the 50 draws. That is, show that $\Pr(K > 25) < .01$ in a binomial process with $n = 50$ and $p = \frac{1}{3}$.

(b) Now assume that there are two red marbles in the urn. Then show that the probability is less than .01 that a red marble would be drawn on fewer than 25 of the 50 draws.

(c) Suppose the 50 draws are executed and a red marble is drawn 29 times. What does this experience indicate about the number of red marbles in the urn? Why?

2. Suppose 900 randomly selected voters are questioned as to whether they favor a certain candidate. Let p represent the fraction of all voters favoring the candidate. Show that the favor of 476 or more of the 900 voters justifies the rejection of the null hypothesis $p \leq .5$ at a .05 level of significance. That is, show that in a binomial process with $n = 900$ and $p \leq .05$, $\Pr(K \geq 476) < .05$.

3. In the following, consider binomial processes with n as given and p in the prescribed range. In each case find the smallest value of k_0, or approximately the smallest, satisfying the stated condition. Then interpret the information as describing a hypothesis test, identifying the null hypothesis H_0, the critical outcome C, and the level of significance.

(a) $n = 100$, $p \leq .5$, $\Pr(K \geq k_0) < .025$.
(b) $n = 400$, $p \leq .8$, $\Pr(K \geq k_0) < .05$.
(c) $n = 400$, $p \leq .8$, $\Pr(K \geq k_0) < .025$.
(d) $n = 600$, $p \leq .4$, $\Pr(K \geq k_0) < .05$.
(e) $n = 600$, $p \leq .4$, $\Pr(K \geq k_0) < .01$.
(f) $n = 1000$, $p \leq .9$, $\Pr(K \geq k_0) < .025$.

4. With respect to the conditions below, follow the directions given in Prob. 3, except that the value of k_0 should be the largest, or approximately the largest, satisying the condition.

(a) $n = 400$, $p \geq .8$, $\Pr(K \leq k_0) < .05$.
(b) $n = 400$, $p \geq .2$, $\Pr(K \leq k_0) < .01$.
(c) $n = 600$, $p \geq .4$, $\Pr(K \leq k_0) < .025$.
(d) $n = 10{,}000$, $p \geq .5$, $\Pr(K \leq k_0) < .01$.
(e) $n = 10{,}000$, $p \geq .9$, $\Pr(K \leq k_0) < .01$.
(f) $n = 500$, $p \geq \frac{2}{3}$, $\Pr(K \leq k_0) < .025$.

5. (a) The flashbulb buyer, wanting some assurance that the probability of a certain type of bulb firing is more than 90%, asks his statistician to design a test of the null hypothesis $H_0: p \leq .9$, at a .025 level of significance with $n = 400$. Determine the appropriate critical outcome $C: K \geq k_0$.

(b) Same problem as part (a), except the desired level of significance is .01.

(c) Again, same as part (a), except the desired level of significance is .001. [Given: $\mathfrak{N}(3.1) = .4990$.]

6. Before a certain new drug can be allowed on the market, evidence must be presented that it does not produce a particular adverse side effect in more than 5% of the overall population. Suppose the drug is tested on 1000 randomly chosen patients. How many of them can suffer the side effect with it still being proclaimed that the desired evidence has been obtained at a .01 level of significance?

7. A student of the powers of the mind, Jason claims to be gifted with ESP (extrasensory perception). As a test, a friend shuffles a conventional deck of cards, draws a card at random, and concentrates on its color, red or black; out of sight in an adjoining room, Jason declares his perception of the card's color. Suppose the experiment is repeated 100 times, and Jason declares the correct color on 69 of the trials. Should we accept Jason's claim to powers of ESP?

8. (a) A competitor in a board game seems to be throwing the advantageous "six" on his die more often than the "laws of probability" would seem to suggest. His next 72 throws are quietly monitored, and "six" is found to occur 17 times. Does this entitle us to reject, at a .05 level of significance, the assumption that there is nothing peculiar about his throw?
   (b) Suppose he had thrown a "six" 22 times out of the 72 trials. At what level of significance may we reject the "nothing peculiar" hypothesis?

9. As soon as the polls close at 9:00 p.m., representatives of a television network start to phone in random returns from around the state. The table below shows the total number of voters and the number voting for Candidate A as compiled at different times.

Time	*Number of Voters*	*Votes for A*
9:17	1,600	827
9:23	3,600	1857
9:31	6,400	3312
9:36	10,000	5179
9:42	14,400	7441

If the network operates at a .001 level of significance, at what time do they project Candidate A as the winner of the election? [Given: $\mathfrak{N}(3.1) = .4990$.]

5-2. Confidence Intervals

Our first statement about probability, early in Chapter 2, was to the effect that "The probability of a coin falling Heads is one-half, because in the long run it should fall Heads about half the time." Now we are ready to tighten the language.

Suppose a coin is flipped 10,000 times. What can we say about the fraction of times the coin falls Heads?

Of course, all we can say absolutely is that the fraction of Heads is somewhere between 0 and 1, inclusive, but that's academic. What we do say is that the fraction is very likely to be fairly close to one-half. Let's be more explicit.

Naturally, we interpret the many flips of the coin as a binomial process with $n = 10{,}000$ and $p = .5$. For the random variable K—total number of Heads—the mean μ and standard deviation σ are as follows:

$$\mu = np = 10{,}000 \times .5 = 5000,$$

$$\sigma = \sqrt{npq} = \sqrt{10{,}000 \times .5 \times .5} = \frac{\sqrt{10{,}000}}{2} = \frac{100}{2} = 50.$$

We recall from the previous section that the area under the normal curve between $x = 0$ and $x = 2$ is a little more than .475. Therefore, the area between $x_1 = -2$ and $x_2 = +2$ is a little more than .95. For the binomial process at hand, the k values corresponding to $x_1 = -2$ and $x_2 = +2$ are

$$k_1 = \mu + x_1 \times \sigma = 5000 - 2 \times 50 = 4900,$$

$$k_2 = \mu + x_2 \times \sigma = 5000 + 2 \times 50 = 5100.$$

As indicated in Fig. 5-4, we conclude that for the binomial process with $n = 10{,}000$ and $p = .5$

$$\Pr(4900 < K < 5100) > .95. \tag{1}$$

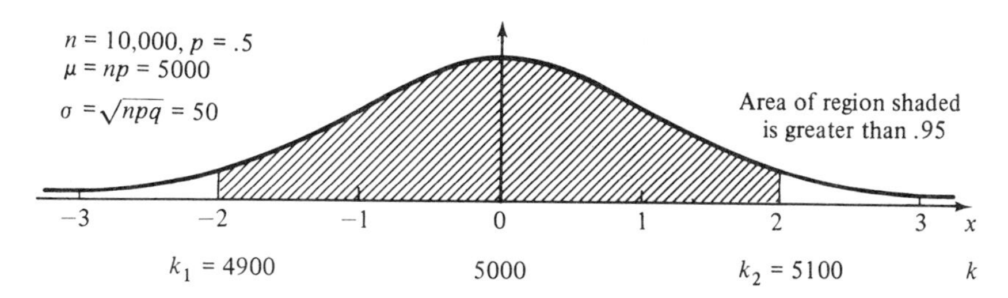

Figure 5-4.

We are concerned with the fraction of Heads in 10,000 flips of a coin. Since K represents the total number of Heads, the fraction of Heads corresponds to the random variable K/n, where $n = 10{,}000$. Inasmuch as

$$\text{“}4900 < K < 5100\text{”} \quad \text{and} \quad \text{“}.49 < \frac{K}{n} < .51\text{”}$$

are one and the same event, we rewrite Eq. (1) in the form

$$\Pr\left(.49 < \frac{K}{n} < .51\right) < .95.$$

Initially we said that the fraction of Heads in the 10,000 flips is very likely to be fairly close to one-half. Now we say that there are better than 19 chances in 20 that the fraction of Heads is between .49 and .51.

Let's imagine that we convey, in plain language, the conclusion in (1) above to a friend who is suspicious by nature of all "theoretical" assertions. He just can't believe that there are 19 chances in 20 that 10,000 flips of a coin will produce between 4900 and 5100 Heads. Not inclined to invest his time in theoretical arguments, he takes the direct approach and spends two long evenings flipping a coin 10,000 times. And he triumphantly reports that he recorded 5140 Heads. Dismissing the possibility that his experience was an unlikely one occurring by chance, our difficult friend insists that he has proven the theory to be wrong.

For the sake of argument let's go along with our friend and not try to explain away his results on the grounds that unlikely outcomes do sometimes occur. Instead, we entertain the possibility that there was a bias in his manner of flipping the coin, or that the coin is unbalanced, making it more likely to fall Heads than Tails. In short, we assume that the probability p of the coin falling Heads is not necessarily one-half. Now let's use his experience to see what it suggests about the value of p.

The conclusion expressed by the inequality (1) can be generalized to apply to any binomial process for which the normal curve provides a good approximation. An extension of the argument that gave 4900 and 5100 in the specific case produces the general result

$$\Pr(\mu - 2\sigma < K < \mu + 2\sigma) > .95. \tag{2}$$

For given n, the value of the standard deviation, $\sigma = \sqrt{npq}$, depends on the values of p and $q = 1 - p$. However, for fixed n it is not difficult to show that the largest value of σ is produced by $p = q = \frac{1}{2}$, in which case

$$\sigma = \sqrt{n \times \frac{1}{2} \times \frac{1}{2}} = \frac{\sqrt{n}}{2}.$$

Thus, regardless of the value of p, the standard deviation σ is less than or equal to $\sqrt{n}/2$.

It follows that the event

$$E_1\colon \mu - \sqrt{n} < K < \mu + \sqrt{n}$$

necessarily includes all values in the range of K which satisfy the event

$$E_2\colon \mu - 2\sigma < K < \mu + 2\sigma.$$

Therefore the probability of E_1 is not less than that of E_2. Thus we can replace 2σ by $\sqrt{n}$ in Eq. (2) without destroying the inequality. Making this substitution and setting $\mu = np$, we deduce from (2) that

$$\Pr(np - \sqrt{n} < K < np + \sqrt{n}) > .95. \tag{3}$$

The following manipulation of the inequalities defining the event in (3) gives us an equivalent characterization:

$$np - \sqrt{n} < K, \qquad K < np + \sqrt{n},$$

$$p < \frac{K + \sqrt{n}}{n}, \qquad \frac{K - \sqrt{n}}{n} < p.$$

Therefore inequality (3) may be rewritten in the form

$$\Pr\left(\frac{K-\sqrt{n}}{n} < p < \frac{K+\sqrt{n}}{n}\right) > .95. \tag{4}$$

Comment: The subtle danger here is that one reads (4) as asserting something about the probability that p lies in a certain interval. Not so. The underlying structure is a binomial process with fixed values of n and p. The inequality (4) speaks only of the probability of the event determined by requiring the random variable K to satisfy a certain pair of simultaneous inequalities.

Let's return to the argument with our friend. Rather than insisting that there was something unusual about his experience, we suggest that his coin and his technique of flipping are biased in such a way that the probability p of Heads on each flip is greater than one-half. We further suggest that the true probability p is such that his result, the event $K = 5140$, is in line with what one would expect in 95% of the cases. To be more precise, 5140 is a value of the random variable K satisfying the requirement stated in (4); that is, p satisfies the inequalities

$$\frac{5140-\sqrt{10{,}000}}{10{,}000} < p < \frac{5140+\sqrt{10{,}000}}{10{,}000},$$

$$\frac{5140-100}{10{,}000} < p < \frac{5140+100}{10{,}000},$$

$$.504 < p < .524.$$

We saw that for $p = .5$ our friends's result is outside the range that had been identified as having probability greater than .95. But if we assume that p is in the interval from .504 to .524, then his result is in the analogous range of values of K for which the probability is greater than .95.

Now let's apply the inequality (4) to a flashbulb variation:

The buyer, contemplating the purchase of a large quantity of flashbulbs, directs his assistant to test 100 bulbs under marginal conditions to get some idea of the probability of a bulb being good. The assistant performs the test and finds that 65 of the 100 bulbs are good.

We view the assistant's test as a binomial process with $n = 100$ and p unknown. Regardless of what the true value of p is, the inequality (4) tells us that the probability is greater than .95 that the observed value of the random variable K satisfies the conditions

$$\frac{K-\sqrt{n}}{n} < p < \frac{K+\sqrt{n}}{n}. \tag{5}$$

Here we have $n = 100$, and the observed value of K is 65. If we reject the possibility that the value of p is such that the observed value of K is not within the 95% range given in (5), we have

$$\frac{65 - \sqrt{100}}{100} < p < \frac{65 + \sqrt{100}}{100},$$

$$.55 < p < .75.$$

This interval, .55 to .75, computed from the result of the assistant's test, is said to be a *95% confidence interval* for the value of p.

In a binomial process of n trials, each value k in the range of the random variable K determines a *95% confidence interval* for the value of p according to the conditions in (5):

$$\frac{k - \sqrt{n}}{n} < p < \frac{k + \sqrt{n}}{n}.$$

The underlying reasoning assures us that the probability is greater than .95 that the observed value, k, is such that the corresponding 95% confidence interval includes the true value of p.

In effect, the computation of a 95% confidence interval is a two-sided hypothesis test. In the assistant's test in which the 95% confidence interval for the value of p is $.55 < p < .75$, we are rejecting the hypothesis $p \leq .55$ at a .025 level of significance because the observed value $k = 65$ is so large, and we are rejecting the hypothesis $p \geq .75$, also at a .025 level of significance, because the observed value $k = 65$ is so small.

We have chosen to focus on the 95% confidence interval, but the same line of reasoning can be applied to determine 90% confidence intervals, or 99% confidence intervals, or what have you. Of course, the higher the degree of confidence demanded, the larger the interval. For example, in the assistant's test, for which the 95% confidence interval is given by $.55 < p < .75$, one finds that the 99% confidence interval is $.52 < p < .78$.

Example: A poll of 1600 randomly selected voters shows that 824 support Candidate A.

Problem: Determine a 95% confidence interval for p, where p is the fraction of all voters supporting Candidate A.

Solution: The voter poll is interpreted as a binomial process with $n = 1600$. On each trial "success" corresponds to "the voter supports Candidate A"; thus the probability of success is p, the fraction of voters supporting the candidate.

The observed value of the random variable K is $k = 824$. The corresponding 95% confidence interval for the value of p is determined as follows:

$$\frac{k - \sqrt{n}}{n} < p < \frac{k + \sqrt{n}}{n},$$

$$\frac{824 - \sqrt{1600}}{1600} < p < \frac{824 + \sqrt{1600}}{1600},$$

$$\frac{824 - 40}{1600} < p < \frac{824 + 40}{1600},$$

$$\frac{784}{1600} < p < \frac{864}{1600},$$

$$.49 < p < .54$$

Based on a 95% confidence interval, the pollster reports that the poll indicates that Candidate A enjoys the support of between 49% and 54% of the voters.

5-2 EXERCISES

1. In the interest of getting an estimate of the probability p that a certain thumbtack will come to rest with its point down, the tack is flipped 400 times. The point-down state is found to occur on 280 of the flips. Use the inequality

$$\Pr\left(\frac{K - \sqrt{n}}{n} < p < \frac{K + \sqrt{n}}{n}\right) > .95$$

to find a 95% confidence interval for p.

2. Gail has a large urn containing thousands of marbles, some white and the rest black. At random and with replacement, she draws 100 times, getting a white marble in 58 of the draws. Find a 95% confidence interval for the fraction p of white marbles in the urn.

3. A random check of 900 households in a large city shows that 510 have pianos in the house. Considering all of the households in the city, find a 95% confidence interval for the fraction p having pianos.

4. In a binomial process with $n = 400$ and p unknown, find 95% confidence intervals for the value of p corresponding to the following values in the range of K: (a) $k = 264$, (b) $k = 140$, (c) $k = 222$.

5. In a binomial process with $n = 2500$ and p unknown, find 95% confidence intervals for p corresponding to the following observed number of successes: (a) $k = 760$, (b) $k = 1261$, (c) $k = 1776$.

6. Just after the polls close, a quick check shows that 212 of 400 randomly selected voters have voted for Candidate A. After all the votes are tallied, it is found that A has won 49.6% of the votes. Does the 95%

confidence interval derived from the check of the 400 voters include the true value of p?

7. In a certain process the probability of success is .4. Consider the following cases of n independent trials producing k successes, with n and k as given. In each case determine whether the 95% confidence interval includes the true value $p = .4$.
   (a) $n = 100$, $k = 48$.
   (b) $n = 1000$, $k = 430$.
   (c) $n = 10{,}000$, $k = 4120$.
   (d) $n = 100{,}000$, $k = 40{,}400$.

8. An insurance company has prepared a mail advertisement for a new Major Medical Policy. Their market research specialist estimates that 35% of those receiving the ad will be interested enough to return the "want more information" card enclosed. As a test, the advertisement is sent to 1000 individuals, and 387 of them return the enclosed card. Does the specialist's estimate of 35% return fall inside the 95% confidence interval determined by the test?

9. On the average, 50% of those visiting a certain gift shop make a purchase. Suppose 584 visitors make a purchase one day. We want to estimate the total number of visitors that day. *Question*: Given a binomial process of n trials with $p = .5$, for what values of n does the observed value $k = 584$ fall within two standard deviations of the mean?

10. Show that for $q = 1 - p$,

$$npq = \frac{n}{4} - n\left(p - \frac{1}{2}\right)^2.$$

Use this result to prove the assertion in the text: In a binomial process of n trials the standard deviation σ is less than or equal to $\sqrt{n}/2$.

5-3. Errors, Statistical and Logical

Before looking into the more objective errors that can occur in hypothesis testing and computations of confidence intervals, we call attention to the danger of a transgression universally forbidden in science. A fundamental dictum of the *scientific method* is that conjectures, not conclusions, are to be drawn from information obtained more by accident than design. In the present context this means that the only worthy statistical inferences to be obtained from a data-gathering experiment are those that pertain to questions that had been formulated in advance of the experiment.

We recall the illustration of the previous section in which the assistant tests 100 flashbulbs with the stated purpose of getting an idea of the value of the probability p that a bulb fires under marginal conditions. He observes that 65 of the bulbs fire. Extracting the lower bound of the 95% confidence interval, this result entitles the assistant to draw the inference that p satisfies the inequality $p > .55$, at a .025 level of significance. The object of the experiment was to test the value of p, and it is true that the probability of getting as many as 65 good bulbs in a sample of 100 is less than .025 if it is assumed that the inequality $p > .55$ is not true. The inference drawn is based on these facts.

But suppose that in testing the 100 bulbs the assistant happens to observe that two of the bulbs crack. On the basis of this observation, the assistant takes as a null hypothesis the assumption that the probability p of a bulb cracking is less than or equal to .002. It can be shown that the probability of getting as many as two cracked bulbs in a lot of 100 is less than .02 if the null hypothesis $p \leq .002$ is true.

Question: Since the assistant did observe two cracked bulbs among the 100 tested, is he entitled to reject the null hypothesis $p \leq .002$ at a .02 level of significance?

The answer is "no," simply because the experiment was not designed to test the probability of a bulb cracking. What undermines the "bulb-cracking" hypothesis test—which was formulated after the fact—is the possibility that the assistant took note of the result because there was something curious about it. Even if the probability of a bulb cracking is about .002, there is still a probability of almost .02 that two or more crack in a test of 100 bulbs. Thus perhaps the assistant was simply reacting to an extraordinary result, the likes of which occurs only about one time in fifty. To say the very least, the .02 level of significance loses its significance.

The scientific method dictates that if the assistant wants to say something substantial about the probability of a bulb cracking, he should design a new experiment with the stated purposes including a test of this probability. The results of the new experiment may then entitle him to draw a statistical inference about its value.

More subtle than a violation of the scientific method is the danger of misinterpreting the meaning of the results of hypothesis testing and confidence interval computations. Taking the easier vehicle for discussion, we'll focus on the confidence interval.

In the example of the previous section, the support of 824 voters among 1600 voters polled produces the 95% confidence interval

$$.49 < p < .54$$

for the fraction p favoring Candidate A from among all the voters.

We recall that the confidence interval is based on the result

$$\Pr\left(\frac{K - \sqrt{n}}{n} < p < \frac{K + \sqrt{n}}{n}\right) > .95,$$

where $n = 1600$ and the observed value of the random variable K is $k = 824$. This being the case, it is understandable that one might interpret the derived confidence interval in the following language:

> The probability is greater than .95 that the value of p lies between .49 and .54.

Looking at this statement, one may get the mistaken idea that we are allowing for a number of undecided voters who will, in effect, decide their preferences by flipping coins, and the probability is greater than .95 that the coin flips will place the fraction p of voters favoring Candidate A between .49 and .54. This is not a correct reading. As has been stressed before, the model assumes that p is fixed—either p lies in the interval or it does not.

But there is a larger setting in which the interpretation of the 95% confidence interval above makes sense. What we do is imagine that a great many polls of 1600 voters each are taken, where a fixed fraction p of all the voters support Candidate A. Each poll produces a 95% confidence interval for the value of p. Then in the long run it is to be expected that more than 95% of the polls produce an interval that includes the true value of p. The particular statement "The value of p lies between .49 and .54" may now be viewed as one of a hypothetical class of statements, more than 95% of which are true.

In short, it is not the particular statistical inference—$.49 < p < .54$—that has a probability greater than .95 of being correct; rather it is the mechanism that produces the inference that has a probability greater than .95 of producing a true statement.

The Objective Errors

Now we turn to the more obvious errors that may occur in hypothesis testing and confidence interval computations. To keep the discussion simple, we return to the hypothesis test in which the candidate's aide polls 400 voters with the aim of obtaining evidence that the fraction p of supporters among the voters is greater than one-half. Recall that if the number of supporters among the 400 sampled is 220 or more, the null hypothesis, $H_0: p \leq .5$, is rejected at a .025 level of significance.

The underlying structure is a binomial process with $n = 400$ and p fixed but unknown. For each value of p between 0 and 1 the critical outcome C has a certain probability α. In Sec. 5-1 we established that if p satisfies the null hypothesis, i.e., $p \leq .5$, then the value of $\alpha = \Pr(C)$ is less than .025, the level of significance.

In Fig. 5-5 the hypothesis test is summarized by what might be called a partial probability tree. The noncritical outcome, $K < 220$, is expressed as $\bar{C}$, in the usual notation of set complement. It is conventional to identify the negation of the null hypothesis—in this case $p > .5$—as the *alternative hypothesis*. In the initial stage of the tree the alternative hypothesis is symbolized as H_a. As the legend reminds us, the null hypothesis H_0 is rejected if the critical outcome C occurs; this action may equally well be characterized as acceptance of the alternative hypothesis H_a. As the legend also shows, there are two types of possible errors, Type I and Type II.

The painful path in the tree is inescapable. On the uppermost path, where the null hypothesis is true and the outcome is critical, we find ourselves rejecting the truth. This is the Type I error. In the voter poll this path corresponds to the case where the actual fraction of supporters is less than .5, but by chance the tested sample of 400 voters does produce 220 or more supporters. The candidate is given to feel that he enjoys the support of more than half the voters, but in this he is wrong.

We do not assign branch probabilities in the first stage of the tree in Fig. 5-5. Indeed our attitude has been that either H_0 is true or false—we just don't

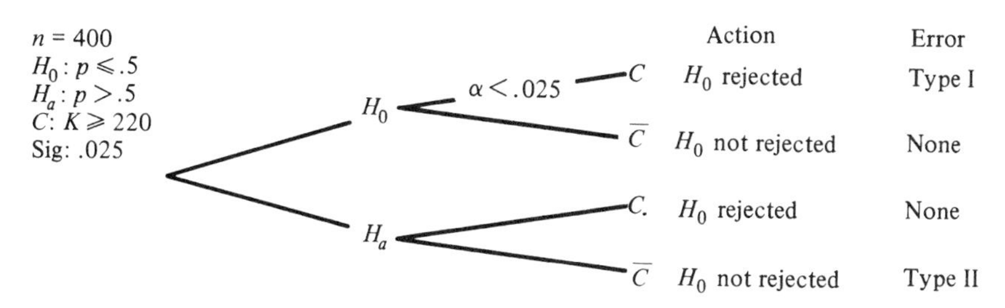

Figure 5-5 Summary of Hypothesis Text.

know which. Thus the branch probability leading to the point H_0 is either 1 or 0. A less strict interpretation of the situation might allow other values for this branch probability, but the value must still be less than or equal to one. In any case, the probability of the uppermost path is less than the level of significance, .025, which is an upper bound on the branch probability α. In short, the probability of a Type I error is less than the level of significance of the test.

As observed in the previous section, the computation of a 95% confidence interval can be viewed as a two-sided hypothesis test, each at a .025 level of significance. There is a probability of less than .025 of committing a Type I error on either side. Thus the probability that neither occurs, i.e., the probability that the 95% confidence interval includes the true value of p, is greater than .95.

The line of action corresponding to the lowermost path in the tree of Fig. 5-5—in effect, not accepting the alternative hypothesis when in fact it is true—is identified as a Type II error. There's not much that can be said here about the probability of a Type II error. In the example, p satisfies the alternative hypothesis if it is only slightly larger than .5. In this case the probability of the noncritical outcome is greater than .97. About as much as we can say is that the probability of a Type II error is less than .98, which isn't saying much.

Secondary Hypotheses

In spirit, the Type II error underlies an important elaboration on hypothesis testing. For an illustration we recall the flashbulb example in Sec. 5-1 in which the buyer tests 150 bulbs in the hope of obtaining assurance that the probability p of a bulb firing is greater than .6. The null hypothesis, $H_0: p \leq .6$, is rejected at a .05 level of significance if $K \geq 101$, i.e., if 101 or more of the 150 bulbs fire. The test is summarized in Fig. 5-6.

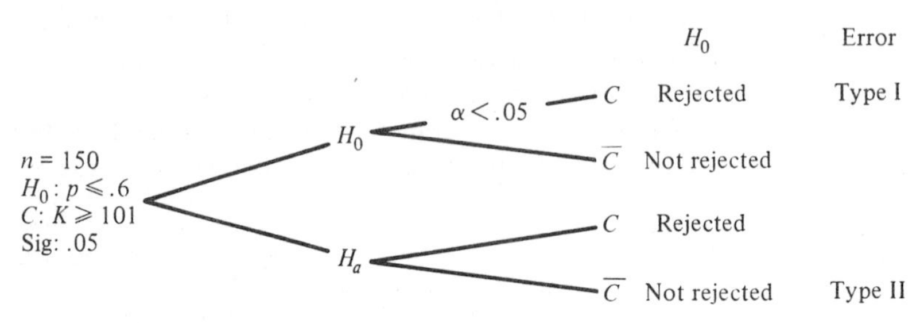

Figure 5-6 Summary of Test of 150 Flashbulbs.

Let's suppose that only 99 of the 150 bulbs fire. Since the critical outcome C does not occur, the buyer regards the test as inconclusive and does not reject the null hypothesis, and he presumably looks for another supplier of flashbulbs.

In the tree of Fig. 5-6 the inconclusive outcome corresponds to the two paths leading to $\bar{C}$. The buyer refuses the lot because he fears that the upper of these two paths represents the truth; i.e., H_0 holds—less than 60% of the bulbs are good in general. But 99 of the 150 bulbs, or 66%, were good. Thus the flashbulb manufacturer would seem to be on solid ground in arguing that it is more likely the lower of the two paths leading to $\bar{C}$ that represents the truth. That is, the manufacturer argues that the buyer is guilty of a Type II error in not rejecting the null hypothesis.

Let's see how the manufacturer's interest can be taken into account. We suppose that quality control at the flashbulb factory is designed to assure that at least 70% of the bulbs are good by the stringent criterion being imposed. Thus in giving the buyer a sample lot to test, the manufacturer would like some assurance that his bulbs will pass the test, if the advertised figure of 70% is correct. Since the buyer requires only that 60% are good, there should be little difficulty in designing a test that will satisfy both parties.

The prospective buyer and the manufacturer get together and decide to test 400 flashbulbs. If the probability of a bulb being good is .6, as the buyer fears, then the expected number of good bulbs in the sample of 400 is 240. On the other hand, if the probability is .7, as the manufacturer asserts, then the expected number of good bulbs is 280.

Splitting the difference, the buyer agrees to purchase a large lot of bulbs if at least 260 in the sample of 400 are good.

From the buyer's point of view, this is a hypothesis test with $n = 400$. The null hypothesis is $H_0: p \leq .6$, and the critical outcome is $C: K \geq 260$.

To find the level of significance, we consider the extreme value, $p = .6$. For a binomial process with $n = 400$ and $p = .6$, the mean μ and standard deviation σ are, respectively,

$$\mu = np = 400 \times .6 = 240,$$

$$\sigma = \sqrt{npq} = \sqrt{400 \times .6 \times .4} = \sqrt{96} = 9.80.$$

The x value corresponding to $k_0 = 260$ is

$$x_0 = \frac{k_0 - \mu}{\sigma} = \frac{260 - 240}{9.80} = \frac{20}{9.80} = 2.04.$$

Thus for $n = 400$ and $p = .6$ the value of $\Pr(K \geq 260)$ is approximately equal to the area under the normal curve to the right of $x_0 = 2.04$. As we've repeatedly had occasion to observe, the area to the right of $x = 2$ is a little less than .025, and likewise for $x_0 = 2.04$. Therefore,

$$\Pr(C) = \Pr(K \geq 260) < .025.$$

The inequality remains valid if $p < .6$. Thus if the null hypothesis H_0: $p \leq .6$ is true, the probability of the critical outcome $C: K \geq 260$ is less than .025. Thus, insofar as the buyer is concerned, the hypothesis test is at a .025 level of significance.

To consider the test from the manufacturer's perspective, we let H_2 represent the statement "The probability of a bulb being good is at least .7":

$$H_2: p \geq .7.$$

We identify H_2 as the *secondary hypothesis*.

The manufacturer's concern is with the prospect of the bulbs failing the test if H_2 is true. Thus he isn't happy unless the probability of the noncritical outcome $\bar{C}: K \leq 259$ is small if it is assumed that the secondary hypothesis $H_2: p \geq .7$ is true.

To compute an upper bound on $\Pr(\bar{C})$ for $p \geq .7$ we consider the extreme case, $p = .7$. With $n = 400$, the mean and standard deviation are as follows:

$$\mu = np = 400 \times .7 = 280,$$

$$\sigma = \sqrt{npq} = \sqrt{400 \times .7 \times .3} = \sqrt{84} = 9.17.$$

For $k_0 = 259$, we have

$$x_0 = \frac{k_0 - \mu}{\sigma} = \frac{259 - 280}{9.17} = \frac{-21}{9.17} = -2.29.$$

The value of $\Pr(K \leq 259)$ is about equal to the area under the normal curve up to $x_0 = -2.29$, as shown in Fig. 5-7.

From Table 4-2, $\mathfrak{N}(2.29) = .487$. Therefore, for $n = 400$ and $p = .7$, $\Pr(K \leq 259) \doteq .5 - .487 = .013$. If p is more than .7, the probability of "$K \leq 259$" will be even less. Allowing for inaccuracy, we conclude that for $p \geq .7$ the probability that the critical outcome does not occur is less than .015. The double-sided hypothesis test is summarized in Fig. 5-8.

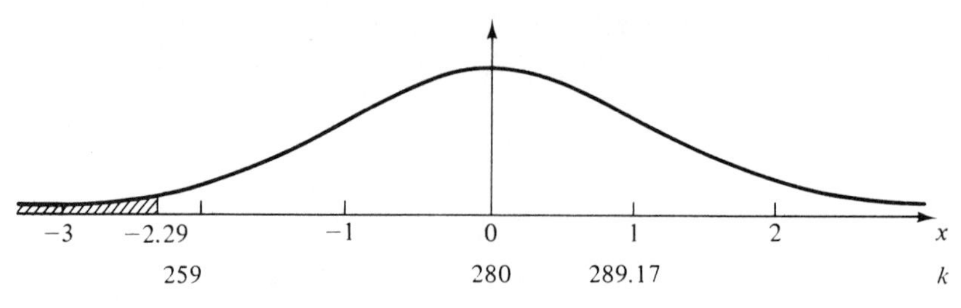

Figure 5-7.

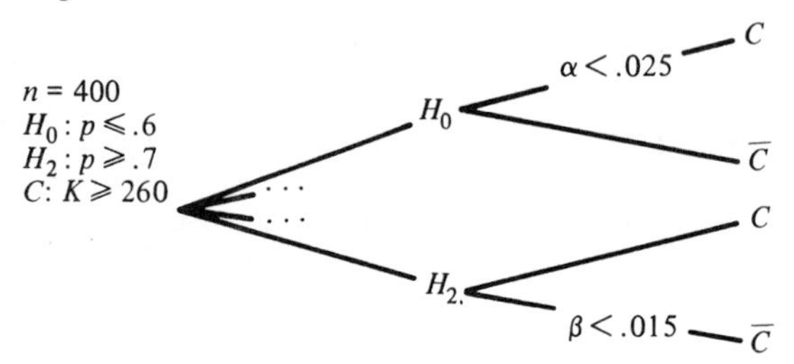

Figure 5-8.

As the tree of Fig. 5-8 shows, the probabilities of the errors feared, respectively, by prospective buyer and manufacturer are small, less than .025 and .015, respectively. Thus both parties should be happy with the test.

5-3 EXERCISES

1. Helen has a large urn containing thousands of marbles, a fraction p of which are white. Helen suspects that $p > .8$. To test this suspicion, she draws $n = 400$ times, at random and with replacement, and she announces the rejection of the null hypothesis $H_0: p \leq .8$ if 336 or more of the 400 draws produce a white marble. In one line, Helen's hypothesis test is as follows:

$$n = 400; \qquad H_0: p \leq .8; \qquad C: K \geq 336.$$

(a) Suppose the actual fraction of white marbles is $p = .784$, and suppose a white marble is drawn in $k = 324$ of the 400 trials. Does Helen's test give rise to a Type I error?

(b) Is a Type I error committed if the true fraction is $p = .824$ and the observed value of K is $k = 319$?

(c) Same question as part (b), with $p = .797$ and $k = 339$.

2. Consider a hypothesis test with essential ingredients as follows:

$$n = 100; \qquad H_0: p \geq .6; \qquad C: K \leq 51; \qquad \text{Sig: } .05.$$

Suppose the true value of p is as given below and that the observed value of K is the given k. In each case, determine whether the hypothesis test gives rise to a Type I error, and also whether it gives rise to a Type II error.

(a) $p = .602$, $k = 53$. (b) $p = .602$, $k = 50$.
(c) $p = .584$, $k = 50$. (d) $p = .584$, $k = 53$.

3. A manufacturer of radios requires "on-off" switches of a type offered by an outside supplier. For the probability p of a switch meeting a certain criterion, the manufacturer demands that $p > .5$, and the supplier asserts that $p > .6$. A sample of 400 switches is tested, with the agreement that the manufacturer will do business with the supplier if 220 or more of the 400 switches meet the criterion set. The two-sided test, with null hypothesis H_0, secondary hypothesis H_2, and critical outcome C is as follows:

$$n = 400; \qquad H_0: p \leq .5; \qquad H_2: p > .6; \qquad C: K \geq 220.$$

Show that if the null hypothesis H_0 is true, then the probability of the critical outcome C is less than .05. Also show that if the secondary hypothesis H_2 is true, then the probability is less than .05 that the critical outcome C does not occur.

4. The buyer wants assurance that the probability p of a certain type of flashbulb firing is greater than .8, and the manufacturer insists that p is at least .9. A random sample of 225 flashbulbs is tested, with the lot passing the test if at least 193 bulbs fire. Find upper bounds on the probability of the lot passing the test if $p \leq .8$, and on the probability of the lot failing the test if $p \geq .9$.

5. As in Prob. 4, the flashbulb buyer demands that $p > .8$, where p is the probability of a bulb being good, and the manufacturer insists that $p \geq .9$. Design a hypothesis test with $n = 400$ such that the probability of the critical outcome C: $K \geq k_0$ is less than .01 if the null hypothesis H_0: $p \leq .8$ is true, and the probability of the noncritical outcome $\bar{C}$: $K < k_0$ is less than .01 if the secondary hypothesis H_2: $p \geq .9$ is true. That is, find k_0 such that in a binomial process of $n = 400$ trials

$$\Pr(K \geq k_0) < .01 \qquad \text{if } p \leq .8$$

and

$$\Pr(K < k_0) < .01 \qquad \text{if } p \geq .9.$$

6. Suppose the buyer of a certain product demands that the probability p of an item being substandard satisfies $p < .4$, and a prospective supplier insists that $p < .2$ for his product. Design a two-sided hypothesis test with $n = 100$. Find an upper bound on the probability of the product passing the test if the buyer's fears are well founded, that is, if $p \geq .4$; also find an upper bound on the probability of the product failing the test if the supplier is correct in his assertion that $p < .2$.

7. Suppose your competitor in a board game has just thrown the advantageous "six" on his die five times in a row. Since this is so unlikely with a fair die, are you entitled to draw the inference that there's something peculiar about his throw?

8. Suppose a wholesaler is checking the freshness of 100 eggs drawn at random from a large shipment, and he observes that 28 of the 100 eggs have double yolks. Is the wholesaler entitled to draw the inference that, at a .025 level of significance, more than 20% of the eggs in the shipment have double yolks?

9. There are five members on the town council. Two are chosen at random and questioned about their position on a proposed ordinance.

(a) What is the probability that both members questioned favor the ordinance if it is assumed that only two of the five members favor the ordinance?

(b) Suppose each of the members determines his position by flipping a coin. What is the probability that three or more members favor the ordinance, given that the two questioned both favor the ordinance?

10. For a certain mathematics textbook it is well established that only 90% of the answers given in the book are correct. Suppose the answer to problem 17 on page 144 is given as $5\sqrt{\pi} - 2$. Assuming you're not able to work problem 17 on page 144, does it make sense to say that the probability is .9 that the correct answer is $5\sqrt{\pi} - 2$?

PART TWO

MATRICES

6

GAMES AND MATRICES

Our attention shifts to the fascinating subject of games. Although the mathematical theory is very modern, the games themselves must have enriched the most primitive of cultures.

> Surely the caveman teased his child, challenging, in some long-lost tongue, "Which hand is it in?," while holding out two clenched fists, one concealing some delectable treat. And that society is deprived indeed which has not devised some variation of "paper, scissors and rock": "Scissors cut the paper, paper covers the rock, and rock smashes the scissors."

The study of games is a bridge between probability, which has occupied us in the first half of the book, and matrices, which will be the main focus of the second half. Matrices form the very ground on which many important questions of mathematics are formulated. As we shall see in Chapter 8, game theory itself has a much wider applicability than one would suspect on first encounter.

6-1. Examples of Games

We begin with the simplest game imaginable:

On the count of three, two boys each extend either one or two fingers. If they match, one and one or two and two, the one boy wins; if they don't match, the other boy wins.

For reasons soon to be clear, we call the one boy "R" and the other "C." The analysis of their "Two-Finger Matching Game" is immediate:

The "Two-Finger Matching Game"

Possibilities R	C	*Result*
I	I	R wins
I	II	C wins
II	I	C wins
II	II	R wins

The square table below conveys the same analysis.

		C	
		I	II
R	I	R wins	C wins
	II	C wins	R wins

Now let's say what it is that R and C win in the various cases, again with a square table:

		C	
		I	II
R	I	C pays R 4 cents	R pays C 2 cents
	II	R pays C 10 cents	C pays R 7 cents

To simplify the table, we adopt the convention that the entry is the amount C pays R. For the cases in which R pays C, we enter a negative amount. For example, if R plays II and C plays I, the *payoff* is -10, meaning R pays C 10 cents. Now the table becomes

		C: I	C: II
R	I	4	−2
R	II	−10	7

Entry is number of cents C pays R.

The mental gymnastics of R and C, common to the sort of games we'll be considering, are easy to imagine:

Thinks C: Since I stand to win more in column I than in II—10 cents rather than 2—and lose less—4 cents rather than 7—I had better play I.

But R is one jump ahead of C: Since C surely prefers column I, I had better play row I and win the 4 cents.

Then C has second thoughts: But R must know that I'm attracted to column I. So he'll probably play row I in hope of winning the 4 cents. Maybe I had better go with column II and settle for the 2 cents.

And R's second thoughts: Since C knows that I know that he ought to prefer column I, he'll probably play column II. So I had better play row II.

Third thoughts for C: But R knows that I know that he knows that I ought to prefer column I. . . .

Etc., etc.

And that's what we mean by a game.

We'll return to the Two-Finger Matching Game shortly. As we shall see, the game definitely favors one of the two players. The reader should try to anticipate the result.

In popular usage, the term "game" embraces many enterprises outside the scope of our restricted use of the word. For example, the game of chess does not qualify as a game in the present context. (But there is a more general mathematical theory that is applicable to chess.) On the other hand, there are many situations which may not look like games which can be treated via game-theoretic models.

Interest will be confined to situations fitting the following description:

1. Each of two players has two or more options he can select, their choices being revealed simultaneously.
2. For each pair of options there is a corresponding numerical payoff which one of the players receives at the expense of the other.

The "Game of Politics" may not seem to fit this model, but politicians may indulge in our kind of game from time to time. The following story is overly simplified—and purely hypothetical, of course:

Tomorrow's statewide election between candidates R and C looks like a standoff; it all seems to hinge on the thus-far uncommitted voters.

Both candidates have reserved time on several TV stations for a final appeal. Candidate R has taped two different speeches: Speech I is the high-minded approach, while Speech II details the moral turpitude of that culprit *C*. But *R* knows his adversary and can well imagine three different approaches, I, II, and III, that *C* may take in his final appeal. And *R* knows the territory, so he is prepared to estimate the probability that he will gain the support of the typical uncommitted voter who hears his Speech I and also hears *C* taking Approach I, and similarly for the other combinations.

These probabilities are presented in the following table:

Candidates Game

		C's Approaches		
		I	II	*III*
R's Speeches	I	.54	.47	.55
	II	.47	.52	.46

Entry is the probability of the voter deciding for *R*.

In the Candidates Game *R* has two options and *C* has three. The probabilities tabulated can be interpreted as the fraction of voters that *C* is awarding to *R* under the respective combinations of options. The problem for *R* is to decide which speech to give. Or should he give different speeches on different stations?

For another example, consider the "Game of Football," which, in the large, is not to be resolved by any abstract mathematical model. But there are games within the game. Again we simplify.

Team *R* has the ball, and the quarterback contemplates whether to call Play I, Play II, or Play III. He knows that the opponent, Team *C*, will try to surprise him with one of two possible defense setups, and he can estimate the number of yards likely to be gained under each combination of Play and Defense. His estimates are shown in the table below:

Play Selection Game

		C's Defenses	
		I	II
R's Offenses	I	1	3
	II	6	1
	III	0	4

Entry is the number of yards that team *R* can expect to gain.

The Payoff Matrix

From the earlier description, it should be clear that a given game can always be conveyed by a rectangular table.

By agreement, the numerical entry in the table is always to represent the amount player C, who chooses the column, pays player R.

The labels R and C for the players, and I, II, etc., for their respective options, are common to all games and can be left implicit in the table representing the game. The essential information is given by the rectangular array of numbers inside.

To illustrate, recall the Two-Finger Matching Game previously represented by the table

		C I	C II
R	I	4	−2
	II	−10	7

Leaving the legends implicit, we convey the game with the following two-by-two array of numbers:

$$\begin{bmatrix} 4 & -2 \\ -10 & 7 \end{bmatrix}.$$

Two-Finger Matching Game

Similarly, the Candidates Game and the Play Selection Game are represented by the arrays

$$\begin{bmatrix} .54 & .47 & .55 \\ .47 & .52 & .46 \end{bmatrix}, \qquad \begin{bmatrix} 1 & 3 \\ 6 & 1 \\ 0 & 4 \end{bmatrix}.$$

Candidates Game   Play Selection Game

In general, a rectangular array of numbers is called a *matrix*. If the matrix has n rows and k columns we identify it as an $n \times k$ *matrix* (read "*n by k* matrix").

The Two-Finger Matching Game has been represented by a 2×2 matrix; the Candidates Game and Play Selection Game have been conveyed by 2×3 and 3×2 matrices, respectively. These matrices are said to be the *payoff matrices* corresponding to the respective games. Every game of the type described has its payoff matrix; we refer to such games as *matrix games.*

Row Strategies and Expectations

Let's return the universal players, R and C, to their Two-Finger Matching Game with payoff matrix

$$\begin{bmatrix} 4 & -2 \\ -10 & 7 \end{bmatrix}.$$

Suppose R and C have agreed to play the game a large number of times, and R decides in advance to play in a random fashion, with one row as likely as the other on each play. In other words, R plays I with probability .5, and he plays II with probability .5. The assignment of probabilities to the respective rows in the payoff matrix is called a *row strategy*. Just as the game itself is represented by a matrix, a row strategy is represented by a matrix of one row whose successive entries are the probabilities of playing the successive rows. In particular, the strategy of playing Row I with probability .5 and Row II with probability .5 is represented by the matrix

$$[.5 \quad .5].$$

Looking forward to the many plays of the game, we think of the row strategy [.5 .5] as an option available to R. In effect, the strategy [.5 .5] gives R a new row in the game. The Column I entry in this new row is the expected payoff for those cases in which player C selects I, and analogously for the Column II entry.

Confining attention to those cases in which C plays I, which has entries 4 and -10, and assuming that R employs the equally likely strategy [.5 .5], the *payoff* is the random variable described by the simple table

Payoff	4	−10
Prob	.5	.5

The expected value is

$$4 \times .5 + (-10) \times .5 = -3.$$

Similarly, the expected payoff for those cases in which C plays II is

$$(-2) \times .5 + 7 \times .5 = 2.5.$$

The payoffs corresponding to the respective columns are represented in a row matrix, i.e., a matrix consisting of a single row. The result is expressed as follows:

In the game with payoff matrix

$$\begin{bmatrix} 4 & -2 \\ -10 & 7 \end{bmatrix}$$

the row strategy

$$[.5 \quad .5]$$

produces the (row) *expectation*

$$[-3 \quad 2.5].$$

Now let's have R use a little imagination. Looking at the relative entries, he senses that Row I seems like a better option in general than Row II. So R decides to flip two coins and play I unless both coins land Tails. In other words, R employs strategy [.75 .25].

To compute the expectation produced by the row strategy [.75 .25], we consider separately the cases in which C plays I and II. If C plays I, the expected payoff is

$$.75 \times 4 + .25 \times (-10) = .5,$$

and if C plays II, the expected payoff is

$$.75 \times (-2) + .25 \times 7 = .25.$$

Thus the row strategy [.75 .25] produces the expectation

$$[.5 \quad .25].$$

This shows that if R employs strategy [.75 .25], then C's best strategy is to play Column II all the time. And even then, R can expect to win $\frac{1}{4}$ cent per game, on the average in the long run. We say *the game favors R*. In fairness, R should pay C at least $\frac{1}{4}$ cent per game for the privilege of playing.

Column Strategies and Expectations

To console C, and to better appreciate the general applicability of the notions of strategy and expectation, we consider another game.

R has two options, and C has three. Depending on the combination selected, the number of dollars that C pays R is given by the following 2×3 payoff matrix:

$$\begin{bmatrix} 4 & -3 & 5 \\ -3 & 2 & -4 \end{bmatrix}.$$

Thus, for instance, if R plays II and C plays III, then R pays 4 dollars to C.

Putting the shoe on the other foot, we have C contemplate the wisdom of a certain strategy. With a good sense of the worthiness of the respective columns, C decides to play Column II with probability .6 and to play each of the other columns with probability .2. This assignment of probabilities is *a column strategy*, and, as one should expect, a column strategy is represented as a column matrix. In particular, the probabilities decided upon by C comprise the column strategy

$$\begin{bmatrix} .2 \\ .6 \\ .2 \end{bmatrix}.$$

We proceed as in the row-oriented analysis. The column strategy is viewed as determining an option for player C in contemplating many plays of the game—in effect, a new column in the game. The (column) expectation associated with the strategy is the 2×1 column matrix whose Row I entry is the expected payoff for those cases in which player R selects I, and analogously for the Row II entry.

Let's find the expectation produced by the column strategy $\begin{bmatrix} .2 \\ .6 \\ .2 \end{bmatrix}$. The entries in the first row of the payoff matrix are 4, −3, and 5, in that order. For those cases in which R plays I the payoff is the random variable

Payoff	4	−3	5
Prob	.2	.6	.2

The expected value is

$$4 \times .2 + (-3) \times .6 + 5 \times .2 = 0.$$

Similarly, the expected payoff for those cases in which R plays II is

$$(-3) \times .2 + 2 \times .6 + (-4) \times .2 = -.2.$$

We express the result in a manner perfectly analogous with the row-oriented situation:

In the game with payoff matrix

$$\begin{bmatrix} 4 & -3 & 5 \\ -3 & 2 & -4 \end{bmatrix}$$

the column strategy

$$\begin{bmatrix} .2 \\ .6 \\ .2 \end{bmatrix}$$

produces the (column) expectation

$$\begin{bmatrix} 0 \\ -.2 \end{bmatrix}.$$

We note that if C uses column strategy $\begin{bmatrix} .2 \\ .6 \\ .2 \end{bmatrix}$, then the best R can do is use row strategy $[1 \quad 0]$, i.e., play I all the time, in which case the expected payoff is 0. For this reason we proclaim that *the game does not favor R*. This is not to say that the game necessarily favors C—it might be a *fair* game, favoring neither player. But if we have C play Column II with a probability slightly larger than .6 and the other columns with equal probability, we find that R can expect to lose in the long run, no matter how he plays. So in fact the game does favor C.

6-1 EXERCISES

1. On the count of three, Tom and Bill each extend one, two, or three fingers. The total number of fingers shown is the number of pennies one player pays the other. If this number is even, Tom pays Bill, and if odd, Bill pays Tom.

(a) Set up the payoff matrix for the game, identifying Tom as the row player R and Bill as the column player C.
(b) Change the identifications, i.e., let Tom be C and Bill be R, and construct the payoff matrix.

2. For a simple card game, Joe retains the 7 of Spades, the 8 of Spades, and the 4 of Hearts, and Jim has the 2 of Spades and the 5 of Hearts. At a signal each displays one of his cards. If the suits match, Joe pays Jim the difference in the face values of the cards; otherwise, Jim pays Joe the difference.
(a) Construct the payoff matrix with Joe as R and Jim as C.
(b) Construct the payoff matrix with Jim as R and Joe as C.
(c) Let Jim and Joe exchange their little decks of cards and then play by the old rules, Jim still winning if the suits match. Set up the payoff matrix with Joe as R and Jim as C.

3. Recall the Two-Finger Matching Game with payoff matrix

$$\begin{bmatrix} 4 & -2 \\ -10 & 7 \end{bmatrix}.$$

(a) Suppose R uses strategy $[.8 \quad .2]$; i.e., he plays I with probability .8 and II with probability .2. Find the corresponding expectation $[r_1 \quad r_2]$, where r_1 and r_2 are the expected payoffs for those cases in which C plays Columns I and II, respectively.
(b) Suppose C employs strategy $\begin{bmatrix} .4 \\ .6 \end{bmatrix}$. Find the corresponding expectation $\begin{bmatrix} c_1 \\ c_2 \end{bmatrix}$.

4. Recall the Play Selection Game with the following payoff matrix:

$$\begin{bmatrix} 1 & 3 \\ 6 & 1 \\ 0 & 4 \end{bmatrix}.$$

(a) Find the expectations produced by the following row strategies:

(a_1) $[.1 \quad .4 \quad .5]$,
(a_2) $[0 \quad 0 \quad 1]$,
(a_3) $[\frac{2}{3} \quad \frac{1}{3} \quad 0]$.

(b) Find the expectations produced by the following column strategies:

$$(b_1)\begin{bmatrix} .5 \\ .5 \end{bmatrix}, \quad (b_2)\begin{bmatrix} 0 \\ 1 \end{bmatrix}, \quad (b_3)\begin{bmatrix} .7 \\ .3 \end{bmatrix}.$$

5. Consider the game with payoff matrix

$$\begin{bmatrix} 3 & 0 & -4 \\ -2 & 1 & 3 \end{bmatrix}.$$

(a) Find the expectations produced by the following row strategies:

(a_1) $[.5 \quad .5]$,
(a_2) $[.4 \quad .6]$,
(a_3) $[\frac{5}{12} \quad \frac{7}{12}]$.

(b) Find the expectations produced by the following column strategies:

$$(b_1)\begin{bmatrix} 1/3 \\ 1/3 \\ 1/3 \end{bmatrix}, \quad (b_2)\begin{bmatrix} .5 \\ .1 \\ .4 \end{bmatrix}, \quad (b_3)\begin{bmatrix} 7/12 \\ 0 \\ 5/12 \end{bmatrix}.$$

6. In the text it was established that the game with payoff matrix

$$\begin{bmatrix} 4 & -3 & 5 \\ -3 & 2 & -4 \end{bmatrix}$$

does not favor the row player R. Show that the game actually does favor player C by considering the column strategy

$$\begin{bmatrix} .19 \\ .62 \\ .19 \end{bmatrix}.$$

7. Consider the game with payoff matrix

$$\begin{bmatrix} 1 & -3 \\ -2 & 5 \end{bmatrix}.$$

(a) Find the expectations produced by the column strategy $\begin{bmatrix} 3/4 \\ 1/4 \end{bmatrix}$.

(b) Show that the game favors the column player by exhibiting a strategy $\begin{bmatrix} p \\ q \end{bmatrix}$ for which both entries in the expectation are negative.

8. Consider the game with payoff matrix

$$\begin{bmatrix} 3 & -9 \\ -1 & 4 \end{bmatrix}.$$

Show that the game favors the row player by exhibiting a row strategy $[p \quad q]$ which produces an expectation in which both entries are positive. (Examine the rows critically, and use a little trial and error.)

9. On the count of three, Mike and Matt each extend one, two, or three fingers. If they show the same number of fingers, Mike pays Matt 1 cent. If one player shows one finger and the other shows three, then Mike pays Matt 6 cents. In all other cases Matt pays Mike 2 cents.

(a) Construct a payoff matrix. Then venture a guess as to which of the players is favored.

(b) Suppose Mike throws a die, plays two fingers if the die shows four or more, and plays one, two, or three fingers when the die shows one, two, or three, respectively. Represent this strategy as a matrix, and compute the corresponding expectation. Then reexamine your guess in part (a).

6-2. The Problem of Game Theory

The analysis of a matrix game is best approached in a spirit of pessimism, always worrying about the worst possible consequences of the strategy contemplated. As we shall see, the critical strategies to which this attitude leads have a validity which cannot be denied by the most speculative of players.

Enter R, one of those hard-luck characters who can invoke a cloudburst simply by not carrying his umbrella. Against his better judgment, R finds himself drawn into a game which translates into the following payoff matrix, identified as G:

$$G = \begin{bmatrix} 3 & -2 & 0 & 1 \\ 1 & 2 & 2 & -1 \\ -3 & 3 & -1 & 2 \end{bmatrix}.$$

The gloomy analysis of pessimistic R is straightforward. Thinks he,

Sure as I play Row I, C will play Column II, and I'll lose 2 points.
But if I play II, C will probably play IV, and I'll lose 1 point.
And if I play III, it's a safe bet that C will play I, and I'll lose 3 points.

So R concludes that he had better play row II where he has the least to lose.

It's too bad that R doesn't mix a little sporting spirit with his chronic pessimism, for then he might roll a die until he gets a 1, 2, or 3 and play the corresponding row. Then the chances that he would lose at all are only 1 in 3 regardless of which column C selects, because there's only one negative entry in each column.

To see the long-run effect of this "equally likely" strategy for R, $[\frac{1}{3} \ \frac{1}{3} \ \frac{1}{3}]$, let's look at the columns of the payoff matrix G and find the expectation. The expected payoffs are

$$\text{Column I expected payoff:} \quad \frac{1}{3} \times 3 + \frac{1}{3} \times 1 + \frac{1}{3} \times (-3) = \frac{1}{3},$$

$$\text{Column II expected payoff:} \quad \frac{1}{3} \times (-2) + \frac{1}{3} \times 2 + \frac{1}{3} \times 3 = 1,$$

$$\text{Column III expected payoff:} \quad \frac{1}{3} \times 0 + \frac{1}{3} \times 2 + \frac{1}{3} \times (-1) = \frac{1}{3},$$

$$\text{Column IV expected payoff:} \quad \frac{1}{3} \times 1 + \frac{1}{3} \times (-1) + \frac{1}{3} \times 2 = \frac{2}{3}.$$

We see that in the game G, i.e., the game with payoff matrix G, the row strategy $[\frac{1}{3} \ \frac{1}{3} \ \frac{1}{3}]$ produces the expectation $[\frac{1}{3} \ 1 \ \frac{1}{3} \ \frac{2}{3}]$. Even pessimistic

R should see that the game is actually in his favor. His grumbling now isn't very convincing:

> Sure as I use strategy $[\frac{1}{3} \ \ \frac{1}{3} \ \ \frac{1}{3}]$, C will stick to Columns I and III, and I can expect to win only about $\frac{1}{3}$ of a point per game, on the average in the long run.

The key value that player R associates with any strategy is the minimum entry in the expectation. For the strategy $[\frac{1}{3} \ \ \frac{1}{3} \ \ \frac{1}{3}]$, with expectation $[\frac{1}{3} \ \ 1 \ \ \frac{1}{3} \ \ \frac{2}{3}]$, the minimum entry is $\frac{1}{3}$. For R, this is the worst payoff that can occur, on the average in the long run, as a consequence of the strategy.

His spirits picking up, R takes a critical look at the payoff matrix and decides that, in general, Row II looks like a better option than Rows I and III. To gauge the effect of giving added weight to Row II, R considers the strategy

$$[.3 \quad .4 \quad .3].$$

The expected payoffs corresponding to the different columns are

$$\begin{aligned}
&\text{Column I:} && .3 \times 3 + .4 \times 1 + .3 \times (-3) = .4,\\
&\text{Column II:} && .3 \times (-2) + .4 \times 2 + .3 \times 3 = 1.1,\\
&\text{Column III:} && .3 \times 0 + .4 \times 2 + .2 \times (-1) = .5,\\
&\text{Column IV:} && .3 \times 1 + .4 \times (-1) + .3 \times 2 = .5,
\end{aligned}$$

Thus row strategy $[.3 \quad .4 \quad .3]$ produces the expectation $[.4 \quad 1.1 \quad .5 \quad .5]$, which has minimum value .4. R notes that the worst possible expected payoff, .4, for the new strategy, $[.3 \quad .4 \quad .3]$, is greater than the worst payoff, $\frac{1}{3}$, for the equally likely strategy, $[\frac{1}{3} \ \ \frac{1}{3} \ \ \frac{1}{3}]$. So R identifies $[.3 \quad .4 \quad .3]$ as a *better* strategy than $[\frac{1}{3} \ \ \frac{1}{3} \ \ \frac{1}{3}]$. And, human nature being as it is, player R will not rest until he is convinced that he has found the very best strategy.

Optimal Strategies

As we've seen, in the game G above R can guarantee himself an expected payoff of at least .4 by employing strategy $[.3 \quad .4 \quad .3]$.

> The question is: How far up can R push the minimum expected payoff? For instance, does there exist a row strategy for which the minimum in the expectation is, say, more than 1?

Here, some insight may be gained by considering the effect of the "equally likely" column strategy

$$\begin{bmatrix} .25 \\ .25 \\ .25 \\ .25 \end{bmatrix}.$$

The expectation produced by this column strategy is found by considering the rows of the matrix G, one at a time. The expected payoffs are

$$\text{Row I:} \quad .25 \times 3 + .25 \times (-2) + .25 \times 0 + .25 \times 1 = .5$$

$$\text{Row II:} \quad .25 \times 1 + .25 \times 2 + .25 \times 2 + .25 \times (-1) = 1.0$$

$$\text{Row III:} \quad .25 \times (-3) + .25 \times 3 + .25 \times (-1) + .25 \times 2 = .25$$

Thus column strategy $\begin{bmatrix} .25 \\ .25 \\ .25 \\ .25 \end{bmatrix}$ produces the expectation $\begin{bmatrix} .5 \\ 1.0 \\ .25 \end{bmatrix}$.

We see that if player C employs the equally likely strategy, he is assured that the expected payoff is 1 or less. Therefore, regardless of what strategy R employs, R cannot guarantee an expected payoff of more than 1.

In plain language: No matter how R plays, he can't be sure of winning more than 1 (long-run average, of course) simply because C can keep him from winning more than 1 by using the equally likely strategy.

Here, the key value "1" associated with the column strategy is the maximum value in the expectation produced. For player C, who wants the payoff to be as negative as possible (where 1 is "more negative" than 2), the maximum in the expectation is the worst possible expected payoff associated with the strategy.

Just as R wants to know how far up he can push the minimum in the expectation by a prudent choice of row strategy, so C would like to know how far down he can push the maximum in the expectation by a choice of column strategy. We've seen R push up to .4, and C push down to 1. Of course, R can't push up higher than C can push down—R can't be sure of winning more than C can keep him from winning. The question is: To what extent can the gap between .4 and 1 be closed?

By now, the reader probably anticipates the answer: *The gap can be closed completely*. Which is to say, R has a row strategy for which the minimum in the expectation is precisely equal to the maximum in the expectation produced by a certain column strategy. The most convincing way to establish this fact for any given game is by exhibiting the two strategies. This we shall proceed to do for the game under consideration, postponing for the present any hint of how these strategies were found.

We assert that in the game G there is no better strategy for R than $[\frac{3}{8} \ \ \frac{3}{8} \ \ \frac{1}{4}]$. Player R should flip three coins and play Row I if he gets two Heads and a Tail, Row II if he gets two Tails and a Head, and Row III if he gets three Heads or three Tails.

One by one we consider the columns of the matrix

$$G = \begin{bmatrix} 3 & -2 & 0 & 1 \\ 1 & 2 & 2 & -1 \\ -3 & 3 & -1 & 2 \end{bmatrix}.$$

The expected payoffs corresponding to row strategy $[\frac{3}{8} \ \ \frac{3}{8} \ \ \frac{1}{4}]$ are

$$\text{Column I:} \quad \frac{3}{8} \times 3 + \frac{3}{8} \times 1 + \frac{1}{4} \times (-3) = \frac{3}{4},$$

$$\text{Column II:} \quad \frac{3}{8} \times (-2) + \frac{3}{8} \times 2 + \frac{1}{4} \times 3 = \frac{3}{4},$$

$$\text{Column III:} \quad \frac{3}{8} \times 0 + \frac{3}{8} \times 2 + \frac{1}{4} \times (-1) = \frac{1}{2},$$

$$\text{Column IV:} \quad \frac{3}{8} \times 1 + \frac{3}{8} \times (-1) + \frac{1}{4} \times 2 = \frac{1}{2}.$$

Thus strategy $[\frac{3}{8} \;\; \frac{3}{8} \;\; \frac{1}{4}]$ produces the expectation $[\frac{3}{4} \;\; \frac{3}{4} \;\; \frac{1}{2} \;\; \frac{1}{2}]$. The minimum in the expectation—the worst expected payoff for R—is $\frac{1}{2}$. If R employs this strategy, the best C can do is stick with Columns III and IV, in which case R can expect to win about $\frac{1}{2}$ point per play, on the average in the long run.

To show that R cannot guarantee an expected payoff greater than $\frac{1}{2}$, let's have C play Columns III and IV each with probability $\frac{1}{2}$ and never play the other columns. In other words, C employs strategy

$$\begin{bmatrix} 0 \\ 0 \\ \frac{1}{2} \\ \frac{1}{2} \end{bmatrix}.$$

For this column strategy the entries in the expectation are

$$\text{Row I:} \quad \frac{1}{2} \times 0 + \frac{1}{2} \times 1 = \frac{1}{2},$$

$$\text{Row II:} \quad \frac{1}{2} \times 2 + \frac{1}{2} \times (-1) = \frac{1}{2},$$

$$\text{Row III:} \quad \frac{1}{2} \times (-1) + \frac{1}{2} \times 2 = \frac{1}{2}.$$

We see that column strategy $\begin{bmatrix} 0 \\ 0 \\ \frac{1}{2} \\ \frac{1}{2} \end{bmatrix}$ produces the expectation $\begin{bmatrix} \frac{1}{2} \\ \frac{1}{2} \\ \frac{1}{2} \end{bmatrix}$. The maximum value—in fact, each value—is $\frac{1}{2}$. With this strategy C is assured that the expected payoff is at most $\frac{1}{2}$. Therefore R must despair of finding a strategy which guarantees a payoff of more than $\frac{1}{2}$. But, as we've seen, row strategy $[\frac{3}{8} \;\; \frac{3}{8} \;\; \frac{1}{4}]$ produces an expected payoff of at least $\frac{1}{2}$. R can do no better than employ this strategy.

We say that $[\frac{3}{8} \;\; \frac{3}{8} \;\; \frac{1}{4}]$ is an *optimal row strategy* in the game G because the minimum in the expectation it produces is as large as possible. We know that this is true because the minimum in the row expectation is equal to the maximum in the column expectation produced by column strategy $\begin{bmatrix} 0 \\ 0 \\ \frac{1}{2} \\ \frac{1}{2} \end{bmatrix}$. The common value of minimum and maximum, $\frac{1}{2}$, is said to be *the value of the game*.

The argument is equally valid from the column player's point of view. The strategy $\begin{bmatrix} 0 \\ 0 \\ \frac{1}{2} \\ \frac{1}{2} \end{bmatrix}$ is an *optimal column strategy* because the maximum in the expectation it produces is the least possible. If player C employs this optimal column strategy, the worst possible payoff is $\frac{1}{2}$; that is, he will do no worse than lose about $\frac{1}{2}$ point per game, on the average in the long run. No other column strategy can guarantee a better payoff for player C simply because player R, by using his optimal row strategy, can assure that the expected payoff is not less than $\frac{1}{2}$.

The same line of reasoning applies to all matrix games.

For any given game the problem is to find optimal row and column strategies. The "optimality" is established by showing that the minimum in the expectation produced by the optimal row strategy is equal to the maximum in the expectation produced by the optimal column strategy. The common value of minimum and maximum is said to be the value of the game.

Of course we haven't yet given any directions for finding optimal strategies; in fact, we've given no argument that strategies satisfying the above test for optimality must exist. But the verification of optimality is straightforward once the strategies are produced.

Example: Consider the game with payoff matrix

$$G = \begin{bmatrix} 1 & -2 \\ -4 & 5 \\ -2 & 1 \end{bmatrix}.$$

Verify that

$$\begin{bmatrix} \frac{3}{4} & \frac{1}{4} & 0 \end{bmatrix}$$

is an optimal row strategy and that

$$\begin{bmatrix} \frac{7}{12} \\ \frac{5}{12} \end{bmatrix}$$

is an optimal column strategy. And find the value of the game.

Solution: The entries in the expectation produced by row strategy $[\frac{3}{4} \quad \frac{1}{4} \quad 0]$ are

$$\text{Column I:} \quad \frac{3}{4} \times 1 + \frac{1}{4} \times (-4) = -\frac{1}{4},$$

$$\text{Column II:} \quad \frac{3}{4} \times (-2) + \frac{1}{4} \times 5 = -\frac{1}{4}.$$

Thus strategy $[\frac{3}{4} \quad \frac{1}{4} \quad 0]$ produces the expectation $[-\frac{1}{4} \quad -\frac{1}{4}]$. The minimum is $-\frac{1}{4}$.

For column strategy $\begin{bmatrix} \frac{7}{12} \\ \frac{5}{12} \end{bmatrix}$ the entries in the expectation are

$$\text{Row I:} \quad \frac{7}{12} \times 1 + \frac{5}{12} \times (-2) = -\frac{3}{12} = -\frac{1}{4},$$

$$\text{Row II:} \quad \frac{7}{12} \times (-4) + \frac{5}{12} \times 5 = -\frac{3}{12} = -\frac{1}{4},$$

$$\text{Row III:} \quad \frac{7}{12} \times (-2) + \frac{5}{12} \times 1 = -\frac{9}{12} = -\frac{3}{4}.$$

We see that the maximum in the expectation for strategy $\begin{bmatrix} \frac{7}{12} \\ \frac{5}{12} \end{bmatrix}$ is $-\frac{1}{4}$, which is equal to the minimum in the expectation for strategy $[\frac{3}{4} \;\; \frac{1}{4} \;\; 0]$. Therefore $[\frac{3}{4} \;\; \frac{1}{4} \;\; 0]$ is an optimal row strategy, and $\begin{bmatrix} \frac{7}{12} \\ \frac{5}{12} \end{bmatrix}$ is an optimal column strategy. The value of the game is $-\frac{1}{4}$.

6-2 EXERCISES

1. Consider the game with payoff matrix

$$\begin{bmatrix} 3 & -3 & 1 \\ -4 & 3 & -1 \end{bmatrix}.$$

(a) Find the minimum in the expectations produced by the following row strategies:

(a_1) [.4 .6].
(a_2) [.5 .5].
(a_3) [.6 .4].

(b) Which of the three row strategies in part (a) is most attractive to player R, in that the minimum in the expectation is the greatest?

2. Consider the game with payoff matrix

$$\begin{bmatrix} 3 & -3 & 1 \\ -4 & 3 & -1 \\ 2 & -2 & 2 \end{bmatrix}.$$

(a) Find the maximum in the expectations produced by the following column strategies:

$$(a_1) \begin{bmatrix} .4 \\ .5 \\ .1 \end{bmatrix}, \quad (a_2) \begin{bmatrix} .5 \\ .5 \\ 0 \end{bmatrix}, \quad \text{and} \quad (a_3) \begin{bmatrix} .4 \\ .6 \\ 0 \end{bmatrix}.$$

(b) Which of the three column strategies in part (a) is most attractive to player C, in that the maximum in the expectation is the least?

3. Consider the game with payoff matrix

$$\begin{bmatrix} 1 & -3 \\ -2 & 5 \\ 3 & -7 \end{bmatrix}.$$

(a) Find the minimum in the expectation produced by the following row strategies:

(a_1) $[.1 \quad .6 \quad .3]$,
(a_2) $[0 \quad .6 \quad .4]$.

(b) Find the maximum in the expectation produced by the following column strategies:

(b_1) $\begin{bmatrix} \frac{2}{3} \\ \frac{1}{3} \end{bmatrix}$, ($b_2$) $\begin{bmatrix} .7 \\ .3 \end{bmatrix}$.

(c) Answer the following questions, giving reasons:

(c_1) Is there a strategy for C that guarantees an expected payoff in his favor?
(c_2) Is there a strategy for R that guarantees an expected payoff in his favor?
(c_3) Is there a strategy for R that guarantees an expected payoff of greater than .1 point per play?

4. Consider the game with payoff matrix

$$\begin{bmatrix} 1 & -3 \\ -2 & 4 \end{bmatrix}.$$

Verify that $[.6 \quad .4]$ is an optimal row strategy and that $\begin{bmatrix} .7 \\ .3 \end{bmatrix}$ is an optimal column strategy; that is, show that the minimum in the expectation produced by row strategy $[.6 \quad .4]$ is equal to the maximum in the expectation produced by column strategy $\begin{bmatrix} .7 \\ .3 \end{bmatrix}$. Also, find the value of the game.

5. Consider the game with payoff matrix

$$\begin{bmatrix} -2 & -3 & 4 \\ 0 & -2 & 1 \\ 3 & 2 & -1 \end{bmatrix}.$$

Verify that

$$[.3 \quad 0 \quad .7]$$

is an optimal row strategy and that

$$\begin{bmatrix} 0 \\ .5 \\ .5 \end{bmatrix}$$

is an optimal column strategy. And find the value of the game.

6. Consider the game with payoff matrix

$$\begin{bmatrix} 2 & -4 & 4 & -3 \\ -3 & 7 & -5 & 4 \end{bmatrix}.$$

Verify that

$$\begin{bmatrix} \frac{7}{12} & \frac{5}{12} \end{bmatrix}$$

is an optimal row strategy and that

$$\begin{bmatrix} \frac{7}{12} \\ 0 \\ 0 \\ \frac{5}{12} \end{bmatrix}$$

is an optimal column strategy. And find the value of the game.

7. For the game of Prob. 2 above, show that

$$[0 \quad \tfrac{4}{11} \quad \tfrac{7}{11}]$$

is an optimal row strategy and that

$$\begin{bmatrix} \frac{5}{11} \\ \frac{6}{11} \\ 0 \end{bmatrix}$$

is an optimal column strategy.

8. For the game of Prob. 3 above, show that

$$[0 \quad \tfrac{10}{17} \quad \tfrac{7}{17}]$$

is an optimal row strategy and that

$$\begin{bmatrix} \frac{12}{17} \\ \frac{5}{17} \end{bmatrix}$$

is an optimal column strategy.

9. Recall the game G discussed in the text. The payoff matrix is

$$G = \begin{bmatrix} 3 & -2 & 0 & 1 \\ 1 & 2 & 2 & -1 \\ -3 & 3 & -1 & 2 \end{bmatrix}.$$

As was shown, $[\frac{3}{8} \quad \frac{3}{8} \quad \frac{1}{4}]$ is an optimal row strategy, and

$$\begin{bmatrix} 0 \\ 0 \\ \frac{1}{2} \\ \frac{1}{2} \end{bmatrix}$$

is an optimal column strategy. Show that

$$\left[\tfrac{6}{14} \quad \tfrac{5}{14} \quad \tfrac{3}{14}\right]$$

is also an optimal row strategy for this game. (This example shows that optimal strategies are not necessarily unique.)

6-3. Strictly Determined Games

On the occasion of its introduction early in the chapter, the "Two-Finger Matching Game" was billed as "the simplest game imaginable." Considering the generality of matrix games, this assertion doesn't hold up. Actually there are games which by any ordinary use of the word "game" should not be called "games" at all. Although they do illustrate the analysis of the previous section, their resolution is self-evident. For example,

Players R and C each have two options. The amount C pays R in the various combinations of choices is represented by the following payoff matrix:

$$G = \begin{bmatrix} 4 & -2 \\ -10 & -7 \end{bmatrix}.$$

It doesn't take much imagination to supply the inescapable reasoning preceding this game.

Thinks R: I'd be out of my mind to play Row II, because Row I gives me a better payoff in both columns.

And thinks C: R would be out of his mind to play Row II. Therefore, since he will surely play Row I, I will play Column II and win the 2 points.

No second thoughts here. We might try to help R by changing the lower left-hand entry in the matrix G from -10 to $+10$: This way he wins, rather than loses, the 10 points. The payoff matrix becomes

$$\begin{bmatrix} 4 & -2 \\ 10 & -7 \end{bmatrix}.$$

But this adjustment doesn't help R at all. Now C is better off playing Column II regardless of what R does, and R knows it. So the best R can do is play Row I, and he still loses the 2 points.

No matter how we change the lower left-hand entry in G, the simple-minded reasoning applies. Leaving the value of this entry in question, the payoff matrix is of the form

$$\begin{bmatrix} 4 & -2 \\ ? & -7 \end{bmatrix}.$$

Now suppose the missing value is less than or equal to 4, e.g.,

$$\begin{bmatrix} 4 & -2 \\ 4 & -7 \end{bmatrix} \quad \text{or} \quad \begin{bmatrix} 4 & -2 \\ 0 & -7 \end{bmatrix} \quad \text{or} \quad \begin{bmatrix} 4 & -2 \\ -10 & -7 \end{bmatrix}.$$

Then, regardless of the column chosen by C, Row I presents at least as good a payoff for R as does Row II. The best interests of R are served by playing Row I, as both players must surely recognize. As before, R should play I, and C should play II.

On the other hand, suppose the missing entry is greater than or equal to -7, e.g.,

$$\begin{bmatrix} 4 & -2 \\ 10 & -7 \end{bmatrix} \text{ or } \begin{bmatrix} 4 & -2 \\ 0 & -7 \end{bmatrix} \text{ or } \begin{bmatrix} 4 & -2 \\ -7 & -7 \end{bmatrix}.$$

Now the best interests of C are served by choosing Column II, as both players know. And again, R plays I and C plays II.

In short, the best interests of both players are served by playing the -2 in the upper right-hand corner of the matrix unless the missing entry is simultaneously greater than 4 and less than -7, and of such there are none.

The property that makes the game G and its variations yield to such straightforward reasoning is the fact that the entry in the upper right-hand corner of the matrix is the minimum value in its row and the maximum in its column. We say that this entry is a *minimax* entry. The simple argument above can be extended to any game represented by a 2×2 matrix containing a minimax entry; the general conclusion is that the players can not do better than play the row and column of that entry.

In the previous section we spoke of optimal strategies for matrix games in general, and we set forth a criterion for verifying their optimality. Let's look at the game G in this technical setting. Recall the payoff matrix:

$$G = \begin{bmatrix} 4 & -2 \\ -10 & -7 \end{bmatrix}.$$

We've decided that the best interests of R are served by playing Row I all the time. This row strategy is given by the matrix $[1 \quad 0]$. Unless the word "optimal" is being misused, $[1 \quad 0]$ should be an optimal row strategy.

There should be little hesitation in producing the expectation for row strategy $[1 \quad 0]$. Since R always plays Row I, the expected payoffs for the different columns of G are just the Row I entries. That is, strategy $[1 \quad 0]$ simply extracts the first row of G as its expectation. The more formal argument is almost as immediate. The respective entries in the expectation for strategy $[1 \quad 0]$ are

$$\text{Column I:} \quad 1 \times 4 + 0 \times (-10) = 4,$$

$$\text{Column II:} \quad 1 \times (-2) + 0 \times (-7) = -2.$$

Thus strategy $[1 \quad 0]$ produces expectation $[4 \quad -2]$, the first row of the payoff matrix G.

For player C, who should play Column II all the time, the indicated optimal strategy is $\begin{bmatrix} 0 \\ 1 \end{bmatrix}$. Following the treatment accorded strategy $[1 \quad 0]$, the expectation produced by $\begin{bmatrix} 0 \\ 1 \end{bmatrix}$ is $\begin{bmatrix} -2 \\ -7 \end{bmatrix}$, the second column of G.

To verify optimal strategies we show that the minimum in the expectation for the optimal row strategy is equal to the maximum in the expectation for the optimal column strategy. Considering the effect of strategies $[1 \quad 0]$ and $\begin{bmatrix} 0 \\ 1 \end{bmatrix}$ in 2×2 matrix games, we conclude that these are optimal strategies precisely when the minimum in Row I is equal to the maximum in Column II—in other words, when there is a minimax entry in Row I and Column II.

Thus there is no conflict between the conclusion of the simple argument for resolving the particular game G and the technical analysis. It is the minimax property of the entry in Row I and Column II that underlies the original argument that the players should select this entry, and it is precisely this property that verifies $[1 \quad 0]$ and $\begin{bmatrix} 0 \\ 1 \end{bmatrix}$ as optimal strategies.

The General Case

The simple argument that was given for the resolution of a 2×2 matrix game with a minimax entry loses its simplicity if we try to extend it to larger payoff matrices with minimax entries. But there is no difficulty in generalizing the more formal argument. For example, consider the game represented by the following 3×4 payoff matrix:

$$G = \begin{bmatrix} 3 & -2 & 0 & -1 \\ 1 & 0 & 2 & -1 \\ -3 & 3 & -1 & -2 \end{bmatrix}.$$

Here it cannot be said that a particular row is the best choice for player R regardless of the column chosen by player C, nor is there a best column by the analogous criterion. But the -1 in Row II and Column IV is a minimax entry—the minimum in its row and the maximum in its column. This being the case, it follows by the criterion of the previous section that an optimal strategy for R is to play Row II all the time and an optimal strategy for C is always to play Column IV.

The expectation produced by row strategy $[0 \quad 1 \quad 0]$—R always plays Row II—is $[1 \quad 0 \quad 2 \quad -1]$, the second row of the payoff matrix G. Similarly, the expectation for column strategy $\begin{bmatrix} 0 \\ 0 \\ 0 \\ 1 \end{bmatrix}$ is $\begin{bmatrix} -1 \\ -1 \\ -2 \end{bmatrix}$, the fourth column of G. Therefore, $[0 \quad 1 \quad 0]$ is an optimal row strategy because the minimum in the expectation $[1 \quad 0 \quad 2 \quad -1]$ is -1, which is equal to the maximum in the expectation $\begin{bmatrix} -1 \\ -1 \\ -2 \end{bmatrix}$ for column strategy $\begin{bmatrix} 0 \\ 0 \\ 0 \\ 1 \end{bmatrix}$, an optimal column strategy. The value of the game is -1.

We know that player R has no strategy that guarantees a higher expected payoff than -1, because by sticking to Column IV player C assures that the payoff is at most -1. Analogously, C has no strategy that guarantees a better, i.e., lesser, expected payoff than -1. And this is the meaning that we have attached to optimal strategies and the value of the game.

The argument adapts to any game whose payoff matrix contains a minimax entry. (Such entries are sometimes called *saddle points*, in analogy with the geometrical nature of the center point of a saddle.)

A matrix game is said to be *strictly determined* if its payoff matrix contains a minimax entry, i.e., an entry which is simultaneously the minimum in its row and the maximum in its column.

For a strictly determined game an optimal row strategy is the row matrix containing 1 in the position corresponding to the row of the minimax entry and zeros in the other positions. Such a strategy extracts the row of the minimax as its expectation. Similarly, an optimal column strategy is the column matrix containing 1 in the position corresponding to the column of the minimax, and zeros elsewhere; its expectation is the column of the minimax. The value of the minimax entry is the value of the game.

Comment: A payoff matrix may contain several minimax entries, but they would necessarily have the same value. For convenience, we have spoken above of *the* minimax entry on the assumption that a particular one has been so identified.

Strategies, row or column, consisting of a single entry 1 with other entries 0 are called *pure strategies.* Strategies not of this simple form are called *mixed strategies.*

Example: Consider the games corresponding to the following payoff matrices. In each case, determine whether the game is strictly determined, and if so, exhibit optimal strategies, verify, and state the value of the game.

(a) $G_a = \begin{bmatrix} 3 & 0 \\ -1 & 2 \end{bmatrix}$. (b) $G_b = \begin{bmatrix} 1 & -3 & 2 \\ -5 & 2 & 3 \end{bmatrix}$.

(c) $G_c = \begin{bmatrix} 3 & -2 & 3 & -2 & -1 \\ -3 & 1 & 2 & -3 & 0 \\ 0 & 1 & -5 & -2 & -10 \end{bmatrix}$.

(d) $G_d = \begin{bmatrix} 3 & 1 \\ 2 & 1 \end{bmatrix}$.

Solution: (a) The game G_a is not strictly determined because neither of the row minima, 0 and -1, are column maxima.

(b) Again, neither of the row minima, -3 and -5, are column maxima. Therefore, G_b is not strictly determined.

(c) The game G_c is strictly determined because the -2 in Row I and Column IV is a minimax entry. The pure strategy $[1 \quad 0 \quad 0]$ is an optimal row strategy, and the pure strategy $\begin{bmatrix} 0 \\ 0 \\ 0 \\ 1 \\ 0 \end{bmatrix}$ is an optimal column strategy.

Verification: Strategy $[1 \quad 0 \quad 0]$ produces expectation

$$[3 \quad -2 \quad 3 \quad -2 \quad -1].$$

Strategy $\begin{bmatrix} 0 \\ 0 \\ 0 \\ 1 \\ 0 \end{bmatrix}$ produces expectation $\begin{bmatrix} -2 \\ -3 \\ -2 \end{bmatrix}$.

The minimum in the row expectation is -2, the same as the maximum in the column expectation. Therefore the strategies are optimal. The value of the game is -2.

(d) The game G_d is strictly determined. There are two minimax entries—the values 1 in Column II. An optimal row strategy is $[1 \quad 0]$, and an optimal column strategy is $\begin{bmatrix} 0 \\ 1 \end{bmatrix}$.

Verification: Strategy $[1 \quad 0]$ produces expectation $[3 \quad 1]$, which has minimum value 1. Strategy $\begin{bmatrix} 0 \\ 1 \end{bmatrix}$ produces expectation $\begin{bmatrix} 1 \\ 1 \end{bmatrix}$, which has maximum value 1.

Therefore the strategies are optimal. The value of the game is 1.

(For extra credit: $[0 \quad 1]$ is also an optimal row strategy in the game G_d, as one may easily verify. For more extra credit: Every row strategy $[p \quad q]$, where $p \geq 0, q \geq 0$, and $p + q = 1$, is an optimal row strategy in this game.)

6-3 EXERCISES

1. Inspect the following 2×2 payoff matrices to ascertain whether the game is strictly determined, i.e., whether there is a minimax entry in the matrix. If so, give a down-to-earth argument that dictates how the players should play. Also, find optimal row and column strategies, and verify by showing that the minimum in the row expectation is equal to the maximum in the column expectation. State the value of the game.

(a) $\begin{bmatrix} -3 & -5 \\ -2 & 1 \end{bmatrix}$. (b) $\begin{bmatrix} 3 & -2 \\ 1 & 2 \end{bmatrix}$.

(c) $\begin{bmatrix} 1 & 0 \\ 2 & -2 \end{bmatrix}$. (d) $\begin{bmatrix} 0 & 2 \\ -1 & 1 \end{bmatrix}$.

2. For the following payoff matrices, all representing strictly determined games, exhibit optimal row and column strategies, and verify. State the value of the game.

(a) $\begin{bmatrix} 2 & -2 & -3 \\ -1 & 5 & -1 \end{bmatrix}$.

(b) $\begin{bmatrix} 1 & -4 & -3 & 5 \\ 2 & 3 & 2 & 3 \\ -2 & 5 & 0 & -7 \end{bmatrix}$.

(c) $\begin{bmatrix} -2 & 3 & -1 & -3 \\ -10 & 5 & 15 & -4 \end{bmatrix}$.

3. Inspect each of the following payoff matrices to ascertain whether the game is strictly determined. If so, exhibit optimal strategies, verify, and state the value of the game.

(a) $\begin{bmatrix} -1 & 2 & -1 & 0 \\ 1 & 3 & -2 & -4 \end{bmatrix}$.

(b) $\begin{bmatrix} 1 & -3 & 2 \\ 3 & 4 & -1 \\ -4 & -1 & 5 \end{bmatrix}$.

(c) $\begin{bmatrix} 7 & 5 & 8 & 6 \\ 7 & 9 & 8 & 7 \\ 8 & 7 & 6 & 7 \end{bmatrix}$.

4. Follow the directions in Prob. 3 for the following payoff matrices:

(a) $\begin{bmatrix} 2 & 1 & -1 \\ 1 & 1 & 1 \\ -1 & 1 & 5 \end{bmatrix}$.

(b) $\begin{bmatrix} -2 & -1 & -4 & -1 \\ -1 & -3 & -2 & -1 \end{bmatrix}$.

(c) $\begin{bmatrix} 4 & 7 & 5 \\ 3 & 4 & 5 \\ 8 & 7 & 9 \end{bmatrix}$.

5. For the following payoff matrix representing a strictly determined game, exhibit several optimal column strategies, and verify each:

$$\begin{bmatrix} 1 & -1 & -1 & 1 \\ 1 & 1 & 2 & 1 \\ 2 & 1 & -2 & -1 \end{bmatrix}.$$

6-4. A Formal Look at Matrices

We have identified matrices as "rectangular arrays" of numbers. For example,

$$G = \begin{bmatrix} 4 & -2 & 3 \\ -10 & 7 & -8 \end{bmatrix}$$

is a 2×3 matrix, meaning G has two rows and three columns.

As a matter of convention we shall represent matrices, other than single rows or single columns, by capital letters; the entries are represented by the lowercase forms of the letters with a double subscript to indicate first the row and then the column of that entry. Thus the general 2×3 matrix is of the form

$$M = \begin{bmatrix} m_{11} & m_{12} & m_{13} \\ m_{21} & m_{22} & m_{23} \end{bmatrix}$$

or

$$G = \begin{bmatrix} g_{11} & g_{12} & g_{13} \\ g_{21} & g_{22} & g_{23} \end{bmatrix}.$$

For the particular matrix G above, we have

$$g_{11} = 4, \quad g_{12} = -2, \quad g_{13} = 3,$$
$$g_{21} = -10, \quad g_{22} = 7, \quad g_{23} = -8.$$

The position taken in Chapter 3 was that mathematical objects are to be assembled from the primitive concepts of set, function, and number. The "double-subscripting" convention for the entries anticipates how the concept of a matrix may be elevated from the intuitive characterization of "rectangular array" to the purely mathematical level. What we do is interpret the value of the entry in the matrix as a function of the pair of integers composing its double subscript.

Let's look again at the 2×3 matrix G:

$$G = \begin{bmatrix} 4 & -2 & 3 \\ -10 & 7 & -8 \end{bmatrix}.$$

In this form we would like to accept G as a purely mathematical entity. But how do we defend it as such in terms of the primitive concepts of set, function, and number? The answer is that the rectangular array is to be interpreted as a way of communicating a function whose domain is the following set of ordered pairs of positive integers:

$$\{(1, 1), (1, 2), (1, 3), (2, 1), (2, 2), (2, 3)\}.$$

As a function, G associates $(1, 1)$ with $g_{11} = 4$, $(1, 2)$ with $g_{12} = -2$, and so on.

We can even talk about the "rows" and "columns" of the matrix G with no pangs of mathematical conscience. Fixing the first entry in the double subscript, the "first row" of the matrix G is interpreted as the function with domain

$$\{(1, 1), (1, 2), (1, 3)\}.$$

The values assigned are as before: (1, 1) is associated with $g_{11} = 4$, (1, 2) with $g_{12} = -2$, and (1, 3) with $g_{13} = 3$. The "second row" is interpreted in the same way, and so also with the "columns." For example, the "second column" of G is interpreted as the function with domain

$$\{(1, 2), (2, 2)\}.$$

Here, of course, (1, 2) is associated with $g_{12} = -2$, and (2, 2) with $g_{22} = 7$.

Turning to the general case, an arbitrary $n \times k$ matrix may be expressed in the form

$$M = \begin{bmatrix} m_{11} & m_{12} & \cdots & m_{1k} \\ m_{21} & m_{22} & \cdots & m_{2k} \\ \vdots & \vdots & & \vdots \\ m_{n1} & m_{n2} & \cdots & m_{nk} \end{bmatrix}.$$

Exactly the same rectangular array is conveyed more efficiently in the concise form

$$M = [m_{ij}], \qquad \begin{array}{l} i = 1, 2, \ldots, n, \\ j = 1, 2, \ldots, k. \end{array}$$

The following definition of a matrix, with its rows and columns, formalizes the discussion of the matrix G above.

Definition: Given positive integers n and k, an $n \times k$ *matrix* is a function mapping the set of ordered pairs of the form (i, j), where $i = 1, 2, \ldots, n$ and $j = 1, 2, \ldots, k$, into the set of real numbers.

The *first row* of the matrix is the function realized by restricting the domain to the set

$$\{(1, 1), (1, 2), \ldots, (1, k)\},$$

the value assigned each $(1, j)$ being the same as in the larger function defining the entire matrix. And in general for $i = 1, 2, \ldots, n$, the *ith row* is the function realized by restricting the domain to the set

$$\{(i, 1), (i, 2), \ldots, (i, k)\},$$

the value assigned each (i, j) being the same as before.

Analogously, for $j = 1, 2, \ldots, k$, the *jth column* of the matrix is the function realized by restricting the domain to the set

$$\{(1, j), (2, j), \ldots, (n, j)\},$$

again with the value assigned each (i, j) being the same as in the original function defining the matrix.

The aim of this definition is to establish that rectangular arrays of numbers, with their rows and columns, can be regarded as legitimate mathematical objects in terms of the requirements set forth in Chapter 3. The definition will not be too visible in the subsequent discussion.

Multiplication of Matrices

To motivate the rules for multiplying matrices, we return to the analysis of games. However, matrix multiplication will not be restricted to products meaningful in a game-theoretic setting.

Consider the game with payoff matrix

$$G = \begin{bmatrix} 4 & -2 & 3 \\ -10 & 7 & -8 \end{bmatrix}.$$

Suppose player R decides to play Row I with probability .8 and Row II with probability .2. Introducing the symbol $\mathbf{r}$ to represent a row strategy, we have

$$\mathbf{r} = [.8 \quad .2].$$

The expected payoffs corresponding to the respective columns of G are

$$\begin{aligned} &\text{Column I:} && .8 \times 4 + .2 \times (-10) = 1.2, \\ &\text{Column II:} && .8 \times (-2) + .2 \times 7 = -.2, \\ &\text{Column III:} && .8 \times 3 + .2 \times (-8) = .8. \end{aligned}$$

Thus row strategy $\mathbf{r} = [.8 \quad .2]$ produces the expectation

$$[1.2 \quad -.2 \quad .8].$$

As matrix multiplication will be defined, the expectation produced by a row strategy is the product of the row strategy times the payoff matrix. In particular, we shall have

$$\mathbf{r}G = [.8 \quad .2]\begin{bmatrix} 4 & -2 & 3 \\ -10 & 7 & -8 \end{bmatrix} = [1.2 \quad -.2 \quad .8].$$

We obtained the entries in the expectation by considering the columns of G one at a time. We also approach multiplication of matrices by considering first the case of a single row times a single column. To this end we extract the columns of G as column matrices and the entries in the expectation as 1×1 matrices. The first stage in the definition of matrix multiplication is designed to give the products

$$[.8 \quad .2]\begin{bmatrix} 4 \\ -10 \end{bmatrix} = [1.2], \qquad [.8 \quad .2]\begin{bmatrix} -2 \\ 7 \end{bmatrix} = [-.2], \qquad [.8 \quad .2]\begin{bmatrix} 3 \\ -8 \end{bmatrix} = [.8].$$

The crucial observation is that the respective values in the expectation are obtained by multiplying the successive entries in the row-strategy matrix times the successive entries in the corresponding column and adding.

Accordingly, we multiply a row matrix times a column matrix of the same length by multiplying successive entries in the two matrices and adding. The product is a 1×1 matrix.

Definition: Given a positive integer n, let A be a $1 \times n$ row matrix:

$$A = [a_1 \quad a_2 \quad \dots \quad a_n];$$

and let B an $n \times 1$ column matrix:

$$B = \begin{bmatrix} b_1 \\ b_2 \\ \cdot \\ \cdot \\ \cdot \\ b_n \end{bmatrix}.$$

Then the product

$$AB$$

is defined to be the 1×1 matrix whose single entry is the value

$$a_1b_1 + a_2b_2 + \cdots + a_nb_n.$$

Examples:

$$[.8 \quad .2]\begin{bmatrix} 4 \\ -10 \end{bmatrix} = [.8 \times 4 + .2 \times (-10)] = [1.2].$$

$$[.8 \quad .2]\begin{bmatrix} -2 \\ 7 \end{bmatrix} = [.8 \times (-2) + .2 \times 7] = [-.2].$$

$$[.2 \quad .5 \quad .3]\begin{bmatrix} -10 \\ 7 \\ -5 \end{bmatrix} = [0].$$

And, with no game-theoretic interpretation intended,

$$[2 \quad 5 \quad -3 \quad 0]\begin{bmatrix} 4 \\ -2 \\ -1 \\ 7 \end{bmatrix} = [1].$$

With respect to the analysis of games, this definition assures that each entry in the expectation is given by the product of the row strategy times the column matrix determined by the corresponding column of the payoff matrix. The expectation as a row matrix is realized by repetitions of this basic step. In informal language,

Let $\mathbf{r}$ be a $1 \times n$ row matrix and let G be an $n \times k$ matrix. Then the product $\mathbf{r}G$ is the $1 \times k$ row matrix whose jth entry is the product of the row matrix $\mathbf{r}$ times the column matrix given by the jth column of G.

In short, we simply look at the columns of G one at a time and apply the "row $\times$ column" definition to find the successive entries in the product. By this rule we obtain the product anticipated earlier:

$$[.8 \quad .2]\begin{bmatrix} 4 & -2 & 3 \\ -10 & 7 & -8 \end{bmatrix} = [1.2 \quad -.2 \quad .8].$$

In general, for a row strategy **r** applied to a game with payoff matrix G the expectation is the matrix product $\mathbf{r}G$.

Example: In the game with payoff matrix

$$G = \begin{bmatrix} -10 & 4 \\ 7 & -7 \\ -5 & 9 \end{bmatrix},$$

find the expectation produced by row strategy

$$\mathbf{r} = [.2 \quad .5 \quad .3].$$

Solution: The computation demanded is given by the matrix product

$$\mathbf{r}G = [.2 \quad .5 \quad .3]\begin{bmatrix} -10 & 4 \\ 7 & -7 \\ -5 & 9 \end{bmatrix} = [0 \quad 0].$$

The expectation is $[0 \quad 0]$.

But we do not confine ourselves to the analysis of games.

Example:

$$[2 \quad 5 \quad -3 \quad 0]\begin{bmatrix} 4 & 3 & 6 \\ -2 & 0 & -2 \\ -1 & 2 & 3 \\ 7 & 5 & 0 \end{bmatrix} = [1 \quad 0 \quad -7].$$

The full definition of matrix multiplication involves repetition of the basic step:

Row on the left times column of the same length on the right.

For matrices A and B the product AB is defined only if the rows of A and the columns of B are of the same length. The entry in a particular row and column of AB is the product, as given by the "row $\times$ column" definition, of the corresponding row of A times the corresponding column of B.

In advance of the formal definition we shall illustrate with a couple of examples. The informal description above should suffice to show how these products are obtained.

Example:

$$\begin{bmatrix} 3 & -1 & 2 \\ 0 & -4 & 5 \end{bmatrix} \begin{bmatrix} 2 \\ -3 \\ 4 \end{bmatrix} = \begin{bmatrix} 17 \\ 32 \end{bmatrix}.$$

Here, for instance, the "32" in the product is the second row of the left factor times the column on the right.

Example:

$$\begin{bmatrix} 3 & -1 & 2 \\ 0 & -4 & 5 \end{bmatrix} \begin{bmatrix} 2 & 1 & -1 \\ -3 & 0 & 3 \\ 4 & 1 & 6 \end{bmatrix} = \begin{bmatrix} 17 & 5 & 6 \\ 32 & 5 & 18 \end{bmatrix}.$$

Here the "18" in the second row and third column of the product is obtained using the second row of the left factor and the third column of the right factor.

Definition: Given positive integers m, n, and p, let A be an $m \times n$ matrix:

$$A = [a_{ij}], \quad \begin{matrix} i = 1, 2, \ldots, m, \\ j = 1, 2, \ldots, n; \end{matrix}$$

and let B be an $n \times p$ matrix:

$$B = [b_{jk}], \quad \begin{matrix} j = 1, 2, \ldots, n, \\ k = 1, 2, \ldots, p. \end{matrix}$$

Then the product AB is defined to be the $m \times p$ matrix

$$AB = C = [c_{ik}], \quad \begin{matrix} i = 1, 2, \ldots, m, \\ k = 1, 2, \ldots, p, \end{matrix}$$

where the entry c_{ik} in the ith row and kth column of C is given by

$$c_{ik} = a_{i1}b_{1k} + a_{i2}b_{2k} + \cdots + a_{in}b_{nk}.$$

Examples: Compute the products AB and BA, where defined, for the following pairs of matrices:

(a) $A = \begin{bmatrix} 3 \\ -1 \end{bmatrix}$, $B = [2 \quad -5]$.

(b) $A = \begin{bmatrix} 1 & 0 & 2 \\ -3 & 4 & 0 \end{bmatrix}$, $B = \begin{bmatrix} -3 & 2 \\ 0 & 1 \end{bmatrix}$.

Solutions: (a)

$$AB = \begin{bmatrix} 3 \\ -1 \end{bmatrix} [2 \quad -5] = \begin{bmatrix} 6 & -15 \\ -2 & 5 \end{bmatrix},$$

$$BA = [2 \quad -5] \begin{bmatrix} 3 \\ -1 \end{bmatrix} = [11].$$

(b) AB is not defined because the rows of A are not of the same length as the columns of B.

$$BA = \begin{bmatrix} -3 & 2 \\ 0 & 1 \end{bmatrix} \begin{bmatrix} 1 & 0 & 2 \\ -3 & 4 & 0 \end{bmatrix} = \begin{bmatrix} -9 & 8 & -6 \\ -3 & 4 & 0 \end{bmatrix}.$$

Application to Column Strategies

Our motivation has been to obtain the expectation produced by a row strategy $\mathbf{r}$ in a game with payoff matrix G as the matrix product $\mathbf{r}G$. Nothing has been said about column strategies and column expectations. Happily, matrix multiplication serves the column player as well as the row player.

We return to the game with payoff matrix

$$G = \begin{bmatrix} 4 & -2 & 3 \\ -10 & 7 & -8 \end{bmatrix}.$$

Suppose player C decides to play Column I with probability .1, Column II with probability .5, and Column III with probability .4. Introducing the symbol $\mathbf{c}$ to represent a column strategy,

$$\mathbf{c} = \begin{bmatrix} .1 \\ .5 \\ .4 \end{bmatrix}.$$

The expected payoffs for the respective rows of G are

$$\text{Row I:} \quad 4 \times .1 + (-2) \times .5 + 3 \times .4 = .6,$$

$$\text{Row II:} \quad (-10) \times .1 + 7 \times .5 + (-8) \times .4 = -.7.$$

Thus column strategy $\mathbf{c}$ produces expectation

$$\begin{bmatrix} .6 \\ -.7 \end{bmatrix}.$$

It should be clear from their form that the Row I and Row II computations above have a natural place as matrix products. Indeed, the arithmetic involved in computing the expectation is precisely that involved in the computation of the product $G\mathbf{c}$. We have

$$G\mathbf{c} = \begin{bmatrix} 4 & -2 & 3 \\ -10 & 7 & -8 \end{bmatrix} \begin{bmatrix} .1 \\ .5 \\ .4 \end{bmatrix} = \begin{bmatrix} .6 \\ -.7 \end{bmatrix}.$$

And so it is in general. Given a game with payoff matrix G, the expectation produced by a column strategy $\mathbf{c}$ is precisely the matrix product $G\mathbf{c}$.

Example: Consider the game with payoff matrix

$$G = \begin{bmatrix} -10 & 4 \\ 7 & -7 \\ -5 & 9 \end{bmatrix}.$$

Find the expectation produced by column strategy

$$\mathbf{c} = \begin{bmatrix} \frac{4}{7} \\ \frac{3}{7} \end{bmatrix}.$$

Solution: The expectation is given by the matrix product

$$G\mathbf{c} = \begin{bmatrix} -10 & 4 \\ 7 & -7 \\ -5 & 9 \end{bmatrix} \begin{bmatrix} \frac{4}{7} \\ \frac{3}{7} \end{bmatrix} = \begin{bmatrix} -4 \\ 1 \\ 1 \end{bmatrix}.$$

6-4 EXERCISES

1. Compute the product AB for the following pairs of matrices:

(a) $A = [2 \quad 0 \quad 1]$, $B = \begin{bmatrix} 3 \\ 5 \\ 2 \end{bmatrix}$.

(b) $A = [2 \quad 0 \quad 1]$, $B = \begin{bmatrix} 3 & -1 \\ 5 & 0 \\ 2 & 2 \end{bmatrix}$.

(c) $A = \begin{bmatrix} 2 & 0 & 1 \\ 1 & -1 & 3 \end{bmatrix}$, $B = \begin{bmatrix} 3 \\ 5 \\ 2 \end{bmatrix}$.

2. Compute the products AB and BA, where defined, for the following pairs of matrices:

(a) $A = \begin{bmatrix} 1 & -1 & 2 \\ 0 & 3 & -1 \end{bmatrix}$, $B = \begin{bmatrix} 3 & 4 \\ -2 & 5 \end{bmatrix}$.

(b) $A = \begin{bmatrix} 1 & -1 & 2 \\ 0 & 3 & -1 \end{bmatrix}$, $B = \begin{bmatrix} 3 & 4 \\ -2 & 5 \\ 1 & -1 \end{bmatrix}$.

3. Compute the products AB and BA, where defined, for the following pairs of matrices:

(a) $A = \begin{bmatrix} 3 & 4 \\ 1 & 1 \\ 2 & 0 \end{bmatrix}$, $B = \begin{bmatrix} 1 & 2 & 3 \\ 3 & -2 & 1 \end{bmatrix}$.

(b) $A = \begin{bmatrix} -2 & 0 \\ 0 & 1 \\ 1 & -3 \end{bmatrix}$, $B = \begin{bmatrix} -3 & 4 \\ 3 & -1 \end{bmatrix}$.

4. Consider the game with payoff matrix

$$G = \begin{bmatrix} 2 & -1 \\ -4 & 2 \end{bmatrix}.$$

(a) Employ matrix multiplication to compute the expectation $\mathbf{r}G$ produced by the following row strategies:

(a_1) $\mathbf{r} = [.6 \quad .4]$,
(a_2) $\mathbf{r} = [\frac{2}{3} \quad \frac{1}{3}]$.

(b) Employ matrix multiplication to compute the expectation $G\mathbf{c}$ produced by the following column strategies:

(b_1) $\mathbf{c} = \begin{bmatrix} .5 \\ .5 \end{bmatrix}$, ($b_2$) $\mathbf{c} = \begin{bmatrix} \frac{1}{3} \\ \frac{2}{3} \end{bmatrix}$.

5. Consider the game with payoff matrix

$$G = \begin{bmatrix} -4 & 6 & 0 \\ 5 & -7 & -1 \end{bmatrix}.$$

(a) Compute the expectation produced by the following row strategies:

(a_1) $\mathbf{r} = [1 \quad 0]$,
(a_2) $\mathbf{r} = [.5 \quad .5]$,
(a_3) $\mathbf{r} = [\frac{6}{11} \quad \frac{5}{11}]$.

(b) Compute the expectation produced by the following column strategies:

(b_1) $\mathbf{c} = \begin{bmatrix} 0 \\ 1 \\ 0 \end{bmatrix}$, ($b_2$) $\mathbf{c} = \begin{bmatrix} .5 \\ .3 \\ .2 \end{bmatrix}$, ($b_3$) $\mathbf{c} = \begin{bmatrix} \frac{13}{22} \\ \frac{9}{22} \\ 0 \end{bmatrix}$.

6. Consider the following matrices:

$$\mathbf{r} = [.4 \quad .6], \qquad G = \begin{bmatrix} 5 & -3 & 2 \\ -2 & 2 & -1 \end{bmatrix}, \qquad \mathbf{c} = \begin{bmatrix} .1 \\ .4 \\ .5 \end{bmatrix}.$$

Show that

$$(\mathbf{r}G)\mathbf{c} = \mathbf{r}(G\mathbf{c}).$$

7. Show that

$$(AB)C = A(BC),$$

where

$$A = \begin{bmatrix} 2 & -1 \\ 1 & 3 \end{bmatrix}, \qquad B = \begin{bmatrix} 1 & 0 & 2 \\ -1 & 1 & 0 \end{bmatrix}, \qquad C = \begin{bmatrix} 0 & 3 \\ 2 & -2 \\ -1 & 0 \end{bmatrix}.$$

6-5. Restatement of the Problem of Game Theory

We continue with the game G considered in the previous section:

$$G = \begin{bmatrix} 4 & -2 & 3 \\ -10 & 7 & -8 \end{bmatrix}.$$

As we saw, the row strategy

$$\mathbf{r}_1 = [.8 \quad .2]$$

produces the expectation

$$\mathbf{r}_1 G = [.8 \quad .2]\begin{bmatrix} 4 & -2 & 3 \\ -10 & 7 & -8 \end{bmatrix}$$

$$= [1.2 \quad -.2 \quad .8].$$

And the column strategy

$$\mathbf{c}_1 = \begin{bmatrix} .1 \\ .5 \\ .4 \end{bmatrix}$$

produces the expectation

$$G\mathbf{c}_1 = \begin{bmatrix} 4 & -2 & 3 \\ -10 & 7 & -8 \end{bmatrix}\begin{bmatrix} .1 \\ .5 \\ .4 \end{bmatrix} = \begin{bmatrix} .6 \\ -.7 \end{bmatrix}.$$

In general, a row strategy for the game G is a 1×2 row matrix $\mathbf{r}$ whose entries—the probabilities of playing the respective rows—are nonnegative numbers adding up to one. The expectation produced by $\mathbf{r}$ is the matrix product $\mathbf{r}G$. Analogously, a column strategy $\mathbf{c}$ is a 3×1 column matrix $\mathbf{c}$, with nonnegative entries totaling one, and the expectation produced by $\mathbf{c}$ is the matrix product $G\mathbf{c}$. This mapping of strategies into expectations gives us the mathematical structure of the game.

To place the analysis of matrix games in a mathematical setting we let G represent an arbitrary matrix.

Mathematical Model of a Matrix Game: We identify a matrix game as an $n \times k$ matrix G, accompanied by the following structure:

Each $1 \times n$ row matrix $\mathbf{r}$ of nonnegative entries totaling one is said to be a *row strategy* for the game G. The *expectation* produced by a given row strategy $\mathbf{r}$ is the $1 \times k$ row matrix $\mathbf{r}G$.

Each $k \times 1$ column matrix $\mathbf{c}$ of nonnegative entries totaling one is said to be a *column strategy* for the game G. The *expectation* produced by a given column strategy $\mathbf{c}$ is the $n \times 1$ column matrix $G\mathbf{c}$.

We recall the discussion of Sec. 6-2 in the interest of giving the problem of game theory a proper setting in the mathematical model. In the game

$$G = \begin{bmatrix} 4 & -2 & 3 \\ -10 & 7 & -8 \end{bmatrix}$$

we have seen that row strategy $\mathbf{r}_1 = [.8 \quad .2]$ produces expectation $\mathbf{r}_1 G = [1.2 \quad -.2 \quad .8]$. The key value which player R extracts from this expectation is the minimum entry, $-.2$, because this value represents the worst possible consequence of strategy $\mathbf{r}_1$. Introducing the symbol "min" to represent the minimum entry in a row matrix, we extract this key value in the fashion

$$\min \mathbf{r}_1 G = \min[1.2 \quad -.2 \quad .8] = -.2.$$

For row strategy

$$\mathbf{r}_2 = [.6 \quad .4]$$

we have expectation

$$\mathbf{r}_2 G = [.6 \quad .4]\begin{bmatrix} 4 & -2 & 3 \\ -10 & 7 & -8 \end{bmatrix} = [-1.6 \quad 1.6 \quad -1.4].$$

Here

$$\min \mathbf{r}_2 G = \min[-1.6 \quad 1.6 \quad -1.4] = -1.6.$$

Thus the worst possible consequence of strategy $\mathbf{r}_2 = [.6 \quad .4]$ is that player R loses 1.6 points per play, on the average in the long run. And this is worse than losing .2 points per play, which is the threat imposed by strategy $\mathbf{r}_1 = [.8 \quad .2]$. Therefore, $\mathbf{r}_1$ is identified as a "better" strategy than $\mathbf{r}_2$.

We identify $\mathbf{r}_1$ as a better strategy precisely because

$$\min \mathbf{r}_2 G < \min \mathbf{r}_1 G.$$

The problem is to find an optimal row strategy $\mathbf{r}_0$ for which

$$\min \mathbf{r}G \leq \min \mathbf{r}_0 G$$

for every row strategy $\mathbf{r}$.

From the column player's point of view, the key value associated with strategy

$$\mathbf{c}_1 = \begin{bmatrix} .1 \\ .5 \\ .4 \end{bmatrix}$$

is

$$\max G\mathbf{c}_1 = \max\begin{bmatrix} .6 \\ -.7 \end{bmatrix} = .6.$$

Here the symbol "max" is employed to extract the maximum value from a column matrix.

We see that if player C employs strategy $\mathbf{c}_1$ and player R stays with Row I, then C loses .6 points per play, on the average in the long run, and this is the worst consequence of strategy $\mathbf{c}_1$. Of course, $\mathbf{c}_1$ is better than a second strategy $\mathbf{c}_2$ if the value of max $G\mathbf{c}_2$ is greater than .6, that is, if

$$\max G\mathbf{c}_2 > \max G\mathbf{c}_1.$$

The problem is to find an optimal column strategy $\mathbf{c}_0$ for which

$$\max G\mathbf{c} \geq \max G\mathbf{c}_0$$

for every column strategy $\mathbf{c}$.

Definition: Given an $n \times k$ matrix game G, a row strategy $\mathbf{r}_0$ is said to be an *optimal row strategy* if

$$\min \mathbf{r}G \leq \min \mathbf{r}_0 G$$

for every row strategy $\mathbf{r}$.

And a column strategy $\mathbf{c}_0$ is said to be an *optimal column strategy* if

$$\max G\mathbf{c} \geq \max G\mathbf{c}_0$$

for every column strategy $\mathbf{c}$.

The Problem of Game Theory: Given an $n \times k$ matrix game G, find optimal strategies $\mathbf{r}_0$ and $\mathbf{c}_0$.

According to the informal argument of Sec. 6-2, $\mathbf{r}_0$ and $\mathbf{c}_0$ are established as optimal strategies if it is the case that the minimum in the expectation produced by $\mathbf{r}_0$ is equal to the maximum in the expectation produced by $\mathbf{c}_0$, that is, if

$$\min \mathbf{r}_0 G = \max G\mathbf{c}_0.$$

In our 2×3 game G we may appeal to this argument to show that optimal strategies are

$$\mathbf{r}_0 = [.75 \quad .25] \quad \text{and} \quad \mathbf{c}_0 = \begin{bmatrix} 0 \\ .55 \\ .45 \end{bmatrix}.$$

We have

$$\mathbf{r}_0 G = [.75 \quad .25]\begin{bmatrix} 4 & -2 & 3 \\ -10 & 7 & -8 \end{bmatrix} = [.5 \quad .25 \quad .25],$$

$$G\mathbf{c}_0 = \begin{bmatrix} 4 & -2 & 3 \\ -10 & 7 & -8 \end{bmatrix}\begin{bmatrix} 0 \\ .55 \\ .45 \end{bmatrix} = \begin{bmatrix} .25 \\ .25 \end{bmatrix}.$$

Thus

$$\min \mathbf{r}_0 G = \max G\mathbf{c}_0 = .25.$$

With the strategy $\mathbf{r}_0$ the expected amount that player R wins per play is .25 points, or better. But by employing strategy $\mathbf{c}_0$, player C assures that the expected amount that R wins is not more than .25 points per play. Therefore, $\mathbf{r}_0$ is optimal.

The common value of $\min \mathbf{r}_0 G$ and $\max G\mathbf{c}_0$, in this case .25, is said to be the *value of the game.*

We shall shortly formalize the condition $\min \mathbf{r}_0 G = \max G\mathbf{c}_0$ for establishing optimal strategies as a theorem in the mathematical system. But first let's pick up a loose end.

The Expected Payoff

Earlier we were looking at the effects of strategies $\mathbf{r} = [.8 \quad .2]$ and $\mathbf{c} = \begin{bmatrix} .1 \\ .5 \\ .4 \end{bmatrix}$ in the game

$$G = \begin{bmatrix} 4 & -2 & 3 \\ -10 & 7 & -8 \end{bmatrix}.$$

But we bypassed the effect of the two players simultaneously employing strategies **r** and **c**.

Returning to first principles, we analyze the game G, played with strategies **r** and **c**, via the probability tree of Fig. 6-1.

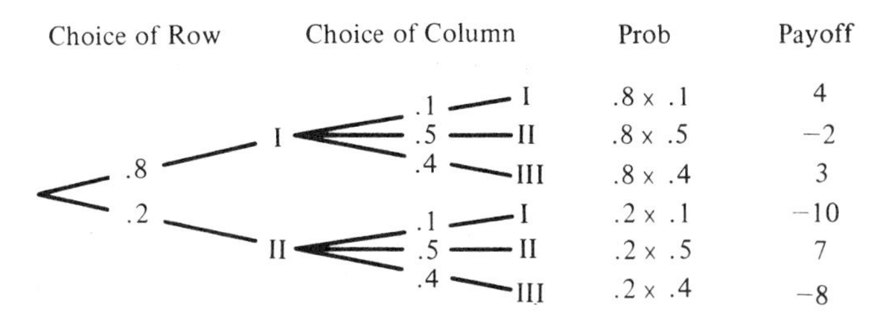

Figure 6-1.

We see that the expected value of the random variable "payoff" is as follows:

$$(.8 \times 4 \times .1) + (.8 \times (-2) \times .5) + (.8 \times 3 \times .4) + (.2 \times (-10) \times .1) + (.2 \times 7 \times .5) + (.2 \times (-8) \times .4) = .34.$$

Inspection of the arithmetic reveals that the computation of the expected payoff is the same as that involved in either of the double matrix multiplications

$$(\mathbf{r}G)\mathbf{c} \quad \text{or} \quad \mathbf{r}(G\mathbf{c}).$$

We have

$$(\mathbf{r}G)\mathbf{c} = \left([.8 \quad .2]\begin{bmatrix} 4 & -2 & 3 \\ -10 & 7 & -8 \end{bmatrix}\right)\begin{bmatrix} .1 \\ .5 \\ .4 \end{bmatrix}$$

$$= [1.2 \quad -.2 \quad .8]\begin{bmatrix} .1 \\ .5 \\ .4 \end{bmatrix} = [.34],$$

$$\mathbf{r}(G\mathbf{c}) = [.8 \quad .2]\left(\begin{bmatrix} 4 & -2 & 3 \\ -10 & 7 & -8 \end{bmatrix}\begin{bmatrix} .1 \\ .5 \\ .4 \end{bmatrix}\right)$$

$$= [.8 \quad .2]\begin{bmatrix} .6 \\ -.7 \end{bmatrix} = [.34]$$

Thus

$$(\mathbf{r}G)\mathbf{c} = \mathbf{r}(G\mathbf{c}) = [.34],$$

where the single entry in the product is .34, the expected payoff.

In advance of any computation we knew that the expected payoff would lie between

$$\min \mathbf{r}G = \min[1.2 \quad -.2 \quad .8] = -.2$$

and

$$\max Gc = \max\begin{bmatrix} .6 \\ -.7 \end{bmatrix} = .6.$$

As has been stressed repeatedly, $-.2$ is the least expected payoff that can result from row strategy **r**, and .6 is the greatest payoff that can result from column strategy **c**. Thus the expected payoff, .34, must lie between these bounds.

The observation that

$$(\mathbf{r}G)\mathbf{c} = \mathbf{r}(G\mathbf{c})$$

is not a peculiarity of the particular matrices **r**, G, and **c** considered. The general argument should be clear if the above manipulations are cast in a more general form. We let

$$\mathbf{r} = [r_1 \quad r_2], \qquad G = \begin{bmatrix} g_{11} & g_{12} & g_{13} \\ g_{21} & g_{23} & g_{23} \end{bmatrix}, \qquad \mathbf{c} = \begin{bmatrix} c_1 \\ c_2 \\ c_3 \end{bmatrix}.$$

The definition of multiplication dictates that the two products $(\mathbf{r}G)\mathbf{c}$ and $\mathbf{r}(G\mathbf{c})$ are both 1×1 matrices. Furthermore, the single entry in either of these products is simply the sum of the six terms of the form

$$r_i g_{ij} c_j,$$

where $i = 1, 2$ and $j = 1, 2, 3$.

Similarly, if G is an $n \times k$ matrix, **r** and **c** being row and column matrices of the proper lengths, then the single entry in $(\mathbf{r}G)\mathbf{c}$, as in $\mathbf{r}(G\mathbf{c})$, is the sum of all terms of the form $r_i g_{ij} c_j$, where $i = 1, 2, \ldots, n$ and $j = 1, 2, \ldots, k$. Thus the two products are equal, and the result may be represented unambiguously as $\mathbf{r}G\mathbf{c}$. [The equality of $(\mathbf{r}G)\mathbf{c}$ and $\mathbf{r}(G\mathbf{c})$ will be embedded in a formal theorem in Sec. 6-7.]

As the argument for the particular game G above suggests, it can be shown that in any matrix game G, played with simultaneous strategies **r** and **c**, the expected payoff is $\mathbf{r}G\mathbf{c}$.

The Duality Theorem

As recalled earlier in the section, and argued at some length in Sec. 6-2, strategies $\mathbf{r}_0$ and $\mathbf{c}_0$ are optimal strategies in a matrix game G if

$$\min \mathbf{r}_0 G = \max G\mathbf{c}_0.$$

We speak of $\mathbf{r}_0$ and $\mathbf{c}_0$ as *dual* strategies.

To this point the arguments have been heuristic, based on the empirical interpretation of the game—R can't win more than C can keep him from winning. The following theorem establishes this criterion for optimality as a result within the mathematical system.

The Duality Theorem of Game Theory: Given an $n \times k$ matrix game G, let $\mathbf{r}_0$ be a row strategy and $\mathbf{c}_0$ be a column strategy such that

$$\min \mathbf{r}_0 G = \max G\mathbf{c}_0.$$

Then $\mathbf{r}_0$ is an optimal row strategy; that is, given any row strategy $\mathbf{r}$,

$$\min \mathbf{r}G \leq \min \mathbf{r}_0 G.$$

And $\mathbf{c}_0$ is an optimal column strategy; that is, given any column strategy $\mathbf{c}$,

$$\max G\mathbf{c} \geq \max G\mathbf{c}_0.$$

In advance of the proof we shall establish an intermediate theorem. The only purpose is to separate the arguments; such theorems are called *lemmas*. The lemma states that the expected payoff $\mathbf{r}G\mathbf{c}$ must lie between the bounds $\min \mathbf{r}G$ and $\max G\mathbf{c}$.

Lemma: With G as in the theorem, let $\mathbf{r}$ and $\mathbf{c}$ be row and column strategies, respectively. And let v be the single entry in the 1×1 matrix $\mathbf{r}G\mathbf{c}$. Then

$$\min \mathbf{r}G \leq v \leq \max G\mathbf{c}.$$

Proof of the Lemma: Let

$$\mathbf{r}G = [a_1 \quad a_2 \quad \dots \quad a_k], \tag{1}$$

and let

$$\min \mathbf{r}G = \min[a_1 \quad a_2 \quad \dots \quad a_k] = a.$$

Since a is the minimum value, it follows that for $i = 1, 2, \dots, k$,

$$a \leq a_i. \tag{2}$$

Let

$$\mathbf{c} = \begin{bmatrix} c_1 \\ c_2 \\ \cdot \\ \cdot \\ \cdot \\ c_k \end{bmatrix},$$

where, as required,

$$c_i \geq 0 \qquad \text{for all } i \tag{3}$$

and

$$c_1 + c_2 + \cdots + c_k = 1. \tag{4}$$

Thus

$$\mathbf{r}G\mathbf{c} = (\mathbf{r}G)\mathbf{c} = [a_1 \quad a_2 \quad \dots \quad a_k] \begin{bmatrix} c_1 \\ c_2 \\ \cdot \\ \cdot \\ \cdot \\ c_k \end{bmatrix}$$

$$= [a_1 c_1 + a_2 c_2 + \cdots + a_k c_k].$$

Since v is the entry in $\mathbf{r}G\mathbf{c}$,

$$v = a_1c_1 + a_2c_2 + \cdots + a_kc_k. \tag{5}$$

From (2) and (3),

$$a \leq a_1 \quad \text{and} \quad c_1 \geq 0.$$

Therefore, since multiplication by the nonegative factor c_1 preserves the "less than or equal" relation,

$$ac_1 \leq a_1c_1.$$

And, in general,

$$ac_1 \leq a_1c_1, ac_2 \leq a_2c_2, \ldots, ac_k \leq a_kc_k. \tag{6}$$

Since the inequality relation is preserved under summation of the respective sides, we conclude from (6) that

$$ac_1 + ac_2 + \cdots + ac_k \leq a_1c_1 + a_2c_2 + \cdots + a_kc_k.$$

Factoring the common a from the left-hand side, we obtain

$$a(c_1 + c_2 + \cdots + c_k) \leq a_1c_1 + a_2c_2 + \cdots + a_kc_k \tag{7}$$

From Eq. (4), the left-hand side in (7) is the number $a = \min \mathbf{r}G$, and from Eq. (5) the right-hand side is the number v. Therefore, (7) asserts that

$$\min \mathbf{r}G \leq v.$$

This establishes one-half of the conclusion of the lemma. The other half is established by a word-for-word adaptation of the argument, interchanging the roles of $\mathbf{r}$ and $\mathbf{c}$ and exchanging min for max.

Proof of the theorem: Now that the lemma has been established, the proof of the theorem is easy. With $\mathbf{r}$ arbitrary and $\mathbf{c} = \mathbf{c}_0$ as in the hypothesis of the theorem, it follows from the lemma that

$$\min \mathbf{r}G \leq \max G\mathbf{c}_0. \tag{8}$$

By hypothesis,

$$\max G\mathbf{c}_0 = \min \mathbf{r}_0G. \tag{9}$$

Therefore, from (8) and (9),

$$\min \mathbf{r}G \leq \min \mathbf{r}_0G.$$

This is the first conclusion in the theorem.

The second conclusion is reached analogously. From the hypothesis recalled in Eq. (9) and the lemma, it follows that

$$\max G\mathbf{c}_0 = \min \mathbf{r}_0G \leq \max G\mathbf{c}$$

or

$$\max G\mathbf{c}_0 \leq \max G\mathbf{c}.$$

Definition: Let G be a matrix game, and let $\mathbf{r}_0$ and $\mathbf{c}_0$ be dual optimal strategies satisfying

$$\min \mathbf{r}_0G = \max G\mathbf{c}_0.$$

Then the common value of min $\mathbf{r}_0 G$ and max $G\mathbf{c}_0$ is said to be the *value of the game G*, denoted as v_G.

6-5 EXERCISES

1. Consider the payoff matrix G, with strategies $\mathbf{r}$ and $\mathbf{c}$, as follows:

$$G = \begin{bmatrix} 2 & -4 \\ -1 & 3 \end{bmatrix}, \qquad \mathbf{r} = [.3 \quad .7], \qquad \mathbf{c} = \begin{bmatrix} .8 \\ .2 \end{bmatrix}.$$

(a) Find the expectations $\mathbf{r}G$ and $G\mathbf{c}$.

(b) Construct a probability tree to represent the game played with the simultaneous strategies $\mathbf{r}$ and $\mathbf{c}$.

(c) From the tree in part (b), compute the expected payoff v.

(d) For the expected payoff v, observe that min $\mathbf{r}G \leq v \leq$ max $G\mathbf{c}$.

2. Let the payoff matrix G, strategies $\mathbf{r}$ and $\mathbf{c}$, be as follows:

$$G = \begin{bmatrix} 2 & -5 \\ -1 & 2 \\ 3 & -8 \end{bmatrix}, \qquad \mathbf{r} = [.3 \quad .6 \quad .1], \qquad \mathbf{c} = \begin{bmatrix} .7 \\ .3 \end{bmatrix}.$$

Follow the directions in Prob. 1.

3. Consider the game with payoff matrix

$$G = \begin{bmatrix} 3 & -5 & 6 \\ -2 & 4 & -5 \end{bmatrix}.$$

(a) For each of the following row strategies, $\mathbf{r}_1$, $\mathbf{r}_2$, and $\mathbf{r}_3$, find min $\mathbf{r}_i G$. Of the three strategies, which represents the best option for player R?

$$\mathbf{r}_1 = [.4 \quad .6],$$
$$\mathbf{r}_2 = [.5 \quad .5],$$
$$\mathbf{r}_3 = [.6 \quad .4].$$

(b) For each of the following column strategies, $\mathbf{c}_1$, $\mathbf{c}_2$, and $\mathbf{c}_3$, find max $G\mathbf{c}_i$. Of the three strategies, which represents the best option for player C?

$$\mathbf{c}_1 = \begin{bmatrix} \frac{1}{3} \\ \frac{1}{3} \\ \frac{1}{3} \end{bmatrix}, \qquad \mathbf{c}_2 = \begin{bmatrix} .1 \\ .5 \\ .4 \end{bmatrix}, \qquad \mathbf{c}_3 = \begin{bmatrix} 0 \\ .5 \\ .5 \end{bmatrix}.$$

4. Continue with the game of Prob. 3:

$$G = \begin{bmatrix} 3 & -5 & 6 \\ -2 & 4 & -5 \end{bmatrix}.$$

It is conjectured that optimal strategies are

$$\mathbf{r}_0 = [.45 \quad .55] \quad \text{and} \quad \mathbf{c}_0 = \begin{bmatrix} 0 \\ .55 \\ .45 \end{bmatrix}.$$

Verify the optimality of these strategies by showing that

$$\min \mathbf{r}_0 G = \max G\mathbf{c}_0.$$

What is the value of the game?

5. For the game of Prob. 2,

$$G = \begin{bmatrix} 2 & -5 \\ -1 & 2 \\ 3 & -8 \end{bmatrix},$$

it is conjectured that optimal strategies are

$$\mathbf{r}_0 = [.3 \quad .7 \quad 0] \quad \text{and} \quad \mathbf{c}_0 = \begin{bmatrix} .7 \\ .3 \end{bmatrix}.$$

Verify by showing that

$$\min \mathbf{r}_0 G = \max G\mathbf{c}_0.$$

What is the value of the game?

6. Let A be a general 1×3 matrix,

$$A = [a_1 \quad a_2 \quad a_3];$$

let B be a general 3×2 matrix,

$$B = \begin{bmatrix} b_{11} & b_{12} \\ b_{21} & b_{22} \\ b_{31} & b_{32} \end{bmatrix};$$

and let C be a general 2×1 matrix,

$$C = \begin{bmatrix} c_1 \\ c_2 \end{bmatrix}.$$

Show that the single entry in

$$(AB)C = A(BC) = ABC$$

is the sum of all products of the form

$$a_i b_{ij} c_j,$$

$i = 1, 2, 3$ and $j = 1, 2$.

7. Let $\mathbf{r}_0$ and $\mathbf{c}_0$ be dual optimal strategies in a game G; i.e.,

$$\min \mathbf{r}_0 G = \max G\mathbf{c}_0.$$

Suppose $\mathbf{r}'_0$ and $\mathbf{c}'_0$ also represent a pair of dual optimal strategies, with

$$\min \mathbf{r}'_0 G = \max G\mathbf{c}'_0.$$

Employ the duality theorem to prove that

$$\min \mathbf{r}_0 G = \min \mathbf{r}'_0 G = \max G\mathbf{c}'_0 = \max G\mathbf{c}_0.$$

(This shows that if optimal strategies exist, then the value of the game v_G is unique.)

8. In the text the details of the proof of the Lemma for the Duality Theorem of Game Theory reached only half of the conclusion. Carry out the steps for the other half to prove that

$$v \leq \max G\mathbf{c}$$

(terminology as in the lemma).

6-6. The Practical Analysis of Games

We know that $\mathbf{r}_0$ and $\mathbf{c}_0$ are optimal strategies if

$$\min \mathbf{r}_0 G = \max G\mathbf{c}_0.$$

But there is a theoretical question which has not been touched, either by the informal arguments of Sec. 6-2 or the mathematical treatment in the previous section:

Given a matrix game G, is it necessarily the case that dual strategies $\mathbf{r}_0$ and $\mathbf{c}_0$ exist such that $\min \mathbf{r}_0 G = \max G\mathbf{c}_0$?

Indeed strategies satisfying this equation must exist for every matrix game, but the argument isn't easy. In the next chapter we shall develop a geometrical interpretation that points the way to a proof of the existence of $\mathbf{r}_0$ and $\mathbf{c}_0$. Of a more practical nature, the geometrical interpretation will make it possible to find optimal strategies in games with $2 \times n$ and $n \times 2$ payoff matrices.

Of course, for a particular game G the existence question is academic if we can actually exhibit strategies $\mathbf{r}_0$ and $\mathbf{c}_0$ for which $\min \mathbf{r}_0 G = \max G\mathbf{c}_0$.

To emphasize the general format of the analysis of a game we now consider a game with a 3×3 payoff matrix G_U, giving no hint of how the dual optimal strategies $\mathbf{r}_0$ and $\mathbf{c}_0$ were found.

Problem: Analyze the game with payoff matrix

$$G_U = \begin{bmatrix} 3 & -5 & -2 \\ -2 & 4 & 3 \\ 7 & -2 & -3 \end{bmatrix}.$$

Solution: Optimal strategies are

$$\mathbf{r}_0 = [0 \;\; \tfrac{2}{3} \;\; \tfrac{1}{3}], \qquad \mathbf{c}_0 = \begin{bmatrix} .4 \\ 0 \\ .6 \end{bmatrix}.$$

Verification:

$$\mathbf{r}_0 G_U = [0 \ \ \tfrac{2}{3} \ \ \tfrac{1}{3}] \begin{bmatrix} 3 & -5 & -2 \\ -2 & 4 & 3 \\ 7 & -2 & 3 \end{bmatrix} = [1 \ \ 2 \ \ 1],$$

$$G_U \mathbf{c}_0 = \begin{bmatrix} 3 & -5 & -2 \\ -2 & 4 & 3 \\ 7 & -2 & -3 \end{bmatrix} \begin{bmatrix} .4 \\ 0 \\ .6 \end{bmatrix} = \begin{bmatrix} 0 \\ 1 \\ 1 \end{bmatrix},$$

$$\min \mathbf{r}_0 G_U = \max G_U \mathbf{c}_0 = 1.$$

Therefore, $\mathbf{r}_0$ and $\mathbf{c}_0$ are dual optimal strategies; the value of the game is $v_{G_U} = 1$.

For reasons that should be clear, a game G is said to *favor player R* if the value of the game v_G is positive, and it *favors player C* if the v_G is negative. If v_G is zero, we say that G is a *fair game.*

Since in the above example the value of the game is

$$v_{G_U} = 1,$$

the game G_U favors R. (The subscript U in G_U stands for "unfair.") By employing strategy $\mathbf{r}_0 = [0 \ \ \tfrac{2}{3} \ \ \tfrac{1}{3}]$, player R is assured of winning about 1 point per game, or more, on the average in the long run. To be fair, R should award C a bonus of 1 point per play of the game.

In general, a game G is converted into a fair game by subtracting the value v_G from each entry in its matrix. Optimal strategies for G qualify as optimal strategies for the converted game.

Let's convert the unfair game G_U into a fair game G_F by subtracting the value of the game, $v_{G_U} = 1$, from each entry in the matrix. The interpretation is that R is giving C a bonus of 1 point per play.

Problem: Analyze the game with payoff matrix

$$G_F = \begin{bmatrix} 2 & -6 & -3 \\ -3 & 3 & 2 \\ 6 & -3 & -4 \end{bmatrix}$$

and show that it is a fair game.

Solution: Optimal strategies are

$$\mathbf{r}_0 = [0 \ \ \tfrac{2}{3} \ \ \tfrac{1}{3}], \qquad \mathbf{c}_0 = \begin{bmatrix} .4 \\ 0 \\ .6 \end{bmatrix}.$$

Verification:

$$\mathbf{r}_0 G_F = [0 \ \ \tfrac{2}{3} \ \ \tfrac{1}{3}] \begin{bmatrix} 2 & -6 & 1 \\ -3 & 3 & 2 \\ 6 & -3 & -4 \end{bmatrix} = [0 \ \ 1 \ \ 0],$$

$$G_F\mathbf{c}_0 = \begin{bmatrix} 2 & -6 & -3 \\ -3 & 3 & 2 \\ 6 & -3 & -4 \end{bmatrix}\begin{bmatrix} .4 \\ 0 \\ .6 \end{bmatrix} = \begin{bmatrix} -1 \\ 0 \\ 0 \end{bmatrix},$$

$$\min \mathbf{r}_0 G_F = \max G_F\mathbf{c}_0 = 0.$$

Therefore, $\mathbf{r}_0$ and $\mathbf{c}_0$ are dual optimal strategies; the value of the game is $v_{G_F} = 0$. Since the value of the game is zero, G_F is a fair game.

Strictly Determined Games

There is one category of games for which dual optimal strategies $\mathbf{r}_0$ and $\mathbf{c}_0$ are immediate. We refer to the strictly determined cases discussed in Sec. 6-3 in which the matrix contains a minimax entry, simultaneously the minimum in its row and the maximum in its column.

Problem: Analyze the game with payoff matrix

$$G = \begin{bmatrix} 2 & -1 & 5 \\ -7 & -2 & 9 \\ 3 & 1 & 2 \\ 4 & 0 & -8 \end{bmatrix}.$$

Solution: Noting that the entry in Row III and Column II is a minimax, it is asserted that the following are optimal strategies:

$$\mathbf{r}_0 = [0 \quad 0 \quad 1 \quad 0], \qquad \mathbf{c}_0 = \begin{bmatrix} 0 \\ 1 \\ 0 \end{bmatrix}.$$

Verification:

$$\mathbf{r}_0 G = [0 \quad 0 \quad 1 \quad 0]\begin{bmatrix} 2 & -1 & 5 \\ -7 & -2 & 9 \\ 3 & 1 & 2 \\ 4 & 0 & -8 \end{bmatrix} = [3 \quad 1 \quad 2],$$

$$G\mathbf{c}_0 = \begin{bmatrix} 2 & -1 & 5 \\ -7 & -2 & 9 \\ 3 & 1 & 2 \\ 4 & 0 & -8 \end{bmatrix}\begin{bmatrix} 0 \\ 1 \\ 0 \end{bmatrix} = \begin{bmatrix} -1 \\ -2 \\ 1 \\ 0 \end{bmatrix},$$

$$\min \mathbf{r}_0 G = \max G\mathbf{c}_0 = 1.$$

Therefore, $\mathbf{r}_0$ and $\mathbf{c}_0$ are dual optimal strategies; the value of the game is $v_G = 1$.

The Analysis of 2 × 2 Games

Beyond the strictly determined cases, the simplest games are those represented by 2×2 payoff matrices. Postponing the revelation of the underlying reasoning until the next chapter, we shall now give a straightforward technique for finding optimal strategies in 2×2 games. For any given game the validity of the procedure is established by verifying that the strategies $\mathbf{r}_0$ and $\mathbf{c}_0$ obtained do satisfy the criterion of the Duality Theorem of Game Theory.

As a first example of the 2×2 technique we return to the "Two-Finger Matching Game" of Sec. 6-1. Here the payoff matrix is

$$G = \begin{bmatrix} 4 & -2 \\ -10 & 7 \end{bmatrix}.$$

By inspection we see that the game G is not strictly determined—neither of the row minima, -2 and -10, is a column maximum. Thus the search for optimal strategies requires a little effort.

The first step is to lift out the two columns of the matrix and, in the obvious manner, compute their "difference":

$$\begin{bmatrix} 4 \\ -10 \end{bmatrix} - \begin{bmatrix} -2 \\ 7 \end{bmatrix} = \begin{bmatrix} 6 \\ -17 \end{bmatrix}.$$

> **Comment:** Note that the "difference" column contains one positive and one negative entry. But suppose both entries had been positive. This would tell us that in each row the entry in the first column is larger than the entry in the second column, which means the column player C can do no better than stick with Column II—the game is strictly determined. If the game is not strictly determined, the difference column must consist of one positive and one negative entry.

The second step is to form a row matrix from the entries in the "difference" column by reversing their order and dropping the minus sign of the negative entry. Thus

$$\begin{bmatrix} 6 \\ -17 \end{bmatrix} \quad \text{produces} \quad [17 \quad 6].$$

Third, we divide each entry in this row matrix by the sum of the two values. This produces the matrix $\mathbf{r}_0$,

$$\mathbf{r}_0 = [\tfrac{17}{23} \quad \tfrac{6}{23}],$$

which we conjecture to be an optimal row strategy.

The candidate for an optimal column strategy $\mathbf{c}_0$ is found in the analogous manner.

Step 1: "Subtract" the rows:

$$[4 \quad -2] - [-10 \quad 7] = [14 \quad -9].$$

(Since the game is not strictly determined, the "difference" row must consist of one positive and one negative entry.)

Step 2: Form a column matrix from the entries in the "difference" row by reversing their order and dropping the single minus sign:

$$\begin{bmatrix} 9 \\ 14 \end{bmatrix}.$$

Step 3: Divide each entry in this column matrix by the sum of the two values to produce the matrix $\mathbf{c}_0$, which we conjecture to be an optimal column strategy:

$$\mathbf{c}_0 = \begin{bmatrix} \frac{9}{23} \\ \frac{14}{23} \end{bmatrix}.$$

To this point, the manipulations are to be viewed as so much scratch work. Now we apply the criterion of the duality theorem to verify that $\mathbf{r}_0$ and $\mathbf{c}_0$ are optimal strategies.

Problem: Analyze the game with payoff matrix

$$G = \begin{bmatrix} 4 & -2 \\ -10 & 7 \end{bmatrix}.$$

Solution: Optimal strategies are

$$\mathbf{r}_0 = [\tfrac{17}{23} \quad \tfrac{6}{23}], \qquad \mathbf{c}_0 = \begin{bmatrix} \frac{9}{23} \\ \frac{14}{23} \end{bmatrix}.$$

Verification:

$$\mathbf{r}_0 G = [\tfrac{17}{23} \quad \tfrac{6}{23}] \begin{bmatrix} 4 & -2 \\ -10 & 7 \end{bmatrix} = [\tfrac{8}{23} \quad \tfrac{8}{23}],$$

$$G\mathbf{c}_0 = \begin{bmatrix} 4 & -2 \\ -10 & 7 \end{bmatrix} \begin{bmatrix} \frac{9}{23} \\ \frac{14}{23} \end{bmatrix} = \begin{bmatrix} \frac{8}{23} \\ \frac{8}{23} \end{bmatrix},$$

$$\min \mathbf{r}_0 G = \max G\mathbf{c}_0 = \tfrac{8}{23}.$$

Therefore, $\mathbf{r}_0$ and $\mathbf{c}_0$ are dual optimal strategies; the value of the game is $v_G = \frac{8}{23}$.

Comment: As the reader may have guessed, the optimal strategies for a 2×2 game which is not strictly determined are those that produce equal entries in the expectations. The "three-step" technique for finding $\mathbf{r}_0$ is no more than a shortcut for solving the algebraic problem

$$\mathbf{r}_0 G = [a \quad a].$$

Now let's analyze a game G_1 which clearly favors C and then convert G_1 into a fair game G_2. The payoff matrix is

$$G_1 = \begin{bmatrix} -1 & -5 \\ -4 & 4 \end{bmatrix}.$$

First, we note that the game G_1 is not strictly determined. We apply the "three-step" method to find candidates for $\mathbf{r}_0$ and $\mathbf{c}_0$.

Search for $\mathbf{r}_0$:

Step 1: $\begin{bmatrix} -1 \\ -4 \end{bmatrix} - \begin{bmatrix} -5 \\ 4 \end{bmatrix} = \begin{bmatrix} 4 \\ -8 \end{bmatrix}.$

Step 2: $[8 \quad 4]$.

Step 3: $\mathbf{r}_0 = [\frac{8}{12} \quad \frac{4}{12}] = [\frac{2}{3} \quad \frac{1}{3}]$.

Search for $\mathbf{c}_0$:

Step 1: $[-1 \quad -5] - [-4 \quad 4] = [3 \quad -9]$.

Step 2: $\begin{bmatrix} 9 \\ 3 \end{bmatrix}.$

Step 3: $\mathbf{c}_0 = \begin{bmatrix} \frac{9}{12} \\ \frac{3}{12} \end{bmatrix} = \begin{bmatrix} \frac{3}{4} \\ \frac{1}{4} \end{bmatrix}.$

The scratch work done, the exercise takes the form

Problem: Analyze the game with payoff matrix

$$G_1 = \begin{bmatrix} -1 & -5 \\ -4 & 4 \end{bmatrix}.$$

Solution: Optimal strategies are

$$\mathbf{r}_0 = [\tfrac{2}{3} \quad \tfrac{1}{3}], \qquad \mathbf{c}_0 = \begin{bmatrix} \frac{3}{4} \\ \frac{1}{4} \end{bmatrix}.$$

Verification:

$$\mathbf{r}_0 G_1 = [\tfrac{2}{3} \quad \tfrac{1}{3}] \begin{bmatrix} -1 & -5 \\ -4 & 4 \end{bmatrix} = [-\tfrac{6}{3} \quad -\tfrac{6}{3}] = [-2 \quad -2],$$

$$G_1 \mathbf{c}_0 = \begin{bmatrix} -1 & -5 \\ -4 & 4 \end{bmatrix} \begin{bmatrix} \frac{3}{4} \\ \frac{1}{4} \end{bmatrix} = \begin{bmatrix} -\frac{8}{4} \\ -\frac{8}{4} \end{bmatrix} = \begin{bmatrix} -2 \\ -2 \end{bmatrix},$$

$$\min \mathbf{r}_0 G_1 = \max G_1 \mathbf{c}_0 = -2.$$

Therefore, $\mathbf{r}_0$ and $\mathbf{c}_0$ are dual optimal strategies; the value of the game is $v_{G_1} = -2$.

To convert G_1 into a fair game G_2, we subtract the value -2 from each entry in the matrix; i.e., we add 2 to each entry. The optimal strategies are unchanged.

Problem: Analyze the game with payoff matrix

$$G_2 = \begin{bmatrix} 1 & -3 \\ -2 & 6 \end{bmatrix}$$

and show that it is a fair game.

Solution: Optimal strategies are

$$\mathbf{r}_0 = [\tfrac{2}{3} \quad \tfrac{1}{3}], \qquad \mathbf{c}_0 = \begin{bmatrix} \tfrac{3}{4} \\ \tfrac{1}{4} \end{bmatrix}.$$

Verification:

$$\mathbf{r}_0 G_2 = [\tfrac{2}{3} \quad \tfrac{1}{3}] \begin{bmatrix} 1 & -3 \\ -2 & 6 \end{bmatrix} = [0 \quad 0],$$

$$G_2 \mathbf{c}_0 = \begin{bmatrix} 1 & -3 \\ -2 & 6 \end{bmatrix} \begin{bmatrix} \tfrac{3}{4} \\ \tfrac{1}{4} \end{bmatrix} = \begin{bmatrix} 0 \\ 0 \end{bmatrix},$$

$$\min \mathbf{r}_0 G_2 = \max G_2 \mathbf{c}_0 = 0.$$

Therefore, $\mathbf{r}_0$ and $\mathbf{c}_0$ are dual optimal strategies; the value of the game is $v_{G_2} = 0$. Consequently, G_2 is a fair game.

6-6 EXERCISES

1. Consider the game with payoff matrix

$$G = \begin{bmatrix} -2 & 1 & 4 \\ 2 & -3 & -2 \\ 0 & 1 & -1 \end{bmatrix}.$$

Verify that

$$\mathbf{r}_0 = [\tfrac{1}{4} \quad \tfrac{1}{4} \quad \tfrac{2}{4}] \quad \text{and} \quad \mathbf{c}_0 = \begin{bmatrix} \tfrac{5}{9} \\ \tfrac{2}{9} \\ \tfrac{2}{9} \end{bmatrix}$$

are optimal strategies, and find the value of the game.

2. Consider the games corresponding to the following 2×2 matrices. Find $\mathbf{r}_0$, $\mathbf{c}_0$, and the value of the game, and verify. Use the three-step method described in the text, unless the game is strictly determined.

(a) $G = \begin{bmatrix} 2 & -1 \\ -3 & 2 \end{bmatrix}$. (b) $G = \begin{bmatrix} 1 & 2 \\ 3 & -6 \end{bmatrix}$.

(c) $G = \begin{bmatrix} -7 & 3 \\ 10 & -5 \end{bmatrix}$. (d) $G = \begin{bmatrix} 1 & 11 \\ 100 & 10 \end{bmatrix}$.

3. Analyze the games corresponding to the following 2×2 matrices; that is, find $\mathbf{r}_0$ and $\mathbf{c}_0$ and verify, concluding with a statement of the value of

the game. (If not strictly determined, try to do the three-step method in your head, noting that the numerator of the first entry in $\mathbf{r}_0$ is the distance between the entries in the second row of G, and analogously for the other entry in $\mathbf{r}_0$ and the two entries in $\mathbf{c}_0$.)

(a) $G = \begin{bmatrix} 1 & -3 \\ -2 & 4 \end{bmatrix}$. (b) $G = \begin{bmatrix} 4 & -5 \\ -10 & 11 \end{bmatrix}$.

(c) $G = \begin{bmatrix} 2 & -4 \\ 0 & 2 \end{bmatrix}$. (d) $G = \begin{bmatrix} 0 & 1 \\ -1 & 0 \end{bmatrix}$.

4. On the count of three, Paul and Steven each extend one or two fingers. If they match one and one, Paul pays Steven 1 cent, and if they match two and two, Paul pays Steven 11 cents. If they don't match, Steven pays Paul 4 cents.
(a) Is this a fair game?
(b) If not, how much should who pay whom in advance of each play to make it fair?

5. Consider the game G_U with payoff matrix

$$G_U = \begin{bmatrix} 1 & -6 \\ -5 & 16 \end{bmatrix}.$$

Convert G_U to a fair game G_F by subtracting a fixed amount from each entry in the matrix G_U. Then prove that G_F is a fair game.

6. Consider the game G with payoff matrix

$$G = \begin{bmatrix} 3 & -11 & 4 \\ -2 & 4 & -1 \end{bmatrix}.$$

(a) Give an argument to the effect that Player C should never play Column III and that therefore an optimal column strategy should be of the form

$$\mathbf{c}_0 = \begin{bmatrix} c_1 \\ c_2 \\ 0 \end{bmatrix}.$$

(b) Ignore Column III and find optimal strategies by the technique for 2×2 matrices. Then present optimal strategies for the game G as given, and verify.

7. Analyze the games corresponding to the following payoff matrices [the matrix in part (b) demands a critical inspection in the spirit of Prob. 6]:

(a) $G = \begin{bmatrix} -4 & 5 & -1 \\ 2 & 1 & 0 \\ 5 & -3 & -2 \end{bmatrix}$. (b) $G = \begin{bmatrix} 2 & 0 & -1 \\ 1 & 2 & -3 \\ -3 & 3 & 2 \end{bmatrix}$.

8. For what value of x does the following matrix represent a fair game?

$$G = \begin{bmatrix} 2 & -3 \\ -8 & x \end{bmatrix}.$$

9. Let the payoff matrix be of the form

$$G = \begin{bmatrix} a & -b \\ -c & d \end{bmatrix},$$

where a, b, c, and d are all positive numbers. Show that if

$$\frac{a}{c} = \frac{b}{d},$$

then G is a fair game, and, conversely, if G is a fair game, then $a/c = b/d$.

6-7. Further Operations with Matrices

In the "scratch work" of searching for optimal strategies for 2×2 games in the previous section we spoke in an offhand way of the "difference" between two row or column matrices. Let's give this operation an official position in the mathematics of matrices. Of course, in the arithmetic of real numbers it is the sum, and not the difference, which is the primary operation, and so it is with matrices. Informally,

Given two matrices, A and B, of the same dimensions, i.e., A and B have the same number of rows and the same number of columns, then the sum $A + B$ is the matrix of the same dimensions realized by adding corresponding entries in A and B. If A and B do not have the same dimensions, the sum is not defined.

In short, the sum is defined "in the obvious manner."

Examples:

$$\begin{bmatrix} 2 & 0 & -5 \\ -3 & 4 & 7 \end{bmatrix} + \begin{bmatrix} -1 & 3 & 1 \\ 2 & -2 & 4 \end{bmatrix} = \begin{bmatrix} 1 & 3 & -4 \\ -1 & 2 & 11 \end{bmatrix}.$$

$$\begin{bmatrix} 3 & -1 \\ 1 & 4 \\ -5 & 0 \end{bmatrix} + \begin{bmatrix} -2 & -1 \\ 2 & -2 \\ 3 & 4 \end{bmatrix} = \begin{bmatrix} 1 & -2 \\ 3 & 2 \\ -2 & 4 \end{bmatrix}.$$

For

$$A = \begin{bmatrix} 2 & 0 & -5 \\ -3 & 4 & 7 \end{bmatrix} \quad \text{and} \quad B = \begin{bmatrix} -2 & -1 \\ 2 & -2 \\ 3 & 4 \end{bmatrix}$$

the sum $A + B$ is not defined because A and B do not have the same dimensions.

Definition: For positive integers, n and k, let

$$A = [a_{ij}] \quad \text{and} \quad B = [b_{ij}]$$

be $n \times k$ matrices. Then

$$C = A + B$$

is the $n \times k$ matrix

$$C = [c_{ij}]$$

such that for $i = 1, 2, \ldots, n$ and $j = 1, 2, \ldots, k$

$$c_{ij} = a_{ij} + b_{ij}.$$

In the arithmetic of real numbers, the operation of subtraction is derived from the operation of addition. For example,

$$5 - 2 = 3$$

because

$$3 + 2 = 5.$$

The same convention gives us the "difference" between matrices. We have

$$\begin{bmatrix} 1 & 3 & -4 \\ -1 & 2 & 11 \end{bmatrix} - \begin{bmatrix} -1 & 3 & 1 \\ 2 & -2 & 4 \end{bmatrix} = \begin{bmatrix} 2 & 0 & -5 \\ -3 & 4 & 7 \end{bmatrix}$$

because

$$\begin{bmatrix} 2 & 0 & -5 \\ -3 & 4 & 7 \end{bmatrix} + \begin{bmatrix} -1 & 3 & 1 \\ 2 & -2 & 4 \end{bmatrix} = \begin{bmatrix} 1 & 3 & -4 \\ -1 & 2 & 11 \end{bmatrix}.$$

The difference $A - B$ can also be defined $A + (-1)B$, but this requires a definition of $(-1)B$. Let's do more, and define what we mean by aB, where a is a real number and B is a matrix.

Definition: Let a be a real number, and let $B = [b_{ij}]$ be an $n \times k$ matrix. Then

$$aB = [c_{ij}]$$

is the $n \times k$ matrix such that for $i = 1, 2, \ldots, n$ and $j = 1, 2, \ldots, k$

$$c_{ij} = a \times b_{ij}.$$

Examples:

$$3\begin{bmatrix} 2 & 0 & -5 \\ -3 & 4 & 7 \end{bmatrix} = \begin{bmatrix} 6 & 0 & -15 \\ -9 & 12 & 21 \end{bmatrix}.$$

$$(-2)\begin{bmatrix} 3 & -1 \\ 1 & 4 \\ -5 & 0 \end{bmatrix} = \begin{bmatrix} -6 & 2 \\ -2 & -8 \\ 10 & 0 \end{bmatrix}.$$

Algebraic Properties

We have three operations to be performed with matrices: (matrix) multiplication, addition, and multiplication by numbers. Worthy of observation are several fundamental algebraic properties, closely related to the laws governing the manipulations of real numbers.

Immediate from the definition above is the fact that multiplication of a matrix B by the number 1 simply produces the matrix B. That is, for all matrices B,

$$1B = B. \tag{1}$$

Immediate also is the fact that if a and b are numbers and C is a matrix, then

$$a(bC) = (ab)C. \tag{2}$$

Other properties bearing on the interplay between the real number operations and the matrix manipulations are the following:

If a and b are numbers and C is a matrix, then

$$(a + b)C = aC + bC. \tag{3}$$

If a is a number and B and C are matrices of the same dimensions, then

$$a(B + C) = aB + aC. \tag{4}$$

If a is a number and B and C are matrices for which the matrix product BC is defined, then

$$a(BC) = (aB)C = B(aC). \tag{5}$$

The next two properties, purely properties of matrix addition, have names derived from the analogous laws for real number addition:

The Associative Law of Matrix Addition: If A, B, and C are matrices of the same dimensions, then

$$A + (B + C) = (A + B) + C. \tag{6}$$

The Commutative Law of Matrix Addition: If A and B are matrices of the same dimensions, then

$$A + B = B + A. \tag{7}$$

Each of the properties (1) through (7) is proved by showing that the entry in the (i, j) position, i.e., in the ith row and the jth column, of the matrix on the one side of the equation is equal to the entry in the same position on the other. For properties (1) through (4) this equality of entries should be clear.

With respect to property (5), suppose B is an $m \times n$ matrix and C is an $n \times p$ matrix. Then by the definitions of the operations, the entry in the (i, j) position in $a(BC)$ on the left in (5) is

$$a(b_{i1}c_{1j} + b_{i2}c_{2j} + \cdots + b_{in}c_{nj}).$$

In $(aB)C$, this entry is

$$(ab_{i1})c_{1j} + (ab_{i2})c_{2j} + \cdots + (ab_{in})c_{nj}.$$

And in $B(aC)$ on the right, we have

$$b_{i1}(ac_{1j}) + b_{i2}(ac_{2j}) + \cdots + b_{2n}(ac_{nj}).$$

The three expressions are established as equal by simply factoring out the common factor a in the latter two.

For the associative law (6) we observe that the entry in the (i, j) position in $A + (B + C)$ on the left is

$$a_{ij} + (b_{ij} + c_{ij}).$$

In $(A + B) + C$ on the right in (6), the entry is

$$(a_{ij} + b_{ij}) + c_{ij}.$$

Thus the associative law for matrix addition follows immediately from the associative law for real number addition: For all real numbers a, b, and c

$$a + (b + c) = (a + b) + c.$$

Similarly for the commutative law (7). In the (i, j) position on the left and right, respectively, we have

$$a_{ij} + b_{ij} = b_{ij} + a_{ij}.$$

By virtue of the associative law, there is no ambiguity in expressing the sum of three matrices as simply

$$A + B + C.$$

Extending the rule, the sum of any number of $n \times k$ matrices is the same regardless of where we insert the parentheses. If $A_1, A_2, \ldots, A_m$ are $n \times k$ matrices, the sum is expressed unambiguously as

$$A_1 + A_2 + \cdots + A_m.$$

The distributive law for real numbers expresses the interplay between the operations of addition and multiplication: For all real numbers a, b, and c

$$a(b + c) = ab + ac.$$

The distributive law carries over to the matrix operation, but its proof requires a little effort.

The Distributive Law: Let A, B, and C be matrices such that the products AB and AC and the sum $B + C$ are defined. Then

$$A(B + C) = AB + AC. \tag{8}$$

To avoid becoming lost in a jungle of symbols, we'll be content to illustrate the argument by considering the case in which A is a general 2×3 matrix and B and C are general 3×2 matrices:

$$A = \begin{bmatrix} a_{11} & a_{12} & a_{13} \\ a_{21} & a_{22} & a_{23} \end{bmatrix}, \qquad B = \begin{bmatrix} b_{11} & b_{12} \\ b_{21} & b_{22} \\ b_{31} & b_{32} \end{bmatrix}, \qquad C = \begin{bmatrix} c_{11} & c_{12} \\ c_{21} & c_{22} \\ c_{31} & c_{32} \end{bmatrix}.$$

Let's consider the entries in the 2nd row and 2nd column—in the (2, 2) position—on either side of Eq. (8). Focusing on the (2, 2) position on the left in Eq. (8),

$$A(B + C) = \begin{bmatrix} \cdot & \cdot & \cdot \\ a_{21} & a_{22} & a_{23} \end{bmatrix} \begin{bmatrix} \cdot & b_{12} + c_{12} \\ \cdot & b_{22} + c_{22} \\ \cdot & b_{32} + c_{32} \end{bmatrix}$$

$$= \begin{bmatrix} \cdot & \cdot \\ \cdot & d_{22} \end{bmatrix},$$

where

$$d_{22} = a_{21}(b_{12} + c_{12}) + a_{22}(b_{22} + c_{22}) + a_{23}(b_{32} + c_{32}).$$

And focusing on this position on the right in Eq. (8),

$$AB + AC = \begin{bmatrix} \cdot & \cdot & \cdot \\ a_{21} & a_{22} & a_{23} \end{bmatrix} \begin{bmatrix} \cdot & b_{12} \\ \cdot & b_{22} \\ \cdot & b_{32} \end{bmatrix} + \begin{bmatrix} \cdot & \cdot & \cdot \\ a_{21} & a_{22} & a_{23} \end{bmatrix} \begin{bmatrix} \cdot & c_{12} \\ \cdot & c_{22} \\ \cdot & c_{32} \end{bmatrix}$$

$$= \begin{bmatrix} \cdot & \cdot \\ \cdot & e_{22} \end{bmatrix} + \begin{bmatrix} \cdot & \cdot \\ \cdot & f_{22} \end{bmatrix} = \begin{bmatrix} \cdot & \cdot \\ \cdot & e_{22} + f_{22} \end{bmatrix},$$

where

$$e_{22} = a_{21}b_{12} + a_{22}b_{22} + a_{23}b_{32}$$

and

$$f_{22} = a_{21}c_{12} + a_{22}c_{22} + a_{23}c_{32}.$$

Thus in the (2, 2) position on the right, we find

$$\begin{aligned} e_{22} + f_{22} &= (a_{21}b_{12} + a_{22}b_{22} + a_{23}b_{32}) + (a_{21}c_{12} + a_{22}c_{22} + a_{23}c_{32}) \\ &= a_{21}(b_{12} + c_{12}) + a_{22}(b_{22} + c_{22}) + a_{23}(b_{32} + c_{32}) \\ &= d_{22}. \end{aligned}$$

As desired, we see that the entry $e_{22} + f_{22}$ in the (2, 2) position on the right in Eq. (8) is equal to the entry d_{22} in this position on the left in Eq. (8).

The same argument works for the other entries on either side in the distributive law (8), and it generalizes with only symbolic complication to matrices A, B, and C of arbitrary dimensions.

There is a companion distributive law that asserts that

$$(A + B)C = AC + BC \tag{9}$$

whenever the matrix combinations are defined. The argument for this property is a straightforward adaptation of the argument for Eq. (8).

In Chapter 8 we shall need the distributive law in the form

$$A(B - C) = AB - AC.$$

This variant is easily established from Eqs. (8) and (5):

$$\begin{aligned} A(B - C) &= A(B + (-1)C) \\ &= AB + A(-1)C \\ &= AB + (-1)(AC) \\ &= AB - AC. \end{aligned}$$

The final property that we'll observe is the extension to matrix multiplication of the associative law of multiplication for real numbers: For all real numbers a, b, and c.

$$(ab)c = a(bc).$$

The Associative Law of Matrix Multiplication: Let A, B, and C be matrices such that the products AB and BC are defined. Then

$$(AB)C = A(BC). \tag{10}$$

We have already encountered this property in a special case. In discussing the expected payoff for a matrix game in Sec. 6-5 we indicated the general argument that

$$(\mathbf{r}G)\mathbf{c} = \mathbf{r}(G\mathbf{c})$$

whenever $\mathbf{r}$ is a row matrix and $\mathbf{c}$ is a column matrix. Which is to say,

$$(AB)C = A(BC)$$

whenever A consists of a single row and C consists of a single column. The argument for the general case is not significantly more complicated. From the definition of matrix multiplication, the entry in the ith row and jth column on either side of Eq. (10) is the single entry in the product

$$(A_iB)C_j = A_i(BC_j),$$

where A_i is the row matrix realized by extracting the ith row of A and C_j is the column matrix realized by extracting the jth column of C. (Compare with the argument for the distributive law above.) Thus, entry for entry, the two sides of Eq. (10) are the same.

Comment: For real numbers we also have the commutative law of multiplication: For all real numbers a and b

$$ab = ba.$$

But it is not true that matrix products AB and BA are necessarily equal. As we saw in the examples in Sec. 6-4, the fact that AB is defined does not even imply that BA is defined. And if AB and BA are both defined, these products are not necessarily of the same dimension. And even if AB and BA are of the same dimensions, they need not be equal. About all we can say is that it is possible for AB and BA to be equal—but it takes some effort to find matrices A and B for which this is true.

Operational View of Matrix Multiplication

In the discussion of strictly determined games in Sec. 6-3, we noted, for example, that the multiplication of the row matrix $\mathbf{r} = [0 \quad 0 \quad 1 \quad 0]$ times a $4 \times m$ matrix B simply extracts the third row of B as the matrix product $\mathbf{r}B$.

Example:

$$[0 \quad 0 \quad 1 \quad 0]\begin{bmatrix} 2 & -1 & 5 \\ -7 & -2 & 9 \\ 3 & 1 & 2 \\ 4 & 0 & -8 \end{bmatrix} = [3 \quad 1 \quad 2].$$

Now that the addition of matrices and multiplication of matrices by numbers have been defined, we can generalize on this observation. Staying with the above example, let's replace the row matrix $[0 \quad 0 \quad 1 \quad 0]$ by the row matrix

$$A = [2 \quad 1 \quad 0 \quad 0].$$

We have

$$AB = [2 \quad 1 \quad 0 \quad 0]\begin{bmatrix} 2 & -1 & 5 \\ -7 & -2 & 9 \\ 3 & 1 & 2 \\ 4 & 0 & -8 \end{bmatrix} = [-3 \quad -4 \quad 19].$$

As the operations have been defined, we can interpret the product $[-3 \quad -4 \quad 19]$ as the sum of the rows of B multiplied, respectively, by the corresponding entries in A:

$$\begin{aligned} AB &= [2 \quad 1 \quad 0 \quad 0]\begin{bmatrix} 2 & -1 & 5 \\ -7 & -2 & 9 \\ 3 & 1 & 2 \\ 4 & 0 & -8 \end{bmatrix} \\ &= 2[2 \quad -1 \quad 5] + 1[-7 \quad -2 \quad 9] + 0[3 \quad 1 \quad 2] + 0[4 \quad 0 \quad -8] \\ &= 2[2 \quad -1 \quad 5] + 1[-7 \quad -2 \quad 9] \\ &= [4 \quad -2 \quad 10] + [-7 \quad -2 \quad 9] = [-3 \quad -4 \quad 19]. \end{aligned}$$

For a matrix product AB in which the left factor A has several rows we simply view the rows of A independently as determining the corresponding rows of AB as combinations of the rows of B. For instance, consider the product

$$AB = \begin{bmatrix} 1 & 0 & 1 \\ 0 & 2 & 0 \\ -1 & 0 & 2 \end{bmatrix}\begin{bmatrix} 0 & 3 \\ 1 & 0 \\ 1 & -2 \end{bmatrix}.$$

In this case the first row of AB is obtained by taking the entries in the first row of A as multipliers of the corresponding rows of B and adding. That is, the first row of AB is

$$1[0 \quad 3] + 0[1 \quad 0] + 1[1 \quad -2].$$

In simple terms, the first row of AB is the sum of the first and third rows of B.

$$\text{First Row of } AB\text{:} \quad [0 \quad 3] + [1 \quad -2] = [1 \quad 1].$$

Similarly, since the second row of A is $[0 \quad 2 \quad 0]$, it follows that the second row of AB is just 2 times the second row of B.

$$\text{Second Row of } AB\text{:} \quad 2[1 \quad 0] = [2 \quad 0].$$

And since the third row of A is $[-1 \quad 0 \quad 2]$, the third row of AB is the negative of the first row of B plus 2 times the third row of B.

$$\text{Third Row of } AB\text{:} \quad -[0 \quad 3] + 2[1 \quad -2] = [2 \quad -7].$$

Thus we have the product

$$AB = \begin{bmatrix} 1 & 0 & 1 \\ 0 & 2 & 0 \\ -1 & 0 & 2 \end{bmatrix} \begin{bmatrix} 0 & 3 \\ 1 & 0 \\ 1 & -2 \end{bmatrix} = \begin{bmatrix} 1 & 1 \\ 2 & 0 \\ 2 & -7 \end{bmatrix}.$$

Now let's look at the multiplication from the other side. Just as the rows of A determine the rows of AB as combinations of the rows of B, so also the columns of B determine the columns of AB as combinations of the columns of A. We stay with the example

$$AB = \begin{bmatrix} 1 & 0 & 1 \\ 0 & 2 & 0 \\ -1 & 0 & 2 \end{bmatrix} \begin{bmatrix} 0 & 3 \\ 1 & 0 \\ 1 & -2 \end{bmatrix}.$$

Since the first column of B is

$$\begin{bmatrix} 0 \\ 1 \\ 1 \end{bmatrix},$$

it follows that the first column of AB is just 0 times the first column of A plus 1 times the second column plus 1 times the third column:

$$0\begin{bmatrix} 1 \\ 0 \\ -1 \end{bmatrix} + 1\begin{bmatrix} 0 \\ 2 \\ 0 \end{bmatrix} + 1\begin{bmatrix} 1 \\ 0 \\ 2 \end{bmatrix}.$$

That is, the first column of AB is the sum of the second and third columns of A.

$$\text{First Column of } AB\text{:} \quad \begin{bmatrix} 0 \\ 2 \\ 0 \end{bmatrix} + \begin{bmatrix} 1 \\ 0 \\ 2 \end{bmatrix} = \begin{bmatrix} 1 \\ 2 \\ 2 \end{bmatrix}.$$

Similarly, since the second column of B is

$$\begin{bmatrix} 3 \\ 0 \\ -2 \end{bmatrix},$$

it follows that the second column of AB is 3 times the first column of A plus -2 times the third column of A.

Second Column of AB: $3\begin{bmatrix} 1 \\ 0 \\ -1 \end{bmatrix} - 2\begin{bmatrix} 1 \\ 0 \\ 2 \end{bmatrix} = \begin{bmatrix} 1 \\ 0 \\ -7 \end{bmatrix}.$

And again we have the product

$$AB = \begin{bmatrix} 1 & 0 & 1 \\ 0 & 2 & 0 \\ -1 & 0 & 2 \end{bmatrix}\begin{bmatrix} 0 & 3 \\ 1 & 0 \\ 1 & -2 \end{bmatrix} = \begin{bmatrix} 1 & 1 \\ 2 & 0 \\ 2 & -7 \end{bmatrix}.$$

6-7 EXERCISES

1. Let

$$A = \begin{bmatrix} -2 & 0 & 3 \\ 5 & -1 & -4 \end{bmatrix} \quad \text{and} \quad B = \begin{bmatrix} 1 & -3 & 7 \\ 0 & 4 & -1 \end{bmatrix}.$$

Find (a) $A + B$, (b) $A - B$, (c) $2A + 3B$.

2. Let

$$A = \begin{bmatrix} 3 & -2 \\ 1 & 0 \\ -2 & 5 \end{bmatrix} \quad \text{and} \quad B = \begin{bmatrix} 4 & -2 \\ 1 & 2 \\ 0 & 3 \end{bmatrix}.$$

Find (a) $A + B$, (b) $B - A$, (c) $3A - 2B$.

3. Let

$$A = \begin{bmatrix} 1 & -1 \\ 2 & 0 \end{bmatrix}, \quad B = \begin{bmatrix} 0 & 3 \\ 1 & -2 \end{bmatrix}, \quad C = \begin{bmatrix} 1 & 3 \\ 0 & -1 \end{bmatrix}.$$

Find AB, AC, and $A(B + C)$. And observe that $AB + AC = A(B + C)$.

4. Let

$$A = \begin{bmatrix} 3 & 0 & 2 \\ -1 & 2 & 0 \end{bmatrix}, \quad B = \begin{bmatrix} 0 & 4 & 1 \\ 2 & 0 & -1 \end{bmatrix}, \quad C = \begin{bmatrix} 0 & -2 \\ 1 & 0 \\ 3 & 1 \end{bmatrix}.$$

(a) Compute

$$(A + 2B)C \quad \text{and} \quad AC + 2BC$$

and observe that the resulting matrices are equal.

(b) Compute

$$C(A - B) \quad \text{and} \quad CA - CB$$

and again observe that the resulting matrices are equal.

5. Let

$$A = \begin{bmatrix} 3 & 0 & 2 \\ -1 & 2 & 0 \end{bmatrix}, \quad B = \begin{bmatrix} 1 & -2 \\ 4 & 2 \\ 0 & 3 \end{bmatrix}, \quad C = \begin{bmatrix} 1 & 3 \\ 0 & -1 \end{bmatrix}.$$

Compute AB and $(AB)C$; also compute BC and $A(BC)$; and observe that $(AB)C = A(BC)$.

6. With the matrices A, B, and C of Prob. 5, show that

$$(BC)A = B(CA).$$

7. Let

$$A = \begin{bmatrix} 5 & -2 \\ 4 & 3 \end{bmatrix}, \quad B = \begin{bmatrix} -1 & 0 \\ 2 & 7 \end{bmatrix}, \quad C = \begin{bmatrix} 5 & 2 \\ 3 & 1 \end{bmatrix}.$$

For $i = 1, 2$, let A_i be the row matrix realized by extracting the ith row of A, and for $j = 1, 2$, let C_j be the column matrix realized by extracting the jth column of C.

(a) Compute A_1BC_1, A_1BC_2, A_2BC_1, A_2BC_2.

(b) Compute ABC. And observe that for $i = 1, 2$ and $j = 1, 2$, the entry in row i and column j of ABC is the single entry in A_iBC_j.

8. Without computing the other entries, find the entry in row 2 and column 3 of the product

$$\begin{bmatrix} -9 & 13 & 6 \\ 7 & -4 & 2 \end{bmatrix} \begin{bmatrix} 1 & 0 \\ 2 & 1 \\ -1 & 3 \end{bmatrix} \begin{bmatrix} 10 & -6 & 3 & 17 \\ 7 & -9 & 5 & 0 \end{bmatrix}.$$

9. Let

$$A = \begin{bmatrix} 1 & 2 & -1 \\ -2 & 0 & 1 \end{bmatrix} \quad \text{and} \quad B = \begin{bmatrix} 3 & -1 \\ 1 & 2 \\ 0 & 1 \end{bmatrix}.$$

(a) Compute AB.

(b) Show that the first row of AB is given by the entries in the first row of A as multipliers in a combination of the rows of B as follows:

$$1[3 \quad -1] + 2[1 \quad 2] - 1[0 \quad 1].$$

(c) In the manner of part (b), interpret the second row of AB.

(d) Show that the first column of AB is given by the entries in the first column of B as multipliers in a combination of the columns of A as follows:

$$3\begin{bmatrix} 1 \\ -2 \end{bmatrix} + 1\begin{bmatrix} 2 \\ 0 \end{bmatrix} + 0\begin{bmatrix} -1 \\ 1 \end{bmatrix}.$$

(e) In the manner of part (d), interpret the second column of AB.

10. Let

$$A = \begin{bmatrix} 1 & 0 & 1 \\ 0 & 2 & 0 \end{bmatrix} \quad \text{and} \quad B = \begin{bmatrix} 0 & 0 \\ -1 & 3 \\ 1 & 0 \end{bmatrix}.$$

(a) In the spirit of the "operational perspective," recalled in Prob. 9 above, write the product AB immediately by regarding the rows of A as "operating" on the rows of B. Also, write AB by regarding the columns of B as "operating" on the columns of A.

(b) Follow the directions in part (a) for the product BA.

11. Consider the general 2×3 game with payoff matrix G and strategies $\mathbf{r}$ and $\mathbf{c}$:

$$\mathbf{r} = [r_1 \quad r_2], \qquad G = \begin{bmatrix} g_{11} & g_{12} & g_{13} \\ g_{21} & g_{22} & g_{23} \end{bmatrix}, \qquad \mathbf{c} = \begin{bmatrix} c_1 \\ c_2 \\ c_3 \end{bmatrix}.$$

(a) Show that

$$\mathbf{r}G = r_1[g_{11} \quad g_{12} \quad g_{13}] + r_2[g_{21} \quad g_{22} \quad g_{23}].$$

(b) Show that

$$G\mathbf{c} = c_1\begin{bmatrix} g_{11} \\ g_{21} \end{bmatrix} + c_2\begin{bmatrix} g_{12} \\ g_{22} \end{bmatrix} + c_3\begin{bmatrix} g_{13} \\ g_{23} \end{bmatrix}.$$

7

VECTORS

We've identified the problem of game theory, and we know how to verify solutions once they are found. But we are very limited in our ability to find solutions, and we've seen no evidence that in general a matrix game must have a solution. Considerable insight into these matters can be gained by forming a geometrical picture of the game and the dynamics of solving it. This carries us to the study of *vectors*.

In large measure our treatment of vectors, like the treatment of matrices thus far, will evolve as if the concept had been invented to serve one's intuition in analyzing games. But games form a very small part of the body of important problems served by these concepts.

7-1. The Vectors in $\mathbf{R}^2$

From the beginning we have treated the concept of real number as a primitive notion with which everyone is familiar. We introduce the symbol $\mathbf{R}$ to represent the set of real numbers. The practical criterion for the elements of the set **R** is the geometrical notion of the points on the number line, as shown in Fig. 7-1.

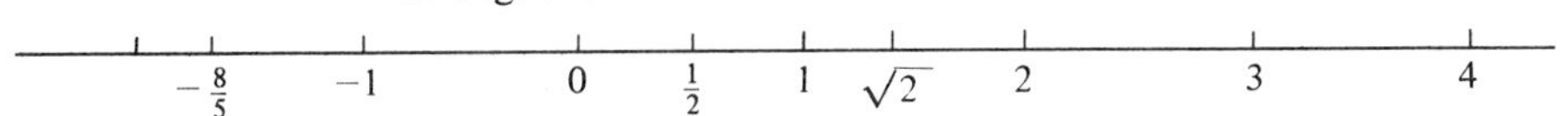

Figure 7-1 The Number Line **R**.

The experience with matrices has shown that it is sometimes fruitful to identify entities that are assembled from several numbers at once. The simplest structure of this variety is the ordered pair of numbers, which is conventionally represented in the form (x, y). These pairs are called *vectors*, and the set of ordered pairs of real numbers is designated by the symbol $\mathbf{R}^2$. Thus (3, 0), (3, 4), (0, 2), and $(-\frac{2}{3}, -1)$ are examples of vectors in $\mathbf{R}^2$.

If we rotate the number line of Fig. 7-1 counterclockwise through a right angle, then the original line together with the new position determines a plane. In a natural way the correspondence between the points on the lines and numbers generates a correspondence between the points in the plane and ordered pairs of numbers. We speak of the plane as the *coordinate plane* $\mathbf{R}^2$; it is depicted in Fig. 7-2, with several vectors plotted in their corresponding positions.

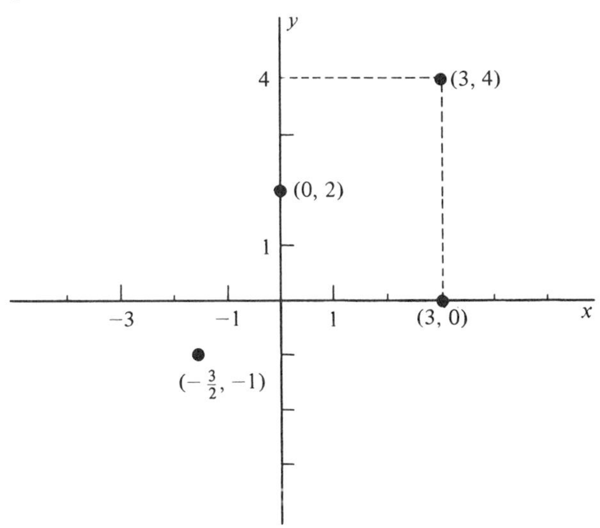

Figure 7-2 The Coordinate Plane $\mathbf{R}^2$.

The first number line, generally pictured horizontal with positive numbers to the right, is conventionally called the *x axis*; the other line, vertical and positive upwards, is called the *y axis*. The vector $(x, y) = (3, 4)$ is plotted as the point where the perpendicular from the point $x = 3$ on the x axis meets the perpendicular from the point $y = 4$ on the y axis. In like manner each vector (x, y) in $\mathbf{R}^2$ is associated with a unique point in the plane and vice versa. The entries x and y in the vector are called the *coordinates* of the point.

Comment: Although the discussion of vectors is dependent on the pictures we draw to represent them, the pictures are not part of the purely mathematical structure. Mathematically, the common perception of points, lines, and planes motivates the formal structure of geometry and assists in understanding and anticipating results, but, like empirical situations, the pictures exist on a level apart. For us, geometry will remain on the informal intuitive level, serving only as a visual aid.

Operations with Vectors

The vectors in $\mathbf{R}^2$ have been identified as ordered pairs of real numbers. Strictly speaking, the designation "vector" should carry with it the capacity to perform two operations: *addition* and *scalar multiplication.*

We add two vectors exactly as one would guess. The sum of the vectors is simply the vector whose coordinates are respectively the sums of the corresponding coordinates. That is,

$$(x_1, y_1) + (x_2, y_2) = (x_1 + x_2, y_1 + y_2).$$

For example,

$$(0, 2) + (-\tfrac{3}{2}, -1) = (-\tfrac{3}{2}, 1),$$
$$(3, 4) + (5, -2) = (8, 2).$$

In the present context, "scalar" is just a fancy word for "real number." The operation of scalar multiplication associates a real number t and a vector (x, y) with the vector (tx, ty), realized by multiplying each coordinate of (x, y) by the factor t:

$$t(x, y) = (tx, ty).$$

For example,

$$\frac{5}{2}(3, 4) = \left(\frac{15}{2}, 10\right),$$
$$-2(5, -2) = (-10, 4).$$

These two operations are reminiscent of the corresponding operations with matrices. One is correct in anticipating that there is a close relation between vectors in $\mathbf{R}^2$ and 1×2 row matrices or 2×1 column matrices. We employ symbols of the form $\mathbf{u}$ and $\mathbf{v}$ to represent vectors; the typographical similarity to $\mathbf{r}$ and $\mathbf{c}$ is intentional.

We have introduced vectors to assist in the study of matrices. (This is a reversal of the roles as they usually evolve in mathematical expositions.) Our empirical problems will always be given matrix models. And formal results will be stated in terms of matrices. We shall not treat vectors with the formality accorded matrices in the previous chapter, so the discussion may sound a little casual at times.

Geometrical Interpretations

The operations of addition and scalar multiplication are easy to interpret in the geometrical representation. To illustrate, we consider the sum

$$(3, 4) + (5, -2) = (8, 2).$$

Figure 7-3 displays the point (3, 4), together with the triangle determined by the *origin*, i.e., the point (0, 0), the point (5, 0) above (5, −2) on the x axis, and (5, −2) itself. As the figure shows, the sum (8, 2) = (3, 4) + (5, −2) is reached by shifting the triangle associated with (5, −2), without distortion or change of direction, so that the vertex originally at the origin is carried to the point (3, 4); the vertex originally at (5, −2) now lies at the point (8, 2).

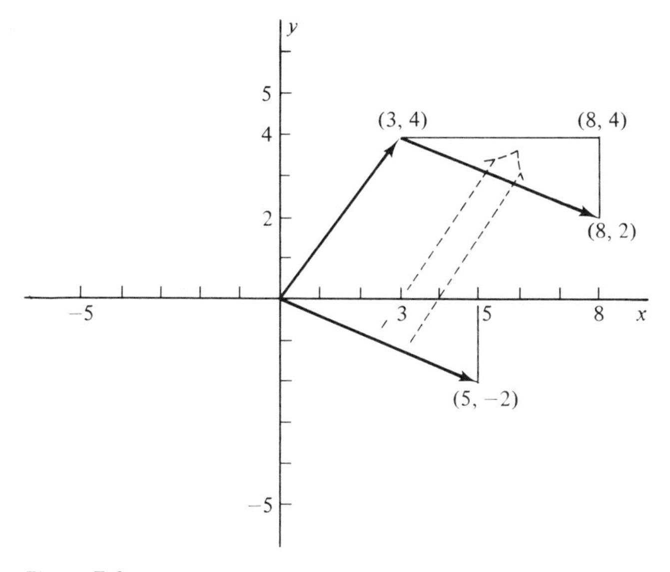

Figure 7-3.

To describe the construction in another way, we associate with the point (3, 4) the directed line segment from the origin to (3, 4), and similarly for the point (5, −2). These "rays" are shown in Fig. 7-3. To arrive at the sum (8, 2) = (3, 4) + (5, −2), the (5, −2) ray is moved, without change in length or direction, until its base coincides with (3, 4); its tip now lies at (8, 2).

In general, a vector $\mathbf{v} = (x, y)$ in $\mathbf{R}^2$ is identified not only with a particular point in the coordinate plane but also with the ray from the origin to that point, which ray continues to represent the vector if it is moved without change of length or direction. The geometrical construction for the sum of two vectors can be described as follows:

Given two vectors, $\mathbf{u}$ and $\mathbf{v}$, we reach the sum $\mathbf{u} + \mathbf{v}$ by moving the "$\mathbf{v}$ ray" until its base coincides with the point $\mathbf{u}$. The tip of the displaced "$\mathbf{v}$ ray" is then at the point $\mathbf{u} + \mathbf{v}$. See Fig. 7-4.

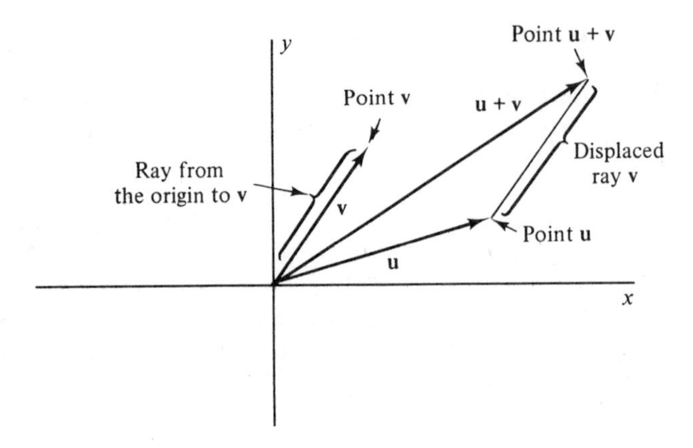

Figure 7-4.

Comment: As has been observed, there are several geometrical interpretations of a vector $\mathbf{v} = (x, y)$ in $\mathbf{R}^2$. The primary interpretation is as a point in the coordinate plane. But it is often helpful to envision the vector as the ray from the origin to that point, or as any ray of the same length and direction.

The earliest historical identification of the vector concept corresponds to that of a directed line segment free to move without change of length or direction. In elementary physics a "vector" is identified as that which has magnitude and direction, such as force or velocity. Geometrically, such phenomena may be represented as directed line segments, whether in the plane or in three-dimensional space.

Scalar multiplication of a vector by a real number is also easy to interpret. In Fig. 7-5 are displayed several vectors of the form $t(5, -2)$.

From the interpretation of addition we reach the point

$$2(5, -2) = (10, -4) = (5, -2) + (5, -2)$$

by proceeding from the origin along a succession of two rays. Both of these

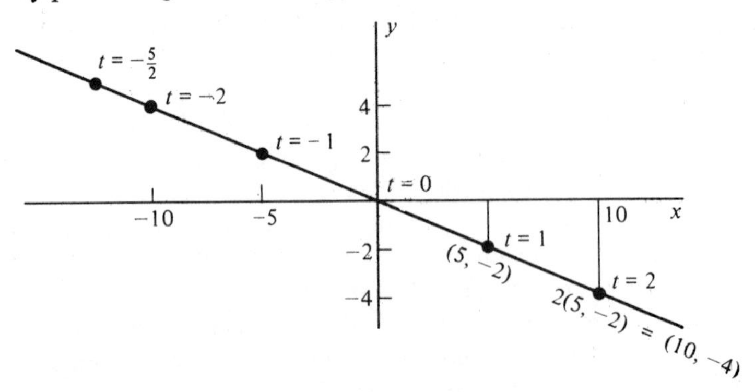

Figure 7-5 Points of the Form t (5, −2).

rays correspond to the vector $(5, -2)$; thus they have the same length and direction. It follows that the point $(10, -4) = 2(5, -2)$ is on the line from the origin in the direction of the point $(5, -2)$, with $(10, -4)$ being twice as far from the origin as is $(5, -2)$.

From a slightly different point of view, the point $(10, -4)$ is reached by doubling all three sides of the triangle determined by the origin $(0, 0)$, the point $(5, 0)$, and the point $(5, -2)$. Of course, this construction also places $(10, -4)$ on the line from the origin to $(5, -2)$, twice as far out.

A generalization of the argument places all points of the form $t(5, -2)$ on the line determined by the origin and $(5, -2)$. Conversely, each point on this line is of the form $t(5, -2)$ for some value of t. For instance, $(0, 0)$ corresponds to $t = 0$, $(5, -2)$ to $t = 1$, $(10, -4)$ to $t = 2$, and $(-5, 2)$ to $t = -1$.

As suggested in Fig. 7-5, these "t values" determine a uniform scale of numbers along the line; that is, relative to each other, the numbers are positioned as on the primitive number line of Fig. 7-1. The variable t is called the *coefficient* of the point $(5, -2)$; it is also identified as the *parameter* determining the line.

The same argument applies to any point (x_0, y_0), where $(x_0, y_0) \neq (0, 0)$. That is, the vectors of the form $t(x_0, y_0)$ correspond to the points on the line determined by $(0, 0)$ and (x_0, y_0). Furthermore, the values of the coefficient t determine a uniform number scale on this line with $t = 0$ at $(0, 0)$ and $t = 1$ at (x_0, y_0).

Of course, the only vector of the form $t(0, 0)$ is $(0, 0)$ itself.

7-1 EXERCISES

1. Compute the following vectors in $\mathbf{R}^2$.
(a) $(1, -3) + (-2, 5)$.
(b) $(-4)(1.5, -2.2)$.
(c) $2(-3, 2) + 5(1, -7)$.

2. (a) Let

$$\mathbf{u}_1 = (2, -1), \quad \mathbf{u}_2 = (4, -2), \quad \text{and} \quad \mathbf{u}_3 = (0, 3).$$

Compute

$$\mathbf{u}_1 + (\mathbf{u}_2 + \mathbf{u}_3) \quad \text{and} \quad (\mathbf{u}_1 + \mathbf{u}_2) + \mathbf{u}_3,$$

and observe that the resulting sums are equal.

(b) Consider the general case

$$\mathbf{u}_1 = (x_1, y_1), \qquad \mathbf{u}_2 = (x_2, y_2), \qquad \mathbf{u}_3 = (x_3, y_3).$$

Show that

$$\mathbf{u}_1 + (\mathbf{u}_2 + \mathbf{u}_3) = (\mathbf{u}_1 + \mathbf{u}_2) + \mathbf{u}_3.$$

(This result demonstrates the *associative law* of vector addition.

Unambiguously we express the sum of the three vectors as

$$\mathbf{u}_1 + \mathbf{u}_2 + \mathbf{u}_3,$$

and similarly for the sum of any number of vectors.)

3. Plot the following vectors in the coordinate plane $\mathbf{R}^2$: $(1, -3)$, $(-2, 5)$, $(2.5, 3.4)$, $(-1.5, -3.75)$.

4. (a) In the coordinate plane $\mathbf{R}^2$, plot the following three vectors:

$$\mathbf{u} = (3, -1), \quad \mathbf{v} = (-1, 5), \quad \text{and} \quad \mathbf{u} + \mathbf{v} = (2, 4).$$

(b) On the graph of part (a), construct the rays from the origin $(0, 0)$ to the points $\mathbf{u}$ and $\mathbf{v}$, respectively. Also construct the ray from $\mathbf{u}$ to $\mathbf{u} + \mathbf{v}$, and observe that it is equal in length and parallel to the ray from $(0, 0)$ to $\mathbf{v}$.

5. Consider the vectors

$$\mathbf{u} = (2, 1), \qquad \mathbf{v} = (-3, 3), \qquad \mathbf{w} = (-2, -3).$$

(a) Plot the vectors $\mathbf{u}$, $\mathbf{v}$, and $\mathbf{w}$ in the coordinate plane, and construct the rays from the origin to the respective points.

(b) Plot the vector $\mathbf{u} + \mathbf{v}$ on the graph of part (a), and construct the ray from $\mathbf{u}$ to $\mathbf{u} + \mathbf{v}$.

(c) Plot the vector $\mathbf{u} + \mathbf{v} + \mathbf{w}$ on the graph of part (a), and construct the ray from $\mathbf{u} + \mathbf{v}$ to $\mathbf{u} + \mathbf{v} + \mathbf{w}$.

(d) In terms of the rays representing $\mathbf{u}$, $\mathbf{v}$, and $\mathbf{w}$, respectively, what is the geometrical interpretation of the triple sum $\mathbf{u} + \mathbf{v} + \mathbf{w}$?

6. In the coordinate plane, plot the position of the vectors of the form $t(4, 3)$ for the following values of t: $t = -2$, $t = \frac{3}{4}$, $t = 0$, $t = 1$, $t = \frac{3}{2}$. Construct the line containing these points.

7. In the coordinate plane, construct the line of points of the form $t\mathbf{v}$ for the following vectors $\mathbf{v}$:

(a) $\mathbf{v} = (3, -3)$.

(b) $\mathbf{v} = (-2, 5)$.

(c) $\mathbf{v} = (2, 1)$.

8. Plot the vectors of the form

$$(4, -1) + t(1, 2)$$

for the following values of t: $t = 0$, $t = 1$, $t = 2$, $t = -1$, $t = -2.5$. Observe that these vectors lie on a line. What general result does this construction suggest?

7-2. Lines and Polygons

We've introduced the two vector operations, addition and scalar multiplication, and we've seen their geometrical interpretations. Now let's combine the operations. We consider

$$(3, 4) + t(5, -2).$$

As we saw in Fig. 7-3, the sum $(3, 4) + (5, -2)$ is reached by taking the ray from the origin to the point $(5, -2)$ and moving it 3 units to the right and up 4 units. And in Fig. 7-5 we saw that the points of the form $t(5, -2)$ comprise the line determined by $(0, 0)$ and $(5, -2)$, with the "t values" constituting a uniform scale, $t = 0$ at $(0, 0)$ and $t = 1$ at $(5, -2)$.

Putting the two constructions together, it should be clear that points of the form

$$(3, 4) + t(5, -2)$$

are realized by moving the entire line of points $t(5, -2)$ to the right 3 units and up 4 units. After this shift the "t values," which have been carried along with the line, still constitute a uniform scale; $t = 0$ now falls at $(3, 4)$ and $t = 1$ at $(8, 2)$. This construction is displayed in Fig. 7-6.

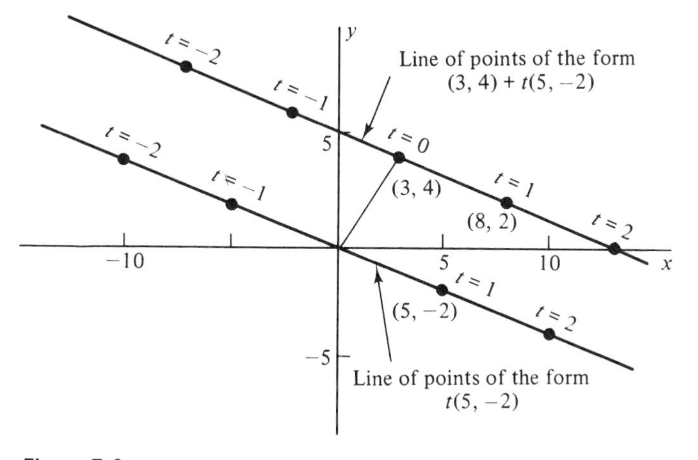

Figure 7-6.

We see that the line determined by the points $(3, 4)$ and $(8, 2)$ is the set of points of the form

$$(x, y) = (3, 4) + t(5, -2).$$

Since $(3, 4)$ and $(8, 2)$ are the points of interest, we remove $(5, -2)$ from the characterization by recalling that

$$(3, 4) + (5, -2) = (8, 2).$$

Thus with the obvious interpretation of subtraction,

$$(5, -2) = (8, 2) - (3, 4).$$

The line of points through $(3, 4)$ and $(8, 2)$ can thus be identified as the set of points (x, y) of the form

$$(x, y) = (3, 4) + t[(8, 2) - (3, 4)].$$

The similarity between the operations on vectors and the operations on numbers justifies the following manipulations of the equation representing the line:

$$\begin{aligned}(x, y) &= (3, 4) + t[(8, 2) - (3, 4)] \\ &= (3, 4) + t(8, 2) - t(3, 4) \\ &= (1 - t)(3, 4) + t(8, 2).\end{aligned}$$

The argument applies for any two points in $\mathbf{R}^2$.

Given two vectors $\mathbf{u}$ and $\mathbf{v}$ in $\mathbf{R}^2$, the vectors (x, y) on the line passing through the points $\mathbf{u}$ and $\mathbf{v}$ are precisely those of the form

$$(x, y) = (1 - t)\mathbf{u} + t\mathbf{v}.$$

Furthermore, the values of the parameter t determine a uniform number scale on this line with $t = 0$ at $\mathbf{u}$ and $t = 1$ at $\mathbf{v}$.

Example: (a) Characterize the vectors in the coordinate plane which lie on the line determined by the points

$$\mathbf{u} = (10, -7) \quad \text{and} \quad \mathbf{v} = (-6, 1).$$

(b) Find the point on the line in part (a) which is one-fourth of the way from $\mathbf{u}$ to $\mathbf{v}$.

Solution: (a) The points on the line determined by $\mathbf{u}$ and $\mathbf{v}$ are those of the form

$$\begin{aligned}(x, y) &= (1 - t)\mathbf{u} + t\mathbf{v} \\ &= (1 - t)(10, -7) + t(-6, 1),\end{aligned}$$

where t is a real number.

(b) Since the value of the parameter t determines a uniform scale on the line, with $t = 0$ at $\mathbf{u}$ and $t = 1$ at $\mathbf{v}$, the desired point corresponds to the value $t = \frac{1}{4}$. We have

$$\begin{aligned}(1 - t)\mathbf{u} + t\mathbf{v} &= (1 - \tfrac{1}{4})(10, -7) + \tfrac{1}{4}(-6, 1) \\ &= \tfrac{3}{4}(10, -7) + \tfrac{1}{4}(-6, 1) \\ &= (\tfrac{30}{4}, -\tfrac{21}{4}) + (-\tfrac{6}{4}, \tfrac{1}{4}) \\ &= (\tfrac{24}{4}, -\tfrac{20}{4}) = (6, -5).\end{aligned}$$

Thus $(6, -5)$ is one-fourth of the way from $\mathbf{u} = (10, -7)$ to $\mathbf{v} = (-6, 1)$. See Fig. 7-7.

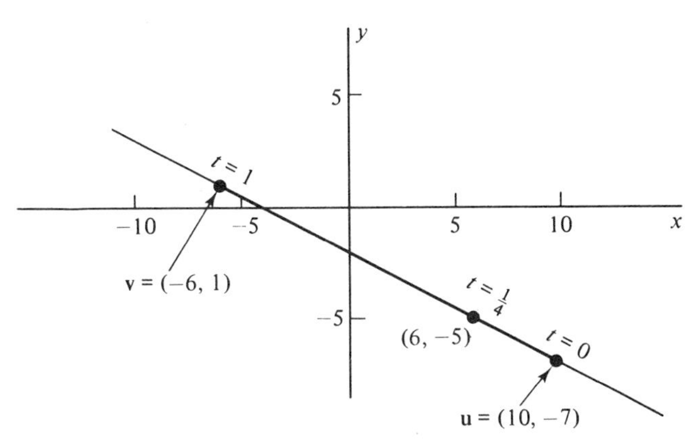

Figure 7-7.

Convex Polygons

We have seen that the line determined by the points (3, 4) and (8, 2) consists of all points of the form

$$(x, y) = (1 - t)(3, 4) + t(8, 2).$$

Furthermore, the values of the parameter t constitute a uniform scale on this line with $t = 0$ at (3, 4) and $t = 1$ at (8, 2). In particular, if t is between 0 and 1, inclusive, then the point (x, y) is on the *line segment* with end points at (3, 4) and (8, 2).

An equivalent characterization of the line segment can be realized by setting

$$c_1 = (1 - t) \quad \text{and} \quad c_2 = t.$$

Of course, $c_1 + c_2 = 1$, and the requirement $0 \leq t \leq 1$ is equivalent to requiring that c_1 and c_2 both be nonnegative, It follows that the line segment with end points (3, 4) and (8, 2) is the set of points of the form

$$(x, y) = c_1(3, 4) + c_2(8, 2),$$

where c_1 and c_2 are nonnegative numbers totaling one.

> **Comment:** As the language "nonnegative numbers totaling one" should suggest, we are laying ground for a probabilistic interpretation. Probability reenters in the next section as we return to the analysis of matrix games.

For example, consider $c_1 = c_2 = \frac{1}{2}$. With the aforementioned vectors we have

$$(x, y) = \tfrac{1}{2}(3, 4) + \tfrac{1}{2}(8, 2) = (5.5, 3).$$

As shown in Fig. 7-8, this point is the midpoint of the line segment determined by (3, 4) and (8, 2).

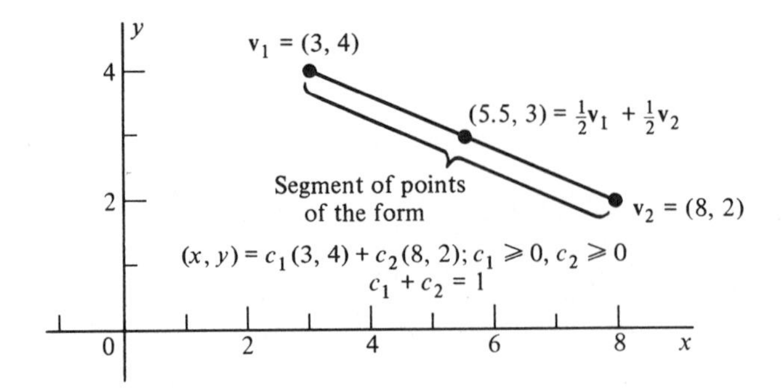

Figure 7-8.

The line segment between any two points can be characterized in this manner.

Given two vectors $\mathbf{v}_1$ and $\mathbf{v}_2$ in $\mathbf{R}^2$, the line segment with end points $\mathbf{v}_1$ and $\mathbf{v}_2$ is the set of all vectors of the form

$$\mathbf{v} = c_1\mathbf{v}_1 + c_2\mathbf{v}_2,$$

where c_1 and c_2 are nonnegative numbers totaling one.

Now let's progress from a combination of two vectors to a combination of three vectors, $\mathbf{v}_1$, $\mathbf{v}_2$, and $\mathbf{v}_3$. Our concern is with the geometrical interpretation of the set of vectors of the form

$$\mathbf{v} = c_1\mathbf{v}_1 + c_2\mathbf{v}_2 + c_3\mathbf{v}_3, \tag{1}$$

where c_1, c_2, and c_3 are nonnegative numbers totaling one. The pleasing result is that this set is precisely the set of points in, or on the boundary of, the triangle with vertices $\mathbf{v}_1$, $\mathbf{v}_2$, and $\mathbf{v}_3$. Let's see why this is true.

In Fig. 7-9 we depict a triangle with particular points $\mathbf{v}_1$, $\mathbf{v}_2$, and $\mathbf{v}_3$ as vertices. However, the argument that follows is valid for any set of three vectors as the vertices.

As the figure indicates, we can identify a point $\mathbf{v}$ as belonging to the triangle if $\mathbf{v}$ is on a line segment with end points $\mathbf{u}$ and $\mathbf{v}_3$, where $\mathbf{u}$ itself is on the line segment with end points $\mathbf{v}_1$ and $\mathbf{v}_2$.

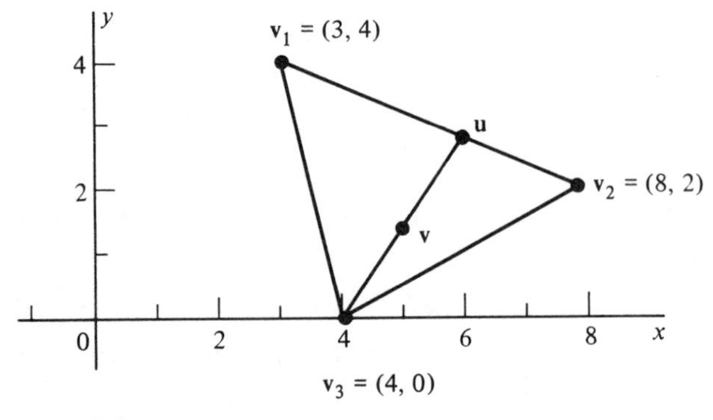

Figure 7-9.

For example, a point $\mathbf{v}$ of the triangle is obtained by setting

$$\mathbf{u} = .4\mathbf{v}_1 + .6\mathbf{v}_2 \tag{2}$$

and

$$\mathbf{v} = .5\mathbf{u} + .5\mathbf{v}_3. \tag{3}$$

Combining the two equations,

$$\begin{aligned} \mathbf{v} &= .5(.4\mathbf{v}_1 + .6\mathbf{v}_2) + .5\mathbf{v}_3 \\ &= .2\mathbf{v}_1 + .3\mathbf{v}_2 + .5\mathbf{v}_3 \\ &= c_1\mathbf{v}_1 + c_2\mathbf{v}_2 + c_3\mathbf{v}_3, \end{aligned} \tag{4}$$

where

$$c_1 = .2, \quad c_2 = .3, \quad \text{and} \quad c_3 = .5$$

are nonnegative numbers totaling one. Thus $\mathbf{v}$ is of the form of Eq. (1).

On the other hand, suppose we start with a vector $\mathbf{v}$ of the form of Eq. (1). Consider

$$\mathbf{v} = \tfrac{1}{6}\mathbf{v}_1 + \tfrac{2}{6}\mathbf{v}_2 + \tfrac{3}{6}\mathbf{v}_3$$

Here, we observe that $\mathbf{v}$ may be written in the form

$$\mathbf{v} = \tfrac{3}{6}(\tfrac{1}{3}\mathbf{v}_1 + \tfrac{2}{3}\mathbf{v}_2) + \tfrac{3}{6}\mathbf{v}_3.$$

In other words,

$$\mathbf{v} = \tfrac{3}{6}\mathbf{u} + \tfrac{3}{6}\mathbf{v}_3 = \tfrac{1}{2}\mathbf{u} + \tfrac{1}{2}\mathbf{v}_3, \tag{5}$$

where

$$\mathbf{u} = \tfrac{1}{3}\mathbf{v}_1 + \tfrac{2}{3}\mathbf{v}_2.$$

We see that $\mathbf{u}$ is on the line segment with end points $\mathbf{v}_1$ and $\mathbf{v}_2$. From Eq. (5), $\mathbf{v}$ is on the line segment with end points $\mathbf{u}$ and $\mathbf{v}_3$. This, of course, is precisely the construction that locates $\mathbf{v}$ as a point of the triangle with vertices $\mathbf{v}_1$, $\mathbf{v}_2$, and $\mathbf{v}_3$.

For the general three-point case, let

$$\mathbf{v} = c_1\mathbf{v}_1 + c_2\mathbf{v}_2 + c_3\mathbf{v}_3, \tag{6}$$

where the c_i are positive and total one. We group the first two terms as follows:

$$\mathbf{v} = (c_1 + c_2)\left\{\frac{c_1}{c_1 + c_2}\mathbf{v}_1 + \frac{c_2}{c_1 + c_2}\mathbf{v}_2\right\} + c_3\mathbf{v}_3. \tag{7}$$

Here

$$\mathbf{u} = \frac{c_1}{c_1 + c_2}\mathbf{v}_1 + \frac{c_2}{c_1 + c_2}\mathbf{v}_2 \tag{8}$$

is on the line segment with end points $\mathbf{v}_1$ and $\mathbf{v}_2$. From Eqs. (7) and (8)

$$\mathbf{v} = (c_1 + c_2)\mathbf{u} + c_3\mathbf{v}_3. \tag{9}$$

Thus $\mathbf{v}$ is on the line segment with end points $\mathbf{u}$ and $\mathbf{v}_3$; therefore, $\mathbf{v}$ is in the triangle with vertices $\mathbf{v}_1$, $\mathbf{v}_2$, and $\mathbf{v}_3$.

In the other direction, we start with a general point $\mathbf{v}$ of the triangle, identified by a generalization of Eqs. (2) and (3), and we imitate the steps leading from Eqs. (2) and (3) to Eq. (4) to show that $\mathbf{v}$ is of the form given in

Eqs. (1) and (6). In effect, we argue backwards from Eq. (9) to Eq. (6). The conclusion is that points $\mathbf{v}$ of the triangle are of the form given in Eq. (1) and vice versa.

The "four-point" argument is concerned with the geometrical character of the set of points of the form

$$\mathbf{v} = c_1\mathbf{v}_1 + c_2\mathbf{v}_2 + c_3\mathbf{v}_3 + c_4\mathbf{v}_4, \tag{10}$$

where the $\mathbf{v}_i$ are fixed points and the c_i are nonnegative numbers totaling one. Shown in Fig. 7-10 is a quadrilateral with particular vertices $\mathbf{v}_1$, $\mathbf{v}_2$, $\mathbf{v}_3$, and $\mathbf{v}_4$.

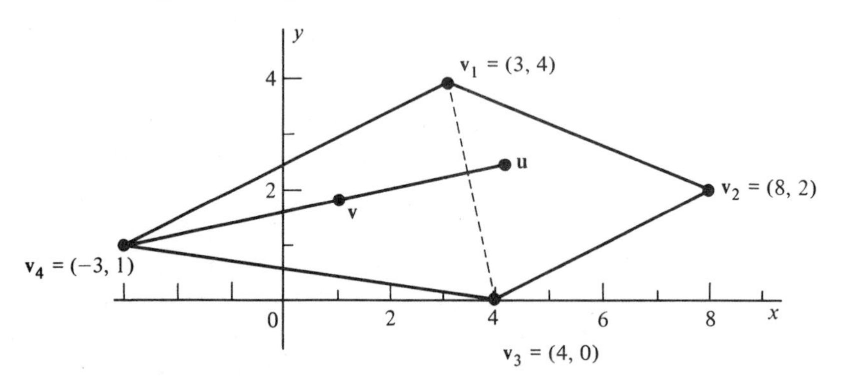

Figure 7-10.

Similar to the previous case, a point $\mathbf{v}$ of the quadrilateral is identified by the fact that it is on a line segment with end points $\mathbf{u}$ and $\mathbf{v}_4$, where $\mathbf{u}$ is a point of the triangle with vertices $\mathbf{v}_1$, $\mathbf{v}_2$, and $\mathbf{v}_3$, as indicated in the figure. An extension of the previous argument establishes that the set of such points $\mathbf{v}$ is the set of vectors satisfying Eq. (10).

In using the word "quadrilateral" we have been assuming that the point $\mathbf{v}_4$ does not itself belong to the triangle with vertices $\mathbf{v}_1$, $\mathbf{v}_2$, and $\mathbf{v}_3$. But suppose $\mathbf{v}_4$ does belong to this triangle; then the criterion above identifying the "quadrilateral" simply identifies again the points belonging to the triangle with vertices $\mathbf{v}_1$, $\mathbf{v}_2$, and $\mathbf{v}_3$.

The argument can be extended to an arbitrary number of points. Typical of the general case is the polygon shown in Fig. 7-11. Here we have indicated ten points, $\mathbf{v}_1, \mathbf{v}_2, \ldots, \mathbf{v}_{10}$. The construction of the polygon can be visualized as if the ten points had been completely enclosed by a stretched rubber band that has then contracted as tightly as possible about posts mounted at the points.

The polygon of Fig. 7-11 is identified as the *convex polygon* generated by the points $\mathbf{v}_1, \mathbf{v}_2, \ldots, \mathbf{v}_{10}$. In general, the convex polygon generated by points $\mathbf{v}_1, \mathbf{v}_2, \ldots, \mathbf{v}_n$ is determined by the following three properties:

1. Each of the generating points $\mathbf{v}_i$ is a point of the polygon. (Here, we take "of the polygon" to mean the following: in the set consisting of the boundary and interior of the polygon.)

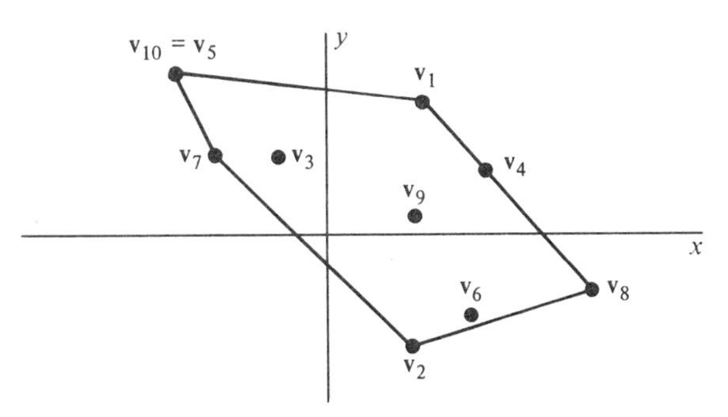

Figure 7-11.

2. Each point on the boundary of the polygon is on a line segment whose end points are in the generating set $\{\mathbf{v}_1, \mathbf{v}_2, \ldots, \mathbf{v}_n\}$.
3. The polygon is *convex*. This means that if **u** and **v** are any two points of the polygon, then the line segment with end points **u** and **v** consists entirely of points of the polygon. In short, a line cannot leave and then reenter the polygon.

It should be noted that this general characterization does not exclude cases in which the "polygon" is no more than a line segment. In fact, if all the $\mathbf{v}_i$ are one and the same, the "polygon" degenerates to a single point.

It can be shown that the points of the convex polygon generated by the points $\mathbf{v}_1, \mathbf{v}_2, \ldots, \mathbf{v}_n$ are precisely those which can be written in the form

$$\mathbf{v} = c_1\mathbf{v}_1 + c_2\mathbf{v}_2 + \cdots + c_n\mathbf{v}_n,$$

where the c_i are nonnegative numbers totaling one.

Example: Consider the following set of five vectors:

$$\mathbf{v}_1 = (0, 2), \quad \mathbf{v}_2 = (-3, 3), \quad \mathbf{v}_3 = (1, 0),$$
$$\mathbf{v}_4 = (-2, 1), \quad \mathbf{v}_5 = (5, -3).$$

(a) In the coordinate plane $\mathbf{R}^2$, construct the convex polygon generated by this set.

(b)–(e) For c_1, c_2, c_3, c_4, and c_5 as given, find

$$\mathbf{v} = c_1\mathbf{v}_1 + c_2\mathbf{v}_2 + c_3\mathbf{v}_3 + c_4\mathbf{v}_4 + c_5\mathbf{v}_5$$

and plot these points in the construction of part (a):

(b) $c_1 = 0, c_2 = 0, c_3 = 0, c_4 = .5, c_5 = .5.$
(c) $c_1 = .2, c_2 = .2, c_3 = .2, c_4 = .2, c_5 = .2.$
(d) $c_1 = 0, c_2 = 1, c_3 = 0, c_4 = 0, c_5 = 0.$
(e) $c_1 = .3, c_2 = .4, c_3 = 0, c_4 = .3, c_5 = 0.$

Solution: (a) The convex polygon generated by the given vectors is the quadrilateral shown in Fig. 7-12.

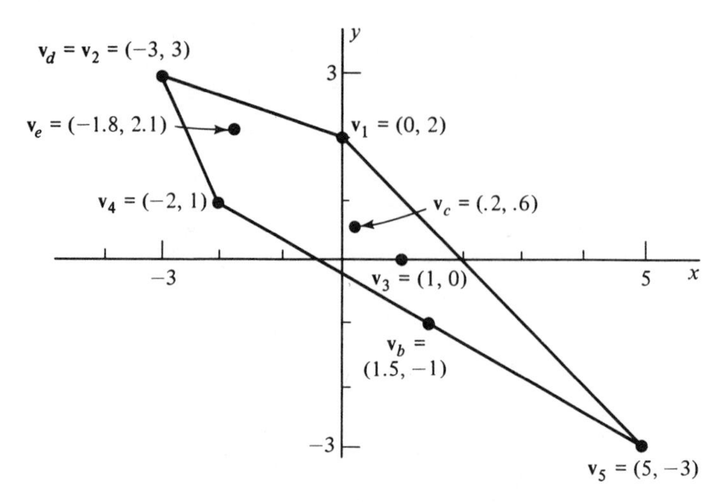

Figure 7-12.

(b) $\mathbf{v}_b = .5\mathbf{v}_4 + .5\mathbf{v}_5 = .5(-2, 1) + .5(5, -3) = (1.5, -1)$.

(c) $\mathbf{v}_c = .2(0, 2) + .2(-3, 3) + .2(1, 0) + .2(-2, 1) + .2(5, -3) = (.2, .6)$.

(d) $\mathbf{v}_d = \mathbf{v}_2 = (-3, 3)$.

(e) $\mathbf{v}_e = .3(0, 2) + .4(-3, 3) + .3(-2, 1) = (-1.8, 2.1)$.

These points are shown in Fig. 7-12.

7-2 EXERCISES

1. In the coordinate plane $\mathbf{R}^2$, plot the line of points of the form

$$(x, y) = (1 - t)(3, 2) + t(-2, 1).$$

On this line label the points corresponding to $t = -1$, $t = 0$, $t = \frac{1}{2}$, $t = 1$, $t = 2$.

2. Let $\mathbf{u} = (5, -3)$ and $\mathbf{v} = (-7, 3)$.

(a) Graph the line segment

$$(x, y) = (1 - t)(5, -3) + t(-7, 3)$$

for $0 \leq t \leq 1$.

(b) Find the coordinates of the point midway between $\mathbf{u}$ and $\mathbf{v}$.

(c) Find the coordinates of the point two-thirds of the way from $\mathbf{u}$ to $\mathbf{v}$. Show this point on the graph in part (a).

3. (a) Graph the line segment with end points

$$\mathbf{u} = (-4, -1) \quad \text{and} \quad \mathbf{v} = (6, 4).$$

(b) Show that the point (2, 2) is on the line segment in part (a) by finding c_1 and c_2 such that

$$(2, 2) = c_1(-4, -1) + c_2(6, 4),$$

where $c_1 \geq 0$, $c_2 \geq 0$, and $c_1 + c_2 = 1$.

4. In the coordinate plane, sketch the set of vectors of the form

$$(x, y) = c_1(5, -4) + c_2(-3, 2),$$

where c_1 and c_2 are nonnegative numbers totaling one.

5. (a) In the coordinate plane, indicate the set of all vectors of the form

$$(x, y) = c_1(5, -4) + c_2(-3, 2) + c_3(1, 1),$$

where the c_i are nonnegative numbers totaling one.

(b) On the graph in part (a), show the points corresponding to the following values of the c_i:

(b_1) $c_1 = 0$, $c_2 = 1$, $c_3 = 0$.
(b_2) $c_1 = \frac{1}{3}$, $c_2 = \frac{2}{3}$, $c_3 = 0$.
(b_3) $c_1 = .4$, $c_2 = .3$, $c_3 = .3$.

6. Let

$$\mathbf{u}_1 = (-3, -2), \quad \mathbf{u}_2 = (2, 0), \quad \text{and} \quad \mathbf{u}_3 = (-1, 2).$$

(a) In the coordinate plane, plot the set of all vectors of the form

$$\mathbf{v} = c_1\mathbf{u}_1 + c_2\mathbf{u}_2 + c_3\mathbf{u}_3,$$

where the c_i are nonnegative numbers totaling one.

(b) On the graph in part (a), show the point corresponding to

$$c_1 = c_2 = c_3 = \tfrac{1}{3}.$$

(c) Give an argument to establish that, relative to the triangle in part (a), the point in part (b) is two-thirds of the way from each vertex to the midpoint of the opposite side.

7. In the coordinate plane, plot the convex polygon generated by the points

$$\mathbf{v}_1 = (-4, -1), \qquad \mathbf{v}_2 = (-1, 1), \qquad \mathbf{v}_3 = (0, 3), \qquad \mathbf{v}_4 = (3, -2).$$

8. (a) Plot the convex polygon consisting of points of the form

$$(x, y) = c_1(-3, 1) + c_2(4, 2) + c_3(-1, 4) + c_4(2, -3),$$

where the c_i are nonnegative numbers totaling one.

(b) On the graph in part (a), show the points corresponding to the following values of the c_i:

(b_1) $c_1 = .5$, $c_2 = ,5$. $c_3 = 0$, $c_4 = 0$.
(b_2) $c_1 = \frac{1}{3}$, $c_2 = \frac{1}{3}$, $c_3 = 0$, $c_4 = \frac{1}{3}$.
(b_3) $c_1 = c_2 = c_3 = c_4 = .25$.

9. Plot the convex polygon generated by the following set of points:

$$\{(1, 2), (-2, 1), (2, -1), (2, 1), (5, 2), (3, 0), (0, 4)\}.$$

10. Set an appropriate scale for the coordinate axes of the plane $\mathbf{R}^2$, and plot the convex polygon generated by the following set of points:

$$\{(-5, 5), (-6, 12), (16, -18), (6, -5), (-21, 18)\}.$$

11. Setting a suitable scale for the axes and displaying the appropriate region in the coordinate plane, construct a prominent picture of the convex polygon generated by the following set:

$$\{(4.8, 6.1), (5.0, 6.0), (5.1, 5.7), (5.0, 5.9)\}.$$

7-3. The Analysis of 2 × n Games

To put the geometrical perspective to work, we return to the analysis of matrix games. In the present section we shall develop a practical procedure for solving games with $2 \times n$ payoff matrices, though it will take a little longer to reach a full appreciation of why the procedure works.

For purposes of illustration we consider the game G corresponding to the following 2×5 payoff matrix:

$$G = \begin{bmatrix} -5 & 3 & -1 & 1 & -3 \\ 3 & -3 & 2 & -1 & 1 \end{bmatrix}.$$

On any given play the column player C selects one of the five columns. In the inescapable fashion the individual columns are identified with corresponding vectors in $\mathbf{R}^2$:

$$\text{I:} \quad \begin{bmatrix} -5 \\ 3 \end{bmatrix} \text{ is identified with } (-5, 3) \text{ in } \mathbf{R}^2.$$

$$\text{II:} \quad \begin{bmatrix} 3 \\ -3 \end{bmatrix} \text{ is identified with } (3, -3) \text{ in } \mathbf{R}^2.$$

And the remaining three columns are identified with the vectors

$$\text{III: } (-1, 2); \qquad \text{IV: } (1, -1); \qquad \text{V: } (-3, 1).$$

The "long-run" options available to player C are the column expectations $G\mathbf{c}$, where the column strategy $\mathbf{c}$ is a 5×1 column matrix of nonnegative entries totaling one. For example, C may employ strategy

$$\mathbf{c} = \begin{bmatrix} .1 \\ .3 \\ .1 \\ .1 \\ .4 \end{bmatrix}.$$

That is, C plays I with probability .1, II with probability .3, and so on. The expected payoffs for the cases in which R plays rows I and II, respectively, are the entries in the expectation

$$Gc = \begin{bmatrix} -5 & 3 & -1 & 1 & -3 \\ 3 & -3 & 2 & -1 & 1 \end{bmatrix} \begin{bmatrix} .1 \\ .3 \\ .1 \\ .1 \\ .4 \end{bmatrix} = \begin{bmatrix} -.8 \\ -.1 \end{bmatrix}.$$

(We see that the game G favors C.)

Needless to say, the expectation

$$Gc = \begin{bmatrix} -.8 \\ -.1 \end{bmatrix}$$

is identified with the vector $(-.8, -.1)$ in $\mathbf{R}^2$. And for each column strategy **c**, the expectation $G\mathbf{c}$ identifies in like manner with a vector in $\mathbf{R}^2$. What we want to do is to form a picture in the coordinate plane of the set of all column expectations $G\mathbf{c}$.

In general, a column strategy is of the form

$$\mathbf{c} = \begin{bmatrix} c_1 \\ c_2 \\ c_3 \\ c_4 \\ c_5 \end{bmatrix},$$

where the c_i are nonnegative numbers totaling one. The expectation produced is

$$G\mathbf{c} = \begin{bmatrix} -5 & 3 & -1 & 1 & -3 \\ 3 & -3 & 2 & -1 & 1 \end{bmatrix} \begin{bmatrix} c_1 \\ c_2 \\ c_3 \\ c_4 \\ c_5 \end{bmatrix}.$$

Recalling the "operational view" of Sec. 6-7, we may write the product $G\mathbf{c}$ as

$$G\mathbf{c} = c_1\begin{bmatrix} -5 \\ -3 \end{bmatrix} + c_2\begin{bmatrix} 3 \\ -3 \end{bmatrix} + c_3\begin{bmatrix} -1 \\ 2 \end{bmatrix} + c_4\begin{bmatrix} 1 \\ -1 \end{bmatrix} + c_5\begin{bmatrix} -3 \\ 1 \end{bmatrix}.$$

Identifying 2×1 matrices as vectors in $\mathbf{R}^2$, we see that the set of column expectations identifies with the set of all vectors of the form

$$c_1(-5, 3) + c_2(3, -3) + c_3(-1, 2) + c_4(1, -1) + c_5(-3, 1),$$

where the c_i are nonnegative numbers totaling one.

From the observations of the previous section, the conclusion is that the set of all column expectations $G\mathbf{c}$ may be viewed as the points of the convex polygon generated by the five points $(-5, 3)$, $(3, -3)$, $(-1, 2)$, $(1, -1)$, and $(-3, 1)$, these five points being precisely the vectors which correspond to the individual columns of the matrix G.

Thus to form the picture of the set of column expectations we plot the columns of the matrix G as points in the coordinate plane, and then construct the convex polygon generated by these points. This *polygon of expectations* is shown in Fig. 7-13.

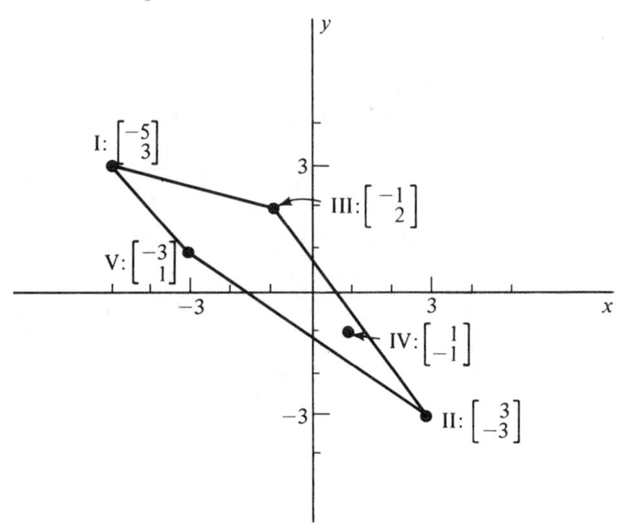

Figure 7-13 The Polygon of Expectations.

The argument extends to any game represented by a $2 \times n$ payoff matrix.

Given a $2 \times n$ payoff matrix G, the set of all column expectations $G\mathbf{c}$, where $\mathbf{c}$ is an $n \times 1$ column matrix of nonnegative entries totaling one, is represented by the convex polygon generated by the points in the plane $\mathbf{R}^2$ corresponding to the columns of the matrix G. This polygon is called the *polygon of (column) expectations* for the game G.

Comment: Of course, if the matrix G has only two columns, or if the columns all lie on the same line in the coordinate plane, then the "polygon" of expectations is just a line segment. And if all of the columns are exactly the same, the "polygon" degenerates to a single point.

The Optimal Column Strategy

In Fig. 7-13 we've given player C a picture of the expectations available to him by a choice of column strategy in the particular 2×5 matrix game G. Now let's show him how to recognize the optimal expectation $\mathbf{u}_0$ and how to find the optimal strategy $\mathbf{c}_0$ that produces it.

We recall that C bases his evaluation of a strategy $\mathbf{c}$ on max $G\mathbf{c}$, which represents the worst possible expected payoff associated with $\mathbf{c}$. For instance, if

$$G\mathbf{c}_1 = \begin{bmatrix} -1 \\ 1 \end{bmatrix},$$

then $\max G\mathbf{c}_1 = 1$; C loses one point per play, on the average in the long run, if R sticks to Row II. An optimal column strategy $\mathbf{c}_0$ is one for which

$$\max G\mathbf{c}_0 \leq \max G\mathbf{c}$$

for all column strategies $\mathbf{c}$.

Since the set of expectations $G\mathbf{c}$ are the points of the polygon of expectations shown in Fig. 7-13, the challenge is to locate a point $\mathbf{u}_0$ in the polygon such that

$$\max \mathbf{u}_0 \leq \max \mathbf{u}$$

for all points $\mathbf{u}$ of the polygon.

To bring geometry to bear on the search for an optimal expectation $\mathbf{u}_0$ we consider first the simple inequality

$$\max \mathbf{u} \leq 0.$$

The points $\mathbf{u} = \begin{bmatrix} x \\ y \end{bmatrix}$ satisfying this inequality are those for which both coordinates, x and y, are zero or negative. For a graphical picture we view the x and y axes as dividing the coordinate plane into four quadrants. The set of points $\mathbf{u}$ satisfying $\max \mathbf{u} \leq 0$ comprises the boundary and interior of the lower left-hand quadrant, which we refer to as the *negative quadrant* (commonly called the *third quadrant*). The negative quadrant is shaded in Fig. 7-14.

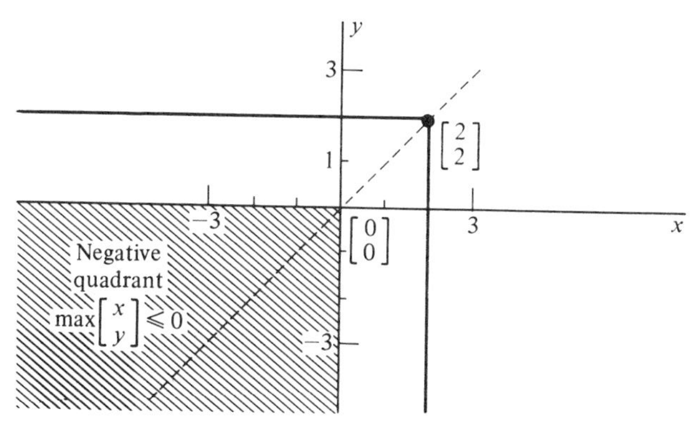

Figure 7-14.

In Fig. 7-14 we have dashed in the *main diagonal* of the plane, the line of points of the form $\begin{bmatrix} x \\ x \end{bmatrix}$. From the point $\begin{bmatrix} 2 \\ 2 \end{bmatrix}$ on the diagonal two half-lines have been constructed heading in the negative directions for the coordinates x and y, respectively. We think of the half-lines together with the points below and to the left as a displaced position of the negative quadrant—the vertex having traveled along the diagonal from $\begin{bmatrix} 0 \\ 0 \end{bmatrix}$ to $\begin{bmatrix} 2 \\ 2 \end{bmatrix}$, pulling the entire quadrant with it. The displaced quadrant with vertex at $\begin{bmatrix} 2 \\ 2 \end{bmatrix}$ represents precisely the set of points for which

$$\max \mathbf{u} \leq 2.$$

After all, if

$$\max \begin{bmatrix} x \\ y \end{bmatrix} \leq 0,$$

then

$$\max \left(\begin{bmatrix} x \\ y \end{bmatrix} + \begin{bmatrix} 2 \\ 2 \end{bmatrix} \right) \leq 2$$

and conversely. Thus we obtain the graph of

$$\max \mathbf{u} \leq 2$$

by adding $\begin{bmatrix} 2 \\ 2 \end{bmatrix}$ to every point of the graph of

$$\max \mathbf{u} \leq 0.$$

The addition of $\begin{bmatrix} 2 \\ 2 \end{bmatrix}$ to the points of the negative quadrant simply "slides" the quadrant up the main diagonal to the point $\begin{bmatrix} 2 \\ 2 \end{bmatrix}$.

Now we are ready to locate the optimal expectation $\mathbf{u}_0$ in the game G at hand. Figure 7-15 shows the polygon of expectations together with the displaced negative quadrant in such a position that there is a point in the intersection of the boundaries of the polygon and quadrant, but there is no point of the polygon in the interior of the quadrant.

With respect to the polygon of expectations, we say that the displaced quadrant is in its *critical position* in Fig. 7-15. It will be argued that the point of intersection $\mathbf{u}_0$ is the optimal expectation.

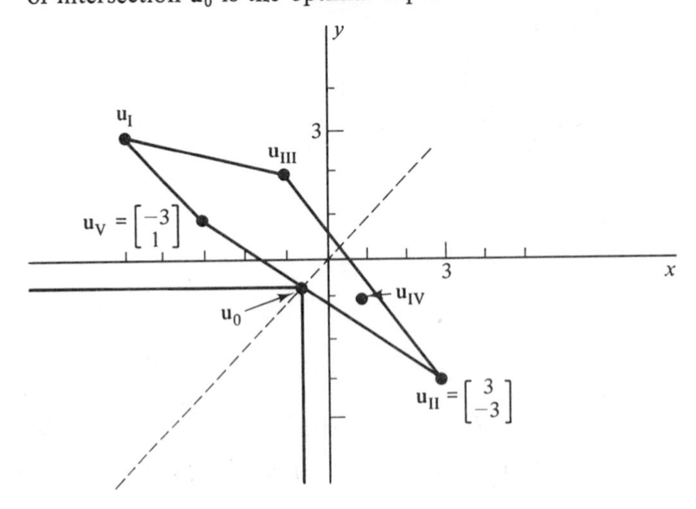

Figure 7-15.

A dynamic view of the construction is as follows:

We start with the negative quadrant displaced well down the main diagonal, and then we slide the quadrant up the diagonal until it first touches the polygon of expectations. A point of first encounter is an optimal expectation $\mathbf{u}_0 = G\mathbf{c}_0$, where $\mathbf{c}_0$ is an optimal column strategy.

The point $\mathbf{u}_0$ of Fig. 7-15 is on the boundary of the displaced quadrant with vertex at a point $\begin{bmatrix} x_0 \\ x_0 \end{bmatrix}$ on the diagonal. (In fact, $\mathbf{u}_0 = \begin{bmatrix} x_0 \\ x_0 \end{bmatrix}$, but in other games this is not necessarily the case.) Thus

$$\max \mathbf{u}_0 = x_0.$$

Since none of the points $\mathbf{u}$ of the polygon of expectations lie in the interior of the displaced quadrant, we know that

$$\max \mathbf{u} \geq x_0$$

for all points $\mathbf{u}$ of the polygon. Thus the construction places $\mathbf{u}_0$ as a point of the polygon of expectations, with

$$x_0 = \max \mathbf{u}_0 \leq \max \mathbf{u}$$

for all points $\mathbf{u}$ of the polygon. This is precisely the condition that qualifies $\mathbf{u}_0$ as an optimal expectation.

To find the optimal strategy $\mathbf{c}_0$ which produces $\mathbf{u}_0$, we note from Fig. 7-15 that $\mathbf{u}_0$ lies on the line segment between the points $\mathbf{u}_{\mathrm{II}}$ and $\mathbf{u}_{\mathrm{V}}$, corresponding to the second and fifth columns of the matrix G. From the discussion of the previous section it follows that there is a value p, where $0 \leq p \leq 1$, such that

$$\mathbf{u}_0 = p\mathbf{u}_{\mathrm{II}} + (1 - p)\mathbf{u}_{\mathrm{V}} \tag{1}$$

$$= p\begin{bmatrix} 3 \\ -3 \end{bmatrix} + (1 - p)\begin{bmatrix} -3 \\ 1 \end{bmatrix} \tag{2}$$

$$= \begin{bmatrix} 3p + (-3)(1 - p) \\ (-3)p + (1 - p) \end{bmatrix}. \tag{3}$$

We also note from Fig. 7-15 that $\mathbf{u}_0$ lies on the main diagonal; thus the two coordinates of $\mathbf{u}_0$ are equal. So we equate the two coordinates of the point in Eq. (3) and solve for p:

$$\begin{aligned} 3p + (-3)(1 - p) &= (-3)p + (1 - p), \\ 3p - 3 + 3p &= -3p + 1 - p, \\ 6p - 3 &= -4p + 1, \\ 10p &= 4, \\ p &= .4, \end{aligned}$$

Thus

$$1 - p = .6.$$

Substituting the values of p and $1 - p$ into Eqs. (1) and (2),

$$\mathbf{u}_0 = .4u_{\text{II}} + .6u_{\text{V}} = .4\begin{bmatrix} 3 \\ -3 \end{bmatrix} + .6\begin{bmatrix} -3 \\ 1 \end{bmatrix},$$

or in terms of all of the columns of the matrix G,

$$\mathbf{u}_0 = 0\begin{bmatrix} -5 \\ 3 \end{bmatrix} + .4\begin{bmatrix} 3 \\ -3 \end{bmatrix} + 0\begin{bmatrix} -1 \\ 2 \end{bmatrix} + 0\begin{bmatrix} 1 \\ -1 \end{bmatrix} + .6\begin{bmatrix} -3 \\ 1 \end{bmatrix}$$

$$= \begin{bmatrix} -5 & 3 & -1 & 1 & -3 \\ 3 & -3 & 2 & -1 & 1 \end{bmatrix}\begin{bmatrix} 0 \\ .4 \\ 0 \\ 0 \\ .6 \end{bmatrix}.$$

From the geometrical argument, $\mathbf{u}_0$ is the optimal column expectation, and we've located $\mathbf{u}_0$ as the expectation produced by the column strategy

$$\mathbf{c}_0 = \begin{bmatrix} 0 \\ .4 \\ 0 \\ 0 \\ .6 \end{bmatrix}.$$

Thus $\mathbf{c}_0$ is the optimal column strategy.

The "sliding quadrant" argument may be applied to any game with a $2 \times n$ payoff matrix. It should be clear that in its critical position the sliding quadrant will always intersect the polygon of expectations at a point on the boundary of the polygon. As observed in the previous section, every point on the boundary of a convex polygon is on a line segment with generating points as endpoints. Just as we located $\mathbf{u}_0$ above as on the line segment with end points $\mathbf{u}_{\text{II}}$ and $\mathbf{u}_{\text{V}}$, so in any game an optimal column expectation is on a line segment with end points corresponding to columns of the payoff matrix. And just as the optimal column strategy $\mathbf{c}_0$ above confines itself to Columns II and V, so in every $2 \times n$ matrix game an optimal column strategy confines itself to the columns which are the end points of the segment containing it.

The conclusion is that in every $2 \times n$ matrix game G there is an optimal column strategy $\mathbf{c}_0$ in which player C confines his play to two of the columns. The sliding quadrant enables us to spot these critical columns. (Of course, if the game is strictly determined, C can confine himself to only one of the columns. As we shall see shortly in an example, the sliding quadrant also conveys this fact.)

The Optimal Row Strategy

Once the critical columns in a $2 \times n$ game have been identified by the sliding quadrant, we can ignore the other columns—they are never to be played. In effect, this reduces the $2 \times n$ game to a 2×2 game.

In the particular 2×5 game G of Fig. 7-15, the sliding quadrant spots Columns II and V as the critical columns. Extracting these columns, the game reduces to the following 2×2 matrix G' (G prime):

$$G' = \begin{bmatrix} 3 & -3 \\ -3 & 1 \end{bmatrix}.$$

Now we obtain optimal row and column strategies, $\mathbf{r}_0'$ and $\mathbf{c}_0'$, for the reduced game G' by the three-step technique of Sec. 6-6.

Search for $\mathbf{r}_0'$:

1. $\begin{bmatrix} 3 \\ -3 \end{bmatrix} - \begin{bmatrix} -3 \\ 1 \end{bmatrix} = \begin{bmatrix} 6 \\ -4 \end{bmatrix}.$
2. $[4 \quad 6].$
3. $\mathbf{r}_0' = [.4 \quad .6].$

Search for $\mathbf{c}_0'$:

1. $[3 \quad -3] - [-3 \quad 1] = [6 \quad -4].$
2. $\begin{bmatrix} 4 \\ 6 \end{bmatrix}.$
3. $\mathbf{c}_0' = \begin{bmatrix} .4 \\ .6 \end{bmatrix}.$

The three-step technique for finding $\mathbf{c}_0'$ is simply an organization of the algebra that earlier led us to the values $p = .4$ and $1 - p = .6$.

With respect to the original game, the assertion is that the optimal strategy for C is to play Column II with probability .4, Column V with probability .6, and never to play the other columns. Thus the optimal strategy $\mathbf{c}_0'$ for the reduced game is converted to an optimal strategy $\mathbf{c}_0$ for the original game by supplying zeros as the probabilities of playing the ignored columns, I, III, and IV. Since the optimal row strategy $\mathbf{r}_0'$ for the reduced game is based on the only columns that C plays in his optimal strategy, it follows that $\mathbf{r}_0'$ is itself an optimal row strategy for the original game.

Now we are ready for the full analysis:

Problem: Analyze the game with payoff matrix

$$G = \begin{bmatrix} -5 & 3 & -1 & 1 & -3 \\ 3 & -3 & 2 & -1 & 1 \end{bmatrix}.$$

Solution: Optimal strategies are

$$\mathbf{r}_0 = [.4 \quad .6], \qquad \mathbf{c}_0 = \begin{bmatrix} 0 \\ .4 \\ 0 \\ 0 \\ .6 \end{bmatrix}.$$

Verification:

$$\mathbf{r}_0 G = [.4 \quad .6]\begin{bmatrix} -5 & 3 & -1 & 1 & -3 \\ 3 & -3 & 2 & -1 & 1 \end{bmatrix}$$
$$= [-.2 \quad -.6 \quad .8 \quad -.2 \quad -.6]$$

$$G\mathbf{c}_0 = \begin{bmatrix} -5 & 3 & -1 & 1 & -3 \\ 3 & -3 & 2 & -1 & 1 \end{bmatrix}\begin{bmatrix} 0 \\ .4 \\ 0 \\ 0 \\ .6 \end{bmatrix} = \begin{bmatrix} -.6 \\ -.6 \end{bmatrix}.$$

$$\min \mathbf{r}_0 G = \max G\mathbf{c}_0 = -.6.$$

Therefore, by the Duality Theorem of Game Theory, $\mathbf{r}_0$ and $\mathbf{c}_0$ are optimal strategies; the value of the game is $\mathbf{v}_G = -.6$.

The following examples should serve to clarify the general procedure.

Examples: Analyze the games corresponding to the following payoff matrices:

(a) $G = \begin{bmatrix} -1 & 5 & 0 & -5 \\ 2 & -2 & 1 & 3 \end{bmatrix}.$

(b) $G = \begin{bmatrix} -5 & 2 & -3 \\ 3 & 0 & -1 \end{bmatrix}.$

(c) $G = \begin{bmatrix} -4 & 5 & -10 \\ 1 & -5 & 5 \end{bmatrix}.$

Solutions: (a) The polygon of expectations with the sliding quadrant in critical position is shown in Fig. 7-16(a).

We see that the optimal expectation lies on the line segment between the points representing Columns II and IV. Since the optimal column strategy $\mathbf{c}_0$ confines itself to these two columns, we delete the other columns of the matrix G and reduce to the 2×2 matrix

$$G' = \begin{bmatrix} 5 & -5 \\ -2 & 3 \end{bmatrix}.$$

By the three-step method optimal strategies for G' are

$$\mathbf{r}_0' = [\tfrac{5}{15} \quad \tfrac{10}{15}] = [\tfrac{1}{3} \quad \tfrac{2}{3}], \qquad \mathbf{c}_0' = \begin{bmatrix} \tfrac{8}{15} \\ \tfrac{7}{15} \end{bmatrix}.$$

Since the entries in $\mathbf{c}_0'$ represent the probabilities with which Columns II and IV are played, we adapt $\mathbf{c}_0'$ to the original game and assert that optimal strategies are

$$\mathbf{r}_0 = [\tfrac{1}{3} \quad \tfrac{2}{3}], \qquad \mathbf{c}_0 = \begin{bmatrix} 0 \\ \tfrac{8}{15} \\ 0 \\ \tfrac{7}{15} \end{bmatrix}.$$

Verification: (a)

$$\mathbf{r}_0 G = [\tfrac{1}{3} \quad \tfrac{2}{3}] \begin{bmatrix} -1 & 5 & 0 & -5 \\ 2 & -2 & 1 & 3 \end{bmatrix} = [\tfrac{3}{3} \quad \tfrac{1}{3} \quad \tfrac{2}{3} \quad \tfrac{1}{3}],$$

$$G\mathbf{c}_0 = \begin{bmatrix} -1 & 5 & 0 & -5 \\ 2 & -2 & 1 & 3 \end{bmatrix} \begin{bmatrix} 0 \\ \tfrac{8}{15} \\ 0 \\ \tfrac{7}{15} \end{bmatrix} = \begin{bmatrix} \tfrac{5}{15} \\ \tfrac{5}{15} \end{bmatrix} = \begin{bmatrix} \tfrac{1}{3} \\ \tfrac{1}{3} \end{bmatrix},$$

$$\min \mathbf{r}_0 G = \max G\mathbf{c}_0 = \tfrac{1}{3}.$$

Thus $\mathbf{r}_0$ and $\mathbf{c}_0$ are optimal strategies; the value of the game is $v_G = \frac{1}{3}$.

(b) A quick inspection shows that this game is strictly determined. Optimal strategies are

$$\mathbf{r}_0 = [0 \quad 1], \qquad \mathbf{c}_0 = \begin{bmatrix} 0 \\ 0 \\ 1 \end{bmatrix}.$$

We have

$$\min(\mathbf{r}_0 G) = \max(G\mathbf{c}_0) = -1.$$

The value of the game is $v_G = -1$.

However, just to see what happens, we depict in Fig. 7-16(b) the polygon of expectations with sliding quadrant in critical position.

We see from the figure that the optimal expectation $\mathbf{u}_0$ is the point corresponding to the third column of the matrix G. The strategy $\mathbf{c}_0$ producing $\mathbf{u}_0$ is

$$\mathbf{c}_0 = \begin{bmatrix} 0 \\ 0 \\ 1 \end{bmatrix};$$

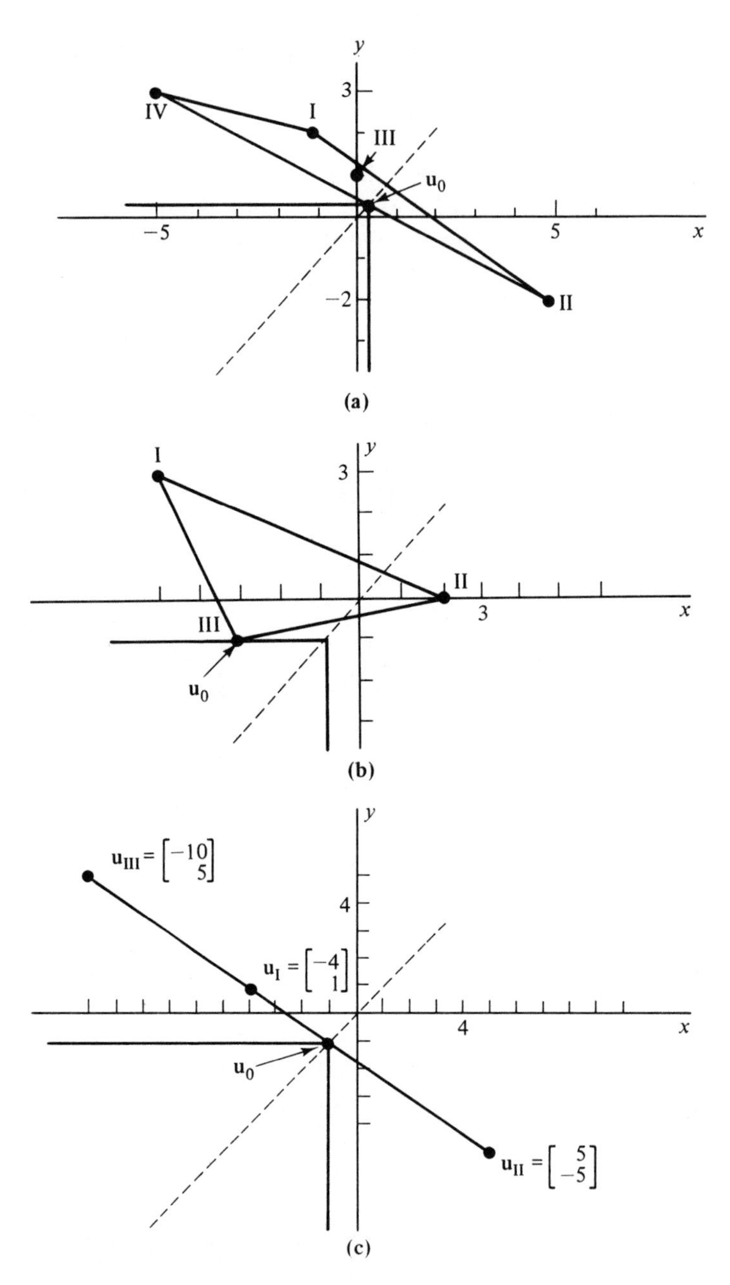

Figure 7-16.

i.e., player C plays Column III all the time. Player R, knowing this fact, simply looks at the third column and, to minimize damage, chooses

$$\mathbf{r}_0 = [0 \quad 1].$$

(c) The "polygon" of expectations with sliding quadrant in critical position is shown in Fig. 7-16(c).

In this case, the three columns of the matrix G all lie on the same line. But let's be more innocent.

Just looking at the picture it's hard to tell whether the optimal expectation $\mathbf{u}_0$ lies between $\mathbf{u}_{\mathrm{I}}$ and $\mathbf{u}_{\mathrm{II}}$ or between $\mathbf{u}_{\mathrm{II}}$ and $\mathbf{u}_{\mathrm{III}}$; possibly both conditions hold.

A straightforward approach is to guess that $\mathbf{u}_0$ lies between $\mathbf{u}_{\mathrm{I}}$ and $\mathbf{u}_{\mathrm{II}}$ and proceed as in part (a). If the optimal strategies satisfy the criterion for verification, we guessed correctly. If the optimal strategies fail the test, we can be sure that $\mathbf{u}_0$ lies between $\mathbf{u}_{\mathrm{II}}$ and $\mathbf{u}_{\mathrm{III}}$, and again we proceed as in part (a). The details are left to the reader.

7-3 EXERCISES

1. Consider the game represented by the following 2 × 3 payoff matrix:

$$G = \begin{bmatrix} -4 & 3 & -1 \\ 3 & -3 & 2 \end{bmatrix}.$$

(a) Construct the polygon of expectations.

(b) Locate in the polygon the expectation $\mathbf{u} = G\mathbf{c}$ for the following column strategy:

$$\mathbf{c} = \begin{bmatrix} .4 \\ .4 \\ .2 \end{bmatrix}.$$

2. Consider the game corresponding to the matrix

$$G = \begin{bmatrix} 3 & 1 & -4 & 2 \\ -6 & -3 & 4 & -1 \end{bmatrix}.$$

(a) Construct the polygon of expectations:

(b) Locate in the polygon the expectations produced by the following strategies:

$$\text{(i)}\quad \mathbf{c}_1 = \begin{bmatrix} .5 \\ 0 \\ .5 \\ 0 \end{bmatrix}, \qquad \text{(ii)}\quad \mathbf{c}_2 = \begin{bmatrix} .4 \\ .2 \\ 0 \\ .4 \end{bmatrix}, \qquad \text{(iii)}\quad \mathbf{c}_3 = \begin{bmatrix} \frac{1}{4} \\ \frac{1}{4} \\ \frac{1}{4} \\ \frac{1}{4} \end{bmatrix}.$$

3. Consider the game corresponding to the matrix

$$G = \begin{bmatrix} -2 & 3 & -5 \\ 2 & -1 & 3 \end{bmatrix}.$$

(a) Construct the polygon of expectations with sliding quadrant in critical position. Encircle the point of contact corresponding to the optimal expectation $\mathbf{u}_0$.

(b) Identify the two critical columns of G involved in the optimal strategy $\mathbf{c}_0$ for which $\mathbf{u}_0 = G\mathbf{c}_0$.

(c) Reduce G to a 2×2 matrix G' by removing the noncritical column. Then use the three-step method to obtain optimal strategies $\mathbf{r}_0'$ and $\mathbf{c}_0'$ for G'.

(d) Let $\mathbf{r}_0 = \mathbf{r}_0'$, and adapt $\mathbf{c}_0'$ to an optimal strategy $\mathbf{c}_0$ for the original game G by appropriate insertion of a zero. Verify that $\mathbf{r}_0$ and $\mathbf{c}_0$ are optimal strategies for G by showing that

$$\min \mathbf{r}_0 G = \max G\mathbf{c}_0.$$

4. Following the steps outlined in Prob. 3, analyze the games corresponding to the following payoff matrices. That is, find optimal strategies $\mathbf{r}_0$ and $\mathbf{c}_0$ and the value of the game, and verify.

(a) $G = \begin{bmatrix} 4 & -3 & 5 \\ -6 & 2 & -7 \end{bmatrix}$, (b) $G = \begin{bmatrix} -3 & 3 & 4 & -1 \\ 3 & -1 & -5 & 1 \end{bmatrix}$.

5. Analyze the games corresponding to the following matrices:

(a) $G = \begin{bmatrix} 1 & -6 & 7 \\ -1 & 5 & -6 \end{bmatrix}$.

(b) $G = \begin{bmatrix} -1 & -2 & 3 \\ 1 & 3 & 1 \end{bmatrix}$.

(c) $G = \begin{bmatrix} -3 & 0 & 4 & -1 & 6 & -2 \\ 3 & 1 & -2 & 2 & -3 & 2 \end{bmatrix}$.

6. Recall the Politicians Game, introduced as an illustration early in Chapter 6. The payoff matrix is as follows:

$$G = \begin{bmatrix} .54 & .47 & .55 \\ .47 & .52 & .46 \end{bmatrix}.$$

Find optimal strategies $\mathbf{r}_0$ and $\mathbf{c}_0$, and the value of the game. With respect to the empirical interpretation in Sec. 6-1, what are the implications for candidates R and C?

7. (a) Complete part (c) of the example in the text, i.e., analyze the game with payoff matrix

$$G = \begin{bmatrix} -4 & 5 & -10 \\ 1 & -5 & 5 \end{bmatrix}.$$

(b) There are many optimal column strategies for this game. Find three, verifying in each case.

7-4. Functionals on $\mathbf{R}^2$

By identifying 2×1 column matrices with vectors in $\mathbf{R}^2$ we've seen a geometrical argument for locating an optimal column expectation in a $2 \times n$ game and for finding the corresponding optimal column strategy. The geometrical construction has not yet been interpreted as affording a view of an optimal row strategy $\mathbf{r}_0$. Since the 2×1 column matrices have been represented as points in the coordinate plane $\mathbf{R}^2$, what we need to complete the picture is a way of visualizing 1×2 row matrices.

For the present, it is advantageous for us to continue to identify a 2×1 matrix $\begin{bmatrix} x \\ y \end{bmatrix}$ as being essentially the same thing as the vector (x, y) in $\mathbf{R}^2$. Insofar as addition and scalar multiplication are concerned, the two are manipulated by the same rules, so one may rightly feel that the difference between 2×1 matrices and vectors in $\mathbf{R}^2$ is largely typographical. But the same observation applies for the 1×2 matrix $[a \quad b]$ and the vector (a, b). However, 1×2 matrices and 2×1 matrices, as they were formally defined in the previous chapter, are different kinds of objects, and this difference is important. We do want to distinguish between $\begin{bmatrix} 3 \\ 4 \end{bmatrix}$ and $[3 \quad 4]$. Thus in relaxing the distinction between the column matrix $\begin{bmatrix} 3 \\ 4 \end{bmatrix}$ and the vector $(3, 4)$, we do not in the same context relax the distinction between the row matrix $[3 \quad 4]$ and the vector $(3, 4)$.

Via matrix multiplication, a fixed row matrix $[a \quad b]$ can be interpreted as a function mapping the set of column matrices $\begin{bmatrix} x \\ y \end{bmatrix}$ into the set of 1×1 matrices. For instance, $[3 \quad 4]$ maps $\begin{bmatrix} 5 \\ -2 \end{bmatrix}$ into the matrix $[7]$, because

$$[3 \quad 4]\begin{bmatrix} 5 \\ -2 \end{bmatrix} = [7].$$

To give the row matrix $[3 \quad 4]$ a role in the vector space $\mathbf{R}^2$, we identify a 1×1 matrix $[t]$ with the real number t; for instance, the matrix $[7]$ is identified with the number 7. We've already identified a 2×1 matrix $\begin{bmatrix} x \\ y \end{bmatrix}$ with the vector (x, y). Since $[3 \quad 4]$ has been interpreted as a function mapping 2×1 matrices into 1×1 matrices, it is natural that we also identify $[3 \quad 4]$ as a function mapping vectors in $\mathbf{R}^2$ into real numbers. Above we saw that the function $[3 \quad 4]$ maps the matrix $\begin{bmatrix} 5 \\ -2 \end{bmatrix}$ into the matrix $[7]$; with the new identification we say that the function $[3 \quad 4]$ maps the vector $(5, -2)$ into the number 7.

In general, a 1×2 row matrix $[a \quad b]$ is interpreted as the function which maps each vector (x, y) into the real number $ax + by$, because

$$[a \quad b]\begin{bmatrix} x \\ y \end{bmatrix} = [ax + by].$$

Functions of this form, represented by 1×2 matrices, are called *functionals* on $\mathbf{R}^2$.

The functional, which we consider here in a simple setting, is an important concept in mathematics. Out of respect for the larger setting, we should observe that the functionals on $\mathbf{R}^2$ are precisely the *linear* mappings of $\mathbf{R}^2$ into $\mathbf{R}$. To say that a function F mapping $\mathbf{R}^2$ into $\mathbf{R}$ is linear means

1. For each pair of vectors $\mathbf{u}$, $\mathbf{v}$ in $\mathbf{R}^2$,

$$F(\mathbf{u} + \mathbf{v}) = F(\mathbf{u}) + F(\mathbf{v}).$$

2. For each vector $\mathbf{v}$ in $\mathbf{R}^2$ and each scalar c,

$$F(c\mathbf{v}) = cF(\mathbf{v}).$$

In short, we can add vectors or multiply by scalars before we map or afterwards—the result will be the same.

It is not difficult to show that every linear function mapping $\mathbf{R}^2$ into $\mathbf{R}$ may be expressed, as above, by a 1×2 row matrix and vice versa.

Example: For each vector $\mathbf{v}$ in $\mathbf{R}^2$ let $F(\mathbf{v})$ be the value assigned $\mathbf{v}$ by the functional $[-2 \quad 5]$.

(a) Find $F(\mathbf{v})$ for the following vectors:

(a_1) $\mathbf{v} = (1, 2)$.
(a_2) $\mathbf{v} = (-3, 0)$.
(a_3) $\mathbf{v} = (4, -1)$.

(b) For the following vectors $\mathbf{u}$ and $\mathbf{v}$, show that

$$F(\mathbf{u} + \mathbf{v}) = F(\mathbf{u}) + F(\mathbf{v}).$$

(b_1) $\mathbf{u} = (1, 2)$ and $\mathbf{v} = (4, -1)$.
(b_2) $\mathbf{u} = (-3, 0)$ and $\mathbf{v} = (3, 1)$.

(c) For the following scalars c and vectors $\mathbf{v}$, show that

$$F(c\mathbf{v}) = cF(\mathbf{v}).$$

(c_1) $c = -3$ and $\mathbf{v} = (1, 2)$.
(c_2) $c = 1.5$ and $\mathbf{v} = (4, -1)$.

Solution: (a) By definition the functional $[-2 \quad 5]$ assigns each vector $\mathbf{v} = (x, y)$ in $\mathbf{R}^2$ the value

$$-2x + 5y.$$

Thus

$$F(\mathbf{v}) = F(x, y) = -2x + 5y.$$

[*Note:* For economy of symbolism we'll write $F(x, y)$ rather than $F((x, y))$.]

In particular,

(a₁) $F(1, 2) = (-2) \times 1 + 5 \times 2 = -2 + 10 = 8.$
(a₂) $F(-3, 0) = (-2) \times (-3) + 5 \times 0 = 6.$
(a₃) $F(4, -1) = -13.$

(b₁) From part (a),

$$F(\mathbf{u}) + F(\mathbf{v}) = F(1, 2) + F(4, -1) = 8 + (-13) = -5.$$

To find $F(\mathbf{u} + \mathbf{v})$, we first compute

$$\mathbf{u} + \mathbf{v} = (1, 2) + (4, -1) = (5, 1).$$

We have

$$F(\mathbf{u} + \mathbf{v}) = F(5, 1) = (-2) \times 5 + 5 \times 1 = -5.$$

As was to be shown,

$$F(\mathbf{u} + \mathbf{v}) = F(\mathbf{u}) + F(\mathbf{v}) = -5.$$

(b₂) Here,

$$F(\mathbf{u}) = F(-3, 0) = 6,$$
$$F(\mathbf{v}) = F(3, 1) = -1.$$

Thus

$$F(\mathbf{u}) + F(\mathbf{v}) = 6 + (-1) = 5.$$

Also,

$$F(\mathbf{u} + \mathbf{v}) = F((-3, 0) + (3, 1)) = F(0, 1) = (-2) \times 0 + 5 \times 1 = 5.$$

Therefore,

$$F(\mathbf{u} + \mathbf{v}) = F(\mathbf{u}) + F(\mathbf{v}) = 5.$$

(c₁) From the definition of the functional $[-2 \quad 5]$,

$$F(c\mathbf{v}) = F(-3(1, 2)) = F(-3, -6) = -24,$$

which we observe to be equal to

$$cF(\mathbf{v}) = -3F(1, 2) = (-3) \times 8 = -24.$$

(c₂) Similarly,

$$F(\mathbf{cv}) = F(1.5(4, -1)) = F(6, -1.5) = -19.5$$

and

$$cF(\mathbf{v}) = 1.5F(4, -1) = 1.5 \times (-13) = -19.5.$$

Geometrical Representation of Functionals

On an ordinary geographical map, a *contour* is an overlying curve connecting points of equal elevation. The geometrical representation of functionals on

$\mathbf{R}^2$ is analogous. For a sense of the behavior of the functional $[a \quad b]$, we connect points for which the values assigned by $[a \quad b]$ are the same.

For example, the *zero contour* of the functional $[3 \quad 4]$ is the set of all points (x, y) in $\mathbf{R}^2$ for which

$$3x + 4y = 0$$

or

$$y = -\tfrac{3}{4}x.$$

Thus the points of the zero contour of $[3 \quad 4]$ are those of the form

$$(x, y) = \left(x, -\frac{3}{4}x\right) = -\frac{x}{4}(-4, 3).$$

Setting $t = -(x/4)$ we conclude that (x, y) is on the zero contour of $[3 \quad 4]$ provided

$$(x, y) = t(-4, 3)$$

for some scalar t.

From the discussion in Sec. 7-1, we recall that the set of points of the form $t(-4, 3)$ comprises the line determined by the origin, $(0, 0)$, and the point $(-4, 3)$. We see that this line is precisely the set of vectors mapped into the value 0 by the functional $[3 \quad 4]$.

Similarly, the "7 contour" of the functional $[3 \quad 4]$ is the set of points (x, y) in $\mathbf{R}^2$ for which

$$3x + 4y = 7.$$

Observing that $(x_0, y_0) = (1, 1)$ is such a point, we write the above equation in the equivalent forms

$$3x + 4y = 3x_0 + 4y_0 = 3 \times 1 + 4 \times 1,$$

$$3(x - 1) + 4(y - 1) = 0,$$

$$(y - 1) = -\frac{3}{4}(x - 1).$$

Thus (x, y) is on the "7 contour" of the functional $[3 \quad 4]$ provided

$$(x - 1, y - 1) = \left(x - 1, -\frac{3}{4}(x - 1)\right)$$

$$= -\frac{x - 1}{4}(-4, 3).$$

Here we set $t = -\dfrac{x - 1}{4}$, and conclude that (x, y) is on the "7 contour" of $[3 \quad 4]$ whenever

$$(x - 1, y - 1) = (x, y) - (1, 1) = t(-4, 3)$$

or

$$(x, y) = (1, 1) + t(-4, 3)$$

for some scalar t.

Again, from the earlier discussion we recall that the vectors of the form

$$(1, 1) + t(-4, 3)$$

comprise the line containing (1, 1) in the direction of the line through (0, 0) and (−4, 3).

The argument generalizes to show that for any value k, the "k contour" of the functional [3 4], i.e., the set of vectors assigned the value k by the function, consists of those (x, y) satisfying

$$(x, y) = (x_0, y_0) + t(-4, 3),$$

where (x_0, y_0) is any particular vector satisfying

$$3x_0 + 4y_0 = k.$$

Geometrically, these points comprise the line containing (x_0, y_0) in the direction of the line through (0, 0) and (−4, 3).

Thus the contours of [3 4], which we'll call the *level lines* of [3 4], constitute a family of parallel lines. Several of the level lines of [3 4] are shown in Fig. 7-17.

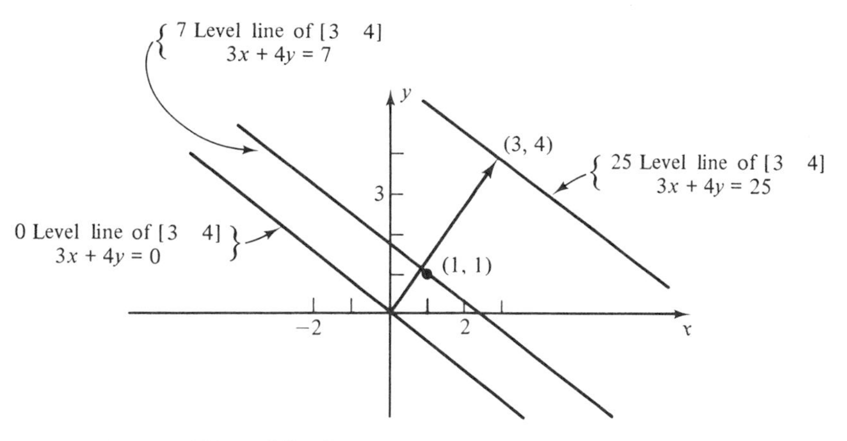

Figure 7-17 Level Lines of [3 4].

As pleasing as the geometrical picture in Fig. 7-17 is, we can say more. In the figure we have indicated the ray from the origin to the point (3, 4). As the figure suggests, the parallel level lines of the functional [3 4] are perpendicular to the ray from (0, 0) to the point (3, 4). Let's see why this is true.

In Fig. 7-18 we observe the effect of rotating the ray from (0, 0) to (3, 4) counterclockwise through a right angle.

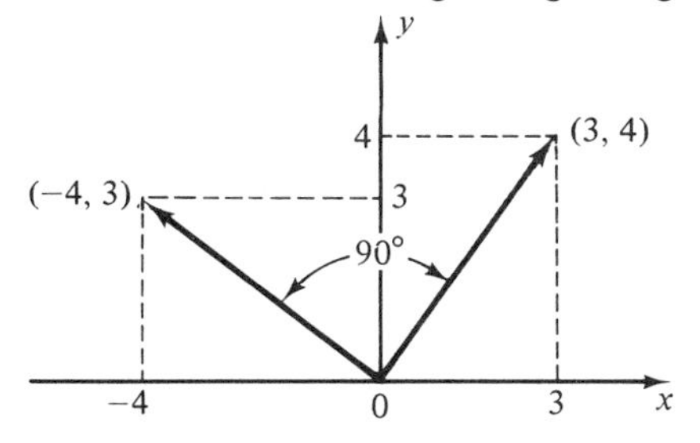

Figure 7-18 Rotation of (3, 4).

Just as the y axis was originally pictured as the new position of the x axis after such a rotation, so also the perpendicular from (3, 4) to the point 3 on the x axis is rotated into a perpendicular to the point 3 on the y axis. Thus the y coordinate of the rotated position is 3. Similarly, the perpendicular from (3, 4) to the point 4 on the y axis is rotated into a perpendicular to the point -4 on the x axis. Thus the x coordinate of the rotated position is -4. In short, a counterclockwise rotation through a right angle carries the point (3, 4) into the point $(-4, 3)$.

It follows that the line of points through the origin perpendicular to the line determined by the origin and (3, 4) is precisely the set of points (x, y) of the form

$$(x, y) = t(-4, 3).$$

As we observed, this line is precisely the zero level line of the functional $[3 \quad 4]$.

Adding a fixed vector (x_0, y_0) to each of the points of the zero level line determines the set of points (x, y) of the form

$$(x, y) = (x_0, y_0) + t(-4, 3).$$

As we've also observed, this is the set of points for which the functional $[3 \quad 4]$ assigns the value $3x_0 + 4y_0$. Geometrically, the addition of the fixed vector (x_0, y_0) simply displaces the line with no change of direction, the origin (0, 0) being carried to the point (x_0, y_0). In its displaced position the line is, of course, still perpendicular to the line determined by (0, 0) and (3, 4).

The argument generalizes to apply to any functional $[a \quad b]$ where a and b are not both equal to 0.

For any fixed value k, the set of points (x, y) mapped by the functional $[a \quad b]$ into k constitutes a line. This line of points (x, y), for which $ax + by = k$, is called the k *level line* of $[a \quad b]$. These level lines form the family of parallel lines perpendicular to the line containing (0, 0) and the point (a, b).

Of course, if $[a \quad b] = [0 \quad 0]$, then the entire plane $\mathbf{R}^2$ is mapped into zero.

Example: In the coordinate plane $\mathbf{R}^2$, construct the k level lines of the functional $[-2 \quad 5]$ for the following values of k:

$$k = -10, \qquad k = 0, \qquad k = 10, \qquad k = 20.$$

Solution: As a first step, the ray from the origin (0, 0) to the point $(-2, 5)$ is drawn. The respective level lines may then be constructed perpendicular to the direction of this ray through particular points assigned the desired value k by the functional $[-2 \quad 5]$. The points selected are as follows:

$$\begin{aligned} k &= -10\colon & (x_0, y_0) &= (5, 0), \\ k &= 0\colon & (x_0, y_0) &= (0, 0), \\ k &= 10\colon & (x_0, y_0) &= (0, 2), \\ k &= 20\colon & (x_0, y_0) &= (0, 4). \end{aligned}$$

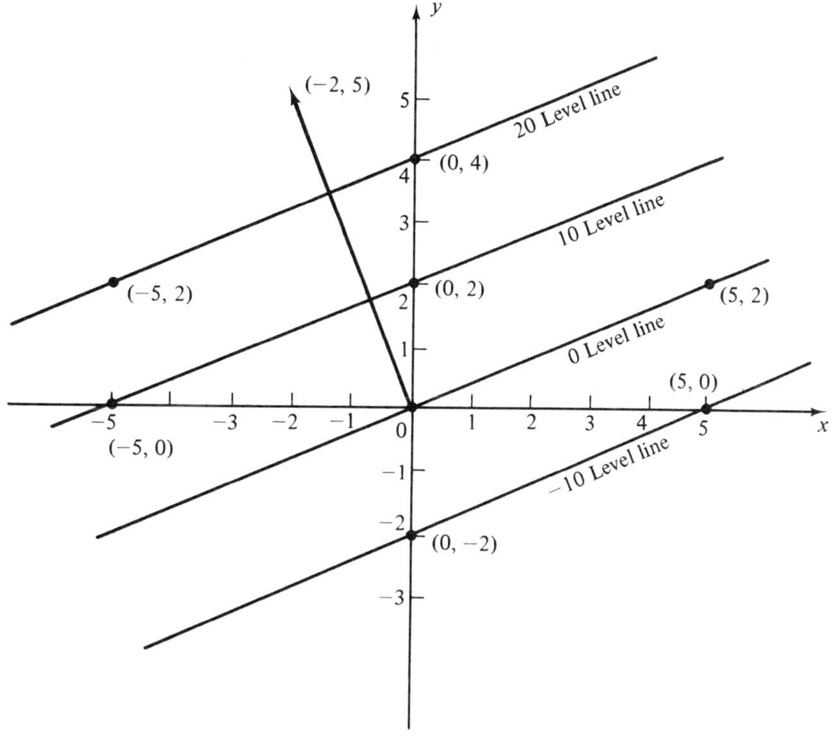

Figure 7-19 Level Lines of the Functional [−2 5].

The construction is shown in Fig. 7-19.

An alternative technique is to plot two distinct points on each level line. In addition to the vectors (x_0, y_0) selected above, a second point on each level line is indicated in Fig. 7-19.

7-4 EXERCISES

1. Consider the functional [2 −1], mapping each vector (x, y) in $\mathbf{R}^2$ into the number

$$2x - y.$$

For each of the following vectors, find the value assigned by [2 −1]:
(a) (4, −2). (b) (2, −1). (c) (1, 2).
(d) (3, 1). (e) (3, 6). (f) (0, −5).

2. For each of the following vectors, find the value assigned by the functional [25 60]:
(a) (0, 10). (b) (12, 5).
(c) (10, 0). (d) (0, 0).

3. For each vector $\mathbf{v}$ in $\mathbf{R}^2$, let $F(\mathbf{v})$ be the value assigned $\mathbf{v}$ by the functional $[3 \quad -2]$.
   (a) For the following vectors $\mathbf{u}$ and $\mathbf{v}$, show that
   $$F(\mathbf{u} + \mathbf{v}) = F(\mathbf{u}) + F(\mathbf{v}).$$
   (a_1) $\mathbf{u} = (4, -1)$ and $\mathbf{v} = (2, 3)$.
   (a_2) $\mathbf{u} = (10, 25)$ and $\mathbf{v} = (15, 5)$.
   (b) For the following scalars c and vectors $\mathbf{v}$, show that
   $$F(c\mathbf{v}) = cF(\mathbf{v}).$$
   (b_1) $c = 5$ and $\mathbf{v} = (3, 1)$.
   (b_2) $c = \frac{1}{3}$ and $\mathbf{v} = (-3, 6)$.

4. (a) Show that the functional $[5 \quad 2]$ assigns the same value to each of the following vectors: $(0, 8)$, $(2, 3)$, $(3.2, 0)$, $(6, -7)$.
   (b) Plot the vectors of part (a) in the coordinate plane $\mathbf{R}^2$. Then construct the line containing these points, and observe that it is perpendicular to the line containing $(0, 0)$ and the point $(5, 2)$.

5. (a) Noting that the functional $[5 \quad 3]$ assigns the value 12 to $(0, 4)$ and also to $(2.4, 0)$, construct the 12 level line of $[5 \quad 3]$. Observe that this line is perpendicular to the line determined by $(0, 0)$ and $(5, 3)$.
   (b) On the graph in part (a), construct the zero level line of the functional $[5 \quad 3]$.

6. In the coordinate plane $\mathbf{R}^2$, construct the k level lines of the functional $[1 \quad 3]$ for the following values of k: $k = -3$, $k = 0$, $k = 6$.

7. In the coordinate plane $\mathbf{R}^2$, construct several level lines of the functional $[-3 \quad 2]$. Indicate the value the functional assumes on each of these lines.

8. (a) Find the point (a, a) where the 10 level line of the functional $[2 \quad 3]$ intersects the main diagonal of the coordinate plane.
   (b) Find the point (b, b) where the level line of the functional $[2 \quad 3]$ through the point $(7, -8)$ crosses the main diagonal.

9. For each of the following pairs of vectors $\mathbf{u}$ and $\mathbf{v}$, find a functional other than $[0 \quad 0]$ that assigns the same value to $\mathbf{u}$ and $\mathbf{v}$.
   (a) $\mathbf{u} = (2, 0)$ and $\mathbf{v} = (0, 2)$.
   (b) $\mathbf{u} = (0, 0)$ and $\mathbf{v} = (2, 3)$.
   (c) $\mathbf{u} = (10, 20)$ and $\mathbf{v} = (25, 30)$.

10. Consider an arbitrary functional $[a \quad b]$ mapping $\mathbf{R}^2$ into $\mathbf{R}$. For each $\mathbf{v}$ in $\mathbf{R}^2$, let $F(\mathbf{v})$ be the value assigned $\mathbf{v}$ by $[a \quad b]$.
    (a) Let $\mathbf{v}_1$ and $\mathbf{v}_2$ be any two vectors,
    $$\mathbf{v}_1 = (x_1, y_1), \qquad \mathbf{v}_2 = (x_2, y_2).$$
    Show that
    $$F(\mathbf{v}_1 + \mathbf{v}_2) = F(\mathbf{v}_1) + F(\mathbf{v}_2).$$
    (b) Show that for each scalar c and each vector $\mathbf{v}$,
    $$F(c\mathbf{v}) = cF(\mathbf{v}).$$

7-5. Row Strategies as Functionals

Returning to $2 \times n$ matrix games, we recall the game G considered in Sec. 7-3:

$$G = \begin{bmatrix} -5 & 3 & -1 & 1 & -3 \\ 3 & -3 & 2 & -1 & 1 \end{bmatrix}.$$

As we have seen, the column expectations may be viewed as the points of the polygon of expectations. The optimal expectation $\mathbf{u}_0$ corresponds to the point where the displaced negative quadrant, sliding up the main diagonal, first touches the polygon. From inspection of the sliding quadrant in critical position we are able to identify the critical columns in the matrix; it is then a straightforward matter to compute the optimal column strategy $\mathbf{c}_0$ such that $\mathbf{u}_0 = G\mathbf{c}_0$.

Now let's consider the geometrical interpretation of the game G from the row player's point of view. To illustrate, we focus on the row strategy

$$\mathbf{r}_1 = [.7 \quad .3].$$

The expectation produced is the matrix product

$$\mathbf{r}_1 G = [.7 \quad .3] \begin{bmatrix} -5 & 3 & -1 & 1 & -3 \\ 3 & -3 & 2 & -1 & 1 \end{bmatrix}$$
$$= [-2.6 \quad 1.2 \quad -.1 \quad .4 \quad -1.8].$$

From the development in the previous section the entries in the expectation $\mathbf{r}_1 G$ may be viewed as the values assigned by the functional $[.7 \quad .3]$ to the vectors, $\mathbf{u}_{\text{I}}$, $\mathbf{u}_{\text{II}}$, etc., corresponding to the columns of the matrix G. That is, $[.7 \quad .3]$ maps $\mathbf{u}_{\text{I}} = (-5, 3)$ into -2.6, maps $\mathbf{u}_{\text{II}} = (3, -3)$ into 1.2, and so on.

For a geometrical picture of the entries in $\mathbf{r}_1 G$, we appeal to the "level-lines" interpretation of the functional $[.7 \quad .3]$. In Fig. 7-20 the points $\mathbf{u}_i$

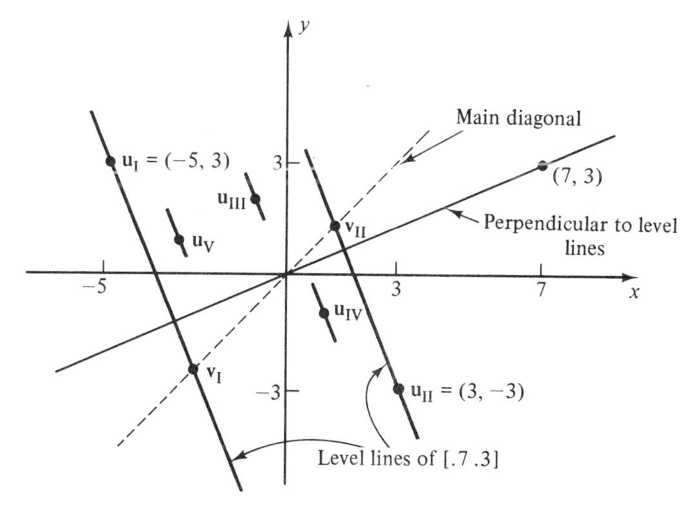

Figure 7-20.

corresponding to the columns of G are plotted in the coordinate plane $\mathbf{R}^2$. The level lines of $[.7 \quad .3]$ through the points $\mathbf{u}_{\mathrm{I}}$ and $\mathbf{u}_{\mathrm{II}}$ are shown; the level lines through the other points $\mathbf{u}_i$ are suggested by shorter segments. From the earlier discussion we know that these lines are perpendicular to the line containing (0, 0) and (.7, .3), which, of course, is the same as the line containing (0, 0) and (7, 3). [And it's more convenient to plot (7, 3) than (.7, .3).] Finally, the main diagonal is dashed in, and the points where the level lines through $\mathbf{u}_{\mathrm{I}}$ and $\mathbf{u}_{\mathrm{II}}$ intersect the diagonal are labeled $\mathbf{v}_{\mathrm{I}}$ and $\mathbf{v}_{\mathrm{II}}$.

Let's look at the point $\mathbf{v}_{\mathrm{I}}$, where the level line of the functional $[.7 \quad .3]$ through $\mathbf{u}_{\mathrm{I}} = (-5, 3)$ intersects the main diagonal. Since $\mathbf{v}_{\mathrm{I}}$ is on the diagonal, it is of the form

$$\mathbf{v}_{\mathrm{I}} = (d_1, d_1).$$

And since $\mathbf{v}_{\mathrm{I}}$ is on the level line of $[.7 \quad .3]$ containing $\mathbf{u}_{\mathrm{I}} = (-5, 3)$, the value assigned by the functional $[.7 \quad .3]$ is the same for $\mathbf{v}_{\mathrm{I}}$ and $\mathbf{u}_{\mathrm{I}}$. The value assigned $\mathbf{v}_{\mathrm{I}} = (d_1, d_1)$ is just

$$.7 \times d_1 + .3 \times d_1 = d_1,$$

the common coordinate of $\mathbf{v}_{\mathrm{I}}$, and the value assigned $\mathbf{u}_{\mathrm{I}} = (-5, 3)$ is

$$.7 \times (-5) + .3 \times 3 = -3.5 + .9 = -2.6,$$

the first entry in the expectation $\mathbf{r}_1 G$. Since the assigned values are equal, it follows that

$$d_1 = -2.6.$$

We see that the Column I entry in the expectation $\mathbf{r}_1 G$ is the common coordinate of the point $\mathbf{v}_1 = (-2.6, -2.6)$, where the main diagonal is intersected by the level line of the functional $[.7 \quad .3]$ passing through $\mathbf{u}_{\mathrm{I}}$.

The same argument applies for the other entries in the expectation $\mathbf{r}_1 G$. For example, the second column in G corresponds to $\mathbf{u}_{\mathrm{II}} = (3, -3)$. Accordingly, the second entry in the expectation $\mathbf{r}_1 G$ is the value assigned $(3, -3)$ by the functional $[.7 \quad .3]$,

$$.7 \times 3 + .3 \times (-3) = 1.2.$$

And the point in Fig. 7-20 where the level line of $[.7 \quad .3]$ through $\mathbf{u}_{\mathrm{II}}$ crosses the diagonal is

$$\mathbf{v}_{\mathrm{II}} = (1.2, 1.2).$$

In line with the general philosophy of matrix games, the evaluation of the strategy $\mathbf{r}_1$ by player R is based on the value of $\min \mathbf{r}_1 G$:

$$\min \mathbf{r}_1 G = \min[-2.6 \quad 1.2 \quad -.1 \quad .4 \quad -1.8] = -2.6.$$

It is this value, -2.6, which represents the worst possible consequence of employing strategy $\mathbf{r}_1$.

In looking at the level lines of $\mathbf{r}_1 = [.7 \quad .3]$ passing through the points corresponding to the columns of the matrix G, player R focuses on the line that intersects farthest down the main diagonal. From Fig. 7-20 we see that this is the line passing through $\mathbf{u}_1$. The point where it crosses the diagonal is $\mathbf{v}_1 = (-2.6, -2.6)$. The common coordinate of $\mathbf{v}_1$, -2.6, is $\min \mathbf{r}_1 G$.

We have been looking at a particular game G and a particular row strategy $\mathbf{r}_1$; now let's consider the general case. For any $2 \times n$ matrix game a row strategy is a matrix of the form $\mathbf{r} = [a \quad b]$, where a and b are nonnegative numbers totaling one. We interpret $\mathbf{r}$ as a functional mapping the vectors in $\mathbf{R}^2$ into real numbers.

As we know, the level lines of the functional $\mathbf{r} = [a \quad b]$ are perpendicular to the line determined by (0, 0) and the point (a, b). Since a and b are both nonnegative, the line through (0, 0) and (a, b) runs from the negative quadrant into the *positive quadrant*. [The "positive quadrant" is, of course, the set of all (x, y) such that $x \geq 0$ and $y \geq 0$.] But if the perpendicular to the level line runs from the negative into the positive quadrant, then the level line through (0, 0) does not enter the interior of these quadrants. This gives us the following geometrical characterization of the zero level lines of row strategies.

The zero level line of a row strategy $\mathbf{r} = [a \quad b]$, where $a \geq 0$, $b \geq 0$, and $a + b = 1$, is a line through (0, 0) that does not enter the interior of the negative quadrant. Conversely, any line through (0, 0) not entering the interior of the negative quadrant is the zero level line of some row strategy.

Now consider the d level line of a row strategy $\mathbf{r} = [a \quad b]$, where d is a fixed number. Since $a + b = 1$, it follows that the point (d, d) on the main diagonal lies on this d level line because

$$a \times d + b \times d = (a + b) \times d = d.$$

The "sliding quadrant" now reenters the discussion. From the characterization of the zero level line of $[a \quad b]$, it follows that the d level line does not enter the interior of the displaced negative quadrant with vertex at (d, d). Indeed, any line through the vertex (d, d) which does not enter the displaced negative quadrant is the d level line for some row strategy. Figure 7-21 illustrates the construction.

The point (x, y) shown on the d level line of $[a \quad b]$ in Fig. 7-21 may be interpreted as a column in the $2 \times n$ payoff matrix for some game. The entry in the corresponding column of the expectation produced by the row strategy

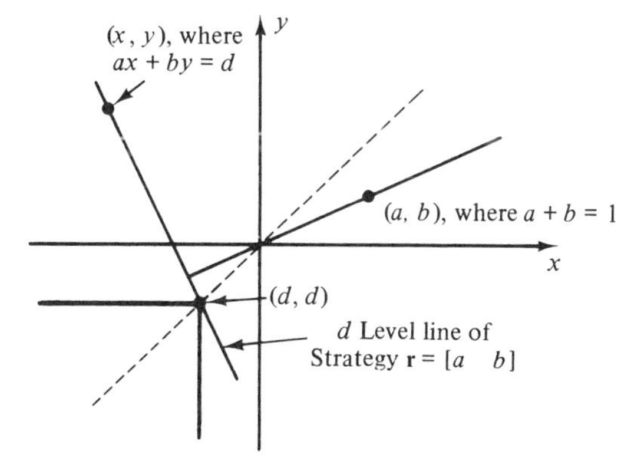

Figure 7-21.

$\mathbf{r} = [a \quad b]$ is, of course, $ax + by$, and since (x, y) is on the d level line of $[a \quad b]$, this entry in the expectation is

$$ax + by = d.$$

The Optimal Row Strategy

Now that we are able to form a geometrical picture of a row strategy in a $2 \times n$ matrix game, we can show player R how to detect the optimal row strategy. The picture is derived from the one constructed for finding the optimal column strategy. To illustrate, we return to the 2×5 matrix game G considered earlier:

$$G = \begin{bmatrix} -5 & 3 & -1 & 1 & -3 \\ 3 & -3 & 2 & -1 & 1 \end{bmatrix}.$$

Figure 7-22 displays again the polygon of (column) expectations for the game G. The sliding quadrant is shown in critical position, just touching the polygon at the point $\mathbf{u}_0 = (d_0, d_0)$ on the diagonal. From Sec. 7-3, we recognize $\mathbf{u}_0$ as the optimal column expectation. An important new ingredient has been added in the figure—we have constructed a line separating quadrant from polygon. This line we identify as the *critical line*.

We'll now argue that the critical line in Fig. 7-22 is a level line for the optimal row strategy.

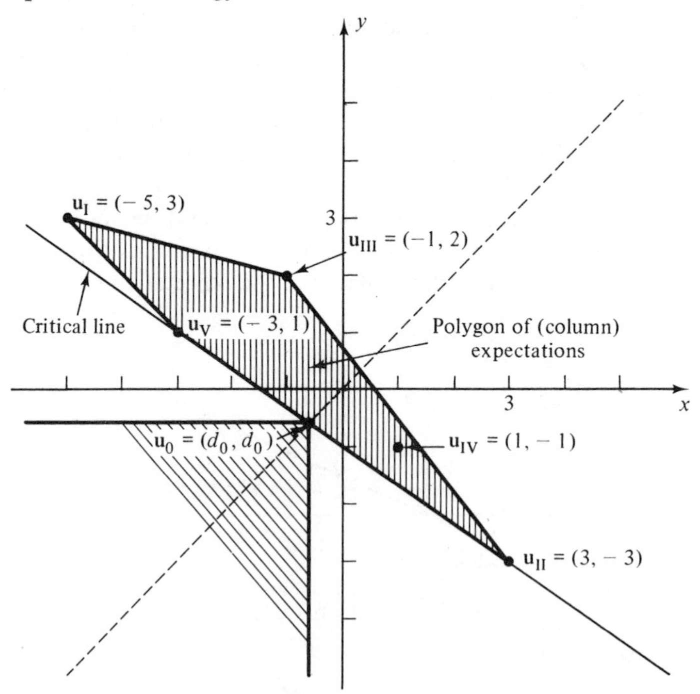

Figure 7-22.

First, the critical line was constructed so that it crosses the main diagonal at a point $\mathbf{u}_0 = (d_0, d_0)$ and does not enter the interior of the displaced negative quadrant with vertex at (d_0, d_0). From the earlier argument this assures that the critical line is a level line for some row strategy $\mathbf{r}_0 = [a_0 \quad b_0]$. Furthermore, the value of the functional $\mathbf{r}_0$ for points on this line is precisely d_0.

Second, each level line of $[a_0 \quad b_0]$ through a point $\mathbf{u}_i$ corresponding to a column of the matrix G is either the critical line itself or a line parallel to it which intersects the main diagonal above (d_0, d_0). The common coordinates of these points on the diagonal are precisely the entries in the expectation $\mathbf{r}_0 G$. Therefore, all of the entries in the expectation are greater than or equal to d_0. Furthermore, there is at least one point $\mathbf{u}_i$ (actually there are two in Fig. 7-22: $\mathbf{u}_{\text{II}}$ and $\mathbf{u}_{\text{V}}$) which lies on the critical line and thus is assigned value d_0 by the functional $r_0 = [a_0 \quad b_0]$. Consequently, the minimum value in the expectation is

$$\min \mathbf{r}_0 G = d_0. \tag{1}$$

To complete the argument we recall from the development in Sec. 7-3 that the quadrant, sliding up from below, first encounters the polygon of column expectations at a point $\mathbf{u}_0$ corresponding to a (column) expectation $\mathbf{u}_0 = G\mathbf{c}_0$. And if (d_0, d_0) is the vertex of the quadrant in this critical position, then

$$\max G\mathbf{c}_0 = d_0. \tag{2}$$

Comparing Eqs. (1) and (2), we see that $\mathbf{r}_0$ and $\mathbf{c}_0$ are qualified as optimal row and column strategies by the criterion of the Duality Theorem of Game Theory:

$$\min \mathbf{r}_0 G = \max G\mathbf{c}_0 = d_0.$$

The value of the game is d_0, the common coordinate of the vertex of the displaced quadrant in critical position.

Now let's find the optimal row strategy $\mathbf{r}_0 = [a_0 \quad b_0]$ which has the critical line as a level line. From inspection of Fig. 7-22 we see that $\mathbf{u}_{\text{II}} = (3, -3)$ and $\mathbf{u}_{\text{V}} = (-3, 1)$ both lie on this line. Since the line is a level line of the functional $[a_0 \quad b_0]$, the values assigned these points are equal:

$$a_0 \times 3 + b_0 \times (-3) = a_0 \times (-3) + b_0 \times 1,$$

$$3a_0 - 3b_0 = -3a_0 + b_0,$$

$$6a_0 = 4b_0. \tag{3}$$

But since $[a_0 \quad b_0]$ is a row strategy,

$$a_0 + b_0 = 1,$$

$$b_0 = 1 - a_0. \tag{4}$$

Substituting Eq. (4) into Eq. (3),

$$6a_0 = 4b_0 = 4(1 - a_0) = 4 - 4a_0,$$

$$10a_0 = 4,$$

$$a_0 = .4.$$

Thus from Eq. (4)

$$b_0 = 1 - a_0 = 1 - .4 = .6.$$

Therefore, the optimal row strategy is

$$\mathbf{r}_0 = [a_0 \quad b_0] = [.4 \quad .6],$$

which is precisely as verified in Sec. 7-3.

Comment: The three-step method for finding $\mathbf{r}_0$ is no more than an organization of the above algebraic steps leading to $[a_0 \quad b_0]$. The efficient technique for finding optimal strategies $\mathbf{r}_0$ and $\mathbf{c}_0$ is still the method of Sec. 7-3. But earlier we had no argument that the procedure will yield strategies $\mathbf{r}_0$ and $\mathbf{c}_0$ that satisfy the criterion of the duality theorem. Now we do.

The argument extends to any $2 \times n$ matrix game G. The polygon of (column) expectations and the displaced quadrant in critical position can always be separated by a critical line. The boundaries of polygon and quadrant intersect in a point $\mathbf{u}_0$, which may or may not be the same point as the vertex (d_0, d_0) of the quadrant. (Recall the strictly determined game in Sec. 7-3.) Necessarily the critical line contains the point of intersection $\mathbf{u}_0$, the quadrant vertex (d_0, d_0), and also a point $\mathbf{u}_i$ corresponding to a column of G.

From Sec. 7-3 we know that $\mathbf{u}_0$ is an expectation of the form $G\mathbf{c}_0$ for some column strategy $\mathbf{c}_0$. Furthermore, the common coordinate d_0 of the vertex of the quadrant satisfies

$$d_0 = \max G\mathbf{c}_0. \tag{5}$$

From the discussion in the present section the critical line is a level line for some row strategy $\mathbf{r}_0$. The level lines of $\mathbf{r}_0$ through the points corresponding to the columns of G all intersect the main diagonal at or above the vertex (d_0, d_0) of the displaced quadrant. Since at least one of these column points $\mathbf{u}_i$ lies on the critical line, and thus is assigned value d_0 by the functional $\mathbf{r}_0$, we conclude that

$$d_0 = \min \mathbf{r}_0 G. \tag{6}$$

From Eqs. (5) and (6) we see that $\mathbf{r}_0$ and $\mathbf{c}_0$ are optimal strategies by the criterion of the duality theorem, and d_0 is the value of the game.

Examples: For games corresponding to the following payoff matrices, construct the polygon of expectations with displaced quadrant in critical position and display the critical line. For each game, find and verify optimal strategies $\mathbf{r}_0$ and $\mathbf{c}_0$, and exhibit the value of the game.

(a) $G_a = \begin{bmatrix} 2 & -3 & 1 & -2 \\ -1 & 2 & 0 & 1 \end{bmatrix}.$

(b) $G_b = \begin{bmatrix} -1 & -3 & -2 \\ 0 & 1 & -1 \end{bmatrix}.$

(c) $G_c = \begin{bmatrix} 4 & 1 & -1 \\ -1 & 1 & 3 \end{bmatrix}.$

Solutions: (a) Polygon, quadrant, and critical line are shown in Fig. 7-23(a).

From the figure we see that the critical columns are I and IV. Deleting the other columns, we reduce to the 2×2 matrix

$$G' = \begin{bmatrix} 2 & -2 \\ -1 & 1 \end{bmatrix}.$$

To compress the algebra, we apply the three-step method to find optimal strategies $\mathbf{r}'_0$ and $\mathbf{c}'_0$:

$$\mathbf{r}'_0 = [\tfrac{2}{6} \quad \tfrac{4}{6}] = [\tfrac{1}{3} \quad \tfrac{2}{3}], \qquad \mathbf{c}'_0 = \begin{bmatrix} \frac{3}{6} \\ \frac{3}{6} \end{bmatrix} = \begin{bmatrix} \frac{1}{2} \\ \frac{1}{2} \end{bmatrix}.$$

These strategies are adapted to the original game, and verified as follows:

$$\mathbf{r}_0 G_a = [\tfrac{1}{3} \quad \tfrac{2}{3}] \begin{bmatrix} 2 & -3 & 1 & -2 \\ -1 & 2 & 0 & 1 \end{bmatrix} = [0 \quad \tfrac{1}{3} \quad \tfrac{1}{3} \quad 0],$$

$$G_a \mathbf{c}_0 = \begin{bmatrix} 2 & -3 & 1 & -2 \\ -1 & 2 & 0 & 1 \end{bmatrix} \begin{bmatrix} \frac{1}{2} \\ 0 \\ 0 \\ \frac{1}{2} \end{bmatrix} = \begin{bmatrix} 0 \\ 0 \end{bmatrix}.$$

We have

$$\min \mathbf{r}_0 G_a = \max G_a \mathbf{c}_0 = 0.$$

Therefore,

$$\mathbf{r}_0 = [\tfrac{1}{3} \quad \tfrac{2}{3}] \quad \text{and} \quad \mathbf{c}_0 = \begin{bmatrix} \frac{1}{2} \\ 0 \\ 0 \\ \frac{1}{2} \end{bmatrix}$$

are optimal strategies. The value of the game is 0—it is a fair game.

(b) Polygon, quadrant, and critical line are shown in Fig. 7-23(b).

We see that the quadrant intersects the polygon at the point corresponding to Column III. Thus Column III represents the optimal column expectation; accordingly, the optimal column strategy is

$$\mathbf{c}_0 = \begin{bmatrix} 0 \\ 0 \\ 1 \end{bmatrix}.$$

We also see that the critical line is horizontal. The perpendicular through (0, 0) contains the point (0, 1). Therefore, the optimal row strategy is

$$\mathbf{r}_0 = [0 \quad 1].$$

Verification:

$$\mathbf{r}_0 G_b = [0 \quad 1]\begin{bmatrix} -1 & -3 & -2 \\ 0 & 1 & -1 \end{bmatrix} = [0 \quad 1 \quad -1],$$

$$G_b \mathbf{c}_0 = \begin{bmatrix} -1 & -3 & -2 \\ 0 & 1 & -1 \end{bmatrix}\begin{bmatrix} 0 \\ 0 \\ 1 \end{bmatrix} = \begin{bmatrix} -2 \\ -1 \end{bmatrix},$$

$$\min \mathbf{r}_0 G_b = \max G_b \mathbf{c}_0 = -1.$$

Thus $\mathbf{r}_0$ and $\mathbf{c}_0$ are optimal strategies. The value of the game is -1.

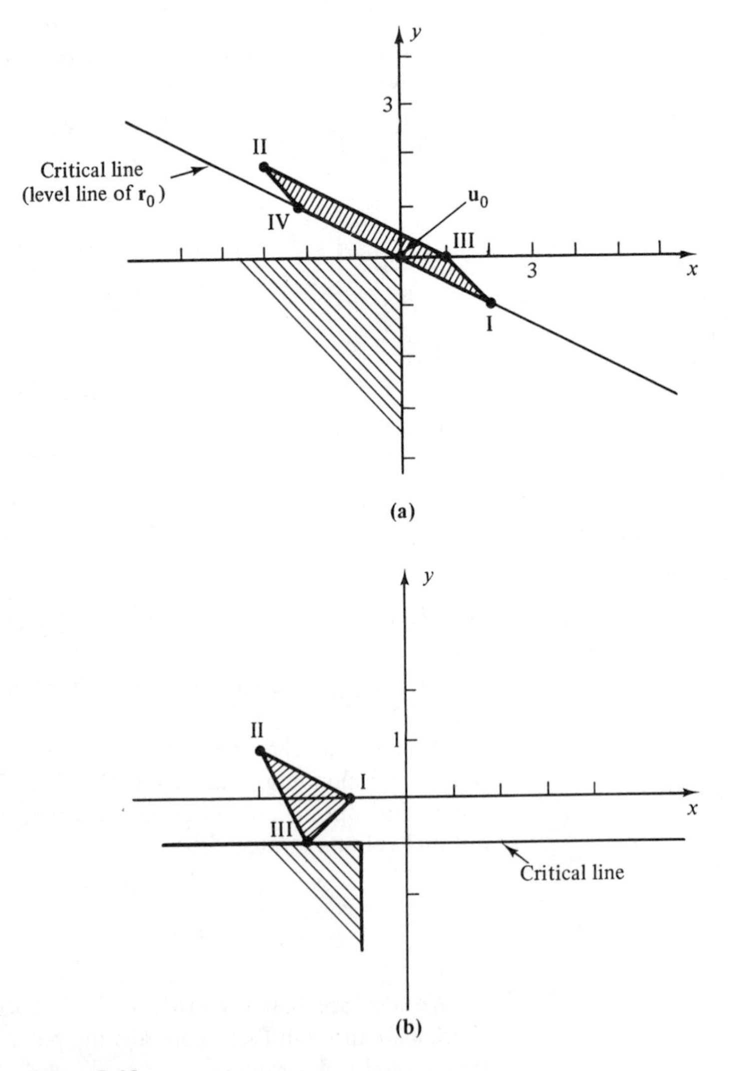

Figure 7-23.

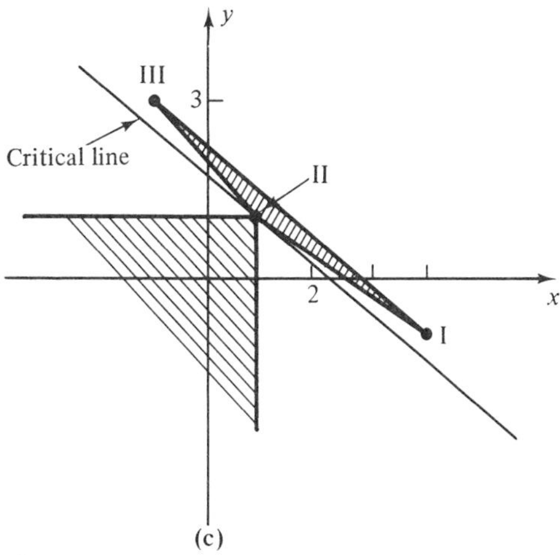

Figure 7-23 Continued.

[Of course, the game G_b is strictly determined. Its analysis requires no picture; but Fig. 7-23(b) is included to emphasize the fact that the pictorial interpretation is always valid.]

(c) As should be clear in Fig. 7-23(c), there are many admissible critical lines entering neither quadrant nor polygon, and hence many possible optimal row strategies. In Fig. 7-23(c) the critical line is taken to be parallel to the line through the points corresponding to Columns I and III.

Since the quadrant intersects the polygon at the point corresponding to Column II, the optimal column strategy is

$$\mathbf{c}_0 = \begin{bmatrix} 0 \\ 1 \\ 0 \end{bmatrix}.$$

The particular critical line selected is parallel to the line through the Column I and III points. Both lines are level lines of the functional $\mathbf{r}_0$, where $\mathbf{r}_0$ is an optimal row strategy. Thus we find $\mathbf{r}_0$ by using the three-step method to find the strategy assigning the same value to Columns I and III. Accordingly, we reduce the matrix to

$$G' = \begin{bmatrix} 4 & -1 \\ -1 & 3 \end{bmatrix}.$$

By the three-step method we find

$$\mathbf{r}_0 = [\tfrac{4}{9} \quad \tfrac{5}{9}].$$

Verification:

$$\mathbf{r}_0 G_c = [\tfrac{4}{9} \quad \tfrac{5}{9}] \begin{bmatrix} 4 & 1 & -1 \\ -1 & 1 & 3 \end{bmatrix} = [\tfrac{11}{9} \quad 1 \quad \tfrac{11}{9}],$$

$$G_c\mathbf{c}_0 = \begin{bmatrix} 4 & 1 & -1 \\ -1 & 1 & 3 \end{bmatrix} \begin{bmatrix} 0 \\ 1 \\ 0 \end{bmatrix} = \begin{bmatrix} 1 \\ 1 \end{bmatrix},$$

$$\min \mathbf{r}_0 G_c = \max G_c\mathbf{c}_0 = 1.$$

Thus $\mathbf{r}_0$ and $\mathbf{c}_0$ are optimal strategies. The value of the game is 1.

(The game G_c illustrates the one type of $2 \times n$ game that would have been troublesome by the method of Sec. 7-3.)

7-5 EXERCISES

1. (a) In the coordinate plane $\mathbf{R}^2$, plot the vectors

$$\mathbf{u}_{\text{I}} = (-2, 1) \quad \text{and} \quad \mathbf{u}_{\text{II}} = (4, -3).$$

On the graph, sketch the line containing (0, 0) and (.6, .4). Then through the point $\mathbf{u}_{\text{I}}$, sketch the level line of the row strategy [.6 .4]. Do the same for $\mathbf{u}_{\text{II}}$.

(b) Solve for the points where the level lines of part (a) cross the main diagonal.

2. Consider the game corresponding to the matrix

$$G = \begin{bmatrix} -2 & 4 & -3 & 2 \\ 1 & -3 & 2 & -3 \end{bmatrix}.$$

Let $\mathbf{r}$ be the row strategy

$$\mathbf{r} = [.6 \quad .4].$$

(a) Find the expectation $\mathbf{r}G$.

(b) Plot the points corresponding to the columns of G in the coordinate plane, and construct the level lines of the functional $\mathbf{r}$ through these points.

(c) Compare the coordinates of the points where these level lines cross the main diagonal with the entries in the expectation $\mathbf{r}G$.

3. Consider the game given by the following 2×4 payoff matrix:

$$G = \begin{bmatrix} -1 & 3 & -2 & 2 \\ 2 & -2 & 4 & -3 \end{bmatrix}.$$

(a) Construct the polygon of (column) expectations with the displaced quadrant in critical position, and display the critical line separating quadrant from polygon.

(b) From the sketch in part (a), identify the critical columns in the matrix G. By the procedure of Sec. 7-3, find optimal strategies $\mathbf{r}_0$ and $\mathbf{c}_0$, and verify.

(c) Show that the critical line of part (a) is a level line of the optimal row strategy $\mathbf{r}_0$.

4. Consider the games corresponding to the following matrices. In each case, construct the polygon of (column) expectations with the displaced quadrant in critical position, and display the critical line. Then find optimal strategies $\mathbf{r}_0$ and $\mathbf{c}_0$, and verify, concluding with a statement of the value of the game.

(a) $G = \begin{bmatrix} 14 & -11 & 9 \\ -16 & 9 & -10 \end{bmatrix}.$

(b) $G = \begin{bmatrix} 3 & -2 & 5 & -1 & 2 \\ -2 & 5 & -5 & 4 & -3 \end{bmatrix}.$

(c) $G = \begin{bmatrix} 2 & 0 & -1 \\ 0 & 1 & -2 \end{bmatrix}.$

5. Analyze the games corresponding to the following matrices. That is, find optimal strategies $\mathbf{r}_0$ and $\mathbf{c}_0$, and verify; state the value of the game.

(a) $G = \begin{bmatrix} 1 & 3 & 1 \\ 1 & 3 & 0 \end{bmatrix}.$ (b) $G = \begin{bmatrix} -5 & 8 & -12 \\ 3 & -10 & 10 \end{bmatrix}.$

6. Find two distinct optimal row strategies for each of the following games:

(a) $G = \begin{bmatrix} -1 & 0 & 1 \\ 1 & 0 & 1 \end{bmatrix}.$ (b) $G = \begin{bmatrix} 10 & 6 & 7 \\ 4 & 10 & 7 \end{bmatrix}.$

7. Consider the game given by the matrix

$$G = \begin{bmatrix} 1 & -1 & 3 \\ 2 & 3 & -1 \end{bmatrix}$$

(a) Construct the polygon of (column) expectations with the displaced quadrant in critical position, and display the critical line.

(b) Purely from inspection of the graph in part (a), guess the critical strategies $\mathbf{r}_0$ and $\mathbf{c}_0$, and the value of the game v_G. Then check your guesses by the criterion of the duality theorem.

7-6. The Dual Perspective

From the empirical interpretation of matrix games it should be apparent that the ability to solve $2 \times n$ games carries with it the ability to solve $n \times 2$ games. Whether a given game corresponds to a $2 \times n$ or an $n \times 2$ payoff matrix is purely a question of which player is identified as the row player and which as the column player.

On a given signal, Ellen extends one, two, or three fingers, while Joan extends one or two fingers. Joan pays Ellen an amount equal to the total number of fingers extended whenever that number is even; if the number is odd, Ellen pays Joan the corresponding amount.

If we identify Ellen as the row player R and Joan as the column player C, the payoff matrix is the 3×2 matrix G_1 below. On the other hand, if Joan is R and Ellen is C, we represent the game with the 2×3 matrix G_2. As always, the respective entries are the amounts the column player pays the row player.

$$G_1 = \begin{bmatrix} 2 & -3 \\ -3 & 4 \\ 4 & -5 \end{bmatrix}, \qquad G_2 = \begin{bmatrix} -2 & 3 & -4 \\ 3 & -4 & 5 \end{bmatrix}.$$

We know how to solve the 2×3 game G_2. One obtains the optimal strategies

$$\mathbf{r}_0 = [\tfrac{9}{16} \quad \tfrac{7}{16}], \qquad \mathbf{c}_0 = \begin{bmatrix} 0 \\ \tfrac{9}{16} \\ \tfrac{7}{16} \end{bmatrix}.$$

We have

$$\min \mathbf{r}_0 G_2 = \max G_2 \mathbf{c}_0 = -\tfrac{1}{16}.$$

Remembering that in the G_2 model of the game Joan is R and Ellen is C, we conclude that the optimal strategy is for Joan to show one finger with probability $\frac{9}{16}$ and two with probability $\frac{7}{16}$. And for Ellen, the optimal strategy is never to show one finger, to show two with probability $\frac{9}{16}$, and to show three with probability $\frac{7}{16}$. The value of the game is $-\frac{1}{16}$; thus Ellen can expect to win about $\frac{1}{16}$ point per play, on the average in the long run.

From the verbal description, the optimal strategies for Ellen and Joan immediately translate into optimal strategies in matrix form for the game as described by the matrix G_1 in which Ellen is R and Joan is C.

Thus optimal strategies for the game with payoff matrix G_1 are

$$\mathbf{r}_0 = [0 \quad \tfrac{9}{16} \quad \tfrac{7}{16}], \qquad \mathbf{c}_0 = \begin{bmatrix} \tfrac{9}{16} \\ \tfrac{7}{16} \end{bmatrix}.$$

We have

$$\min \mathbf{r}_0 G_1 = \max G_1 \mathbf{c}_0 = \tfrac{1}{16}.$$

In the manner indicated by this example, the optimal strategies in any game with an $n \times 2$ payoff matrix G are easily obtained from the analysis of the $2 \times n$ matrix realized by interchanging rows and columns in G and changing all signs. But let's consider the $n \times 2$ matrix in its own right.

We have been identifying a 2×1 column matrix $\begin{bmatrix} x \\ y \end{bmatrix}$ with the vector (x, y) in $\mathbf{R}^2$, and we have identified the 1×2 row matrix $[a \quad b]$ with a functional on $\mathbf{R}^2$. But the vector model for matrices is equally valid the other way around. That is, we can forget the old identifications and regard the row matrix $[x \quad y]$ as essentially the same thing as the vector (x, y) in $\mathbf{R}^2$, in which case it is inescapable that we regard the column matrix $\begin{bmatrix} a \\ b \end{bmatrix}$ as the functional that maps (x, y) into $ax + by$, because

$$[x \quad y]\begin{bmatrix} a \\ b \end{bmatrix} = [ax + by].$$

The entire discussion of the preceding sections can be reconstructed, consistently replacing column matrix by row matrix and vice versa. The $2 \times n$ matrix becomes an $n \times 2$ matrix G, and the polygon of (column) expectations becomes the polygon of (row) expectations $\mathbf{r}G$. Because we continue to interpret the entry in the payoff matrix as the amount C pays R, the crucial value min $G\mathbf{c}$ is replaced by the crucial value max $\mathbf{r}G$. Most important, the displaced negative quadrant becomes the displaced positive quadrant, which, sliding down the diagonal, first touches the polygon of (row) expectations at a point $\mathbf{r}_0G$, where $\mathbf{r}_0$ is an optimal row strategy. In this position, quadrant and polygon can be separated by a critical line which is a level line of an optimal column strategy $\mathbf{c}_0$.

The conclusion is that we can analyze $n \times 2$ games just as we analyze $2 \times n$ games. The rows are plotted in the coordinate plane, and the polygon of (row) expectations is constructed as before. The only difference is that now we slide the displaced positive quadrant down the diagonal, rather than bringing the negative quadrant up. The critical rows of the matrix are detected from this picture, and we proceed as before.

Example: Analyze the game corresponding to the following payoff matrix:

$$G = \begin{bmatrix} 1 & -3 \\ -3 & 4 \\ 5 & -4 \\ -2 & 1 \end{bmatrix}.$$

Solution: In Fig. 7-24 the rows of G are plotted as points in the coordinate plane, and the polygon of (row) expectations is constructed. The displaced positive quadrant is in critical position, and the critical line is shown separating polygon from quadrant.

We see that the optimal expectation lies on the segment between the points corresponding to Rows II and III. Thus we reduce the matrix, deleting the other rows:

$$G' = \begin{bmatrix} -3 & 4 \\ 5 & -4 \end{bmatrix}.$$

By the three-step method, optimal strategies for G' are

$$\mathbf{r}_0' = [\tfrac{9}{16} \quad \tfrac{7}{16}], \qquad \mathbf{c}_0' = \begin{bmatrix} \frac{8}{16} \\ \frac{8}{16} \end{bmatrix} = \begin{bmatrix} \frac{1}{2} \\ \frac{1}{2} \end{bmatrix}.$$

Strategy $\mathbf{r}_0'$ is adapted to the original game by assigning zero as the probability of playing Rows I or IV. Strategy $\mathbf{c}_0'$ is

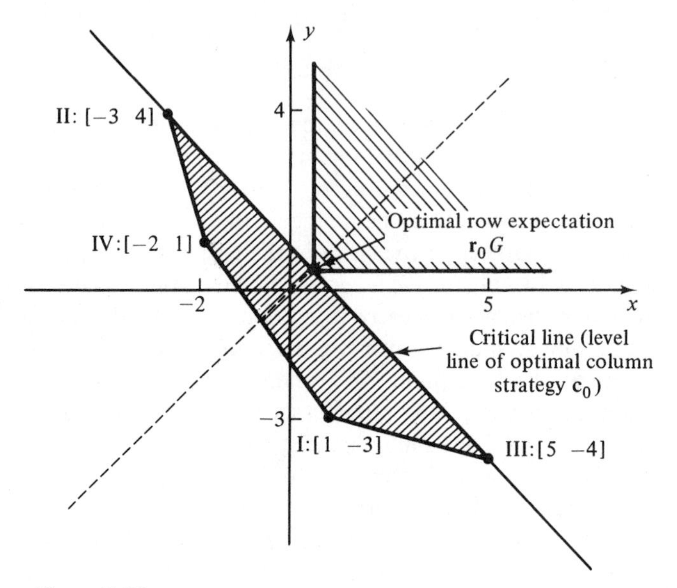

Figure 7-24.

the optimal strategy in the game as given. Thus optimal strategies are

$$\mathbf{r}_0 = [0 \quad \tfrac{9}{16} \quad \tfrac{7}{16} \quad 0], \qquad \mathbf{c}_0 = \begin{bmatrix} \tfrac{1}{2} \\ \tfrac{1}{2} \end{bmatrix}.$$

Verification:

$$\mathbf{r}_0 G = [0 \quad \tfrac{9}{16} \quad \tfrac{7}{16} \quad 0]\begin{bmatrix} 1 & -3 \\ -3 & 4 \\ 5 & -4 \\ -2 & 1 \end{bmatrix} = [\tfrac{8}{16} \quad \tfrac{8}{16}] = [\tfrac{1}{2} \quad \tfrac{1}{2}],$$

$$G\mathbf{c}_0 = \begin{bmatrix} 1 & -3 \\ -3 & 4 \\ 5 & -4 \\ -2 & 1 \end{bmatrix}\begin{bmatrix} \tfrac{1}{2} \\ \tfrac{1}{2} \end{bmatrix} = \begin{bmatrix} -1 \\ \tfrac{1}{2} \\ \tfrac{1}{2} \\ -\tfrac{1}{2} \end{bmatrix},$$

$$\min \mathbf{r}_0 G = \max G\mathbf{c}_0 = \tfrac{1}{2}.$$

Thus $\mathbf{r}_0$ and $\mathbf{c}_0$ are optimal; the value of the game is $\frac{1}{2}$.

7-6 EXERCISES

1. The payoff matrices G_1 and G_2 below represent the same game with identification of row and column players interchanged. Analyze the game G_2 by the $2 \times n$ technique of the previous sections. Then convert the optimal strategies for G_2 to optimal strategies for G_1; verify, and state the value of the game.

$$G_1 = \begin{bmatrix} 1 & -3 \\ -2 & 3 \\ 1 & -2 \\ -4 & 5 \end{bmatrix}, \qquad G_2 = \begin{bmatrix} -1 & 2 & -1 & 4 \\ 3 & -3 & 2 & -5 \end{bmatrix}.$$

2. Convert the 3×2 payoff matrix G_1 below to a 2×3 payoff matrix G_2. Find optimal strategies for G_2, and convert to optimal strategies for G_1. Verify, and state the value of the game.

$$G_1 = \begin{bmatrix} 5 & -3 \\ -5 & 2 \\ 2 & -1 \end{bmatrix}.$$

3. Consider the game given by the following 3×2 matrix:

$$G = \begin{bmatrix} -2 & 1 \\ 2 & -3 \\ -1 & 2 \end{bmatrix}.$$

(a) Adopting the dual perspective, plot the rows of G as points in the coordinate plane. Construct the polygon of (row) expectations with displaced (positive) quadrant in critical position. Display the critical line separating polygon from quadrant.

(b) From the sketch in part (a), identify the critical rows in the matrix G. Then find optimal strategies $\mathbf{r}_0$ and $\mathbf{c}_0$, and verify. State the value of the game.

4. Analyze the games given by the following matrices:

$$\text{(a)}\quad G = \begin{bmatrix} 1 & -2 \\ -3 & 2 \\ 2 & -4 \\ -2 & 3 \end{bmatrix}, \qquad \text{(b)}\quad G = \begin{bmatrix} 1 & 3 \\ 6 & 1 \\ 0 & 4 \end{bmatrix}.$$

5. Analyze the games given by the following matrices:

$$\text{(a)}\quad G = \begin{bmatrix} 3 & -6 \\ -2 & 4 \\ 2 & -5 \\ -1 & 1 \end{bmatrix}, \qquad \text{(b)}\quad G = \begin{bmatrix} 3 & -1 \\ -2 & 1 \\ 5 & -2 \\ -4 & 3 \\ 6 & -3 \end{bmatrix}.$$

6. Analyze the games given by the following matrices:

$$\text{(a)}\quad G = \begin{bmatrix} 5 & -2 & 4 & -5 \\ -4 & 4 & -2 & 3 \end{bmatrix}. \qquad \text{(b)}\quad G = \begin{bmatrix} 5 & -4 \\ -2 & 4 \\ 4 & -2 \\ -5 & 3 \end{bmatrix}.$$

7. Analyze the games given by the following matrices:

(a) $G = \begin{bmatrix} 5 & 4 & 3 & 2 \\ 0 & 4 & 2 & 6 \end{bmatrix}$, $G = \begin{bmatrix} 5 & 0 \\ 4 & 4 \\ 3 & 2 \\ 2 & 6 \end{bmatrix}$.

8. Consider the games given by the following 3×3 matrices. A critical inspection should reveal that one of the rows, or one of the columns, is never played in an optimal strategy. Deleting this row, or column, find optimal strategies for the reduced game. Then convert to optimal strategies for the game as given, verify, and state the value of the game.

(a) $G = \begin{bmatrix} 5 & -3 & 7 \\ -1 & 2 & -3 \\ 4 & -4 & 5 \end{bmatrix}$, (b) $G = \begin{bmatrix} 5 & -1 & -2 \\ -4 & 5 & 4 \\ 2 & -1 & -1 \end{bmatrix}$.

9. Analyze the games given by the following matrices. [In part (c), compare the expectation produced by strategy [.5 .5 0] with Row III.]

(a) $G = \begin{bmatrix} 9 & -5 & 8 & -3 \\ -6 & 4 & -7 & 5 \\ 4 & -3 & 3 & -2 \end{bmatrix}$.

(b) $G = \begin{bmatrix} 1 & 0 & 4 \\ 2 & 4 & 2 \\ 2 & 5 & 1 \end{bmatrix}$.

(c) $G = \begin{bmatrix} 12 & 7 & 2 \\ 2 & 5 & 8 \\ 4 & 5 & 5 \end{bmatrix}$.

7-7. Vectors and Functionals, in General

As we've seen, $2 \times k$ matrix games (and $k \times 2$ games, from the dual perspective) can fruitfully be viewed as problems concerning vectors in the space $\mathbf{R}^2$ and the functionals defined on the space. To generalize from $2 \times k$ games to $n \times k$ games, the setting of the problem must be moved from the vector space $\mathbf{R}^2$ to the vector space $\mathbf{R}^n$. Our remarks about the general space $\mathbf{R}^n$ will be very brief, and—as has been the case with $\mathbf{R}^2$—informal. We'll give a little more attention to the three-dimensional space $\mathbf{R}^3$, its geometry, and the interpretation of $3 \times n$ games.

For $n = 1, 2, 3, \ldots$, the *vector space* $\mathbf{R}^n$ consists of all ordered n-tuples of real numbers of the form

$$\mathbf{v} = (v_1, v_2, \ldots, v_n).$$

Here, $\mathbf{v}$ is a *vector* in $\mathbf{R}^n$ and the *coordinates* v_i are real numbers.

The operations of addition and scalar multiplication are defined as one should expect.

For each pair of vectors in $\mathbf{R}^n$,

$$\mathbf{u} = (u_1, u_2, \ldots, u_n),$$
$$\mathbf{v} = (v_1, v_2, \ldots, v_n),$$

the sum is the vector

$$\mathbf{u} + \mathbf{v} = (u_1 + v_1, u_2 + v_2, \ldots, u_n + v_n).$$

And for each real number t and vector $\mathbf{v}$ in $\mathbf{R}^n$, the operation of scalar multiplication determines the vector,

$$t\mathbf{v} = t(v_1, v_2, \ldots, v_n) = (tv_1, tv_2, \ldots, tv_n).$$

Thus, for example, the following are vectors in $\mathbf{R}^4$:

$$\mathbf{u} = (-2, 1, 0, \tfrac{3}{2}) \quad \text{and} \quad \mathbf{v} = (3, -2, \tfrac{1}{3}, 4).$$

Here we have

$$\mathbf{u} + \mathbf{v} = (1, -1, \tfrac{1}{3}, \tfrac{11}{2})$$

and

$$4\mathbf{u} = (-8, 4, 0, 6).$$

Since the vectors in $\mathbf{R}^n$ and the $n \times 1$ column matrices are alike in structure and enjoy the same rules for addition and scalar multiplication, we may interpret the vector space as a model of the set of column matrices. In effect, we regard the $n \times 1$ column matrix and the corresponding vector in $\mathbf{R}^n$ as being essentially the same things. (This, of course, was our original perspective with $\mathbf{R}^2$; as we witnessed in the previous section, it is subject to change.)

With the $n \times 1$ column matrices identified with the vectors in $\mathbf{R}^n$, it is natural that we identify the $1 \times n$ row matrices as functions mapping $\mathbf{R}^n$ into $\mathbf{R}$, the set of real numbers. For instance, $[1 \quad -3 \quad 5 \quad 4]$ represents the function mapping $\mathbf{R}^4$ into $\mathbf{R}$ that associates each vector

$$\mathbf{v} = (v_1, v_2, v_3, v_4)$$

with the number

$$v_1 - 3v_2 + 5v_3 + 4v_4.$$

Needless to say, this identification derives from the fact that

$$[1 \quad -3 \quad 5 \quad 4]\begin{bmatrix} v_1 \\ v_2 \\ v_3 \\ v_4 \end{bmatrix} = [t],$$

where

$$t = v_1 - 3v_2 + 5v_3 + 4v_4.$$

The function represented by the row matrix [1 −3 5 4] is said to be a *functional* on $\mathbf{R}^4$.

In general, a *functional* on $\mathbf{R}^n$ is a mapping of $\mathbf{R}^n$ into $\mathbf{R}$ that associates each vector

$$\mathbf{v} = (v_1, v_2, \ldots, v_n)$$

with the number

$$a_1v_1 + a_2v_2 + \cdots + a_nv_n,$$

where the a_i are given numbers. With the vectors identified with the corresponding column matrices, the functional is represented by the row matrix

$$[a_1 \quad a_2 \quad \ldots \quad a_n].$$

Having laid some ground for talking about vectors in a less inhibited fashion in the future, we turn our attention to the space $\mathbf{R}^3$.

The Vector Space $\mathbf{R}^3$

A vector $\mathbf{v}$ in the space $\mathbf{R}^3$ is conventionally represented as (x, y, z). For a geometrical picture of the vector (x, y, z) we extend the coordinate plane $\mathbf{R}^2$ to the coordinate space $\mathbf{R}^3$ by erecting a perpendicular "z axis." The correspondence between vector and point in space should be clear from the perspective drawing in Fig. 7-25, which focuses on the particular vector $(3, -4, 5)$.

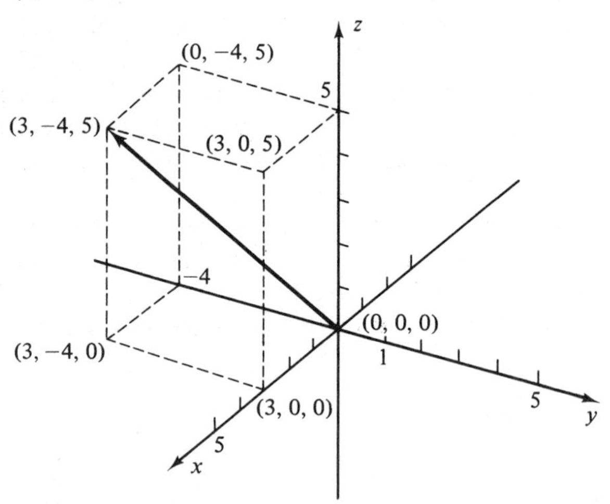

Figure 7-25 The Coordinate Space $\mathbf{R}^3$.

As in the plane $\mathbf{R}^2$, we admit two subsidiary geometrical interpretations of the vector in $\mathbf{R}^3$. In addition to identifying the vector as a point in space, we may think of the vector as represented by the ray from the origin (0, 0, 0) to that point, or as being characterized by any directed line segment of the same length and direction as that ray.

The arguments from the plane adapt to the three-dimensional configuration to produce the same interpretation of vector addition and scalar multiplication. Given any vector $\mathbf{v}_0 = (x_0, y_0, z_0)$, the vectors of the form $t\mathbf{v}_0$, where t is a real number, comprise the line through (0, 0, 0) and (x_0, y_0, z_0), with the "t values" constituting a uniform scale along this line. Also, it can be argued that, given two vectors $\mathbf{v}_1$ and $\mathbf{v}_2$, the point corresponding to the sum $\mathbf{v}_1 + \mathbf{v}_2$ is reached by proceeding from the origin along the "$\mathbf{v}_1$ ray" to the point $\mathbf{v}_1$, and thence along the directed line segment corresponding to $\mathbf{v}_2$ to the point $\mathbf{v}_1 + \mathbf{v}_2$.

Similarly, the vectors of the form

$$\mathbf{v}_1 + t\mathbf{v}_2$$

correspond to the points on the line containing $\mathbf{v}_1$ parallel to the line through (0, 0, 0) and $\mathbf{v}_2$. And the vectors of the form

$$(1 - t)\mathbf{v}_1 + t\mathbf{v}_2$$

constitute the line through the points $\mathbf{v}_1$ and $\mathbf{v}_2$, the "t values" determining a uniform scale with $t = 0$ at $\mathbf{v}_1$ and $t = 1$ at $\mathbf{v}_2$. Again, the interpretations are the same as in $\mathbf{R}^2$; indeed, the arguments supporting these interpretations are essentially the same.

The new notion that $\mathbf{R}^3$ offers is the idea of a *plane* of points. A plane is determined by three distinct points which do not lie on the same line. In particular, a plane through the origin (0, 0, 0) is determined by (0, 0, 0) and points $\mathbf{v}_1$ and $\mathbf{v}_2$, where $\mathbf{v}_1$ is distinct from the origin and $\mathbf{v}_2$ is not a scalar multiple of $\mathbf{v}_1$. From another point of view, a plane through the origin may be identified as consisting of (0, 0, 0) and the set of all vectors whose rays from the origin are perpendicular to the ray from the origin to some fixed vector $\mathbf{v}_0$. In the construction of Fig. 7-26, the plane through the origin perpendicular to the vector (3, −4, 5) is indicated by a circle of points in the plane.

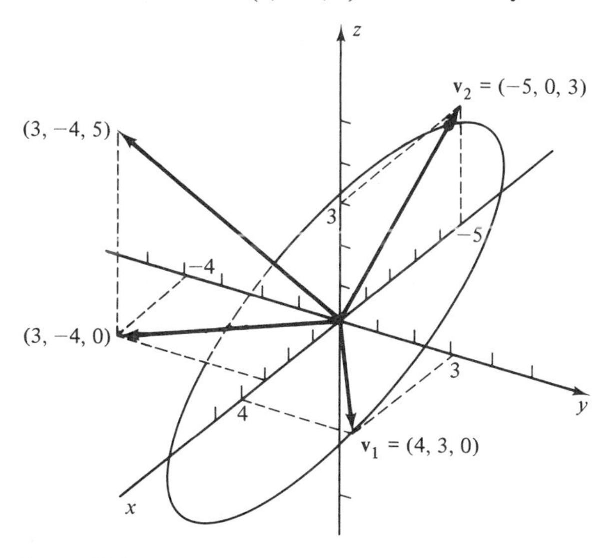

Figure 7-26.

This circle appears as an ellipse on the flat surface of the perspective drawing. Two particular vectors $\mathbf{v}_1$ and $\mathbf{v}_2$, which we shall shortly defend as lying in the plane, are shown. The vector $(3, -4, 0)$ is also included for purposes of discussion.

We turn now to the assertion that the vector $\mathbf{v}_1 = (4, 3, 0)$ lies in the plane through the origin perpendicular to the vector $(3, -4, 5)$. (As the language suggests, we are interpreting the vector geometrically as the ray from the origin. We shall continue for awhile with this interpretation.)

From Sec. 7-4 we know that in the coordinate plane $\mathbf{R}^2$ the vector $(3, -4)$ is perpendicular to the vector $(4, 3)$. In the coordinate space $\mathbf{R}^3$, the set of points of the form $(x, y, 0)$ is a replica of $\mathbf{R}^2$. Thus the vector $(3, -4, 0)$ is perpendicular to the vector $(4, 3, 0)$. Also, the z axis, i.e., the points of the form $(0, 0, z)$, is perpendicular to the plane of points of the form $(x, y, 0)$. Therefore, the vector $(0, 0, 5)$ is also perpendicular to $(4, 3, 0)$.

We see that the plane through the origin perpendicular to $(4, 3, 0)$ (not the plane in the figure) includes the points $(3, -4, 0)$ and $(0, 0, 5)$. It should be clear that if this plane includes these points, it must also include their sum—the addition of vectors by the "succession of rays" is a phenomenon that takes place entirely in a plane. It follows that the plane through the origin perpendicular to $(4, 3, 0)$ contains the vector

$$(3, -4, 0) + (0, 0, 5) = (3, -4, 5).$$

We conclude that the vector $(4, 3, 0)$ is perpendicular to the vector $(3, -4, 5)$.

It is also the case that the vector $\mathbf{v}_2 = (-5, 0, 3)$ is perpendicular to the vector $(3, -4, 5)$. To justify this assertion, we note first that in the plane of vectors of the form $(x, 0, z)$ the vector $(3, 0, 5)$ is perpendicular to the vector $(-5, 0, 3)$ by the argument of Sec. 7-4. Also, $(0, -4, 0)$ is perpendicular to $(-5, 0, 3)$ since the y axis is perpendicular to the plane of points of the form $(x, 0, z)$. Thus the plane through the origin perpendicular to $(-5, 0, 3)$ includes $(3, 0, 5)$ and $(0, -4, 0)$. This plane must also include their sum

$$(3, 0, 5) + (0, -4, 0) = (3, -4, 5).$$

We've now established that the plane through the origin perpendicular to the vector $(3, -4, 5)$ includes the points $\mathbf{v}_1 = (4, 3, 0)$ and $\mathbf{v}_2 = (-5, 0, 3)$ shown in Fig. 7-26.

This concern with the plane through the origin perpendicular to the vector $(3, -4, 5)$ anticipates the interpretation of the functional on $\mathbf{R}^3$ represented by the row matrix $[3 \quad -4 \quad 5]$. The two vectors

$$\mathbf{v}_1 = (4, 3, 0) \quad \text{and} \quad \mathbf{v}_2 = (-5, 0, 3)$$

were selected because they are assigned the value zero by the functional $[3 \quad -4 \quad 5]$. In matrix terms,

$$[3 \quad -4 \quad 5]\begin{bmatrix} 4 \\ 3 \\ 0 \end{bmatrix} = [3 \quad -4 \quad 5]\begin{bmatrix} -5 \\ 0 \\ 3 \end{bmatrix} = [0].$$

Analogous to the experience in $\mathbf{R}^2$, the fact is that this plane—through the origin perpendicular to (3, —4, 5)—consists precisely of the points assigned the value zero by the functional $[3 \quad -4 \quad 5]$; we say that it is the *zero level plane* of the functional. The argument follows.

We observed earlier that, given two points in a plane through the origin, the sum of the points must also lie in the plane. In particular, in the plane at hand, determined by (0, 0, 0), $\mathbf{v}_1 = (4, 3, 0)$ and $\mathbf{v}_2 = (-5, 0, 3)$, the sum $\mathbf{v}_1 + \mathbf{v}_2$ is in the plane. Again the argument is that the "succession of rays" is a phenomenon that takes place entirely in a plane. By this "succession of rays" argument the vector

$$-2\mathbf{v}_1 + 3\mathbf{v}_2$$

also lies in the plane, as indeed does every vector of the form

$$s\mathbf{v}_1 + t\mathbf{v}_2$$

for scalars s and t. Moreover, we can reach any point in the plane by proceeding first in the "$\mathbf{v}_1$ direction" to a point $s\mathbf{v}_1$, and thence in the "$\mathbf{v}_2$ direction" to $s\mathbf{v}_1 + t\mathbf{v}_2$. Thus, in fact, the plane is comprised precisely of points of this form.

It is easy to show that every vector of the form $s\mathbf{v}_1 + t\mathbf{v}_2$, i.e., every point in the plane, is assigned value zero by the functional [3, —4, 5]:

$$\begin{aligned} s\mathbf{v}_1 + t\mathbf{v}_2 &= s(4, 3, 0) + t(-5, 0, 3) \\ &= (4s - 5t, 3s, 3t). \end{aligned}$$

Applying the functional $[3 \quad -4 \quad 5]$,

$$[3 \quad -4 \quad 5]\begin{bmatrix} 4s - 5t \\ 3s \\ 3t \end{bmatrix} = [0s + 0t] = [0].$$

On the other hand, let (x, y, z) be an arbitrary vector assigned value zero by the functional $[3 \quad -4 \quad 5]$. That is, suppose

$$3x - 4y + 5z = 0$$

or

$$x = \frac{1}{3}(4y - 5z) = 4 \times \frac{y}{3} - 5 \times \frac{z}{3}.$$

We may express (x, y, z) in the form

$$\begin{aligned} (x, y, z) &= \left(4 \times \frac{y}{3} - 5 \times \frac{z}{3}, 3 \times \frac{y}{3}, 3 \times \frac{z}{3}\right) \\ &= \frac{y}{3}(4, 3, 0) + \frac{z}{3}(-5, 0, 3). \end{aligned}$$

It follows that

$$(x, y, z) = s\mathbf{v}_1 + t\mathbf{v}_2,$$

where $s = y/3$ and $t = z/3$. This is the form that places (x, y, z) in the plane determined by (0, 0, 0), $\mathbf{v}_1$ and $\mathbf{v}_2$, which, of course, is the plane through the origin perpendicular to the vector (3, −4, 5). We've now qualified the plane as the zero level plane of the functional $[3 \quad -4 \quad 5]$.

The arguments from Sec. 7-4 about level lines of functionals on $\mathbf{R}^2$ adapt to the three-dimensional configuration to lead to the conclusion that the sets of points for which $[3 \quad -4 \quad 5]$ assumes constant values are the parallel planes in $\mathbf{R}^3$ perpendicular to the line containing (0, 0, 0) and (3, −4, 5). Needless to say, we identify these planes as the *level planes* of the functional.

We started with the vector (3, −4, 5). One may start with any vector (a, b, c), distinct from (0, 0, 0), and reconstruct the entire argument. The conclusion is as follows:

Given a functional $[a \quad b \quad c]$ on $\mathbf{R}^3$, a, b, and c not all zero, then for each fixed real number k the set of points (x, y, z) in the coordinate space $\mathbf{R}^3$ comprises a plane, called the *k level plane* of the functional, which is perpendicular to the line containing (0, 0, 0) and the point (a, b, c). Conversely, each plane perpendicular to this line is the k level plane of the functional $[a \quad b \quad c]$ for some value of k.

Of course, if $[a \quad b \quad c] = [0 \quad 0 \quad 0]$, then the entire space $\mathbf{R}^3$ is mapped into zero.

The Analysis of 3 × *n* Games

Inasmuch as the motivation for considering vectors and functionals has been the analysis of matrix games, some remarks are in order about the insight the geometry of $\mathbf{R}^3$ brings to the analysis of $3 \times n$ games. The analogy with the pictorial representation of $2 \times n$ games is perfect, but the actual perspective drawings tend to be too confusing to be of practical value. We'll give a verbal description of the general argument and leave the construction of the spatial configurations to the imagination.

The general $3 \times n$ payoff matrix is of the form

$$G = \begin{bmatrix} g_{11} & g_{12} & \cdots & g_{1n} \\ g_{21} & g_{22} & \cdots & g_{2n} \\ g_{31} & g_{32} & \cdots & g_{3n} \end{bmatrix}.$$

The columns of G are regarded as points in the coordinate space $\mathbf{R}^3$:

$$\mathbf{u}_1 = (g_{11}, g_{21}, g_{31}), \mathbf{u}_2 = (g_{12}, g_{22}, g_{32}), \ldots, \mathbf{u}_n = (g_{1n}, g_{2n}, g_{3n}).$$

For a column strategy $\mathbf{c}$, i.e., an $n \times 1$ column matrix $\mathbf{c}$ of nonnegative entries totaling one, the expectation is as follows (the "operational view"):

$$G\mathbf{c} = \begin{bmatrix} g_{11} & g_{12} & \cdots & g_{1n} \\ g_{21} & g_{22} & \cdots & g_{2n} \\ g_{31} & g_{32} & \cdots & g_{3n} \end{bmatrix} \begin{bmatrix} c_1 \\ c_2 \\ \cdot \\ \cdot \\ \cdot \\ c_n \end{bmatrix} = c_1 \begin{bmatrix} g_{11} \\ g_{22} \\ g_{31} \end{bmatrix} + c_2 \begin{bmatrix} g_{12} \\ g_{22} \\ g_{32} \end{bmatrix} + \cdots + c_n \begin{bmatrix} g_{1n} \\ g_{2n} \\ g_{3n} \end{bmatrix}.$$

As points in $\mathbf{R}^3$, the set of column expectations is the set of points of the form

$$\mathbf{v} = c_1\mathbf{u}_1 + c_2\mathbf{u}_2 + \cdots + c_n\mathbf{u}_n, \tag{1}$$

where the points $\mathbf{u}_i$ correspond to the columns of the matrix G, and the c_i are nonnegative numbers totaling one.

If $n = 2$, i.e., G has only two columns, then the points satisfying Eq. (1) are of the form

$$c_1\mathbf{u}_1 + c_2\mathbf{u}_2,$$

where c_1 and c_2 are nonnegative and $c_1 + c_2 = 1$. As observed earlier, the arguments from $\mathbf{R}^2$ extend to $\mathbf{R}^3$ to show that the points of this form comprise the line segment with end points $\mathbf{u}_1$ and $\mathbf{u}_2$. For $n = 3$ the points of the form of Eq. (1),

$$c_1\mathbf{u}_1 + c_2\mathbf{u}_2 + c_3\mathbf{u}_3,$$

the c_i nonnegative and totaling one, comprise the triangle with vertices $\mathbf{u}_1$, $\mathbf{u}_2$, and $\mathbf{u}_3$—the argument of Sec. 7-2 applies in the new setting. For $n = 4$ we reach a three-dimensional configuration. In this case, points of the form of Eq. (1),

$$c_1\mathbf{u}_1 + c_2\mathbf{u}_2 + c_3\mathbf{u}_3 + c_4\mathbf{u}_4,$$

the c_i nonnegative and totaling one, comprise the tetrahedron with vertices $\mathbf{u}_1$, $\mathbf{u}_2$, $\mathbf{u}_3$, and $\mathbf{u}_4$. (One may think of the tetrahedron as a pyramid having the triangle with vertices $\mathbf{u}_1$, $\mathbf{u}_2$, and $\mathbf{u}_3$ as base and its peak at $\mathbf{u}_4$). In some cases the "tetrahedron" may degenerate to a triangle or a line segment—or to just a point if the four $\mathbf{u}_i$ are all the same.

For the general result, we recall that a set in $\mathbf{R}^2$ is said to be *convex* if the line segment between any two points of the set lies entirely in the set. Exactly the same definition of convex set applies in $\mathbf{R}^3$ (in fact, in $\mathbf{R}^n$ in general).

For the $3 \times n$ game G the set of column expectations $G\mathbf{c}$ is the smallest convex set that contains the points $\mathbf{u}_i$ corresponding to the columns of the matrix G, on the boundary or in the interior. This set—the set of points of the form of Eq. (1)—is called the *polyhedron* of (column) expectations.

Analogous to the polygon of expectations in $\mathbf{R}^2$, the polyhedron of expectations may be visualized as if the points $\mathbf{u}_i$ had been completely enclosed by a stretched spherical surface that has then contracted as tightly as possible against the fixed points $\mathbf{u}_i$.

An optimal column strategy $\mathbf{c}_0$ is determined by the property that

$$\max G\mathbf{c}_0 \leq \max G\mathbf{c}$$

for all column strategies c. The pictorial location of an optimal expectation $G\mathbf{c}_0$ is the analogue of the construction in the $\mathbf{R}^2$ situation. We start with the *negative octant* consisting of points (x, y, z), where $x \leq 0$, $y \leq 0$, and $z \leq 0$, displaced down the main diagonal, $x = y = z$, below the polyhedron of expectations. Then we slide the octant up the diagonal until it first touches the polyhedron of expectations. With the displaced octant in this critical position it can be argued that a point of intersection of octant and polyhedron is an optimal column expectation $G\mathbf{c}_0$. Furthermore, for this point we have

$$\max G\mathbf{c}_0 = d,$$

where d is the common coordinate of the point (d, d, d) on the diagonal serving as vertex of the displaced octant in critical position.

From the row player's point of view, a row strategy $\mathbf{r}$ is a functional $[a \quad b \quad c]$, where a, b, and c are nonnegative numbers totaling one. Such a functional is characterized by the fact that its zero level plane does not enter the interior of the negative octant. By essentially the same argument advanced in Sec. 7-5 for the $\mathbf{R}^2$ situation, the value of the functional on each of its level planes is the common coordinate d of the point (d, d, d) where the main diagonal intersects the plane.

Now we return to the polyhedron of (column) expectations with displaced octant in critical position. Both polyhedron and octant are convex sets, and the only points they have in common are boundary points of both sets. This being the case, there must exist a *critical plane* such that the points of the polygon lie on the plane or to one side of it, while the points of the displaced octant lie on the plane or to the other side. The critical plane is analogous to the critical line in $\mathbf{R}^2$.

Comment: The existence of a plane separating two nonempty convex sets which have only boundary points in common is the mathematical key to the whole analysis. Indeed, this important result, called the *separation principle*, extends to all vector spaces $\mathbf{R}^n$. For $\mathbf{R}^2$ and $\mathbf{R}^3$, we leave the existence of critical line and critical plane on an intuitive geometrical basis—which has been our position all along.

Since the critical plane does not intrude into the interior of the displaced octant, it is a level plane for some row strategy—call it $\mathbf{r}_0$. Necessarily the critical plane contains the vertex (d, d, d) of the displaced octant, and it also must contain at least one of the points corresponding to a column of the matrix G. Level planes of the functional $\mathbf{r}_0$ containing other columns of the matrix intersect the diagonal at or above the point (d, d, d). Since the common coordinate of the point where a level plane intersects the diagonal is the value of the functional for that plane, we conclude that

$$\min \mathbf{r}_0 G = d. \tag{2}$$

As has already been observed, the octant in critical position intersects the polyhedron of expectations at a point of the form $G\mathbf{c}_0$, where $\mathbf{c}_0$ is a column strategy. And since the octant has its vertex at the point (d, d, d) on the diagonal, it follows that

$$\max G\mathbf{c}_0 = d. \tag{3}$$

From Eqs. (2) and (3), we see that $\mathbf{r}_0$ and $\mathbf{c}_0$ are optimal strategies by the Duality Theorem of Game Theory:

$$\min \mathbf{r}_0 G = \max G\mathbf{c}_0 = d.$$

The value of the game is d, the common coordinate of the vertex of the displaced octant in critical position.

Thus we have a plausible geometrical argument that every $3 \times n$ game has optimal strategies satisfying the criterion demanded for verification. Of course, the geometrical argument is no mathematical proof. But the formal mathematical definitions of the underlying concepts—vector, functional, line, plane, convex set—are consistent with the geometrical interpretations. And the "sliding quadrant" and "sliding octant" arguments suggest the steps in a formal proof.

In the formal setting, the proof can be extended to show that every $n \times k$ matrix game does have optimal strategies $\mathbf{r}_0$ and $\mathbf{c}_0$ that satisfy the criterion of the Duality Theorem of Game Theory.

8

LINEAR PROGRAMMING

A major advantage of the mathematical perspective derives from the fact that the concepts and theory developed for serving one problem are often of value in analyzing situations of a seemingly different character. From the general nature of matrices and vectors, it should come as no surprise that these concepts are important in many applications of mathematics. Indeed, one may protest that the treatment of matrices and vectors in the previous two chapters has been overly biased toward serving the rather special problem of matrix games.

It may come as more of a surprise that matrix games themselves are not so special as one might guess. The applications to which we now turn not only put matrices and vectors to work in a new setting, but they also offer a new interpretation for the mathematics of game theory. Left behind in the passage is the empirical concept that has motivated all of our work since Chapter 2—the new problems have nothing to do with probability.

8-1. A Diet Problem

A natural reaction to the problem posed below is that we are shifting from matrix games to a brand new subject. As we shall see, we haven't budged an inch.

After a little research, a hobbyist discovers the following facts relevant to the economics of feeding the fish in his aquarium:

The diet of the fish requires three nutrients, X, Y, and Z, and available on the market are two foods, U and V, containing these nutrients.

A dollar's worth of food U contains a 4-day supply of nutrient X, a 2-day supply of nutrient Y, and a 3-day supply of nutrient Z; and a dollar's worth of food V contains a 3-day supply of X, a 6-day supply of Y, and a 5-day supply of Z.

Question: What is the maximum length of time that the fish's dietary requirements can be met by an expenditure of one dollar on a mixture of foods U and V? Also, how is the dollar divided in achieving this most judicious mixture?

The information is more easily read from the following tabulation:

		Nutrient		
		X	Y	Z
Food	U	4	2	3
	V	3	6	5

Entry is the amount of nutrient, in number of days, supplied by $1 worth of the food.

The process is to divide the one-dollar expenditure between foods U and V. For reasons soon to be apparent, we represent the possibilities as 1×2 row matrices $[u \quad v]$, where u and v are nonnegative numbers totaling one. For instance, the matrix $[.4 \quad .6]$, which we refer to as a "diet," represents the statement "40 cents is spent on U and 60 cents is spent on V."

To develop the mathematical structure, the given data are used to define a function which associates each diet $[u \quad v]$ with a 1×3 "nutritional-content" matrix $[x \quad y \quad z]$, where x is the number of days of nutrient X supplied by u dollars worth of U and v dollars worth of V, and analogously for y and z.

To illustrate, let's find the nutritional-content matrix $[x_1 \quad y_1 \quad z_1]$ corresponding to the diet $[u_1 \quad v_1] = [.4 \quad .6]$.

Since $1 worth of U furnishes a 4-day supply of X, the amount of X, in number of days, supplied by $.40 worth of U is

$$.40 \times 4 = 1.6.$$

Similarly, since \$1 worth of V furnishes a 3-day supply of X, the amount of X supplied by \$.60 worth of V is

$$.60 \times 3 = 1.8.$$

Thus for $[u_1 \quad v_1] = [.4 \quad .6]$, we have

$$x_1 = 1.6 + 1.8 = 3.4.$$

That is, 40 cents worth of U and 60 cents worth of V meets the X requirement for 3.4 days.

Analogously, the amount of Y, in number of days, supplied by the diet $[u_1 \quad v_1] = [.4 \quad .6]$ is

$$y_1 = .4 \times 2 + .6 \times 6 = .8 + 3.6 = 4.4.$$

And the amount of Z furnished by this diet is

$$z_1 = .4 \times 3 + .6 \times 5 = 1.2 + 3.0 = 4.2.$$

Therefore, the diet $[u_1 \quad v_1] = [.4 \quad .6]$, representing the way the dollar is divided between foods U and V, is mapped into the nutritional-content matrix

$$[x_1 \quad y_1 \quad z_1] = [3.4 \quad 4.4 \quad 4.2],$$

which represents the number of days the respective requirements are met.

The computation of $[x_1 \quad y_1 \quad z_1]$ from $[u_1 \quad v_1]$ may sound like a matrix multiplication; indeed it is. A check of the spirit, and the details, of the procedure reveals that the calculation is given by the matrix product

$$[x_1 \quad y_1 \quad z_1] = [u_1 \quad v_1]\begin{bmatrix} 4 & 2 & 3 \\ 3 & 6 & 5 \end{bmatrix} = [.4 \quad .6]\begin{bmatrix} 4 & 2 & 3 \\ 3 & 6 & 5 \end{bmatrix}$$

$$= [3.4 \quad 4.4 \quad 4.2].$$

The computation is based on the 2×3 matrix

$$M = \begin{bmatrix} 4 & 2 & 3 \\ 3 & 6 & 5 \end{bmatrix},$$

which can be viewed as having been lifted directly from the tabulation of the data.

In general, for any diet $[u \quad v]$, where u and v are nonnegative numbers totaling one, the nutritional-content matrix $[x \quad y \quad z]$ is given by the product

$$[x \quad y \quad z] = [u \quad v]M = [u \quad v]\begin{bmatrix} 4 & 2 & 3 \\ 3 & 6 & 5 \end{bmatrix}.$$

With respect to the question posed in the problem, the crucial entry in the matrix

$$[x_1 \quad y_1 \quad z_1] = [3.4 \quad 4.4 \quad 4.2],$$

corresponding to $[u_1 \quad v_1] = [.4 \quad .6]$, is the minimum value $x_1 = 3.4$. All three nutritional requirements are met for 3.4 days, but no longer, because nutrient X is exhausted after 3.4 days.

For contrast, let's test the diet

$$[u_2 \quad v_2] = [.5 \quad .5],$$

corresponding to 50 cents worth of U and 50 cents worth of V. The nutritional-content matrix is

$$[x_2 \quad y_2 \quad z_2] = [u_2 \quad v_2]M = [.5 \quad .5]\begin{bmatrix} 4 & 2 & 3 \\ 3 & 6 & 5 \end{bmatrix}$$
$$= [3.5 \quad 4.0 \quad 4.0].$$

Since

$$3.5 = \min[x_2 \quad y_2 \quad z_2] = \min[3.5 \quad 4.0 \quad 4.0],$$

we conclude that the diet $[u_2 \quad v_2] = [.5 \quad .5]$ sustains the fish for 3.5 days. That's better than 3.4 days, so we identify $[u_2 \quad v_2]$ as a *better* diet than $[u_1 \quad v_1]$. The problem, of course, is to find the best diet.

With this background, it should be apparent that the "Fish-Diet" problem may be formulated as follows:

Problem: Given the 2×3 matrix

$$M = \begin{bmatrix} 4 & 2 & 3 \\ 3 & 6 & 5 \end{bmatrix},$$

find a 1×2 row matrix $\mathbf{r} = [u \quad v]$, where u and v are nonnegative numbers totaling one, for which

$$\min \mathbf{r}M = \min[x \quad y \quad z]$$

is as large as possible.

Mathematically, this is precisely the problem of finding an optimal row strategy $\mathbf{r}_0$ in the game with payoff matrix M. The techniques evolved for solving matrix games can be applied without reservation to solve the new problem. As in the previous chapter, we draw a picture to spot the critical columns in the matrix M and employ the three-step method to find $\mathbf{r}_0$ and $\mathbf{c}_0$. Of course, it's only $\mathbf{r}_0$ that interests us, but we need $\mathbf{c}_0$ to verify the correctness of $\mathbf{r}_0$. The solution is as follows:

Solution: The optimal 1×2 row matrix called for in the matrix statement of the problem is

$$\mathbf{r}_0 = [.6 \quad .4].$$

Verification: We note that

$$\mathbf{r}_0 M = [.6 \quad .4]\begin{bmatrix} 4 & 2 & 3 \\ 3 & 6 & 5 \end{bmatrix} = [3.6 \quad 3.6 \quad 3.8].$$

Now consider the 3×1 column matrix

$$\mathbf{c}_0 = \begin{bmatrix} .8 \\ .2 \\ 0 \end{bmatrix}$$

of nonnegative entries totaling one. Then

$$M\mathbf{c}_0 = \begin{bmatrix} 4 & 2 & 3 \\ 3 & 6 & 5 \end{bmatrix} \begin{bmatrix} .8 \\ .2 \\ 0 \end{bmatrix} = \begin{bmatrix} 3.6 \\ 3.6 \end{bmatrix}.$$

We have

$$\min \mathbf{r}_0 M = \max M\mathbf{c}_0 = 3.6.$$

By the Duality Theorem of Game Theory it follows that if $\mathbf{r} = [u \quad v]$ is any 1×2 row matrix of nonnegative entries totaling one, then

$$\min \mathbf{r}_0 M \geq \min \mathbf{r}M.$$

Therefore, $\mathbf{r}_0 = [.6 \quad .4]$ is the solution of the problem.

In terms of the original statement of the Fish-Diet problem we conclude that the diet $\mathbf{r}_0 = [.6 \quad .4]$ sustains the fish for 3.6 days, and that no diet $\mathbf{r} = [u \quad v]$ will sustain them for longer than this. To answer the original question in its own language:

Answer: The maximum length of time that the fish's dietary requirements can be met by an expenditure of one dollar on a mixture of foods U and V is 3.6 days. This maximum time is realized with 60 cents worth of U and 40 cents worth of V.

Comment: We've see that the Fish-Diet problem is mathematically inseparable from a corresponding problem in game theory. True, game theory involves probability and the diet problem does not, but the probabilistic interpretation is a phenomenon at the empirical level of the analysis, and not the mathematical.

A Production-Scheduling Problem

It may be apparent that the problem stated below enjoys a certain kinship with the "Fish-Diet" problem above. What may not be so apparent is that, in a sense, it is the same problem.

A factory is set up to produce three products X, Y, and Z. Each unit of each product produces a profit of one dollar.

There are two machines, U and V, both of which are employed in the manufacture of each product. A lot of 1000 units of product X requires 4 days' time on machine U and also requires 3 days' time on machine V. A 1000-unit lot of Y requires 2 days on U and 6 days on V, and a 1000-unit lot of Z requires 3 days on U and 5 days on V.

Question: How many units of each product should be manufactured in order to produce a total of 1000 units, and therefore a profit of \$1000, in the least possible time?

Reminiscent of the Fish-Diet problem, the data are presented in the following table:

		Product		
		X	Y	Z
Machine	U	4	2	3
	V	3	6	5

Entry is the number of days on the machine required in the production of a 1000-unit lot of the product.

Here the process is to divide the 1000-unit mixed lot into fractions representing the proportions of products X, Y, and Z contained. We represent the possibilities as 1×3 column matrices $\begin{bmatrix} x \\ y \\ z \end{bmatrix}$, or, for typographical convenience, as vectors (x, y, z) in $\mathbf{R}^3$, where x, y, and z are nonnegative numbers totaling one. For instance, the vector $(x_1, y_1, z_1) = (.3, .5, .2)$ represents a lot consisting of 300 units of X, 500 units of Y, and 200 units of Z. We refer to these vectors, or matrices, as "production schedules."

The next step is to use the given data to define a function which associates each production schedule (x, y, z) with a 2×1 "time-requirement" matrix $\begin{bmatrix} u \\ v \end{bmatrix}$, which may be identified as a vector (u, v) in $\mathbf{R}^2$, where u is the number of days required on machine U to produce the 1000-unit lot represented by (x, y, z) and v is the time required on machine V.

To illustrate, let's find the time-requirement vector (u_1, v_1) corresponding to the production schedule $(x_1, y_1, z_1) = (.3, .5, .2)$.

Since 1000 units of X requires 4 days' time on machine U, the number of days on U required to produce $.3 \times 1000$ units of X is

$$.3 \times 4 = 1.2.$$

Similarly, since 1000 units of Y require 2 days on U, the number of days required by $.5 \times 1000$ units is

$$.5 \times 2 = 1.0.$$

And since 1000 units of Z require 3 days on U, the number of days required by $.2 \times 1000$ units is

$$.2 \times 3 = .6.$$

All in all, the number of days required for machine U to do its part in producing the 1000-unit lot represented by the production schedule $(x_1, y_1, z_1) = (.3, .5, .2)$ is

$$u_1 = .3 \times 4 + .5 \times 2 + .2 \times 3 = 2.8.$$

Analogously, the number of days required on machine V to produce this lot is

$$v_1 = .3 \times 3 + .5 \times 6 + .2 \times 5 = 4.9.$$

Thus the production schedule $(x_1, y_1, z_1) = (.3, .5, .2)$ is associated with the time-requirement vector $(u_1, v_1) = (2.8, 4.9)$.

The reasoning and the arithmetic involved in computing (u_1, v_1) from (x_1, y_1, z_1) is given by the matrix product

$$\begin{bmatrix} u_1 \\ v_1 \end{bmatrix} = \begin{bmatrix} 4 & 2 & 3 \\ 3 & 6 & 5 \end{bmatrix} \begin{bmatrix} x_1 \\ y_1 \\ z_1 \end{bmatrix}$$

$$= \begin{bmatrix} 4 & 2 & 3 \\ 3 & 6 & 5 \end{bmatrix} \begin{bmatrix} .3 \\ .5 \\ .2 \end{bmatrix} = \begin{bmatrix} 2.8 \\ 4.9 \end{bmatrix}.$$

As in the earlier Fish-Diet problem, the computation is based on the matrix

$$M = \begin{bmatrix} 4 & 2 & 3 \\ 3 & 6 & 5 \end{bmatrix},$$

which we again view as being lifted directly from the tabulation of the data.

In general, for any production schedule (x, y, z) of nonnegative entries totaling one, the time-requirement vector (u, v) is given in matrix form by the product

$$\begin{bmatrix} u \\ v \end{bmatrix} = M \begin{bmatrix} x \\ y \\ z \end{bmatrix} = \begin{bmatrix} 4 & 2 & 3 \\ 3 & 6 & 5 \end{bmatrix} \begin{bmatrix} x \\ y \\ z \end{bmatrix}.$$

Comment: Although the matrix M is exactly as in the Fish-Diet problem, one should not confuse the vectors—equivalently column matrices—(x_1, y_1, z_1) and (u_1, v_1) with the row matrices

$$[x_1 \quad y_1 \quad z_1] \quad \text{and} \quad [u_1 \quad v_1]$$

of the earlier problem. They are different kinds of things, and their roles are different. Previously we witnessed the row matrix $[u_1 \quad v_1] = [.4 \quad .6]$ determining the row matrix $[x_1 \quad y_1 \quad z_1] = [3.4 \quad 4.4 \quad 4.2]$; now we've seen the vector $(x_1, y_1, z_1) = (.3, .5, .2)$ determine the vector $(u_1, v_1) = (2.8, 4.9)$.

The problem is to determine how the 1000-unit lot should be mixed so that it can be produced in the least possible time. Since the production schedule $(x_1, y_1, z_1) = (.3, .5, .2)$ requires 2.8 days' time on machine U and 4.9 days' time on machine V, the time required to produce this lot is 4.9 days. We have to wait on machine V while machine U sits idle for 2.1 days. Which is simply to observe that the value of interest in the time-requirement vector (2.8, 4.9) is the maximum of the entries.

And so it is in general. If (x, y, z) is any production schedule, then the number of days required by the factory to produce the corresponding 1000-unit lot is

$$\max(u, v),$$

where (u, v) is the time-requirement vector obtained from (x, y, z). The problem is to find a production schedule (x, y, z) for which $\max(u, v)$ is as small as possible.

With this background the problem is reformulated as follows:

Problem: Given the 2×3 matrix

$$M = \begin{bmatrix} 4 & 2 & 3 \\ 3 & 6 & 5 \end{bmatrix},$$

find a 3×1 column matrix

$$\mathbf{c} = \begin{bmatrix} x \\ y \\ z \end{bmatrix}$$

of nonnegative entries totaling one such that

$$\max M\mathbf{c} = \max \begin{bmatrix} u \\ v \end{bmatrix}$$

is as small as possible.

Of course, this is precisely the problem faced by the column player in analyzing the game with payoff matrix M.

We've already seen that the Fish-Diet problem is mathematically equivalent to the row player's problem in the game corresponding to the matrix M. In that context we analyzed the game, finding optimal strategies $\mathbf{r}_0$ and $\mathbf{c}_0$ and verifying. With respect to the present problem we now see that the optimal column strategy $\mathbf{c}_0$ represents the production schedule that produces a 1000-unit lot in the least possible time.

From the earlier analysis we recall that the solution of the problem is given by

$$\mathbf{c}_0 = \begin{bmatrix} .8 \\ .2 \\ 0 \end{bmatrix}.$$

The time-requirement vector, or matrix, corresponding to this schedule is

$$\begin{bmatrix} u \\ v \end{bmatrix} = M\mathbf{c}_0 = \begin{bmatrix} 4 & 2 & 3 \\ 3 & 6 & 5 \end{bmatrix} \begin{bmatrix} .8 \\ .2 \\ 0 \end{bmatrix} = \begin{bmatrix} 3.6 \\ 3.6 \end{bmatrix}.$$

To answer the question as originally posed,

Answer: To realize a 1000-unit lot, and therefore a \$1000 profit, in the least possible time the factory should produce 800 units of product X, 200 units of Y, and no Z. With this combination the time required is 3.6 days; the factory can do no better.

Comment: We've seen two distinct empirical problems, neither of which involves probability, as being equivalent to the perspectives of the row and column players, respectively, in the game with payoff matrix M. In solving one, we solved the other. But no attempt has been made to interpret empirically the solution of the one problem in the context of the other. One must wonder whether the optimal column strategy $\mathbf{c}_0$, which solved the production-scheduling problem, tells us anything about the economics of feeding those fish. Here we have a case of the mathematics suggesting that there may be an ingredient in the problem not recognized in advance—and an important ingredient it is, as we shall see in Sec. 8-5.

8-1 EXERCISES

1. A diet requires two nutrients, X and Y, and available on the market are two foods, U and V, containing these nutrients. A dollar's worth of food U contains a 4-day supply of X and an 8-day supply of Y, while a dollar's worth of V contains a 7-day supply of X and a 5-day supply of Y.

(a) For how many days are the X and Y requirements, respectively, met by an expenditure of 40 cents on food U and 60 cents on food V?

(b) For nonnegative numbers, u and v, totaling one, let $[u \quad v]$ represent the diet consisting of u dolllars worth of U and v dollars worth of V. Each diet $[u \quad v]$ is associated with a nutritional-content matrix $[x \quad y]$, where x is the number of days the X requirement is met, and analogously for y. Find the matrix $[x \quad y]$ obtained from the following diets:

i) $[u_1 \quad v_1] = [.6 \quad .4]$.
ii) $[u_2 \quad v_2] = [.3 \quad .7]$.
iii) $[u_3 \quad v_3] = [.2 \quad .8]$.

(c) Which of the three diets considered in part (b) meets the dietary requirements for the longest time?

(d) Use the techniques of game theory to determine the optimal diet $[u \quad v]$ which meets the dietary requirements for the longest possible time.

2. A diet requires two nutrients, X and Y, and available are three foods, U, V, and W, containing these nutrients. The number of days the respective nutrients are supplied by a dollar's worth of the respective foods is given in the following table:

		Nutrient	
		X	Y
Food	U	12	0
	V	8	5
	W	0	10

Determine the maximum length of time both requirements can be met by a one-dollar expenditure divided between foods U, V, and W. (Recall the dual perspective of Sec. 7-6.)

3. A factory is set up to produce two products, X and Y. Each unit of each product produces a profit of one dollar. There are two machines, U and V, both of which are employed in the manufacture of each product. The number of days on the respective machines required to produce a 1000-unit lot of the respective products is shown in the following table:

		Product	
		X	Y
Machine	U	4	8
	V	7	5

(a) For nonnegative numbers, x and y, totaling one, let $\begin{bmatrix} x \\ y \end{bmatrix}$ represent the production schedule corresponding to a mixed lot of 1000 units consisting of $1000x$ units of X and $1000y$ units of Y. Each schedule $\begin{bmatrix} x \\ y \end{bmatrix}$ is associated with a time-requirement matrix $\begin{bmatrix} u \\ v \end{bmatrix}$, where u is the number of days required on machine U, and analogously for v. Find the column matrix $\begin{bmatrix} u \\ v \end{bmatrix}$ obtained from the following production schedules:

(i) $\begin{bmatrix} x \\ y \end{bmatrix} = \begin{bmatrix} .6 \\ .4 \end{bmatrix}$. (ii) $\begin{bmatrix} x \\ y \end{bmatrix} = \begin{bmatrix} .3 \\ .7 \end{bmatrix}$. (iii) $\begin{bmatrix} x \\ y \end{bmatrix} = \begin{bmatrix} 1 \\ 0 \end{bmatrix}$.

(b) For which of the three schedules considered in part (a) is the lot of 1000 units produced in the least time?

(c) Find the mixed lot of 1000 units which can be produced in the least possible time. (Compare with Prob. 1.)

4. The diet problem in Prob. 2 is mathematically equivalent to the row player's problem in the game with payoff matrix

$$M = \begin{bmatrix} 12 & 0 \\ 8 & 5 \\ 0 & 10 \end{bmatrix}.$$

State a production-scheduling problem that corresponds to the column player's problem in this game, and solve your problem.

5. A factory is set up to produce three products X, Y, and Z. There are two machines, U and V, both employed in the manufacture of each product. A 5000-unit lot of product X requires 3 days on U and 7 days on V; a 5000-unit lot of Y requires 8 days on U and 2 days on V; and a 5000-unit lot of Z requires 5 days on U and 4 days on V.

(a) How many units of each product should be manufactured in order to produce a total of 5000 units in the least time? And what is the minimum time?

(b) State a diet problem the solution of which is contained in the solution and verification of the production-scheduling problem in part (a). State the solution of your diet problem.

8-2. Restatement of the Diet Problem

The statements of the "Fish-Diet" problem and its "Production-Scheduling" companion in the previous section were contrived to force a perfect fit with the mathematical model that has been evolved for game theory. Normally one does not encounter such an agreeable phrasing of the problem. Consider the following variation:

The fish in a certain aquarium require three nutrients, X, Y, and Z, and available are two foods, U and V, containing these nutrients.

Each package of food U costs \$18 and supplies 18 units of nutrient X, 12 units of Y, and 9 units of Z.

Each package of V costs \$12 and supplies 9 units of X, 24 units of Y, and 10 units of Z.

The yearly requirement is 90 units of X, 120 units of Y, and 60 units of Z.

Question: How many packages of each food should be purchased in order to feed the fish for one year at the least possible cost?

The information is tabulated below:

		Amount of nutrient (in units per pkg)			Cost of food (in dollars per pkg)
		X	*Y*	*Z*	
Food	*U*	18	12	9	18
	V	9	24	10	12
Yearly Requirement (in nutritional units)		90	120	60	

One attack on this problem is to recast it in the language of the Fish-Diet problem of the previous section. As the following development shows, the new problem is essentially the same as the old.

For convenience we assume there are 360 days in a year. (This really doesn't entail an approximation in the result—our "days" are just slightly longer than the standard 24 hours.)

Since \$18 worth of food U contains 18 units of X, one dollar's worth of U contains one unit of X, and since the requirement for 360 days is 90 units, the one unit of X supplied by a dollar's worth of U meets the X requirement for $\frac{360}{90}$, or 4, days. By the same reasoning the dollar's worth of U meets the Y requirement for 2 days and the Z requirement for 3 days. Similarly, a dollar's worth of V meets the X, Y, and Z requirements for 3, 6, and 5 days, respectively. Thus the new Fish-Diet data conform perfectly with the earlier statement.

From the analysis of the previous section, in which the diet problem was observed to be mathematically equivalent to a problem in game theory, we can easily solve the new problem.

It was established that one dollar's worth of a mixture of foods U and V sustains the fish for at most 3.6 days. Thus the fish are sustained for at most 360 days, or one year, by a \$100 expenditure. Since their nutritional requirements cannot be met for longer than a year by a \$100 mixture of foods, it follows that \$100 is the minimum cost for one year's diet.

We recall that the optimal diet for 3.6 days was $[.6 \quad .4]$, 60 cents worth of U and 40 cents worth of V. Multiplying by 100, we conclude that the optimal diet for 360 days consists of \$60 worth of U and \$40 worth of V, which translates into

$$\frac{60}{18} = \frac{10}{3} = 3\frac{1}{3} \text{ packages of food } U,$$

and

$$\frac{40}{12} = \frac{10}{3} = 3\frac{1}{3} \text{ packages of food } V.$$

Therefore, in answer to the question posed,

Answer: The minimum cost diet for one year consists of $3\frac{1}{3}$ packages of U and $3\frac{1}{3}$ packages of V, costing a total of \$100.

Comment: In general we'll not be disturbed by the appearance of fractions in the solution of problems of this type. Here, we simply interpret that 10 packages of U and 10 packages of V should be purchased every three years.

Now that we've seen how the problem can be solved by a game-theoretic interpretation, let's look at it in its own wording.

The process is to consider diets consisting of u packages of U and v packages of V. The possibilities—referred to as "diets"—are represented as row matrices $[u \quad v]$ where u and v are nonnegative numbers. We use the given data to define a function which associates each such $[u \quad v]$ with a "nutritional-content" matrix $[x \quad y \quad z]$, where x is the number of units of nutrient X contained in u packages of U and v packages of V, and analogously for y and z.

For example, consider the diet

$$[u_1 \quad v_1] = [5 \quad 2].$$

Since each package of U contains 18 units of X and each package of V contains 9 units of X, the number of units of X contained in $u_1 = 5$ packages of U together with $v_1 = 2$ packages of V is

$$x_1 = 5 \times 18 + 2 \times 9 = 90 + 18 = 108.$$

By the same argument, the diet $[u_1 \quad v_1] = [5 \quad 2]$ contains

$$y_1 = 5 \times 12 + 2 \times 24 = 108$$

units of Y and

$$z_1 = 5 \times 9 + 2 \times 10 = 65$$

units of Z.

Thus the diet $[u_1 \quad v_1] = [5 \quad 2]$ is associated with the nutritional-content matrix

$$[x_1 \quad y_1 \quad z_1] = [108 \quad 108 \quad 65].$$

The foregoing argument is essentially the same as in the earlier diet problem. The computation of $[x \quad y \quad z]$ from $[u \quad v]$ is given by the matrix product

$$[x \quad y \quad z] = [u \quad v]\begin{bmatrix} 18 & 12 & 9 \\ 9 & 24 & 10 \end{bmatrix},$$

or

$$[x \quad y \quad z] = [u \quad v]M,$$

where

$$M = \begin{bmatrix} 18 & 12 & 9 \\ 9 & 24 & 10 \end{bmatrix}.$$

Here, the matrix M may be viewed as having been lifted directly from the tabulation of the data.

In particular,

$$[x_1 \quad y_1 \quad z_1] = [u_1 \quad v_1]M = [5 \quad 2]\begin{bmatrix} 18 & 12 & 9 \\ 9 & 24 & 10 \end{bmatrix}$$
$$= [108 \quad 108 \quad 65].$$

The yearly nutritional requirement is 90 units of X, 120 units of Y, and 60 units of Z. Since the diet $[u_1 \quad v_1] = [5 \quad 2]$ provides only $y_1 = 108$ units of Y, we announce that it is not *feasible;* that is, it doesn't meet all of the requirements.

On the other hand, consider the diet

$$[u_2 \quad v_2] = [4 \quad 3],$$

4 packages of U and 3 packages of V. The corresponding nutritional-content matrix is

$$\begin{aligned}[x_2 \quad y_2 \quad z_2] = [u_2 \quad v_2]M &= [4 \quad 3]\begin{bmatrix} 18 & 12 & 9 \\ 9 & 24 & 10 \end{bmatrix} \\ &= [99 \quad 120 \quad 66].\end{aligned}$$

Since the diet $[u_2 \quad v_2] = [4 \quad 3]$ supplies 99 units of X, 120 units of Y, and 66 units of Z, which meets the requirements in all three nutrients, we identify the diet as feasible.

To bring the feasibility criterion into the matrix setting we represent the yearly nutritional requirement—90 units of X, 120 units of Y, and 60 units of Z—by the "requirements" matrix

$$A = [90 \quad 120 \quad 60].$$

The matrix A is viewed as having been lifted directly from the bottom row in the tabulation of the data.

With the natural interpretation of inequalities between matrices, we see the feasibility criterion as being satisfied by the diet $[u_2 \quad v_2] = [4 \quad 3]$ because

$$[u_2 \quad v_2]M = [99 \quad 120 \quad 66] \geq [90 \quad 120 \quad 60] = A.$$

The following definition brings the inequality relation into the formal system of matrices:

Definition: Let $H = [h_{ij}]$ and $K = [k_{ij}]$ be $n \times k$ matrices. Then we say that $H \leq K$, and equivalently $K \geq H$, provided

$$h_{ij} \leq k_{ij}$$

for $i = 1, 2, \ldots, n$ and $j = 1, 2, \ldots, k$. That is, $H \leq K$ if each entry in H is less than or equal to the corresponding entry in K.

Comment: Although we mentally read $H \leq K$ as "H is less than or equal to K," it should be noted that for $H = [90 \quad 120 \quad 60]$ and $K = [99 \quad 120 \quad 66]$ it is not the case that H is "less than" K, which, naturally, would require that each entry in H be less than the corresponding entry in K, nor is H equal to K. However, it is the case that $H \leq K$.

Since the nutritional content of any diet $[u \quad v]$ is given by $[u \quad v]M$, and since A represents the yearly requirement, we define a diet $[u \quad v]$ to be feasible if

$$[u \quad v]M \geq A.$$

Having seen that the diet $[u_2 \quad v_2] = [4 \quad 3]$ is feasible, the *cost* of the diet is of interest. Since each package of U costs \$18 and each package of

V costs \$12, the total cost of the diet [4 3] is given by

$$4 \times 18 + 3 \times 12 = 108,$$

or \$108.

To complete the matrix structure of the problem, the cost data is lifted from the column on the right in the tabulation and represented as the matrix

$$B = \begin{bmatrix} 18 \\ 12 \end{bmatrix}.$$

The cost of the diet $[u_2 \quad v_2] = [4 \quad 3]$ is then given by

$$[u_2 \quad v_2]B = [4 \quad 3]\begin{bmatrix} 18 \\ 12 \end{bmatrix} = [108].$$

And, in general, the cost of any diet $[u \quad v]$ is given by

$$[u \quad v]B.$$

The problem is to find the feasible diet $[u \quad v]$ for which the cost is as small as possible. With the background that has been developed we can restate the problem in matrix terms as follows:

Problem: Given the matrices

$$M = \begin{bmatrix} 18 & 12 & 9 \\ 9 & 24 & 10 \end{bmatrix}, \qquad \begin{bmatrix} 18 \\ 12 \end{bmatrix} = B,$$

$$A = [90 \quad 120 \quad 60],$$

find a row matrix $[u \quad v]$ of nonnegative entries which statisfies

$$[u \quad v]M \geq A$$

such that

$$[u \quad v]B$$

is as small as possible.

We have here an example of a minimization problem in *linear programming*. The matrices M, A, and B can be varied to obtain other problems of this type as long as the row matrix A contains the same number of columns as M and the column matrix B has the same number of rows as M.

The Graphical Attack

The matrix model which has been evolved for the new Fish-Diet problem gives us a concise statement of the problem, directly in terms of the given data. Now let's bring the geometrical interpretations of the previous chapter into play to assist in solving the problem.

First, we want to form a picture of the feasible matrices $[u \quad v]$ of nonnegative entries satisfying the *constraint*

$$[u \quad v]M = [u \quad v]\begin{bmatrix} 18 & 12 & 9 \\ 9 & 24 & 10 \end{bmatrix} \geq [90 \quad 120 \quad 60] = A.$$

We isolate the columns and express this requirement in terms of the three simpler matrix inequalities

$$[u \quad v]\begin{bmatrix}18\\9\end{bmatrix} \geq [90], \qquad [u \quad v]\begin{bmatrix}12\\24\end{bmatrix} \geq [120], \qquad [u \quad v]\begin{bmatrix}9\\10\end{bmatrix} \geq [60].$$

As in the "dual perspective" of Sec. 7-6, we identify the 1×2 row matrix $[u \quad v]$ with the vector (u, v) in $\mathbf{R}^2$. This identification carries with it the recognition of a 2×1 column matrix $\begin{bmatrix}a\\b\end{bmatrix}$ as the functional which maps (u, v)—equivalently $[u \quad v]$—into the value

$$au + bv.$$

In particular, we recognize $\begin{bmatrix}18\\9\end{bmatrix}$ as the functional mapping (u, v) into the value

$$18u + 9v.$$

Referring to the requirements imposed by the constraint above, the feasible points (u, v) are those in which u and v are both nonnegative and which satisfy the conditions

1. The value assigned by the functional $\begin{bmatrix}18\\9\end{bmatrix}$ is 90 or more,
2. The value assigned by $\begin{bmatrix}12\\24\end{bmatrix}$ is 120 or more, and
3. The value assigned by $\begin{bmatrix}9\\10\end{bmatrix}$ is 60 or more.

For a geometrical view of the feasible points we recall the "level line" interpretation of functionals on $\mathbf{R}^2$. Sketched in Fig. 8-1 is the level line of the functional $\begin{bmatrix}18\\9\end{bmatrix}$ corresponding to the value 90. Since both coordinates in (u, v) are required to be nonnegative, it is only the positive quadrant of the coordinate plane that is of interest; the points outside the positive quadrant are not relevant and are not to be regarded as part of the graph at all.

We note that the line segment in the figure has end points at the values 5 and 10 on the u and v axes, respectively; in fact, it was these values which

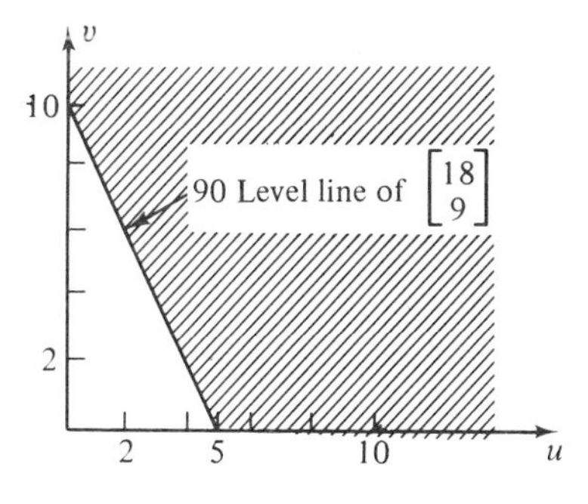

Figure 8-1.

were first computed to determine the segment. The 90 level line of $\begin{bmatrix} 18 \\ 9 \end{bmatrix}$ consists of points (u, v) satisfying

$$18u + 9v = 90.$$

Setting $v = 0$ yields the point (5, 0), and setting $u = 0$ produces the point (0, 10).

In Fig. 8-1 the region above and to the right of the level line has been shaded. It should be clear that this region, including its boundary, is precisely the set of all points (u, v) in the positive quadrant for which the value assigned by the functional $\begin{bmatrix} 18 \\ 9 \end{bmatrix}$ is 90 or more. That is, the shaded region consists of those points (u, v) for which

$$[u \quad v]\begin{bmatrix} 18 \\ 9 \end{bmatrix} \geq [90].$$

These are the points which satisfy the requirement for feasibility imposed by the first column of the constraint $[u \quad v]M \geq A$.

The requirements imposed by the other two columns of the constraint are treated in the analogous manner. In Fig. 8-2 we've sketched all three level lines: the 90 level line of the functional $\begin{bmatrix} 18 \\ 9 \end{bmatrix}$, the 120 level line of $\begin{bmatrix} 12 \\ 24 \end{bmatrix}$, and the 60 level line of $\begin{bmatrix} 9 \\ 10 \end{bmatrix}$. The lines are labeled 1, 2, and 3, respectively, to indicate the corresponding columns in the constraint.

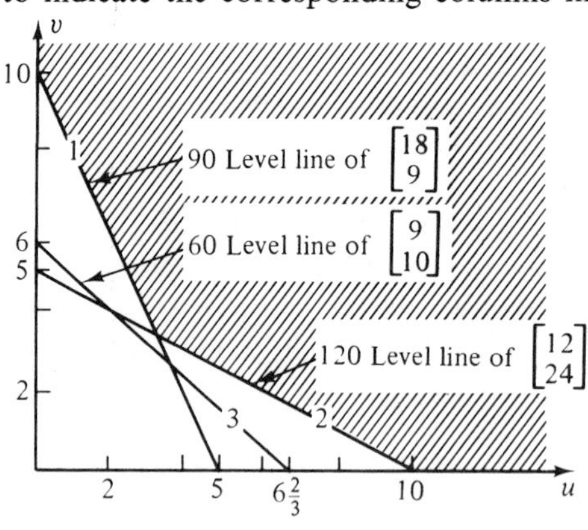

Figure 8-2.

Shaded in the figure is the region consisting of points (u, v), which lie above and to the right of all three level lines. It should be clear that this region, including its boundary, consists of those points (u, v) which simultaneously satisfy the three inequalities

$$[u \quad v]\begin{bmatrix} 18 \\ 9 \end{bmatrix} \geq [90], \qquad [u \quad v]\begin{bmatrix} 12 \\ 24 \end{bmatrix} \geq [120], \qquad [u \quad v]\begin{bmatrix} 9 \\ 10 \end{bmatrix} \geq [60].$$

Or, reassembling the columns in the constraint,

$$[u \quad v]\begin{bmatrix} 18 & 12 & 9 \\ 9 & 24 & 10 \end{bmatrix} \geq [90 \quad 120 \quad 60].$$

In short, the region depicted in Fig. 8-2 represents the set of feasible points (u, v), or feasible matrices $[u \quad v]$, which satisfy the constraint $[u \quad v]M \geq A$. From its manner of construction as the intersection of half-planes, this *region of feasibility* is a convex set, meaning that the line segment between any two points of the region lies entirely in the region.

To solve the problem, we need to pinpoint a feasible point (u, v) for which the value of

$$[u \quad v]\begin{bmatrix} 18 \\ 12 \end{bmatrix}$$

is as small as possible. Which is to say, we want to find a feasible point for which the value assigned by the functional $\begin{bmatrix} 18 \\ 12 \end{bmatrix}$ is minimized.

From the previous chapter we know that the level lines of the functional $\begin{bmatrix} 18 \\ 12 \end{bmatrix}$ are the lines perpendicular to the line through (0, 0) and the point (18, 12). In Fig. 8-3 we show again the region of feasibility, together with several level lines of $\begin{bmatrix} 18 \\ 12 \end{bmatrix}$. (To avoid unnecessary clutter, the shading of the region of feasibility is suppressed to points near the boundary.) The line through (0, 0) and (18, 12) has been determined by the point $(9, 6) = \frac{1}{2}(18, 12)$.

Now we can interpret the minimization problem dynamically. Since the value of the functional $\begin{bmatrix} 18 \\ 12 \end{bmatrix}$ decreases as we follow the family of parallel level lines down the perpendicular through (18, 12) and (0, 0), the problem

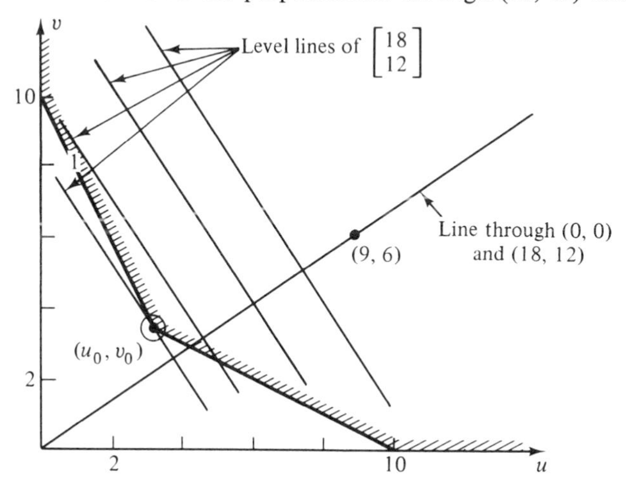

Figure 8-3.

is to spot the point in the region of feasibility such that its level line is as far down as possible. From inspection of Fig. 8-3 we see that this is the point, indicated as (u_0, v_0), where the boundary lines 1 and 2 intersect. As the figure shows, level lines of $\begin{bmatrix} 18 \\ 12 \end{bmatrix}$ lying below the level line through (u_0, v_0) contain no feasible points at all.

Recalling the source of boundary lines 1 and 2, the geometrical evidence of Fig. 8-3 tells us that the feasible point (u, v) for which the value of the functional $\begin{bmatrix} 18 \\ 12 \end{bmatrix}$ is as small as possible is the point (u_0, v_0) which simultaneously satisfies the conditions

$$[u_0 \quad v_0]\begin{bmatrix} 18 \\ 9 \end{bmatrix} = [90] \quad \text{and} \quad [u_0 \quad v_0]\begin{bmatrix} 12 \\ 24 \end{bmatrix} = [120].$$

That is, (u_0, v_0) is the solution of the system of equations

$$18u_0 + 9v_0 = 90, \tag{1}$$

$$12u_0 + 24v_0 = 120. \tag{2}$$

To simplify Eq. (1) is divided by 9 and Eq. (2) by 12:

$$2u_0 + v_0 = 10, \tag{3}$$

$$u_0 + 2v_0 = 10. \tag{4}$$

We multiply Eq. (3) by 2 and retain Eq. (4):

$$4u_0 + 2v_0 = 20, \tag{5}$$

$$u_0 + 2v_0 = 10. \tag{4}$$

Subtracting Eq. (4) from Eq. (5) gives

$$3u_0 = 10 \quad \text{or} \quad u_0 = \tfrac{10}{3} = 3\tfrac{1}{3}.$$

To find v_0, and to check the work, we substitute $u_0 = \frac{10}{3}$ into the original equations:

$$\text{Eq. (1):} \quad 18 \times \tfrac{10}{3} + 9v_0 = 90,$$

$$60 + 9v_0 = 90 \quad \text{or} \quad v_0 = \tfrac{30}{9} = \tfrac{10}{3} = 3\tfrac{1}{3}.$$

$$\text{Eq. (2):} \quad 12 \times \tfrac{10}{3} + 24v_0 = 120,$$

$$40 + 24v_0 = 120 \quad \text{or} \quad v_0 = \tfrac{80}{24} = \tfrac{10}{3} = 3\tfrac{1}{3}.$$

Thus the geometrical attack on the new Fish-Diet problem indicates that the desired solution is given in matrix terms by

$$[u_0 \quad v_0] = [3\tfrac{1}{3} \quad 3\tfrac{1}{3}].$$

This, of course, is precisely the solution obtained earlier by interpreting the problem as being equivalent to a certain matrix game.

8-2 EXERCISES

1. A certain diet requires two nutrients, X and Y, and available on the market are two foods, U and V, containing these nutrients. Each package of U contains 6 units of X and 6 units of Y and costs \$3. Each package of V contains 10 units of X and 6 units of Y and costs \$4. The yearly requirement is 60 units of X and 48 units of Y. For each of the following diets $[u \quad v]$—u packages of U and v packages of V—compute the corresponding nutritional-content matrix $[x \quad y]$. Determine whether $[u \quad v]$ is a feasible diet, and, if so, compute its cost.

(a) $[u \quad v] = [2 \quad 5]$.
(b) $[u \quad v] = [7 \quad 2]$.
(c) $[u \quad v] = [4 \quad 3]$.
(d) $[u \quad v] = [10 \quad 0]$.

2. A diet must contain three nutrients, X, Y, and Z, to be obtained from two foods, U and V. The following tabulation displays the nutritional content and cost per package of the respective foods and also the yearly requirement of the nutrients:

		Amount of nutrient (in units per pkg)			*Cost of food (in dollars per pkg)*
		X	Y	Z	
Food	U	5	8	5	5
	V	10	4	15	5
Yearly Requirement (nutritional units)		60	48	75	

For each of the following diets $[u \quad v]$, compute the nutritional-content matrix $[x \quad y \quad z]$. Determine whether $[u \quad v]$ is a feasible diet; if so, compute its cost.

(a) $[u \quad v] = [2 \quad 7]$.
(b) $[u \quad v] = [3 \quad 6]$.
(c) $[u \quad v] = [4 \quad 4]$.
(d) $[u \quad v] = [6 \quad 3]$.
(e) $[u \quad v] = [8 \quad 2]$.

3. Observe that the cost minimization problem for the diet data of Prob. 1 translates into the following linear programming problem: Given the matrices

$$M = \begin{bmatrix} 6 & 6 \\ 10 & 6 \end{bmatrix}, \quad \begin{bmatrix} 3 \\ 4 \end{bmatrix} = B,$$

$$A = [60 \quad 48],$$

find a matrix $[u \quad v]$ of nonnegative entries satisfying the constraint

$$[u \quad v]M \geq A$$

which minimizes the value of

$$[u \quad v]B.$$

(a) In the positive quadrant of the coordinate plane, sketch the line consisting of points (u, v) such that

$$[u \quad v]\begin{bmatrix} 6 \\ 10 \end{bmatrix} = [60];$$

i.e., sketch the 60 level line of the functional $\begin{bmatrix} 6 \\ 10 \end{bmatrix}$.

(b) On the same graph, sketch the line of points (u, v) for which

$$[u \quad v]\begin{bmatrix} 6 \\ 6 \end{bmatrix} = [48].$$

(c) Shade the region of feasible points (u, v) satisfying the constraint

$$[u \quad v]M = [u \quad v]\begin{bmatrix} 6 & 6 \\ 10 & 6 \end{bmatrix} \geq [60 \quad 48] = A.$$

(d) On the graph, construct the line from $(0, 0)$ through the point $(3, 4)$, and by erecting perpendiculars display several level lines of the functional $\begin{bmatrix} 3 \\ 4 \end{bmatrix}$. From inspection of the picture, identify the feasible point for which the value assigned by $\begin{bmatrix} 3 \\ 4 \end{bmatrix}$ is minimized.

(e) Solve for the critical feasible point identified in part (d), thereby obtaining the matrix $[u \quad v]$, which the graphical evidence suggests as the solution of the minimization argument.

(f) From the solution obtained in part (e), declare the minimum cost diet for the situation in Prob. 1. What is its cost?

4. Observe that the cost minimization problem for the diet data of Prob. 2 translates into the following linear programming problem: Given the matrices

$$M = \begin{bmatrix} 5 & 8 & 5 \\ 10 & 4 & 15 \end{bmatrix}, \qquad \begin{bmatrix} 5 \\ 5 \end{bmatrix} = B,$$

$$A = [60 \quad 48 \quad 75],$$

find a matrix $[u \quad v]$ of nonnegative entries satisfying the constraint

$$[u \quad v]M \geq A$$

which minimizes the value of

$$[u \quad v]B.$$

(a) In the first quadrant of the coordinate plane shade the region of feasible points (u, v) satisfying the constraint

$$[u \quad v]M \geq A.$$

[Follow steps suggested in parts (a), (b), and (c) of Prob. 3.]

(b) On the graph, display several level lines of the functional B. Identify the feasible point which minimizes the value of $[u \quad v]B$, and solve for this point [as in parts (d) and (e) of Prob. 3].

(c) Declare the minimum cost diet for the diet situation in Prob. 2. What is its cost?

5. Consider the following minimization problem in linear programming: Given the matrices

$$M = \begin{bmatrix} 6 & 4 & 2 \\ 2 & 4 & 6 \end{bmatrix}, \qquad \begin{bmatrix} 3 \\ 6 \end{bmatrix} = B,$$

$$A = [12 \quad 16 \quad 12],$$

find a matrix $[u \quad v]$ of nonnegative entries satisfying the constraint

$$[u \quad v]M \geq A$$

which minimizes the value of

$$[u \quad v]B.$$

(a) In the first quadrant of the coordinate plane, shade the region of feasible points (u, v) for which

$$[u \quad v]M \geq A.$$

(b) Sketch several level lines of the functional B, and identify the feasible point for which the value assigned by B is minimized. Solve for this point.

6. With matrices M and A as in Prob. 5 and B as given below, follow the directions in Prob. 5 to find solutions of the respective minimization problems:

$$\text{(a) } B = \begin{bmatrix} 6 \\ 3 \end{bmatrix}, \qquad \text{(b) } B = \begin{bmatrix} 4 \\ 1 \end{bmatrix}, \qquad \text{(c) } B = \begin{bmatrix} 1 \\ 4 \end{bmatrix}.$$

7. A man resolves to buy life insurance for his family. For himself he wants \$24,000 of coverage; for his wife, \$10,000; and for each child, \$5000.

Two plans are available. Each unit of plan U affords \$1000 coverage on the man and \$300 on the wife and each child, all for a premium of \$3 per month. Each unit of plan V affords \$1000 coverage on the man and his wife and \$100 on each child at a premium of \$4 per month.

How many units of each plan should the man buy in order to obtain the desired coverage for the least cost? (First state the problem in matrix terms, as a minimization problem in linear programming. Then apply the graphical technique of the previous problems. For convenience, measure the insurance coverage in units of \$100.)

8-3. The Duality Theorem

The cost minimization problem associated with feeding the fish in the previous section was a restatement in more natural language of the Fish-Diet problem of Sec. 8-1. In its original form we saw that the Fish-Diet problem is mathematically the same as the problem faced by the row player in a certain matrix game. To verify the solution, we considered the column player's problem in the game. As we also saw in Sec. 8-1, the column player's problem translated into a Production-Scheduling problem—how to find the combination of products yielding a given profit in the shortest possible time. The challenge now is to restate the Production-Scheduling problem in the spirit of the restatement of the Fish-Diet problem. The analogy couldn't be better.

A factory has two machines, U and V, with which it can produce three products, X, Y, and Z. A case of X requires 18 hours on machine U and also requires 9 hours of time on machine V and produces a profit of \$90. A case of Y requires 12 hours on U and 24 hours on V and produces a profit of \$120. And a case of Z requires 9 hours on U and 10 hours on V and produces a profit of \$60. Machine U can operate a maximum of 18 hours a day, and machine V can operate a maximum of 12 hours a day.

Question: How many cases (or fractional cases) of X, Y, and Z, respectively, should be manufactured in order to realize the maximum daily profit?

Except for the interpretation, the tabulation of the data is the same as in the restated Fish-Diet problem:

		Product			*Machine capacity (in hours per day)*
		X	Y	Z	
Time Required on Machine (in hours per case of product)	U	18	12	9	18
	V	9	24	10	12
Profit (in dollars per case of product)		90	120	60	

This profit maximization problem is essentially the same as the Production-Scheduling problem of Sec. 8-1. The only adjustments needed are that we measure product output in cases rather than in lots of 1000 and that we measure profit in units of \$360 rather than in units of \$1000. Now, one unit of profit, i.e., \$360, from product X requires 72 hours on machine U and 36 hours on machine V. Given the number of hours that the respective machines can operate, we see that this one unit of profit from X requires 4 days on U and 3 days on V, which is just as in Sec. 8-1, and similarly for the units of profit from products Y and Z.

From the analysis in Sec. 8-1, we know that the minimum time required to realize one unit of profit—now \$360—is 3.6 days. Therefore, the maximum profit for one day is \$100. We also recall from the earlier analysis that eight-tenths of the profit was realized from product X, two-tenths from Y, and none from Z, which tells us that of the maximum daily profit of \$100, \$80 is from X and \$20 from Y. Since a case of X produces a profit of \$90, we conclude that the optimal daily production calls for $\frac{8}{9}$ case of X. And since a case of Y produces a profit of \$120, the optimal daily production calls for $\frac{1}{6}$ case of Y. And no Z.

Thus the reduction of the problem to the matrix game of Sec. 8-1 yields the following answer to the question posed:

Answer: The maximum daily profit is \$100, realized by the production of $\frac{8}{9}$ case of product X, $\frac{1}{6}$ case of product Y, and none of product Z.

The reduction of the profit maximization problem to a matrix game is a rather roundabout approach. Let's try a more direct attack.

We identify a *production schedule*, or just a *schedule*, as a column matrix

$$\mathbf{c} = \begin{bmatrix} x \\ y \\ z \end{bmatrix}$$

of nonegative entries, where x corresponds to the number of cases of X to be produced in one day, and analogously for y and z. Associated with each such schedule is a "time-requirement" matrix $\begin{bmatrix} u \\ v \end{bmatrix}$, where u and v are the numbers of hours required on machines U and V, respectively.

For example, consider the schedule

$$\mathbf{c}_1 = \begin{bmatrix} x_1 \\ y_1 \\ z_1 \end{bmatrix} = \begin{bmatrix} .4 \\ .4 \\ .4 \end{bmatrix}.$$

This schedule corresponds to a daily production of .4 case of each of X, Y, and Z. Since the time required on U is 18 hours for each case of X, 12 hours for each case of Y, and 9 hours for each case of Z, the total number of hours on U required to produce .4 case of each of the three products is

$$.4 \times 18 + .4 \times 12 + .4 \times 9 = 15.6.$$

Similarly, from the time-requirement data for machine V, the total number of hours on V required by schedule $\mathbf{c}_1$ is

$$.4 \times 9 + .4 \times 24 + .4 \times 10 = 17.2.$$

Thus the schedule $\mathbf{c}_1$ produces the time-requirement matrix

$$\begin{bmatrix} u_1 \\ v_1 \end{bmatrix} = \begin{bmatrix} 15.6 \\ 17.2 \end{bmatrix}.$$

Since this schedule calls for 17.2 hours on machine V and since V can operate only 12 hours a day, we announce that schedule $\mathbf{c}_1$ is not *feasible*.

The calculation of the time-requirement matrix is given by the matrix product

$$M\mathbf{c}_1 = M\begin{bmatrix} x_1 \\ y_1 \\ z_1 \end{bmatrix} = \begin{bmatrix} 18 & 12 & 9 \\ 9 & 24 & 10 \end{bmatrix}\begin{bmatrix} .4 \\ .4 \\ .4 \end{bmatrix} = \begin{bmatrix} 15.6 \\ 17.2 \end{bmatrix}.$$

Here the matrix M has been lifted directly from the tabulation of the data. This is the same tabulation that served the restated Fish-Diet problem of the previous section, and the matrix M is the same.

Now let's look at another schedule:

$$\mathbf{c}_2 = \begin{bmatrix} x_2 \\ y_2 \\ z_2 \end{bmatrix} = \begin{bmatrix} .8 \\ .1 \\ .2 \end{bmatrix}.$$

The time-requirement matrix is

$$\begin{bmatrix} u_2 \\ v_2 \end{bmatrix} = M\mathbf{c}_2 = \begin{bmatrix} 18 & 12 & 9 \\ 9 & 24 & 10 \end{bmatrix}\begin{bmatrix} .8 \\ .1 \\ .2 \end{bmatrix} = \begin{bmatrix} 17.4 \\ 11.6 \end{bmatrix}.$$

Since the hours required, 17.4 and 11.6, are within the daily machine capacities, 18 and 12, respectively, we identify $\mathbf{c}_2$ as a feasible schedule.

As in the Fish-Diet problem, we lift the column from the right in the tabulation, forming the column matrix

$$B = \begin{bmatrix} 18 \\ 12 \end{bmatrix}.$$

We see that the schedule $\mathbf{c}_2$ is feasible because

$$\begin{bmatrix} 17.4 \\ 11.6 \end{bmatrix} = M\mathbf{c}_2 \leq B = \begin{bmatrix} 18 \\ 12 \end{bmatrix}.$$

Inasmuch as $\mathbf{c}_2$ is a feasible schedule, the profit produced is of interest. Since the profits per case for the respective products are \$90, \$120, and \$60, it follows that the profit produced by schedule $\mathbf{c}_2$ is \$ 96 because

$$90 \times .8 + 120 \times .1 + 60 \times .2 = 96.$$

Again as in the earlier problem, we lift the row from the bottom in the tabulation, forming the row matrix

$$A = [90 \quad 120 \quad 60].$$

We see that the profit produced by schedule $\mathbf{c}_2$ is given by

$$A\mathbf{c}_2 = [90 \quad 120 \quad 60]\begin{bmatrix} .8 \\ .1 \\ .2 \end{bmatrix} = [96].$$

The problem, of course, is to find the feasible schedule producing the maximum profit. Building on the development of schedules $\mathbf{c}_1$ and $\mathbf{c}_2$ above, the problem may be stated in matrix terms as follows:

The Maximization Problem: Given the matrices

$$M = \begin{bmatrix} 18 & 12 & 9 \\ 9 & 24 & 10 \end{bmatrix}, \qquad \begin{bmatrix} 18 \\ 12 \end{bmatrix} = B,$$

$$A = [90 \quad 120 \quad 60],$$

find a column matrix

$$\mathbf{c} = \begin{bmatrix} x \\ y \\ z \end{bmatrix}$$

of nonnegative entries satisfying the *constraint*

$$M\mathbf{c} \leq B$$

such that the value of

$$A\mathbf{c}$$

is the maximum value possible.

In its new form the Production-Scheduling problem is still the companion of the Fish-Diet problem. We recall from the previous section that the restated Fish-Diet problem assumed the following matrix form, closely related to the form of the maximization problem:

The Minimization Problem: Given the matrices M, A, and B in the maximization problem above, find a row matrix $\mathbf{r} = [u \quad v]$ of nonnegative entries satisfying the constraint

$$\mathbf{r}M \geq A$$

such that the value of

$$\mathbf{r}B$$

is the minimum value possible.

We recall also that the solution of the minimization problem was found to be

$$\mathbf{r}_0 = [\tfrac{10}{3} \quad \tfrac{10}{3}].$$

Let's verify that $\mathbf{r}_0$ satisfies the constraint

$$\mathbf{r}_0 M = [\tfrac{10}{3} \quad \tfrac{10}{3}]\begin{bmatrix} 18 & 12 & 9 \\ 9 & 24 & 10 \end{bmatrix} = [90 \quad 120 \quad 63\tfrac{1}{3}]$$
$$\geq [90 \quad 120 \quad 60] = A. \checkmark$$

We note that the value of $\mathbf{r}_0 B$—the least possible for feasible $\mathbf{r}$—is

$$\mathbf{r}_0 B = [\tfrac{10}{3} \quad \tfrac{10}{3}]\begin{bmatrix} 18 \\ 12 \end{bmatrix} = [100].$$

The solution of the maximization problem was obtained earlier by reducing the problem to a matrix game. In the matrix form of a production schedule the solution is

$$\mathbf{c}_0 = \begin{bmatrix} \frac{8}{9} \\ \frac{1}{6} \\ 0 \end{bmatrix}.$$

Let's verify that $\mathbf{c}_0$ satisfies the constraint of the maximization problem:

$$M\mathbf{c}_0 = \begin{bmatrix} 18 & 12 & 9 \\ 9 & 24 & 10 \end{bmatrix}\begin{bmatrix} \frac{8}{9} \\ \frac{1}{6} \\ 0 \end{bmatrix} = \begin{bmatrix} 18 \\ 12 \end{bmatrix} \leq \begin{bmatrix} 18 \\ 12 \end{bmatrix} = B. \checkmark$$

The value of $A\mathbf{c}_0$—the maximum possible for feasible $\mathbf{c}$—is

$$A\mathbf{c}_0 = [90 \quad 120 \quad 60]\begin{bmatrix} \frac{8}{9} \\ \frac{1}{6} \\ 0 \end{bmatrix} = [100].$$

The crucial fact here is that for $\mathbf{r}_0$ and $\mathbf{c}_0$, feasible in their respective problems, the values of $\mathbf{r}_0 B$ and $A\mathbf{c}_0$ are equal:

$$\mathbf{r}_0 B = [100] = A\mathbf{c}_0.$$

In the Fish-Diet problem we interpret the 100 as the minimum number of dollars required to feed the fish for one year, and in the Production-Scheduling companion, we interpret the 100 as the maximum profit, in number of dollars, that can be realized by one day's production.

The closely related minimization and maximization problems evolved from the problems faced by row and column players, respectively, in a certain matrix game. The optimal diet $\mathbf{r}_0$ and optimal schedule $\mathbf{c}_0$ were adapted from optimal strategies for the game, which strategies were verified in Sec. 8-1. The interesting question is,

How does the criterion for verification of optimal strategies in the underlying game translate into the setting of the derived minimization-maximization problems?

The answer is,

$\mathbf{r}_0$ and $\mathbf{c}_0$ are established as the solutions in the respective problems if they satisfy the constraints, $\mathbf{r}_0 M \geq A$ and $M\mathbf{c}_0 \leq B$ and if $\mathbf{r}_0 B = A\mathbf{c}_0$.

The minimization and maximization problems are identified as *dual problems in linear programming*. The generalization is as follows:

Dual Problems of Linear Programming: Given an $n \times k$ matrix M, a $1 \times k$ row matrix A, and an $n \times 1$ column matrix B,

$$M = \begin{bmatrix} m_{11} & m_{12} & \dots & m_{1k} \\ m_{21} & m_{22} & \dots & m_{2k} \\ \cdot & \cdot & & \cdot \\ \cdot & \cdot & & \cdot \\ \cdot & \cdot & & \\ m_{n1} & m_{n2} & \dots & m_{nk} \end{bmatrix}, \quad \begin{bmatrix} b_1 \\ b_2 \\ \cdot \\ \cdot \\ \cdot \\ b_n \end{bmatrix} = B,$$

$$A = [a_1 \quad a_2 \quad \dots \quad a_k].$$

The Minimization Problem: Find a $1 \times n$ row matrix $\mathbf{r} = [r_1 \quad r_1 \quad \dots \quad r_n]$ of nonnegative entries satisfying the constraint

$$\mathbf{r}M \geq A$$

such that

$$\mathbf{r}B$$

is the minimum value possible.

The Maximization Problem: Find a $k \times 1$ column matrix

$$\mathbf{c} = \begin{bmatrix} c_1 \\ c_2 \\ \cdot \\ \cdot \\ \cdot \\ c_k \end{bmatrix}$$

of nonnegative entries satisfying the constraint

$$M\mathbf{c} \leq B$$

such that

$$A\mathbf{c}$$

is the maximum value possible.

Problem pairs of this type are obtained as generalizations of the problems faced by row and column players in matrix games. The Duality Theorem of Game Theory translates into the following result:

The Duality Theorem of Linear Programming: Given an $n \times k$ matrix M, a $1 \times k$ row matrix A, and an $n \times 1$ column matrix B, let $\mathbf{r}_0$ be a $1 \times n$ row matrix of nonnegative entires satisfying

$$\mathbf{r}_0 M \geq A,$$

and let $\mathbf{c}_0$ be a $k \times 1$ column matrix of nonnegative entries satisfying

$$M\mathbf{c}_0 \leq B.$$

Suppose further that

$$\mathbf{r}_0 B = A\mathbf{c}_0.$$

Then for each $1 \times n$ row matrix $\mathbf{r}$ of nonnegative entries satisfying

$$\mathbf{r}M \geq A$$

we have

$$\mathbf{r}_0 B \leq \mathbf{r}B,$$

And for each $k \times 1$ column matrix $\mathbf{c}$ of nonnegative entries satisfying

$$M\mathbf{c} \leq B$$

we have

$$A\mathbf{c} \leq A\mathbf{c}_0.$$

In short, $\mathbf{r}_0$ and $\mathbf{c}_0$ are solutions of the minimization and maximization problems, respectively.

The proof of the duality theorem is deferred until two sections hence. In the next section we'll consider the practical problem of finding solutions to dual problems. Once conjectured solutions have been obtained, by whatever means, the verification is straightforward. The following example displays the format; the source of the solutions $\mathbf{r}_0$ and $\mathbf{c}_0$ will be revealed in the next section.

Example: Given the matrices

$$M = \begin{bmatrix} 1 & 2 \\ 2 & 2 \\ 2 & 1 \end{bmatrix}, \quad \begin{bmatrix} 20 \\ 26 \\ 22 \end{bmatrix} = B,$$

$$A = [10 \quad 15].$$

The Minimization Problem: Find a 1×3 row matrix $\mathbf{r} = [r_1 \quad r_2 \quad r_3]$ of nonnegative entries satisfying

$$\mathbf{r}M \geq A$$

such that

$$\mathbf{r}B$$

is the minimum value possible.

The Maximization Problem: Find a 2×1 column matrix $\mathbf{c} = \begin{bmatrix} c_1 \\ c_2 \end{bmatrix}$ of nonnegative entries satisfying

$$M\mathbf{c} \leq B$$

such that

$$A\mathbf{c}$$

is the maximum value possible.

Solution: The solution of the minimization problem is

$$\mathbf{r}_0 = [5 \quad 2.5 \quad 0],$$

and the solution of the maximization problem is

$$\mathbf{c}_0 = \begin{bmatrix} 6 \\ 7 \end{bmatrix}.$$

Verification: To check that the constraints are satisfied,

$$\mathbf{r}_0 M = [5 \quad 2.5 \quad 0] \begin{bmatrix} 1 & 2 \\ 2 & 2 \\ 2 & 1 \end{bmatrix} = [10 \quad 15] \geq [10 \quad 15] = A, \checkmark$$

$$M\mathbf{c}_0 = \begin{bmatrix} 1 & 2 \\ 2 & 2 \\ 2 & 1 \end{bmatrix} \begin{bmatrix} 6 \\ 7 \end{bmatrix} = \begin{bmatrix} 20 \\ 26 \\ 19 \end{bmatrix} \leq \begin{bmatrix} 20 \\ 26 \\ 22 \end{bmatrix} = B. \checkmark$$

To check that $\mathbf{r}_0 B = A\mathbf{c}_0$,

$$\mathbf{r}_0 B = [5 \quad 2.5 \quad 0] \begin{bmatrix} 20 \\ 26 \\ 22 \end{bmatrix} = [165],$$

$$A\mathbf{c}_0 = [10 \quad 15] \begin{bmatrix} 6 \\ 7 \end{bmatrix} = [165].$$

Thus $\mathbf{r}_0 B = A\mathbf{c}_0 = [165]. \checkmark\checkmark$

Since $\mathbf{r}_0$ and $\mathbf{c}_0$ satisfy the constraints and since $\mathbf{r}_0 B = A\mathbf{c}_0$, it follows from the duality theorem that $\mathbf{r}_0$ is the solution of the minimization problem and that $\mathbf{c}_0$ is the solution of the maximization problem. Subject to the constraints, the minimum possible value of $\mathbf{r}B$ and the maximum possible value of $A\mathbf{c}$ are both equal to [165].

8-3 EXERCISES

1. A factory has three machines, U, V, and W, with which it can manufacture two products, X and Y. The tabulation below displays the number of hours on each machine required for the production of a unit of the respective products. Also shown is the profit per unit of product and the daily machine capacity.

		Product		*Machine capacity (in hours per day)*
		X	Y	
Time Required on Machine (in hours per unit of product)	U	2	1	16
	V	2	0	14
	W	2	3	24
Profit (in dollars per unit of product)		40	30	

Consider the following production-schedule matrices $\begin{bmatrix} x \\ y \end{bmatrix}$, where x and y are the number of units of X and Y to be produced in one day. In each case, determine whether the schedule is feasible, i.e., whether the machine time required for its production is within the daily capacity. If the schedule is feasible, compute the profit it yields.

(a) $\begin{bmatrix} x \\ y \end{bmatrix} = \begin{bmatrix} 2 \\ 7 \end{bmatrix}$. (b) $\begin{bmatrix} x \\ y \end{bmatrix} = \begin{bmatrix} 3 \\ 6 \end{bmatrix}$. (c) $\begin{bmatrix} x \\ y \end{bmatrix} = \begin{bmatrix} 4 \\ 4 \end{bmatrix}$.

(d) $\begin{bmatrix} x \\ y \end{bmatrix} = \begin{bmatrix} 6 \\ 2 \end{bmatrix}$.

2. Consider the following matrices M, A, and B:

$$M = \begin{bmatrix} 6 & 4 & 2 \\ 2 & 4 & 6 \end{bmatrix}, \quad \begin{bmatrix} 96 \\ 60 \end{bmatrix} = B,$$

$$A = [12 \quad 16 \quad 12].$$

For each of the following 3×1 column matrices $\mathbf{c} = \begin{bmatrix} x \\ y \\ z \end{bmatrix}$, determine whether $\mathbf{c}$ satisfies the constraint $M\mathbf{c} \leq B$, and if so, compute the value of $A\mathbf{c}$:

(a) $\mathbf{c} = \begin{bmatrix} 7 \\ 4 \\ 5 \end{bmatrix}$. (b) $\mathbf{c} = \begin{bmatrix} 15 \\ 0 \\ 5 \end{bmatrix}$. (c) $\mathbf{c} = \begin{bmatrix} 8 \\ 11 \\ 0 \end{bmatrix}$.

3. Given the matrices

$$M = \begin{bmatrix} 3 & 1 \\ 0 & 2 \\ 2 & 2 \end{bmatrix}, \quad \begin{bmatrix} 12 \\ 8 \\ 12 \end{bmatrix} = B,$$

$$A = [30 \quad 20].$$

(a) For the following 1×3 row matrices $\mathbf{r} = [u \quad v \quad w]$, determine whether $\mathbf{r}$ satisfies the constraint $\mathbf{r}M \geq A$, and if so, compute the value of $\mathbf{r}B$:

(i) $\mathbf{r} = [5 \quad 2 \quad 7]$. (ii) $\mathbf{r} = [7 \quad 2 \quad 5]$.

(b) For the following 2×1 column matrices $\mathbf{c} = \begin{bmatrix} x \\ y \end{bmatrix}$, determine whether $\mathbf{c}$ satisfies the constraint $M\mathbf{c} \leq B$, and if so, compute the value of $A\mathbf{c}$:

(i) $\mathbf{c} = \begin{bmatrix} 2 \\ 4 \end{bmatrix}$. (ii) $\mathbf{c} = \begin{bmatrix} 3 \\ 2.5 \end{bmatrix}$.

4. For the matrices M, A, and B of Prob. 3, consider the dual problems of linear programming.

The Minimization Problem: Find a 1×3 row matrix $\mathbf{r} = [u \quad v \quad w]$ of nonnegative entries satisfying the constraint $\mathbf{r}M \geq A$ such that the value of $\mathbf{r}B$ is minimized.

The Maximization Problem: Find a 2×1 column matrix $\mathbf{c} = \begin{bmatrix} x \\ y \end{bmatrix}$ of nonnegative entries satisfying the constraint $M\mathbf{c} \leq B$ such that the value of $A\mathbf{c}$ is maximized.

Appeal to the Duality Theorem of Linear Programming to show that

$$\mathbf{r}_0 = [5 \quad 0 \quad 7.5]$$

is a solution of the minimization problem and that

$$\mathbf{c}_0 = \begin{bmatrix} 3 \\ 3 \end{bmatrix}$$

is a solution of the maximization problem.

5. For the matrices M, A, and B of Prob. 2, state the dual problems of linear programming. Then appeal to the duality theorem to show that solutions to the respective problems are given by

$$\mathbf{r}_0 = [1 \quad 3] \quad \text{and} \quad \mathbf{c}_0 = \begin{bmatrix} 9 \\ 10.5 \\ 0 \end{bmatrix}.$$

6. For the production-scheduling situation in Prob. 1, employ the matrices

$$\mathbf{r}_0 = [15 \quad 0 \quad 5] \quad \text{and} \quad \mathbf{c}_0 = \begin{bmatrix} 6 \\ 4 \end{bmatrix}$$

to deduce that the maximum daily profit is realized by manufacturing 6 units of product X and 4 units of product Y.

7. Given

$$M = \begin{bmatrix} 10 & 7 & 25 \\ 8 & 14 & 10 \end{bmatrix}, \qquad \begin{bmatrix} 250 \\ 80 \end{bmatrix} = B,$$

$$A = [80 \quad 98 \quad 125].$$

(a) By the graphical technique of the previous section, find a row matrix $\mathbf{r} = [u \quad v]$ of nonnegative entries satisfying $\mathbf{r}M \geq A$ such that $\mathbf{r}B$ is minimized.

(b) Given that

$$\mathbf{c}_0 = \begin{bmatrix} 0 \\ 0 \\ 8 \end{bmatrix}$$

is the solution of the dual maximization problem, appeal to the duality theorem to prove that the solution found in part (a) is correct.

8-4. Finding the Solutions

In Sec. 8-2 we developed a graphical technique for solving the minimization problem of linear programming for those cases in which the matrix M has two rows. An obvious variation of the procedure produces a solution of the maximization problem when the matrix M has two columns.

We recall the maximization problem from the example given at the end of the previous section.

Problem: Given

$$M = \begin{bmatrix} 1 & 2 \\ 2 & 2 \\ 2 & 1 \end{bmatrix}, \qquad \begin{bmatrix} 20 \\ 26 \\ 22 \end{bmatrix} = B,$$

$$A = [10 \quad 15],$$

find a 2×1 column matrix $\mathbf{c} = \begin{bmatrix} x \\ y \end{bmatrix}$ of nonnegative entries satisfying the constraint $M\mathbf{c} \leq B$ such that $A\mathbf{c}$ is maximized.

We return to the original perspective of Chapter 7, and identify a 2×1 column matrix $\begin{bmatrix} x \\ y \end{bmatrix}$ with the vector (x, y) in $\mathbf{R}^2$. The individual rows of M are interpreted as functionals on $\mathbf{R}^2$. In terms of these functionals the constraint $M\begin{bmatrix} x \\ y \end{bmatrix} \leq B$ is expressed by the following requirements on the vector (x, y):

1. The value assigned by the functional $[1 \quad 2]$ is 20 or less,
2. The value assigned by $[2 \quad 2]$ is 26 or less, and
3. The value assigned by $[2 \quad 1]$ is 22 or less.

We confine ourselves to the positive quadrant, i.e., the set of (x, y) such that $x \geq 0$ and $y \geq 0$.

The sketch in Fig. 8-4 displays the 20 level line of the functional $[1 \quad 2]$. This line has been determined by the points $(20, 0)$ and $(0, 10)$ on the respec-

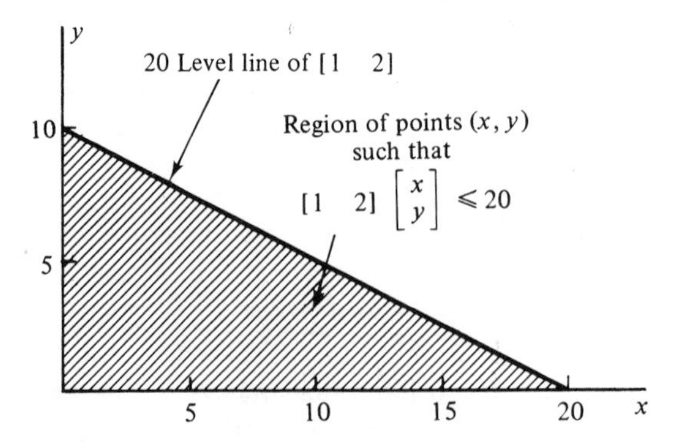

Figure 8-4.

tive axes; these points are easily found by setting $y = 0$ and $x = 0$, respectively, in the equation of the line

$$x + 2y = 20.$$

The region below the line has been shaded. It should be clear that this triangle, interior and boundary, is the set of points (x, y) in the positive quadrant for which the value assigned by $[1 \quad 2]$ is 20 or less, as required by the first row in the constraint $M\begin{bmatrix} x \\ y \end{bmatrix} \leq B$.

In Fig. 8-5 the picture has been extended to include the 26 level line of the functional $[2 \quad 2]$ and the 22 level line of the functional $[2 \quad 1]$. The lines have been labeled 1, 2, and 3, corresponding to the rows of the matrix M from which the functionals are obtained. Shaded in the figure is the convex region consisting of all points (x, y) which simultaneously lie below and to the left of all 3 level lines. This is the *region of feasibility*, whose points are precisely those which satisfy all three rows of the constraint

$$M\begin{bmatrix} x \\ y \end{bmatrix} \leq B.$$

The problem is to pinpoint a feasible point (x, y) for which the value assigned by the functional $A = [10 \quad 15]$ is as large as possible.

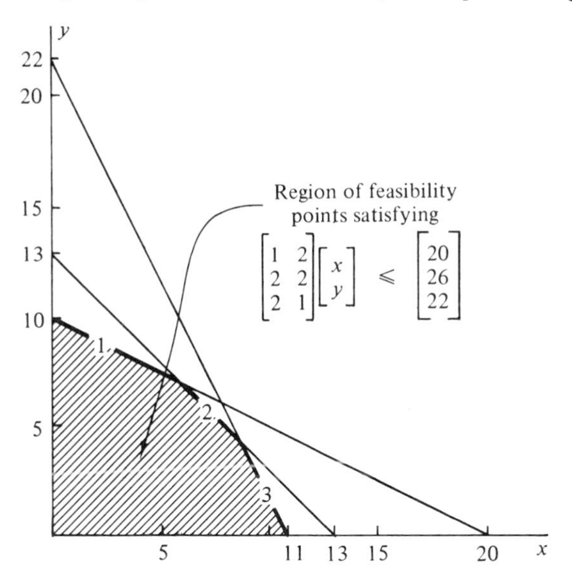

Figure 8-5.

Figure 8-6 displays the region of feasibility together with several level lines of the function $[10 \quad 15]$. The line from $(0, 0)$ through the point $(10, 15)$, to which the level lines are perpendicular, is also shown in the sketch.

Now we interpret the maximization problem dynamically. Since the value of the functional $[10 \quad 15]$ decreases as we move the level line down, the prob-

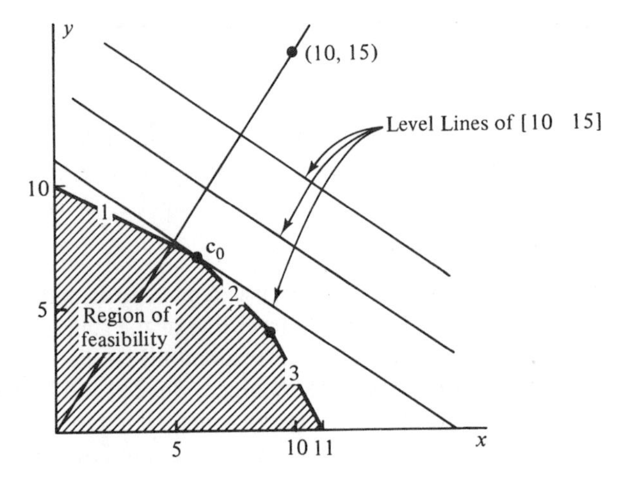

Figure 8-6.

lem is to spot a feasible point $\mathbf{c}_0$ such that there are no feasible points above the level line through $\mathbf{c}_0$. Thus we start with a level line that lies above the region of feasibility and then slide the level line down until it first encounters the region. A point of first encounter must then determine the maximum feasible value of the functional [10 15]. From the sketch in Fig. 8-6 we see that the descending level line first encounters the region of feasibility at the point $\mathbf{c}_0$, where the 20 level line of [1 2] (labeled 1) meets the 26 level line of [2 2] (labeled 2).

Thus the geometrical evidence indicates that the solution $\mathbf{c}_0 = \begin{bmatrix} x \\ y \end{bmatrix}$ of the maximization problem is determined by the simultaneous equations

$$x + 2y = 20,$$
$$2x + 2y = 26,$$

giving

$$x = 6 \quad \text{and} \quad y = 7.$$

Let's check the conjectured solution $\mathbf{c}_0 = \begin{bmatrix} 6 \\ 7 \end{bmatrix}$ in the constraint $M\mathbf{c} \leq B$. We have

$$M\mathbf{c}_0 = \begin{bmatrix} 1 & 2 \\ 2 & 2 \\ 2 & 1 \end{bmatrix} \begin{bmatrix} 6 \\ 7 \end{bmatrix} = \begin{bmatrix} 20 \\ 26 \\ 19 \end{bmatrix} \leq \begin{bmatrix} 20 \\ 26 \\ 22 \end{bmatrix} = B.$$

To verify $\mathbf{c}_0$ as correct by the criterion of the duality theorem we need a solution $\mathbf{r}_0$ to the dual minimization problem:

With M, A, and B as in the above maximization problem, find a row matrix $\mathbf{r} = [u \quad v \quad w]$ of nonnegative entries satisfying $\mathbf{r}M \geq A$ such that $\mathbf{r}B$ is minimized.

The practical procedure for finding $\mathbf{r}_0$ is an adaptation of the game-theoretic technique of Sec. 7-3 in which an optimal row strategy for a $2 \times n$ matrix game is found after reducing the matrix to the two *critical* columns involved in the optimal column strategy.

First, we note that both entries in $\mathbf{c}_0 = \begin{bmatrix} 6 \\ 7 \end{bmatrix}$ are strictly positive. We say that both columns of the matrix M are *critical.*

Second, we note that the inequality $M\mathbf{c}_0 \leq B$ is satisfied with equality in the first two rows and strict inequality in the third. We say that the first two rows of M are *critical* and that the third row is not.

To find $\mathbf{r}_0$, we translate the identification of critical rows and columns of M, as given by $\mathbf{c}_0$, to the dual problem. Since the third row of M is not critical, we look for a solution to the minimization problem of the form $\mathbf{r}_0 = [u \quad v \quad 0]$, and since both columns of M are critical, the constraint $\mathbf{r}_0 M \geq A$ is to be satisfied with equality in both columns.

In short, $\mathbf{r}_0$ is found by setting $\mathbf{r}_0 = [u \quad v \quad 0]$ and $\mathbf{r}_0 M = A$:

$$\mathbf{r}_0 M = [u \quad v \quad 0] \begin{bmatrix} 1 & 2 \\ 2 & 2 \\ 2 & 1 \end{bmatrix} = [10 \quad 15] = A,$$

or

$$u + 2v = 10,$$
$$2u + 2v = 15,$$

giving

$$u = 5 \quad \text{and} \quad v = 2.5.$$

We recall that $\mathbf{r}_0 = [5 \quad 2.5 \quad 0]$ and $\mathbf{c}_0 = \begin{bmatrix} 6 \\ 7 \end{bmatrix}$ are precisely the solutions verified at the end of the previous section, with $\mathbf{r}_0 B = A\mathbf{c}_0 = [165]$. Now let's vary the problem by decreasing the second entry in the matrix A:

Variation 1: Given the matrices M and B as before, let $A = [10 \quad 12]$. Find the solutions $\mathbf{r}_0$ and $\mathbf{c}_0$ of the dual minimization and maximization problems.

Since M and B are unchanged, the region of feasibility for the maximization problem is unchanged. The only adjustment required in Fig. 8-6 is a slight clockwise tilt of the level lines of the functional to be maximized. With this adjustment it is still the case that the level line sliding down from above first encounters the region of feasibility at the point (6, 7). Thus $\mathbf{c}_0 = \begin{bmatrix} 6 \\ 7 \end{bmatrix}$ is again proposed as the solution of the maximization problem. The identification of critical columns and rows of M is also unchanged. We find the dual solution by setting $\mathbf{r}_0 = [u \quad v \quad 0]$ and solving $\mathbf{r}_0 M = A = [10 \quad 12]$.

This gives $\mathbf{r}_0 = [2 \quad 4 \quad 0]$. By the duality theorem one can easily verify that $\mathbf{r}_0 = [2 \quad 4 \quad 0]$ and $\mathbf{c}_0 = \begin{bmatrix} 6 \\ 7 \end{bmatrix}$ solve the dual problems, with $\mathbf{r}_0 B = A\mathbf{c}_0 = [144]$.

The second variation is more interesting:

Variation 2: Given the matrices M and B above, let $A = [10 \quad 10]$. Find the solutions $\mathbf{r}_0$ and $\mathbf{c}_0$ of the dual problems.

Again the region of feasibility is as shown in Fig. 8-6. But now the level line of the functional $A = [10 \quad 10]$ to be maximized encounters the region of feasibility in coincidence with boundary line 2—the 26 level line of the functional $[2 \quad 2]$—corresponding to the second row of the constraint $M\mathbf{c} \leq B$. The end points of this segment are (6, 7), as found before, and (9, 4), the intersection of boundary lines 2 and 3. Any point on this segment qualifies as a solution of the maximization problem.

If we set $\mathbf{c}_0 = \begin{bmatrix} 6 \\ 7 \end{bmatrix}$ as before, we again are led to identify both columns and the first two rows of M as critical. Setting $\mathbf{r}_0 = [u \quad v \quad 0]$ and solving $\mathbf{r}_0 M = A = [10 \quad 10]$, we obtain $\mathbf{r}_0 = [0 \quad 5 \quad 0]$.

If we set $\mathbf{c}_0 = \begin{bmatrix} 9 \\ 4 \end{bmatrix}$, we find that the constraint $M\mathbf{c}_0 \leq B$ is satisfied with a strict inequality in the first row and equality in the other two. Thus we set $\mathbf{r}_0 = [0 \quad v \quad w]$ and solve $\mathbf{r}_0 M = A = [10 \quad 10]$. Again we obtain $\mathbf{r}_0 = [0 \quad 5 \quad 0]$.

But let's look at another point on the segment of first encounter. Midway between $\begin{bmatrix} 6 \\ 7 \end{bmatrix}$ and $\begin{bmatrix} 9 \\ 4 \end{bmatrix}$ we select the solution $\mathbf{c}_0 = \begin{bmatrix} 7.5 \\ 5.5 \end{bmatrix}$. In this case the constraint $M\mathbf{c}_0 \leq B$ is satisfied with equality only in the second row. Therefore, we set $\mathbf{r}_0 = [0 \quad v \quad 0]$. In this case both columns in the matrix equation $\mathbf{r}_0 M = A = [10 \quad 10]$ reduce to the equation $2v = 10$. Once again we obtain $\mathbf{r}_0 = [0 \quad 5 \quad 0]$.

It is a straightforward matter to verify that $\mathbf{r}_0 = [0 \quad 5 \quad 0]$ solves the minimization problem, with the maximization problem solved by $\mathbf{c}_0 = \begin{bmatrix} 6 \\ 7 \end{bmatrix}$ or $\mathbf{c}_0 = \begin{bmatrix} 9 \\ 4 \end{bmatrix}$ or $\mathbf{c}_0 = \begin{bmatrix} 7.5 \\ 5.5 \end{bmatrix}$. In all cases we have $\mathbf{r}_0 B = A\mathbf{c}_0 = [130]$.

The third variation represents the analogue of a strictly determined game.

Variation 3: Given the matrices M and B above, let $A = [10 \quad 4]$. Find the solutions $\mathbf{r}_0$ and $\mathbf{c}_0$ of the dual problems.

The region of feasibility is still as shown in Fig. 8-6. But now the level lines of the functional A have been tilted so that the point of first encounter is the vertex (11, 0) on the x axis. Thus we conjecture that $\mathbf{c}_0 = \begin{bmatrix} 11 \\ 0 \end{bmatrix}$ solves the maximization problem.

Let's check the constraint $M\mathbf{c}_0 \leq B$. We have

$$M\mathbf{c}_0 = \begin{bmatrix} 1 & 2 \\ 2 & 2 \\ 2 & 1 \end{bmatrix} \begin{bmatrix} 11 \\ 0 \end{bmatrix} = \begin{bmatrix} 11 \\ 22 \\ 22 \end{bmatrix} \leq \begin{bmatrix} 20 \\ 26 \\ 22 \end{bmatrix} = B.$$

Since $M\mathbf{c}_0 \leq B$ is satisfied with equality only in the third row, we look for a solution to the minimization problem of the form $\mathbf{r}_0 = [0 \quad 0 \quad w]$.

The new feature of this variation is that the solution $\mathbf{c}_0$ is positive in the first entry and zero in the second. Thus only the first column of M is critical. We translate this fact to the dual problem by requiring that the constraint $\mathbf{r}_0 M \geq A = [10 \quad 4]$ be satisfied with equality in the first column. We have

$$\mathbf{r}_0 M = [0 \quad 0 \quad w] \begin{bmatrix} 1 & 2 \\ 2 & 2 \\ 2 & 1 \end{bmatrix} = [2w \quad w] \geq [10 \quad 4] = A.$$

Requiring equality in the first column,

$$2w = 10 \quad \text{or} \quad w = 5.$$

Thus we conjecture that $\mathbf{r}_0 = [0 \quad 0 \quad 5]$ is a solution to the minimization problem and $\mathbf{c}_0 = \begin{bmatrix} 11 \\ 0 \end{bmatrix}$ is a solution to the maximization problem. Let's verify by the duality theorem:

$$\mathbf{r}_0 M = [0 \quad 0 \quad 5] \begin{bmatrix} 1 & 2 \\ 2 & 2 \\ 2 & 1 \end{bmatrix} = [10 \quad 5] \geq [10 \quad 4] = A, \checkmark$$

$$M\mathbf{c}_0 = \begin{bmatrix} 1 & 2 \\ 2 & 2 \\ 2 & 1 \end{bmatrix} \begin{bmatrix} 11 \\ 0 \end{bmatrix} = \begin{bmatrix} 11 \\ 22 \\ 22 \end{bmatrix} \leq \begin{bmatrix} 20 \\ 26 \\ 22 \end{bmatrix} = B, \checkmark$$

$$\mathbf{r}_0 B = [0 \quad 0 \quad 5] \begin{bmatrix} 20 \\ 26 \\ 22 \end{bmatrix} = [110],$$

$$A\mathbf{c}_0 = [10 \quad 4] \begin{bmatrix} 11 \\ 0 \end{bmatrix} = [110].$$

Thus

$$\mathbf{r}_0 B = A\mathbf{c}_0 = [110]. \checkmark\checkmark$$

Another Example

A practical situation may call for the solution of only one of the dual problems. Of course, the solution of the other problem is needed for verification by the criterion of the duality theorem. As the following example illustrates, it may be advantageous to solve the other problem first.

A factory has two machines, U and V, with which it can manufacture three products, X, Y, and Z. The following tabulation displays the machine time required for the production of one unit of the respective products. Also shown is the profit per unit of product, and the daily capacity of the machines.

		Product			*Machine capacity (in hours per day)*
		X	Y	Z	
Time required (in hours on each machine to produce one unit of product)	U	3	3	2	19
	V	2	0	3	21
Profit (in dollars per unit of product)		120	90	90	

Question: How many units of each of the respective products should be manufactured in order to realize the maximum daily profit?

The matrices of interest are lifted directly from the tabulation of the data:

$$M = \begin{bmatrix} 3 & 3 & 2 \\ 2 & 0 & 3 \end{bmatrix}, \qquad \begin{bmatrix} 19 \\ 21 \end{bmatrix} = B,$$

$$A = [120 \quad 90 \quad 90]\cdot$$

The problem of finding the feasible production schedule yielding the maximum profit is the maximization problem associated with the matrices M, A, and B.

The Maximization Problem: With M, A, and B as above, find a matrix $\mathbf{c} = \begin{bmatrix} x \\ y \\ z \end{bmatrix}$ of nonnegative entries satisfying $M\mathbf{c} \leq B$ such that $A\mathbf{c}$ is maximized.

A graphical attack on this problem would require a construction in three-dimensional space. But since the matrix M has only two rows, the dual minimization problem can be represented in the coordinate plane $\mathbf{R}^2$. And, in the manner of the preceding example, the solution of the dual problem will indicate how the given problem can be solved algebraically. Thus we turn to the dual problem:

The Minimization Problem: With M, A, and B as above, find a matrix $\mathbf{r} = [u \quad v]$ of nonnegative entries satisfying $\mathbf{r}M \geq A$ such that $\mathbf{r}B$ is minimized.

Looking at the individual columns in the constraint $\mathbf{r}M \geq A$, the region of feasibility is defined by the requirements

$$[u \quad v]\begin{bmatrix}3\\2\end{bmatrix} \geq 120, \qquad [u \quad v]\begin{bmatrix}3\\0\end{bmatrix} \geq 90, \qquad [u \quad v]\begin{bmatrix}2\\3\end{bmatrix} \geq 90.$$

In Fig. 8-7 the region of feasibility is shown as the set of points above and to the right of the 120 level line of $\begin{bmatrix}3\\2\end{bmatrix}$ (labeled 1), to the right of the 90 level line of $\begin{bmatrix}3\\0\end{bmatrix}$ (labeled 2), and above and to the right of the 90 level line of $\begin{bmatrix}2\\3\end{bmatrix}$ (labeled 3). The functional to be minimized is $B = \begin{bmatrix}19\\21\end{bmatrix}$; several level lines of this functional are shown.

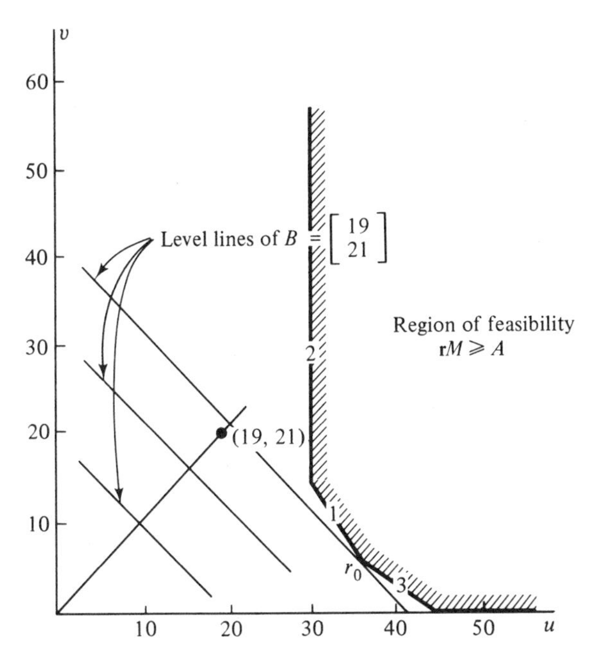

Figure 8-7.

The figure indicates that the level line of $B = \begin{bmatrix}19\\21\end{bmatrix}$, pushed up from below the region of feasibility, first encounters the region at the point $\mathbf{r}_0 = [u \quad v]$, where the 120 level line of $\begin{bmatrix}3\\2\end{bmatrix}$ (labeled 1) meets the 90 level line of $\begin{bmatrix}2\\3\end{bmatrix}$ (labeled 3). That is, $[u \quad v]$ is given by the set of equations

$$3u + 2v = 120,$$
$$2u + 3v = 90.$$

Solving, we obtain,

$$u = 36 \quad \text{and} \quad v = 6.$$

Thus we conjecture that

$$\mathbf{r}_0 = [36 \quad 6]$$

is the solution of the minimization problem.

Let's check that $\mathbf{r}_0$ satisfies the constraint

$$\mathbf{r}_0 M = [36 \quad 6]\begin{bmatrix} 3 & 3 & 2 \\ 2 & 0 & 3 \end{bmatrix} = [120 \quad 108 \quad 90]$$
$$\geq [120 \quad 90 \quad 90] = A, \checkmark$$

Now we use the evidence contained in $\mathbf{r}_0$ to find $\mathbf{c}_0 = \begin{bmatrix} x \\ y \\ z \end{bmatrix}$, the solution of the given maximization problem.

First, since $\mathbf{r}_0$ satisfies the constraint $\mathbf{r}_0 M \geq A$ with equality only in the first and third columns, we look for a solution $\mathbf{c}_0$ of the dual problem of the form $\mathbf{c}_0 = \begin{bmatrix} x \\ 0 \\ z \end{bmatrix}$.

Second, since both entries in $\mathbf{r}_0$ are strictly positive, the dual solution $\mathbf{c}_0$ is to satisfy the constraint $M\mathbf{c}_0 \leq B$ with equality in both rows. Now we have $\mathbf{c}_0$ determined by the equation

$$M\mathbf{c}_0 = \begin{bmatrix} 3 & 3 & 2 \\ 2 & 0 & 3 \end{bmatrix}\begin{bmatrix} x \\ 0 \\ z \end{bmatrix} = \begin{bmatrix} 19 \\ 21 \end{bmatrix} = B,$$

or

$$3x + 2z = 19,$$
$$2x + 3z = 21,$$

giving

$$x = 3 \quad \text{and} \quad z = 5.$$

Thus we conjecture that $\mathbf{c}_0 = \begin{bmatrix} 3 \\ 0 \\ 5 \end{bmatrix}$ is the solution of the maximization problem.

We have already seen that for $\mathbf{r}_0 = [36 \quad 6]$

$$\mathbf{r}_0 M = [120 \quad 108 \quad 90] \geq [120 \quad 90 \quad 90] = A. \checkmark$$

And from the manner in which $\mathbf{c}_0$ is obtained,

$$M\mathbf{c}_0 = \begin{bmatrix} 19 \\ 21 \end{bmatrix} \leq \begin{bmatrix} 19 \\ 21 \end{bmatrix} = B. \checkmark$$

All that remains to verify both solutions by the duality theorem is to confirm that

$$\mathbf{r}_0 B = A\mathbf{c}_0.$$

We have

$$\mathbf{r}_0 B = [36 \quad 6]\begin{bmatrix} 19 \\ 21 \end{bmatrix} = [810],$$

$$A\mathbf{c}_0 = [120 \quad 90 \quad 90]\begin{bmatrix} 3 \\ 0 \\ 5 \end{bmatrix} = [810] = \mathbf{r}_0 B. \checkmark\checkmark$$

Therefore, $\mathbf{r}_0 = [36 \quad 6]$ is the solution of the minimization problem, and $\mathbf{c}_0 = \begin{bmatrix} 3 \\ 0 \\ 5 \end{bmatrix}$ is the solution of the maximization problem.

To answer the question posed,

Answer: The factory should manufacture 3 units of product X, no units of Y, and 5 units of Z in order to realize a maximum daily profit of \$810.

The method in these examples, relying on a geometrical interpretation, can be employed as long as the matrix M has only two rows or two columns. Usually the "critical" identifications derived from the solution of the two dimensional problem will generate a unique solution for the dual problem; if this is not the case, extra care may be demanded in selecting a particular dual solution that satisfies the entire matrix inequality given by the constraint.

The technique extends to three dimensional space—three rows or three columns in M—but it is difficult to construct and to read the pictures. As one might guess, the applications of linear programming in honestly encountered problems generally involve somewhat larger matrices. The diet in an institution will demand many nutrients and there will be many foods to be considered. And a factory will likely have a large variety of products calling for a prudent allocation of the many elements in its resources—machines, labor, floor space, etc. One doesn't try to solve the larger problems with pencil and paper—there are efficient computer algorithms for searching for the optimal points. The geometrical interpretations that we've developed give useful insight into the general method, and our theory covers the general case, but, as far as actually finding solutions, we'll be content with the $2 \times n$ and $n \times 2$ cases.

8-4 EXERCISES

1. Given the matrices

$$M = \begin{bmatrix} 2 & 6 \\ 2 & 2 \\ 4 & 2 \end{bmatrix}, \qquad \begin{bmatrix} 24 \\ 12 \\ 20 \end{bmatrix} = B,$$

$$A = [6 \ \ 10].$$

(a) In the first quadrant of the coordinate plane, sketch the region of feasible points $\mathbf{c} = \begin{bmatrix} x \\ y \end{bmatrix}$ satisfying the constraint

$$M\mathbf{c} \leq B.$$

(b) On the sketch in part (a), display several level lines of the function $A = [6 \quad 10]$.

(c) From inspection of the construction in part (b), locate the point $\mathbf{c}_0$ of the region of feasibility such that the value assigned by the functional $A = [6 \quad 10]$ is maximized.

(d) With $\mathbf{c}_0$ identifying the critical rows and columns of the matrix M, algebraically solve for the solution $\mathbf{r}_0$ of the minimization problem associated with M, A, and B.

(e) Verify by the Duality Theorem of Linear Programming that $\mathbf{r}_0$ and $\mathbf{c}_0$ are solutions of the minimization and maximization problems associated with M, A, and B.

2. With M and B as given in Prob. 1 and A as given below, find the solutions $\mathbf{r}_0$ and $\mathbf{c}_0$ of the dual minimization and maximization problems. Verify by the duality theorem.

(a) $A = [6 \quad 5]$.

(b) $A = [6 \quad 2]$.

3. Given the matrices

$$M = \begin{bmatrix} 1 & 2 & 2 & 1 \\ 2 & 2 & 0 & 4 \end{bmatrix}, \qquad \begin{bmatrix} 6 \\ 4 \end{bmatrix} = B,$$

$$A = [8 \quad 10 \quad 8 \quad 10].$$

(a) In the first quadrant of the coordinate plane, sketch the region of feasible points $\mathbf{r} = [u \quad v]$ satisfying the constraint

$$\mathbf{r}M \geq A.$$

Sketch several level lines of the functional $B = \begin{bmatrix} 6 \\ 4 \end{bmatrix}$, and locate the point $\mathbf{r}_0$ which minimizes the value assigned by the functional B.

(b) With $\mathbf{r}_0$ identifying the critical rows and columns of M, algebraically solve for the solution $\mathbf{c}_0$ of the maximization problem associated with M, A, and B.

(c) Verify by the duality theorem that $\mathbf{r}_0$ and $\mathbf{c}_0$ are the solutions of the minimization and maximization problems associated with M, A, and B.

4. Given the matrices

$$M = \begin{bmatrix} 1 & 2 \\ 2 & 2 \\ 2 & 0 \\ 1 & 4 \end{bmatrix}, \qquad \begin{bmatrix} 8 \\ 10 \\ 8 \\ 10 \end{bmatrix} = B,$$

$$A = [6 \quad 4],$$

find the solutions $\mathbf{r}_0$ and $\mathbf{c}_0$ of the dual minimization and maximization problems associated with the matrices M, A, and B. Verify by the duality theorem. (Compare with Prob. 3.)

5. In her pottery shop, Aleeta has created two distincitve types of pots, much in demand. The larger pot sells for $12 and takes 30 minutes to produce, requiring 2 pounds of clay and 3 ounces of glaze. The smaller pot sells for $7.50, takes 20 minutes to produce, and calls for 1 pound of clay and 4 ounces of glaze. Suppose Aleeta has 150 pounds of clay and 10 quarts (320 ounces) of glaze and she's willing to devote 40 hours to throwing this material into pots.

   The question is, How many large and small pots, respectively, should she produce in order to maximize her income?

   Display the matrices M, A, and B for which this is the maximization problem. Find and verify $\mathbf{r}_0$ and $\mathbf{c}_0$. Then answer the question in words.

6. Given

$$M = \begin{bmatrix} 4 & 5 \\ 0 & 7 \\ 5 & 3 \end{bmatrix}, \quad \begin{bmatrix} 30 \\ 28 \\ 30 \end{bmatrix} = B,$$

$$A = [100 \quad 40],$$

   solve the minimization problem associated with the matrices M, A, and B.

7. To prepare the spaghetti sauce for the annual social, Ellen needs 20 pounds of ground beef and 12 pounds of ground pork. At Store A she can buy the beef for 85 cents a pound and the pork for 79 cents a pound; or she can buy their "Meat Loaf Special," each pack of which costs $1.15 and contains 1 pound of ground beef and $\frac{1}{2}$ pound of ground pork. At Store B, Ellen can buy the beef for 89 cents a pound and the pork for 75 cent a pound, or she can buy the "Combo Special," each pack of which costs 75 cents and contains $\frac{1}{2}$ pound of ground beef and $\frac{1}{2}$ pound of ground pork. How can Ellen obtain the desired 20 pounds of beef and 12 pounds of pork for the least possible cost?

8. A mining company has contracted to deliver 10 tons of Mineral A and 2 tons of Mineral B. They have three mines: From Mine 1, 8% of the ore is extracted as Mineral A and 1% as B; from Mine 2, 4% of the ore is extracted as A and 2% as B; and from Mine 3, 10% of the ore is extracted as A and none as B. How can the company meet its contract of 10 tons of A and 2 tons of B while minimizing the total amount of ore required?

9. The work force at the Peerless Awning Company delivers 1000 man-hours per week. A maximum of 400 of these hours can be spent on the sewing machines. The company has a dependable market for tents, tarps, and trampolines. Each tent requires 3 man-hours of labor including one hour on the sewing machines, and produces a profit to the owners of $21. Each tarp requires 20 minutes of labor including 10 minutes on the sewing machines, and produces a profit of $2.50. Each

trampoline requires 1.5 man-hours including 20 minutes on the sewing machines, and produces a profit of $9. How many of each of the respective products should the company produce for maximum profit in a week?

10. Given the matrices

$$M = \begin{bmatrix} 1 & 3 & 2 \\ 3 & 1 & 2 \end{bmatrix}, \qquad \begin{bmatrix} 6 \\ 6 \end{bmatrix} = B,$$

$$A = [6 \quad 6 \quad 8],$$

(a) Find and verify three distinct solutions $\mathbf{r}_0$ for the minimization problem associated with M, A, and B.

(b) Replace the matrix A with

$$A = [6 \quad 6 \quad 6].$$

Now find and verify three distinct solutions $\mathbf{c}_0$ for the maximization problem associated with M, A, and B.

11. (For a general "urn model" of minimization problems, read "assortment" as "type of urn.") Carol requires 100 red marbles and 100 white marbles. Red and white marbles are available in three assortments. Each bag of assortment U contains 10 red and 15 white marbles and costs 90 cents; each bag of assortment V contains 10 red and 10 white marbles and costs 80 cents; and each bag of assortment W contains 10 red and 5 white marbles and costs 60 cents. How many bags of each assortment should Carol buy in order to get the 100 marbles of each color for the least possible cost? And what is the least cost?

12. (Suggestive of an "urn model" for maximization problems. Compare with Prob. 11 above.) Carol has 100 red and 100 white marbles which she is going to package into bags to sell at the Charity Bazaar. The assortments in the bags are to be of three possible types. Each bag of assortment X is to contain 10 red and 15 white marbles and sell for 90 cents; each bag of assortment Y is to contain 10 red and 10 white marbles and sell for 80 cents; and each bag of assortment Z is to contain 10 red and 5 white marbles and sell for 60 cents. How many bags of each assortment should Carol assemble so that the Bazaar may realize the maximum possible income from the 100 red and 100 white marbles? And what is the maximum income?

8-5. Proof and Interpretation of the Duality Theorem

By now we have a significant investment in the Duality Theorem of Linear Programming. But the only argument that has been offered in its defense is the development of the "Fish-Diet" problem and its "Production-Scheduling" companion to suggest that the new duality theorem is a variation on the Duality Theorem of Game Theory. Let's look at the new theorem in its own right.

It is often the case in mathematics that one can gain insight into a general result by considering a simplified special case. The Duality Theorem of Linear Programming furnishes a beautiful example. Reducing the theorem to its simplest case, we consider the instance in which the matrices involved are 1×1 in dimension—in effect, the matrices are just numbers. So as a warmup to attacking the general theorem, we restate it, substituting real numbers for matrices. In this form it is easy to prove.

Preliminary Theorem: Given real numbers M, A, and B, let r_0 and c_0 be nonnegative real numbers which satisfy the following properties:

$$r_0 M \geq A,$$

$$Mc_0 \leq B,$$

$$r_0 B = Ac_0,$$

Then for each nonnegative real number r such that

$$rM \geq A$$

we have

$$r_0 B \leq rB,$$

and for each nonnegative real number c such that

$$Mc \leq B$$

we have

$$Ac \leq Ac_0.$$

Proof: Suppose r is a nonnegative number such that

$$rM \geq A. \tag{1}$$

Since c_0 is nonnegative, we have

$$rMc_0 \geq Ac_0. \tag{2}$$

Similarly, since $Mc_0 \leq B$ and r is nonnegative,

$$rMc_0 \leq rB. \tag{3}$$

Combining (2) and (3), we have

$$Ac_0 \leq rMc_0 \leq rB, \tag{4}$$

from which it follows that

$$Ac_0 \leq rB. \tag{5}$$

By hypothesis $Ac_0 = r_0 B$. Therefore, from (5)

$$r_0 B \leq rB. \tag{6}$$

This is the first of the desired conclusions.

Now let c be a nonnegative real number such that

$$Mc \leq B.$$

We have

$$r_0 Mc \leq r_0 B = Ac_0. \tag{7}$$

Since $r_0M \geq A$,

$$r_0Mc \geq Ac. \tag{8}$$

From (7) and (8) it follows that

$$Ac \leq r_0Mc \leq Ac_0$$

or

$$Ac \leq Ac_0,$$

which is the second of the desired conclusions.

As we see, the argument for the imitation theorem, with numbers substituted for matrices, is not difficult. Happily, this argument carries over to the larger theorem. All we need to do is to establish that for all of the real number properties on which the argument depends the analogous properties are valid for matrices. Essentially only three properties are involved—let's look at them.

Property 1: The Associative Law. Let S, T, and W be matrices such that the products

$$ST \quad \text{and} \quad TW$$

are defined. Then

$$S(TW) = (ST)W.$$

We recall that the argument for the associative law was given in Sec. 6-7.

Property 2: Preservation of Inequality. Let T and U be matrices of like dimensions such that

$$T \geq U.$$

Let S be a matrix of *nonnegative entries* such that ST, and therefore SU, is defined. Then

$$ST \geq SU.$$

Similarly, if W is a matrix of nonnegative entries such that TW, and therefore UW, is defined, then

$$TW \geq UW.$$

Proof: It is immediate from the definition of inequality for matrices of like dimensions that

$$T \geq U$$

is satisfied precisely when the entries in the difference $T - U$ are all nonnegative. Since the entries in both S and the difference $T - U$ are nonnegative, it follows from the definition of matrix multiplication that the entries in $S(T - U)$ are nonnegative. (This conclusion depends only on the fact that a sum of products of nonnegative numbers is nonnegative.)

From the distributive law of matrix multiplication, given in Sec. 6-7,

$$S(T - U) = ST - SU.$$

Since the entries in $S(T - U)$ and therefore in

$$ST - SU$$

are nonnegative, if follows that

$$ST \geq SU.$$

The other half of the conclusion is established analogously.

Property 3: Transitivity of Inequality. If T, U, and V are matrices of like dimension such that

$$T \leq U \quad \text{and} \quad U \leq V,$$

then

$$T \leq V.$$

Proof: Immediate from the definition of inequality for matrices and the analogous transitivity property for real numbers.

We are now ready to state the general theorem, and prove it, virtually word for word as we proved the preliminary theorem.

The Duality Theorem of Linear Programming: Given an $n \times k$ matrix M, a $1 \times k$ row matrix A, and an $n \times 1$ column matrix B, let $\mathbf{r}_0$ be a $1 \times n$ matrix and $\mathbf{c}_0$ be a $k \times 1$ matrix, both with entries nonnegative, satisfying the following:

$$\mathbf{r}_0 M \geq A,$$

$$M\mathbf{c}_0 \leq B,$$

$$\mathbf{r}_0 B = A\mathbf{c}_0.$$

Then for each $1 \times n$ matrix $\mathbf{r}$ of nonnegative entries such that

$$\mathbf{r}M \geq A$$

we have

$$\mathbf{r}_0 B \leq \mathbf{r}B,$$

and for each $k \times 1$ matrix $\mathbf{c}$ of nonnegative entries such that

$$M\mathbf{c} \leq B$$

we have

$$A\mathbf{c} \leq A\mathbf{c}_0.$$

(In short, $\mathbf{r}_0$ solves the minimization problem and $\mathbf{c}_0$ solves the maximization problem.)

Proof: Suppose $\mathbf{r}$ is a $1 \times n$ matrix of nonnegative entries such that

$$\mathbf{r}M \geq A. \tag{1}$$

Since $\mathbf{c}_0$ is a $k \times 1$ matrix of nonnegative entries and since multiplication by such a matrix preserves inequality, we have

$$(\mathbf{r}M)\mathbf{c}_0 \geq A\mathbf{c}_0. \tag{2}$$

Similarly, since $M\mathbf{c}_0 \leq B$ and since the entries in $\mathbf{r}$ are nonnegative,

$$\mathbf{r}(M\mathbf{c}_0) \leq \mathbf{r}B. \tag{3}$$

Applying the associative law and combining (2) and (3), we have

$$A\mathbf{c}_0 \leq (\mathbf{r}M)\mathbf{c}_0 = \mathbf{r}(M\mathbf{c}_0) \leq \mathbf{r}B. \tag{4}$$

(We recall that $(\mathbf{r}M)\mathbf{c}_0 \geq A\mathbf{c}_0$ and $A\mathbf{c}_0 \leq (\mathbf{r}M)\mathbf{c}_0$ mean the same thing.) From (4) it follows by the transitivity property that

$$A\mathbf{c}_0 \leq \mathbf{r}B. \tag{5}$$

By hypothesis $A\mathbf{c}_0 = \mathbf{r}_0 B$. Therefore, from (5)

$$\mathbf{r}_0 B \leq \mathbf{r}B. \tag{6}$$

This is the first of the desired conclusions.

Now let $\mathbf{c}$ be a $k \times 1$ matrix of nonnegative entries such that

$$M\mathbf{c} \leq B.$$

We have

$$\mathbf{r}_0(M\mathbf{c}) \leq \mathbf{r}_0 B = A\mathbf{c}_0. \tag{7}$$

Since $\mathbf{r}_0 M \geq A$,

$$(\mathbf{r}_0 M)\mathbf{c} \geq A\mathbf{c}. \tag{8}$$

From (7) and (8) it follows that

$$A\mathbf{c} \leq (\mathbf{r}_0 M)\mathbf{c} = \mathbf{r}_0(M\mathbf{c}) \leq A\mathbf{c}_0$$

or

$$A\mathbf{c} \leq A\mathbf{c}_0,$$

which is the second of the desired conclusions.

The Evaluation Question

The duality theorem reveals an intimate connection between the minimization and maximization problems in linear programming. However, in the examples we've seen, such as the "Fish-Diet" problem and its "Production-Scheduling" companion, there has been no visible connection between the empirical interpretations of the two problems. We shall see now how the dual problem may be stated in the framework of the given situation.

We return to the Fish-Diet problem, as restated in Sec. 8-2.

The diet of the fish requires three nutrients, X, Y, and Z, to be obtained from two foods, U and V. The cost and nutritional content per package of food and the yearly nutritional requirement are displayed in the following tabulation:

		Nutrient			Cost
		X	Y	Z	(in dollars per pkg of food)
Amount of Nutrient (in nutritional units, per pkg of food)	U	18	12	9	18
	V	9	24	10	12
Yearly Nutritional Requirement (in nutritional units)		90	120	60	

The problem of minimizing the cost of the year's diet is the minimization problem associated with the matrices

$$M = \begin{bmatrix} 18 & 12 & 9 \\ 9 & 24 & 10 \end{bmatrix}, \qquad \begin{bmatrix} 18 \\ 12 \end{bmatrix} = B,$$

$$A = [90 \quad 120 \quad 60].$$

We recall from the earlier analysis that the solutions of the dual problems are, respectively,

$$\mathbf{r}_0 = [\tfrac{10}{3} \quad \tfrac{10}{3}] \quad \text{and} \quad \mathbf{c}_0 = \begin{bmatrix} \frac{8}{9} \\ \frac{1}{6} \\ 0 \end{bmatrix},$$

Here

$$\mathbf{r}_0 M = [90 \quad 120 \quad 63\tfrac{1}{3}] \geq A, \qquad M\mathbf{c}_0 = \begin{bmatrix} 18 \\ 12 \end{bmatrix} \leq B,$$

and

$$\mathbf{r}_0 B = A\mathbf{c}_0 = [100].$$

Thus the minimum cost diet consists of $\frac{10}{3}$ packages of U and $\frac{10}{3}$ packages of V. The cost of this combination is \$100.

Now that the old problem has been reestablished, we turn to the interesting question,

What does the solution of the dual problem

$$\mathbf{c}_0 = \begin{bmatrix} \frac{8}{9} \\ \frac{1}{6} \\ 0 \end{bmatrix}$$

tell us about the economics of feeding those fish?

In the original statement of the Fish-Diet problem in Sec. 8-1 we were able to identify the hobbyist, who was trying to feed his fish for as long as possible for \$1, as the row player R in a certain matrix game. As the problem has been restated, we have R trying to find the minimum cost combination of packages of foods U and V to sustain his fish for one year. The question: What is the role of the column player C in the Fish-Diet situation? And a reasonable answer:

We may think of C as a hypothetical salesman who is offering buyer R pure nutrients at prices to be determined by the competition from foods U and V and by R's requirements.

Thus we keep the Fish-Diet situation exactly as it is, and we have salesman C enter, offering pure nutrient X at a price of x dollars per unit of nutrient, pure Y at a price of y dollars per unit, and pure Z at a price of z dollars per unit. The problem faced by C is to fix his prices so that the maximum income will be realized from R's purchase of his yearly requirements. The constraint on salesman C is that the prices must not be too high, for then buyer R will be attracted to meet some, or all, of his nutritional requirements by purchases of foods U and/or V.

With x, y, and z as interpreted above we identify the "price schedule" offered by C as a 3×1 column matrix $\mathbf{c} = \begin{bmatrix} x \\ y \\ z \end{bmatrix}$ of nonnegative entries. Buyer R, who we assume is agreeable to doing business with C, accepts or rejects the schedule $\mathbf{c}$ on simple grounds: All that R wants to know is whether the nutrients are overpriced vis-à-vis the packages of foods U and V.

One package of U contains 18 units of X, 12 units of Y, and 9 units of Z. Under schedule $\mathbf{c} = \begin{bmatrix} x \\ y \\ z \end{bmatrix}$ the price in dollars charged by salesman C for the same combination of nutrients is

$$18x + 12y + 9z.$$

Since a package of U costs \$18, R accepts the schedule $\mathbf{c}$ as feasible with respect to U if

$$18x + 12y + 9z \leq 18.$$

Similarly, $\mathbf{c}$ is feasible with respect to V if

$$9x + 24y + 10z \leq 12.$$

In short, the schedule $\mathbf{c} = \begin{bmatrix} x \\ y \\ z \end{bmatrix}$ is feasible if

$$M\mathbf{c} = \begin{bmatrix} 18 & 12 & 9 \\ 9 & 24 & 10 \end{bmatrix} \begin{bmatrix} x \\ y \\ z \end{bmatrix} \leq \begin{bmatrix} 18 \\ 12 \end{bmatrix} = B.$$

Thus we have buyer R accepting the price schedule $\mathbf{c}$ as feasible, and buying all of his nutrients from salesman C, provided

$$M\mathbf{c} \leq B.$$

In furnishing the yearly requirement of 90 units of X, 120 units of Y, and 60 units of Z, the income for C in dollars is

$$90x + 120y + 60z.$$

Thus the challenge for C is to find the feasible schedule $\mathbf{c}$, satisfying the constraint $M\mathbf{c} \leq B$, which maximizes the value of

$$A\mathbf{c} = [90 \quad 120 \quad 60]\begin{bmatrix} x \\ y \\ z \end{bmatrix} = [90x + 120y + 60z].$$

We see that C's problem is precisely the maximization problem associated with matrices M, A, and B. As we already know, the solution is

$$\mathbf{c}_0 = \begin{bmatrix} \frac{8}{9} \\ \frac{1}{6} \\ 0 \end{bmatrix}.$$

Therefore, salesman C should charge $\frac{8}{9}$ dollar per unit of X, $\frac{1}{6}$ dollar per unit of Y, and nothing for Z.

With this optimal price schedule $\mathbf{c}_0$ the yearly income for C is \$100, since

$$A\mathbf{c}_0 = [100].$$

Of course, this \$100 comes from buyer R. And this is exactly the amount that R must pay for the minimal cost diet $\mathbf{r}_0$ assembled from foods U and V, since

$$\mathbf{r}_0 B = [100].$$

The duality theorem asserts that R has solved his problem of finding the minimal-cost feasible combination of foods U and V, and, at the same time, C has solved his problem of finding the maximal-income feasible schedule of prices for pure nutrients precisely when the cost to R is the same whether he buys the foods or the pure nutrients.

In the actual empirical situation there really isn't a salesman C offering pure nutrients at prices he's free to fix. But there is a buyer R, and there's good reason for R to be interested in the problem of hypothetical C. As we've seen, the optimal price schedule in the Fish-Diet situation is

$$\mathbf{c}_0 = \begin{bmatrix} \frac{8}{9} \\ \frac{1}{6} \\ 0 \end{bmatrix}.$$

Even though no one is offering units of nutrients at these prices, R should see himself as paying $\frac{8}{9}$ dollar per unit of X, $\frac{1}{6}$ dollar per unit of Y, and nothing for Z. Of the \$100 that R spends for his yearly requirements of nutrients, \$80 of this should be assigned as the cost of the 90 units of X, \$20 as the cost of the 120 units of Y; and the Z is free.

To confirm the reasonableness of assigning the above prices to the units of nutrients, let's suppose a new food, W, appears on the market. We'll say that each package of W contains 9 units of X, 12 units of Y, and 24 units of Z. If single units of X, Y, and Z are evaluated at $\frac{8}{9}$, $\frac{1}{6}$, and 0 dollars, respectively, then the value of a package of food W is

$$\frac{8}{9} \times 9 + \frac{1}{6} \times 12 + 0 \times 24 = 10,$$

or $10.

The tabulation of the Fish-Diet situation and the matrix statement of the cost minimization problem can be extended to encompass the new food W. If the actual cost of a package of W is more than $10, then it's easy to show that the old solution—$\frac{10}{3}$ packages of U and $\frac{10}{3}$ packages of V and no W—still represents the minimal cost diet. On the other hand, if a package of W costs less than $10, then the minimal cost diet includes some W. If the price is exactly $10, then there is not a unique minimal cost diet, and some W may or may not be included.

The assigned prices, $\frac{8}{9}$ dollar per unit of X, etc.—called *shadow prices*—can be employed to attach a value to any combination of nutrients. If the actual price of the combination exceeds the evaluation, then the minimal cost diet is, as before, restricted to foods U and V. But if the actual price of the combination is less than the evaluation, then the minimal cost diet includes some of the new combination.

There is another sense in which the shadow prices are applicable. Suppose the X requirement increases slightly, say from 90 units per year to 91 units per year. Then it can be shown that the minimal cost for a feasible diet has increased by exactly $\frac{8}{9}$ dollar, the shadow price associated with the one unit of X. And if the X requirement decreases by one-half unit, the minimal cost goes down by $\frac{1}{2} \times \frac{8}{9}$ dollar, and analogously for slight changes in the Y and Z requirements.

We recall that the minimal cost diet $\mathbf{r}_0 = [\frac{10}{3} \;\; \frac{10}{3}]$ provides $63\frac{1}{3}$ units of Z, whereas the year's requirement is only 60 units. So a slight change in the Z requirement doesn't change the minimal cost diet at all, which should explain why the shadow price attached to Z is zero. If we increase the Z requirement from 60 to 64 units, then the old minimal cost diet no longer works, and the cost increases. But the increase from 60 to 64 units is not a "slight" change—it switches the critical vertex in the graphical interpretation of the problem. The use of shadow prices to measure the change in the cost for a change in requirements is valid only for sufficiently small changes.

The same interpretation of shadow prices applies in any cost minimization linear programming problem in which the dual maximization problem has a unique solution. An analogous evaluation applies for maximization problems. For example, consider the Production-Scheduling dual of the Fish-Diet problem, where one unit of product X requires 18 hours on machine U and 9 hours on machine V and produces a profit of $90, etc. Here the interpretation of the entries in the solution $\mathbf{r}_0 = [\frac{10}{3} \;\; \frac{10}{3}]$ of the minimization problem is that 1 hour's worth of time of machine U yields a profit of $\frac{10}{3}$ dollars, and likewise for machine V. If someone will rent the factory an extra machine U for less than $\frac{10}{3}$ dollars per hour, then the operators of the factory should consider buying some time on it.

8-5 EXERCISES

1. A diet requires two nutrients, X and Y; available are two foods, U and V, containing these nutrients. The following table displays the nutritional content and cost per package of food, and the required amounts of nutrients:

		Amount of nutrient (in units per pkg)		*Cost of food (in dollars per pkg)*
		X	Y	
Food	U	1	2	20
	V	2	1	22
Requirements (in nutritional units)		16	14	

(a) State the problem of finding the minimal cost diet meeting the requirements as a minimization problem in linear programming, and state the dual maximization problem. Appeal to the duality theorem to prove that

$$\mathbf{r}_0 = [4 \quad 6] \quad \text{and} \quad \mathbf{c}_0 = \begin{bmatrix} 8 \\ 6 \end{bmatrix}$$

are the solutions to the respective problems.

(b) What is the shadow price of single units of nutrients X and Y, respectively?

(c) Suppose a new food, W, appears on the market, each package of which contains one unit of X and one unit of Y and costs \$15. Taking into account food W, exhibit the matrices invoved in the cost minimization problem and its maximization dual. Adapt the solutions of the original problem, and prove that the minimal cost diet includes no W.

(d) Part (c) shows that one would not be willing to pay \$15 for a package of food W. What is the maximum amount that one would be willing to pay?

(e) Suppose the X requirement is increased from 16 to 16.3 units, the Y requirement remaining at 14 units. Again, consider the cost minimization problem (forget food W) and its dual. Appeal to the duality theorem to prove that

$$\mathbf{r}_0 = [3.9 \quad 6.2] \quad \text{and} \quad \mathbf{c}_0 = \begin{bmatrix} 8 \\ 6 \end{bmatrix}$$

are solutions to the adjusted problems. By how much has the minimal cost of the diet been increased by the increase in the X requirement? What is the relation between this increase in cost and the shadow price of a unit of X?

2. Consider the "Production-Scheduling" dual of the diet problem in Prob. 1 (as originally stated):

A factory employs machines U and V to produce products X and Y. Each unit of X requires 1 hour on U and 2 hours on V and yields a profit of \$16. Each unit of Y requires 2 hours on U and 1 hour on V and yields a profit of \$14. Machine U can operate 20 hours a day, and V can operate 22 hours.

(a) Recalling the solutions given in Prob. 1(a), state the production schedule yielding the maximum daily profit. What is the maximum profit?

(b) Suppose the factory is able to produce a new product, Z, each unit of which requires 2 hours of U and 3 hours of V and yields a profit of \$25. Taking into account product Z, consider the maximum profit problem as a maximization problem in linear programming. Show that the optimal production schedule includes no product Z. What is the minimum profit per unit to be demanded of product Z for it to warrant manufacture?

(c) Suppose the number of hours available daily on machine U is increased from 20 to 20.6, the number of hours of Y remaining at 22. Again, consider the profit maximization problem (ignore product Z) and its dual as linear programming problems. Prove that the solutions to the respective problems are given by

$$\mathbf{r}_0 = [4 \quad 6] \quad \text{and} \quad \mathbf{c}_0 = \begin{bmatrix} 7.8 \\ 6.4 \end{bmatrix}.$$

By how much has the maximum profit been increased by the increase in the time available on machine U? What is the relation between this increase in profit and the u value in the solution $\mathbf{r}_0$ of the dual problem?

3. Consider the linear programming problems associated with the following matrices M, A, and B, where B contains an unknown, p:

$$M = \begin{bmatrix} 3 & 2 \\ 2 & 4 \\ 2 & 3 \end{bmatrix}, \qquad \begin{bmatrix} 24 \\ 24 \\ p \end{bmatrix} = B,$$

$$A = [24 \quad 32].$$

(a) Prove that if $p > 21$, the solution of the minimization problem is given by

$$\mathbf{r}_0 = [4 \quad 6 \quad 0].$$

(b) Interpret the minimization problem as a diet problem involving a food W of unknown cost p. Explain the inequality $p > 21$ in terms of the shadow prices of the nutrients as determined by the other foods.

4. Consider the linear programming problems associated with the following matrices M, A, and B, where A contains an unknown, p. (Note the similarity with Prob. 3.)

$$M = \begin{bmatrix} 3 & 2 & 2 \\ 2 & 4 & 3 \end{bmatrix}, \quad \begin{bmatrix} 24 \\ 24 \end{bmatrix} = B,$$

$$A = [24 \quad 32 \quad p].$$

(a) Show that if $p < 26$, the solution of the maximization problem is of the form

$$\mathbf{c}_0 = \begin{bmatrix} x \\ y \\ 0 \end{bmatrix}.$$

(b) Interpret the maximization problem as a production-scheduling problem involving a product Z of unknown profit. Explain the inequality $p < 26$ in terms of the entries in the solution of the dual minimization problem.

5. (a) Find the solutions $\mathbf{r}_0$ and $\mathbf{c}_0$ of the minimization and maximization problems associated with the following matrices, and verify:

$$M = \begin{bmatrix} 1 & 1 & 2 \\ 3 & 1 & 1 \end{bmatrix}, \quad \begin{bmatrix} 4 \\ 3 \end{bmatrix} = B,$$

$$A = [12 \quad 10 \quad 15].$$

(b) With M and B as in part (a), replace the matrix A by

$$A' = [12 \quad 9.9 \quad 15].$$

Find the solution $\mathbf{r}_0'$ of the new minimization problem, and show that the solution $\mathbf{c}_0$ of part (a) also solves the new maximization problem. What is the relation between the entries in $\mathbf{c}_0$ and the change in the minimum value of $\mathbf{r}B$ as A is replaced by A'?

(c) Repeat part (b) with A replaced by the matrix

$$A'' = [13 \quad 10 \quad 15].$$

In general, the examples and exercises of the preceding sections furnish a ready source of applications of the "shadow price" and the dual "shadow profit" concepts. For instance, in Prob. 7 of Sec. 8-4, what is the price Ellen pays for each pound of beef and each pound of pork, respectively? And in Prob. 9 of the same section, what profit should the owners attach to each man-hour of labor and to each hour on the sewing machines?

8-6. Games and Linear Programming

As the original Fish-Diet problem was stated in Sec. 8-1, it was mathematically equivalent to the row player's problem in a certain matrix game, and the Production-Scheduling dual corresponded to the column player's problem. Let's recall the Fish-Diet problem.

> The fish require nutrients X, Y, and Z, to be obtained from foods U and V. One dollar's worth of U furnishes a 4-day supply of X, a 2-day supply of Y, and a 3-day supply of Z, and one dollar's worth of V furnishes a 3-day supply of X, a 6-day supply of Y, and a 5-day supply of Z.

Question: How should a one-dollar expenditure be divided between foods U and V in order to feed the fish for the longest possible time?

As we saw in Sec. 8-1, the problem is precisely that of finding an optimal row strategy in the game with payoff matrix

$$G = \begin{bmatrix} 4 & 2 & 3 \\ 3 & 6 & 5 \end{bmatrix}.$$

The optimal row strategy was found to be

$$\mathbf{r}_0 = [.6 \quad .4].$$

This corresponds to an optimal diet consisting of 60 cents worth of U and 40 cents worth of V. The nutritional requirements are maintained for 3.6 days since

$$\min \mathbf{r}_0 G = 3.6 = v_G.$$

Here, v_G is the value of the game G.

Now let's stay very close to the above data and reword the problem in language that translates into a minimization problem in linear programming. All we do is interpret a one-day requirement of nutrient as one unit and one dollar's worth of food as one package. The new problem is to find the number of packages of the respective foods that should be bought in order to meet the one-day requirements for the least possible cost.

> The fish require nutrients X, Y, and Z, to be obtained from foods U and V. Each package of U contains 4 units of X, 2 units of Y, and 3 units of Z and costs \$1. Each package of V contains 3 units of X, 6 units of Y, and 5 units of Z and costs \$1. The (daily) requirement is 1 unit of X, 1 unit of Y, and 1 unit of Z.

Problem: How many packages (or fractions of packages) of foods U and V should be purchased in order to meet the nutritional requirements for the least possible cost?

Now we have a conventional minimization problem in linear programming. The matrix G is the same as in the game-theoretic interpretation of the problem. We represent the row matrices of concern as $\mathbf{r}'$, to distinguish them from the row strategies $\mathbf{r}$ in the matrix game.

Given

$$G = \begin{bmatrix} 4 & 2 & 3 \\ 3 & 6 & 5 \end{bmatrix}, \quad \begin{bmatrix} 1 \\ 1 \end{bmatrix} = B,$$

$$A = [1 \quad 1 \quad 1],$$

The Minimization Problem: Find a row matrix $\mathbf{r}' = [u \quad v]$ of nonnegative entries satisfying

$$\mathbf{r}'G \geq A$$

such that $\mathbf{r}'B$ is minimized.

We already know that the fish are maintained for at most 3.6 days for a one-dollar expenditure—the optimal expenditure being 60 cents on U and 40 cents on V, corresponding to the optimal row strategy $\mathbf{r}_0 = [.6 \quad .4]$. It follows that the minimum cost for one day's requirements is 1/3.6 dollar, and this is realized with $(1/3.6) \times .6$ dollars worth of U, or $\frac{1}{6}$ package of U, and $(1/3.6) \times .4$ dollars worth of V, or $\frac{1}{9}$ package of V. Thus the solution of the minimization problem is

$$\mathbf{r}'_0 = \begin{bmatrix} \frac{1}{6} & \frac{1}{9} \end{bmatrix} = \frac{1}{3.6}[.6 \quad .4] = \frac{1}{v_G}\mathbf{r}_0.$$

Here, $\mathbf{r}_0$ is the optimal row strategy in the game with payoff matrix G, and v_G is the value of the game. Furthermore, the minimum cost for meeting the (daily) requirement is given by

$$\mathbf{r}'_0B = \begin{bmatrix} \frac{1}{3.6} \end{bmatrix} = \begin{bmatrix} \frac{1}{v_G} \end{bmatrix}.$$

An analogous development can be applied to the production-scheduling dual of the problem. As stated in Sec. 8-1, the problem was interpreted as calling for an optimal column strategy $\mathbf{c}_0$ in the game with payoff matrix G. This we found to be

$$\mathbf{c}_0 = \begin{bmatrix} .8 \\ .2 \\ 0 \end{bmatrix}.$$

Imitating the way in which the diet problem is translated from game-theoretic to linear programming form, the production-scheduling problem translates into the maximization problem associated with the same matrices G, A, and B.

The Maximization Problem: With G, A, and B as before, find a column matrix $\mathbf{c}' = \begin{bmatrix} x \\ y \\ z \end{bmatrix}$ of nonnegative entries satisfying

$$G\mathbf{c}' \leq B$$

such that $A\mathbf{c}'$ is maximized.

By an argument analogous to the one that led us from $\mathbf{r}_0$ to $\mathbf{r}'_0$, we are led to the conclusion that the solution of the maximization problem is given by

$$\mathbf{c}'_0 = \frac{1}{v_G}\mathbf{c}_0 = \frac{1}{3.6}\begin{bmatrix} .8 \\ .2 \\ 0 \end{bmatrix} = \begin{bmatrix} \frac{2}{9} \\ \frac{1}{18} \\ 0 \end{bmatrix}.$$

Furthermore,

$$A\mathbf{c}'_0 = \left[\frac{1}{3.6}\right] = \left[\frac{1}{v_G}\right].$$

The point of the foregoing development is to show that the analysis of the matrix game G can be embodied in the linear programming problems associated with G, row matrix A, and column matrix B, where all of the entries in A and B are ones. The argument has been heuristic, based on the interpretations of the problems as related diet and and production-scheduling problems. An algebraic argument is not terribly difficult, but is somewhat tiresome.

In general, we could start with a payoff matrix G, give the game an interpretation as diet and dual production-scheduling problems, and repeat the argument, thereby tying the optimal strategies in the game to the solutions of the related linear programming problems involving G and row and column matrices A and B consisting of all ones. But there's a rub. For the matrix G to have a diet interpretation as in the examples we've seen, the entries in G must be nonnegative; for example, it would confuse the argument for food U to contain -4 units of nutrient X. Furthermore, each food should contain a positive amount of at least one nutrient, and each nutrient should be present in at least one food. Thus a reasonable diet interpretation of a payoff matrix G demands that G consist of nonnegative entries, with at least one nonzero entry in each row and in each column. These same properties make for a reasonable production-scheduling interpretation. The formal mathematical result—rendered plausible by the diet and production-scheduling interpretations—is as follows:

Theorem: Let G be an $n \times k$ matrix consisting of nonnegative entries with at least one nonzero entry in each row and each column. Let A be the $1 \times k$ row matrix consisting of all ones, and let B be the $n \times 1$ column matrix consisting of all ones.

Suppose $\mathbf{r}_0$ and $\mathbf{c}_0$ are optimal row and column strategies for the game with G as the payoff matrix, with

$$\min \mathbf{r}_0 G = \max G\mathbf{c}_0 = v_G.$$

Then

$$\mathbf{r}'_0 = \frac{1}{v_G}\mathbf{r}_0 \quad \text{and} \quad \mathbf{c}'_0 = \frac{1}{v_G}\mathbf{c}_0$$

are solutions of the minimization and maximization problems of linear programming associated with the matrices G, A, and B, with

$$\mathbf{r}'_0 B = A\mathbf{c}'_0 = \left[\frac{1}{v_G}\right].$$

Conversely, if $\mathbf{r}'_0$ and $\mathbf{c}'_0$ are solutions of the minimization and maximization problems associated with G, A, and B, with

$$\mathbf{r}'_0 B = A\mathbf{c}'_0 = [d_0]$$

then

$$\mathbf{r}_0 = \frac{1}{d_0}\mathbf{r}'_0 \quad \text{and} \quad \mathbf{c}_0 = \frac{1}{d_0}\mathbf{c}'_0$$

are optimal row and column strategies in the game with payoff matrix G, with

$$\min \mathbf{r}_0 G = \max G\mathbf{c}_0 = \frac{1}{d_0}.$$

Our development may suggest that the usefulness of the result is in translating optimal strategies $\mathbf{r}_0$ and $\mathbf{c}_0$ into solutions $\mathbf{r}'_0$ and $\mathbf{c}'_0$ of linear programming problems. Actually it's the other way around—we analyze the game by first solving the linear programming problems. If the given payoff matrix G is not of the prescribed form, we convert to a matrix G^* by adding a fixed quantity to each entry, just as in Sec. 6-6 we converted an unfair game into a fair game. The addition of a fixed quantity to each entry lifts the value of the game and all entries in the expectations by this amount but does not affect the optimal strategies. Thus the analysis of the converted game is easily adapted to an analysis of the given game.

Example: Analyze the game with payoff matrix

$$G = \begin{bmatrix} -3 & 5 \\ 1 & -3 \\ -1 & 2 \end{bmatrix}.$$

Solution: We convert to a game G^* by adding 3 to all entries in G. The analysis of G^* is embodied in the solutions of the minimization and maximization problems associated with the matrices

$$G^* = \begin{bmatrix} 0 & 8 \\ 4 & 0 \\ 2 & 5 \end{bmatrix}, \quad \begin{bmatrix} 1 \\ 1 \\ 1 \end{bmatrix} = B,$$

$$A = [1 \quad 1].$$

Figure 8-8 displays the set of feasible points $\mathbf{c}' = \begin{bmatrix} x \\ y \end{bmatrix}$, where x and y are nonnegative and

$$G^*\mathbf{c}' \leq B.$$

Several level lines of the functional $A = [1 \quad 1]$ are shown.

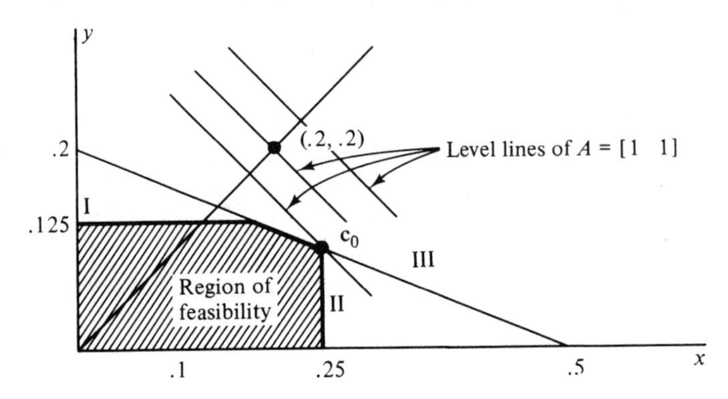

Figure 8-8.

The sketch suggests that the maximum value of the functional A for feasible points is at the intersection of the boundary lines corresponding to Rows II and III of the matrix G^*:

$$4x \qquad = 1,$$

and

$$2x + 5y = 1,$$

giving

$$x = .25 \quad \text{and} \quad y = .1.$$

Thus we conjecture that

$$\mathbf{c}'_0 = \begin{bmatrix} .25 \\ .1 \end{bmatrix}$$

solves the maximization problem. Checking with respect to the constraint,

$$G^*\mathbf{c}'_0 = \begin{bmatrix} 0 & 8 \\ 4 & 0 \\ 2 & 5 \end{bmatrix} \begin{bmatrix} .25 \\ .1 \end{bmatrix} = \begin{bmatrix} .8 \\ 1 \\ 1 \end{bmatrix} \leq \begin{bmatrix} 1 \\ 1 \\ 1 \end{bmatrix} = B. \checkmark$$

Since both entries in $\mathbf{c}'_0$ are strictly positive, we identify both columns of G^* as critical, and since the constraint $G^*\mathbf{c}' \leq B$ is satisfied with equality in Rows II and III, we identify these rows as critical. Reflecting this identification to the dual problem, we look for a solution of the form

$$\mathbf{r}'_0 = [0 \quad v \quad w].$$

with the constraint $\mathbf{r}'_0 G^* \geq A$ satisfied with equality in both columns

$$\mathbf{r}'_0 G^* = [0 \quad v \quad w]\begin{bmatrix} 0 & 8 \\ 4 & 0 \\ 2 & 5 \end{bmatrix} = [1 \quad 1] = A$$

or

$$4v + 2w = 1,$$
$$5w = 1,$$

giving

$$v = .15 \quad \text{and} \quad w = .2.$$

Thus we conjecture that

$$\mathbf{r}'_0 = [0 \quad .15 \quad .2]$$

solves the minimization problem. The verification that

$$\mathbf{r}'_0 G^* \geq A$$

is straightforward, and

$$\mathbf{r}'_0 B = A\mathbf{c}'_0 = [.35].$$

The common value, .35, is just the sum of the entries in both $\mathbf{r}'_0$ and $\mathbf{c}'_0$. This value is the key to translating the solutions of the linear programming problem to optimal strategies in the game.

With respect to the game G^* we know from the theorem that the value of the game is

$$v_{G^*} = \frac{1}{.35} = \frac{20}{7}.$$

Optimal strategies are

$$\mathbf{r}_0 = v_{G^*}\mathbf{r}'_0 = \frac{20}{7}\begin{bmatrix} 0 & \frac{3}{20} & \frac{4}{20} \end{bmatrix} = \begin{bmatrix} 0 & \frac{3}{7} & \frac{4}{7} \end{bmatrix}$$

and

$$\mathbf{c}_0 = v_{G^*}\mathbf{c}'_0 = \frac{20}{7}\begin{bmatrix} \frac{5}{20} \\ \frac{2}{20} \end{bmatrix} = \begin{bmatrix} \frac{5}{7} \\ \frac{2}{7} \end{bmatrix}.$$

The optimal strategies $\mathbf{r}_0$ and $\mathbf{c}_0$ for the converted game G^* are also optimal strategies for the given game G. Since we passed from G to G^* by adding 3 to each entry of G, the value of the game G is

$$v_G = v_{G^*} - 3 = \frac{20}{7} - 3 = -\frac{1}{7}.$$

Conclusion: In the game with payoff matrix

$$G = \begin{bmatrix} -3 & 5 \\ 1 & -3 \\ -1 & 2 \end{bmatrix}$$

optimal strategies are

$$\mathbf{r}_0 = [0 \quad \tfrac{3}{7} \quad \tfrac{4}{7}] \quad \text{and} \quad \mathbf{c}_0 = \begin{bmatrix} \tfrac{5}{7} \\ \tfrac{2}{7} \end{bmatrix}.$$

The value of the game is

$$v_G = -\frac{1}{7}.$$

Verification:

$$\mathbf{r}_0 G = [0 \quad \tfrac{3}{7} \quad \tfrac{4}{7}] \begin{bmatrix} -3 & 5 \\ 1 & -3 \\ -1 & 2 \end{bmatrix} = [-\tfrac{1}{7} \quad -\tfrac{1}{7}],$$

$$G\mathbf{c}_0 = \begin{bmatrix} -3 & 5 \\ 1 & -3 \\ -1 & 2 \end{bmatrix} \begin{bmatrix} \tfrac{5}{7} \\ \tfrac{2}{7} \end{bmatrix} = \begin{bmatrix} -\tfrac{5}{7} \\ -\tfrac{1}{7} \\ -\tfrac{1}{7} \end{bmatrix}.$$

As had been asserted,

$$\min \mathbf{r}_0 G = \max G\mathbf{c}_0 = -\frac{1}{7} = v_G. \checkmark\checkmark$$

Remarks

The analysis of the game with 2×3 payoff matrix G in the above example by the techniques of linear programming was a roundabout method, considerably more involved than the procedure of the previous chapter. But whereas this line of attack may not seem efficient for handling $n \times 2$ or $2 \times k$ payoff matrices, it does have merit in the general case of the $n \times k$ payoff matrix. As mentioned at the end of Sec. 8-4, there are efficient computer techniques for solving general linear programming problems. If one needs to analyze a game with, say, a 9×12 payoff matrix, it is a simple matter to cast the problem into the linear programming format and let the computer do the rest.

Another reason for transforming the game is to emphasize the close relation between game theory and linear programming. As the development of the Fish-Diet problem suggests, every cost minimization diet problem can be reworded so that it becomes, in effect, the row player's problem in a matrix game. With the most recent developments, we see that problems in which the matrices A and B have all of their entries equal to one can be immediately identified with matrix games without any rewording. But any cost minimization diet problem can be put in this form—A and B consisting of all 1's—by simply adjusting the units in which we measure. All we have to do is measure the foods in one-dollar packages and measure the nutrients in units equal to the requirements imposed.

The point is that there is no essential mathematical distinction between minimization problems of the "diet" variety and matrix games as viewed by the row player. Analogously, there is no essential distinction between maximization problems of the "production-scheduling" variety and matrix games as viewed by the column player.

To clarify what is meant by the "diet" and "production-scheduling" varieties of problems, we recall the conditions imposed earlier on a payoff matrix G to allow these interpretations. Conventionally, we've stated the problems in terms of matrices M, A, and B. (For the special interpretations of the present section, we used G or G^* instead of M.) In the diet setting, the entries in M are amounts of nutrient per package of food. A reasonable interpretation requires that the entries in M be nonnegative, with at least one nonzero entry in each row—each food contains some nutrient—and at least one nonzero entry in each column—each nutrient is present somewhere. Also, it is assumed that each nutrient is required in some positive amount, so all of the entries in A should be positive. Finally, each package of food costs something, so all of the entries in B should be positive. The nature of the production-scheduling problem suggests the same requirements on M, A, and B.

For purposes of discussion, let's say that a "simple" linear programming problem is one in which the entries in the matrix M are all nonnegative with at least one nonzero entry in each row and in each column and all of the entries in A and B are positive. By the arguments that have been advanced, the analysis of a "simple" linear programming problem is tantamount to the analysis of a matrix game.

In Chapter 7 a geometrical argument was given to the effect that every matrix game has solutions $\mathbf{r}_0$ and $\mathbf{c}_0$ which satisfy the criterion of the Duality Theorem of Game Theory:

$$\min \mathbf{r}_0 G = \max G\mathbf{c}_0.$$

Thus far, we've seen no argument that every linear programming problem has solutions $\mathbf{r}_0$ and $\mathbf{c}_0$ which satisfy the criterion of the Duality Theorem of Linear Programming:

$$\mathbf{r}_0 B = A\mathbf{c}_0.$$

But now the identification of "simple" linear programming problems with matrix games allows us to carry over the argument from game theory to conclude that every "simple" linear programming problem must have solutions satisfying the criterion of the duality theorem.

It should be noted that the statement of the Duality Theorem of Linear Programming does not impose any assumptions of "simplicity" on the matrices M, A, and B. Indeed, there are many important applications of linear programming, some very easy to state and analyze geometrically, in which the matrices M, A, and B do not satisfy our requirements for "simplicity." The duality theorem still holds, but there may be no guarantee that solutions $\mathbf{r}_0$ and $\mathbf{c}_0$ exist, or, for that matter, that there exist any feasible matrices $\mathbf{r}$ and $\mathbf{c}$ at all. However, it can be shown that whenever either of the solutions

$\mathbf{r}_0$ or $\mathbf{c}_0$ exists, then both must exist, and they must satisfy the criterion of the duality theorem.

8-6 EXERCISES

1. Use the techniques of Chapter 7 to find optimal strategies $\mathbf{r}_0$ and $\mathbf{c}_0$, and the value of the game v_G, in the game with payoff matrix

$$G = \begin{bmatrix} 8 & 6 & 0 \\ 5 & 7 & 9 \end{bmatrix}.$$

Then appeal to the Duality Theorem of Linear Programming to prove that

$$\mathbf{r}_0' = \frac{1}{v_G}\mathbf{r}_0 \quad \text{and} \quad \mathbf{c}_0' = \frac{1}{v_G}\mathbf{c}_0$$

solve the minimization and maximization problems associated with the matrices

$$G = \begin{bmatrix} 8 & 6 & 0 \\ 5 & 7 & 9 \end{bmatrix}, \quad \begin{bmatrix} 1 \\ 1 \end{bmatrix} = B,$$

$$A = [1 \quad 1 \quad 1].$$

2. By first analyzing the game with payoff matrix G, solve the dual problems of linear programming associated with the matrices

$$G = \begin{bmatrix} 3 & 4 \\ 1 & 6 \\ 7 & 2 \end{bmatrix}, \quad \begin{bmatrix} 1 \\ 1 \\ 1 \end{bmatrix} = B,$$

$$A = [1 \quad 1].$$

3. Using linear programming techniques, find the solutions $\mathbf{r}_0'$ and $\mathbf{c}_0'$ of the minimization and maximization problems associated with the following matrices. Let d_0 be the common value in $\mathbf{r}_0'B$ and $A\mathbf{c}_0'$.

$$G = \begin{bmatrix} 3 & 2 & 0 \\ 1 & 3 & 4 \end{bmatrix}, \quad \begin{bmatrix} 1 \\ 1 \end{bmatrix} = B,$$

$$A = [1 \quad 1 \quad 1].$$

Then appeal to the Duality Theorem of Game Theory to prove that

$$\mathbf{r}_0 = \frac{1}{d_0}\mathbf{r}_0' \quad \text{and} \quad \mathbf{c}_0 = \frac{1}{d_0}\mathbf{c}_0'$$

are optimal strategies in the game with payoff matrix G, and show that the value of the game is

$$v_G = \frac{1}{d_0}.$$

4. Follow the directions in Prob. 3 for the following matrices:

$$G = \begin{bmatrix} 3 & 4 \\ 7 & 1 \\ 0 & 7 \\ 2 & 6 \end{bmatrix}, \quad \begin{bmatrix} 1 \\ 1 \\ 1 \\ 1 \end{bmatrix} = B,$$

$$A = [1 \quad 1].$$

5. Consider the game with payoff matrix

$$G = \begin{bmatrix} 7 & -3 \\ -3 & 2 \\ 2 & -1 \end{bmatrix}.$$

(a) Convert the matrix G to a matrix G^* of nonnegative entries by adding a fixed quantity to each entry of G. Solve the dual problems of linear programming associated with the matrices G^*, A, and B, where

$$A = [1 \quad 1] \quad \text{and} \quad B = \begin{bmatrix} 1 \\ 1 \\ 1 \end{bmatrix}.$$

(b) Adapt the solutions $\mathbf{r}_0'$ and $\mathbf{c}_0'$ of the linear programming problems to obtain optimal strategies $\mathbf{r}_0$ and $\mathbf{c}_0$ in the game with payoff matrix G^*. Finally, appeal to the duality theorem to prove that $\mathbf{r}_0$ and $\mathbf{c}_0$ are optimal strategies in the game with the given payoff matrix G, and state the value of the game.

6. Use linear programming techniques to find optimal strategies in the games corresponding to the following payoff matrices. Verify by the Duality Theorem of Game Theory.

(a) $G = \begin{bmatrix} 5 & 0 \\ 4 & 4 \\ 3 & 2 \\ 2 & 6 \end{bmatrix}$. (b) $G = \begin{bmatrix} 1 & -1 & 0 \\ 0 & 0 & 1 \end{bmatrix}$.

7. The diet in a certain institution requires nutrients X and Y, to be obtained from foods U and V. In terms of the minimum daily requirement—MDR—each package of food U contains $\frac{1}{5}$ MDR of nutrient X and $\frac{2}{5}$ MDR of Y, and each package of V contains $\frac{1}{6}$ MDR of X and $\frac{1}{15}$ MDR of Y. A package of U costs \$2 and a package of V costs \$1.

How many packages of the respective foods are needed in order to satisfy the minimum daily requirements for the least possible cost?

Set up the cost minimization problem in the linear programming format, measuring food U in "half-packages" costing one dollar each, and measuring food V and nutrients X and Y in the units given. Observe that your linear programming problem is of the form developed for analyzing a certain 2×2 matrix game. Analyze the game by the methods of Chapter 6. Then adapt your solutions to obtain an answer to the question as stated.

8. Given the matrices

$$M = \begin{bmatrix} 2 & -1 & 4 \\ 1 & 0 & -5 \end{bmatrix}, \quad \begin{bmatrix} 5 \\ -10 \end{bmatrix} = B,$$

$$A = [15 \quad -5 \quad -10],$$

consider the dual problems of linear programming ("nonsimple" variety):

The Minimization Problem: Find a matrix $\mathbf{r} = [u \quad v]$ of nonnegative entries satisfying the constraint $\mathbf{r}M \geq A$ such that $\mathbf{r}B$ is minimized.

The Maximization Problem: Find a matrix $\mathbf{c} = \begin{bmatrix} x \\ y \\ z \end{bmatrix}$ of nonnegative entries satisfying $M\mathbf{c} \leq A$ such that $A\mathbf{c}$ is maximized.

Appeal to the Duality Theorem of Linear Programming to prove that

$$\mathbf{r}_0 = [5 \quad 6]$$

solves the minimization problem, and

$$\mathbf{c}_0 = \begin{bmatrix} 0 \\ 3 \\ 2 \end{bmatrix}$$

solves the maximization problem.

9. A diet requires two nutrients, X and Y, to be obtained from foods U and V. The minimum yearly requirement is 20 units of X and 24 units of Y; also, because of an adverse effect, the maximum tolerable amount of Y in a year is 36 units. Each package of food U contains 2 units of X and 6 units of Y and costs \$5, and each package of V contains 4 units of X and 4 units of Y and costs \$12.

The problem is to find the minimum cost combination of packages of foods U and V that will satisfy the minimum requirements in X and Y without violating the upper bound on the amount of Y.

(a) Observe that the given problem is the minimization problem associated with the matrices

$$M = \begin{bmatrix} 2 & 6 & -6 \\ 4 & 4 & -4 \end{bmatrix}, \quad \begin{bmatrix} 5 \\ 12 \end{bmatrix} = B,$$

$$A = [20 \quad 24 \quad -36].$$

(b) In the first quadrant of the coordinate plane, sketch the region of feasible points $\mathbf{r} = [u \quad v]$ satisfying the constraint

$$\mathbf{r}M \geq A$$

and sketch several level lines of the functional $B = \begin{bmatrix} 5 \\ 12 \end{bmatrix}$. From your picture identify the point $\mathbf{r}_0$ which appears to solve the minimization problem.

(c) As in Sec. 8-4, let $\mathbf{r}_0$ identify the critical rows and columns of M; then algebraically solve for the matrix $\mathbf{c}_0$ to be conjectured as the solution of the dual maximization problem.

(d) Appeal to the duality theorem to prove that $\mathbf{r}_0$ and $\mathbf{c}_0$ are solutions of the dual problems.

(e) Finally, present the solution to the problem originally posed.

10. Given the matrices

$$M = \begin{bmatrix} 1 & -1 \\ -1 & 1 \end{bmatrix}, \quad \begin{bmatrix} 1 \\ -2 \end{bmatrix} = B,$$
$$A = [\ 1 \quad 1].$$

(a) Show that there are no feasible points $\mathbf{r} = [u \quad v]$ for the minimization problem associated with M, A, and B.

(b) Replace the matrix A with

$$A' = [-2 \quad 1].$$

Show that there are feasible points $\mathbf{r} = [u \quad v]$ for the minimization problem associated with M, A', and B but that there is no minimum value of $\mathbf{r}B$ for feasible $\mathbf{r}$.

(c) Replace the matrix B with

$$B' = \begin{bmatrix} 2 \\ -1 \end{bmatrix}.$$

Solve the dual problems associated with M, A', and B'. Verify by the duality theorem.

8-7. Linear Mappings

Permeating the structure of linear programming problems, and indeed underlying the evolution and applications of matrices in general, is the concept of *linear mappings* of vector spaces. Thus far, the only attention given the word "linear" was the observation in Sec. 7-4 that the functionals on the vector space $\mathbf{R}^2$ are the linear mappings of $\mathbf{R}^2$ into $\mathbf{R}$, the set of real numbers. To return to that setting, we consider a piece of a simple production-scheduling problem.

A factory is equipped to produce two products, X and Y. Each unit of X yields a profit of \$18, and each unit of Y yields a profit of \$32.

As has been our custom, a production schedule—x units of X and y units of Y—is identified as the vector (x, y) in the space $\mathbf{R}^2$, or, equivalently, as the 2×1 column matrix $\begin{bmatrix} x \\ y \end{bmatrix}$. Of interest is the function—call it P (for "profit")—that associates each production schedule (x, y) with the corresponding number of dollars in profit, $P(x, y)$. The formula for $P(x, y)$ is, of course, a standard ingredient in problems of this type. There should be little hesitation in recognizing that, in this case,

$$P(x, y) = 18x + 32y,$$

or, in matrix form,

$$P(x, y) = [18 \quad 32]\begin{bmatrix} x \\ y \end{bmatrix}.$$

But let's repeat the argument that underlies the formula for $P(x, y)$, and elaborate a little on it.

> Since each unit of X produces a profit of \$18, it follows that x units of X produce a profit of $18x$ dollars, and since each unit of Y produces a profit of \$32, it follows that y units of Y produce a profit of $32y$ dollars. It also follows that x units of X together with y units of Y, that is, the production schedule (x, y), produce a profit in dollars of
>
> $$P(x, y) = 18x + 32y.$$

The argument may be paraphrased in the following terms (compare closely with the language above):

> Since $P(1, 0) = 18$, it is the nature of the interpretation of $P(x, y)$ that $P(x, 0) = xP(1, 0) = 18x$. And since $P(0, 1) = 32$, it is the nature of the interpretation that $P(0, y) = yP(0, 1) = 32y$. Finally, since $P(x, 0) = 18x$ and $P(0, y) = 32y$, it is the nature of the interpretation that
>
> $$P(x, y) = P(x, 0) + P(0, y) = 18x + 32y.$$

The argument for the form of $P(x, y)$ is based on two properties that the interpretation leads us to assign the function:

1. For any pair of vectors (x_1, y_1) and (x_2, y_2), the function $P(x, y)$ satisfies

$$P((x_1, y_1) + (x_2, y_2)) = P(x_1, y_1) + P(x_2, y_2).$$

2. For any vector (x, y) and any scalar, i.e., real number, s,

$$P(s(x, y)) = sP(x, y).$$

We recall that $P(1, 0) = 18$ and $P(0, 1) = 32$. Applying property 2 to the vector $(1, 0)$ and scalar x, we have

$$P(x, 0) = P(x(1, 0)) = xP(1, 0) = 18x.$$

Similarly,

$$P(0, y) = P(y(0, 1)) = yP(0, 1) = 32y.$$

And applying property 1, we have

$$P(x, y) = P((x, 0) + (0, y)) = P(x, 0) + P(0, y) = 18x + 32y.$$

We've employed properties 1 and 2 only for vectors of special form, but it is easy to show from the deduced form of $P(x, y)$ that the function satisfies the properties in all cases.

For property 1,

$$\begin{aligned} P((x_1, y_1) + (x_2, y_2)) &= P(x_1 + x_2, y_1 + y_2) \\ &= 18(x_1 + x_2) + 32(y_1 + y_2) \\ &= (18x_1 + 32y_1) + (18x_2 + 32y_2) \\ &= P(x_1, y_1) + P(x_2, y_2). \end{aligned}$$

For property 2,

$$\begin{aligned} P(s(x, y)) &= P(sx, xy) \\ &= 18sx + 32sy \\ &= s(18x + 32y) \\ &= sP(x, y). \end{aligned}$$

Any function mapping $\mathbf{R}^2$ into $\mathbf{R}$ satisfying properties 1 and 2 is said to be *linear*. The argument deducing the form of $P(x, y)$ can be applied to any linear mapping of $\mathbf{R}^2$ into $\mathbf{R}$.

Suppose T is a linear mapping $\mathbf{R}^2$ into $\mathbf{R}$. Let $T(1, 0) = a$ and $T(0, 1) = b$. Then from properties 1 and 2 of linearity,

$$\begin{aligned} T(x, y) &= T((x, 0) + (0, y)) = T(x, 0) + T(0, y) \\ &= T(x(1, 0)) + T(y(0, 1)) = xT(1, 0) + yT(0, 1). \end{aligned}$$

Thus

$$T(x, y) = ax + by. \tag{1}$$

By the treatment accorded the particular mapping P above, where $a = 18$ and $b = 32$, it is easy to show that any mapping T of the form in Eq. (1) is a linear mapping. In short, Eq. (1) characterizes the class of all linear mappings of $\mathbf{R}^2$ into $\mathbf{R}$.

In matrix form—identifying the vector (x, y) with the column matrix $\begin{bmatrix} x \\ y \end{bmatrix}$—we recognize the class of linear mappings of $\mathbf{R}^2$ into $\mathbf{R}$ as being precisely the set of functionals on $\mathbf{R}^2$, as defined by row matrices $[a \quad b]$. From Eq. (1),

$$[T(x, y)] = [ax + by] = [a \quad b]\begin{bmatrix} x \\ y \end{bmatrix}.$$

Thus the linear mapping T given by Eq. (1) is the functional

$$T = [a \quad b].$$

Now we pick up the statement of the linear programming situation begun earlier.

> Employed in the manufacture of products X and Y is a machine, machine U. Each unit of X requires 3 hours on U, and each unit of Y requires 2 hours on U.

Let $u(x, y)$ be the number of hours on U required for the production schedule (x, y). Clearly the mapping u is linear:

> Since each unit of X requires 3 hours on U, it follows from the interpretation that x units of X require $3x$ hours. That is, since $u(1, 0) = 3$, it follows that $u(x, 0) = 3x$. Similarly, since $u(0, 1) = 2$, it follows that $u(0, y) = 2y$. Finally, from the interpretation it follows that the number of hours required by the schedule (x, y) is
>
> $$u(x, y) = u(x, 0) + u(0, y) = 3x + 2y.$$

In matrix form, $u(x, y)$ is the (linear) functional $[3 \quad 2]$ since

$$[u] = [3 \quad 2]\begin{bmatrix} x \\ y \end{bmatrix}. \tag{2}$$

Next, a second machine is brought into the picture:

> The manufacture of products X and Y also requires time on machine V: 1 hour for each unit of X and 4 hours for each unit of Y.

We let v be the number of hours on V required for the production schedule $\begin{bmatrix} x \\ y \end{bmatrix}$. Without further ado,

$$[v] = [1 \quad 4]\begin{bmatrix} x \\ y \end{bmatrix}. \tag{3}$$

In Eqs. (2) and (3) we see that the requirements on machines U and V, considered separately, determine two (linear) functionals, u and v. But we do not want to consider the requirements separately. Our attitude in the past has been that the production-schedule vector (x, y)—or matrix $\begin{bmatrix} x \\ y \end{bmatrix}$—determines a machine requirements vector (u, v)—or matrix $\begin{bmatrix} u \\ v \end{bmatrix}$. We view Eqs. (2) and (3) as determining the two rows in a single matrix equation:

$$\begin{bmatrix} u \\ v \end{bmatrix} = \begin{bmatrix} 3 & 2 \\ 1 & 4 \end{bmatrix}\begin{bmatrix} x \\ y \end{bmatrix}. \tag{4}$$

In vector terms, what we have in Eq. (4) is a function mapping the vector space $\mathbf{R}^2$ into itself, the vector (x, y) being associated with the vector (u, v). The important property of this mapping is that the coordinates u and v in the image vector (u, v) are themselves (linear) functionals on the space $\mathbf{R}^2$. The linearity in the individual coordinates extends into a linearity property for the larger mapping. The properties defining a linear mapping of $\mathbf{R}^2$ into $\mathbf{R}^2$ are essentially the same as those defining a linear mapping of $\mathbf{R}^2$ into $\mathbf{R}$:

> A function T mapping $\mathbf{R}^2$ into $\mathbf{R}^2$ is said to be *linear* provided it satisfies the following properties:

1. For any pair of vectors $\mathbf{c}_1 = (x_1, y_1)$ and $\mathbf{c}_2 = (x_2, y_2)$ in $\mathbf{R}^2$,
$$T(\mathbf{c}_1 + \mathbf{c}_2) = T(\mathbf{c}_1) + T(\mathbf{c}_2).$$
2. For any vector $\mathbf{c} = (x, y)$ in $\mathbf{R}^2$ and any scalar s in $\mathbf{R}$,
$$T(s\mathbf{c}) = sT(\mathbf{c}).$$

In short, we can combine vectors by addition and scalar multiplication before we map or after we map—the result is the same.

We've seen that the linear mappings of $\mathbf{R}^2$ into $\mathbf{R}$ are the functionals determined by row matrices $[a \quad b]$. The argument can be extended to establish that the linear mappings of $\mathbf{R}^2$ into $\mathbf{R}^2$ are precisely those determined by 2×2 matrices as in Eq. (4). That is, if T is a linear mapping of $\mathbf{R}^2$ into $\mathbf{R}^2$, then there are numbers a_1, b_1, a_2, and b_2 such that the image vector $(u, v) = T(x, y)$ is determined by the matrix multiplication

$$\begin{bmatrix} u \\ v \end{bmatrix} = \begin{bmatrix} a_1 & b_1 \\ a_2 & b_2 \end{bmatrix} \begin{bmatrix} x \\ y \end{bmatrix}. \tag{5}$$

Conversely, every mapping $T(x, y) = (u, v)$ determined by a 2×2 matrix as in Eq. (5) is a linear mapping. From a purely mathematical point of view, 2×2 matrices are created as the arithmetic devices for expressing linear mappings of $\mathbf{R}^2$ into $\mathbf{R}^2$.

Now we complete the linear programming problem.

The maximum time allowed on machine U is 18 hours a day, and on machine V, 16 hours a day. Find a feasible schedule $\begin{bmatrix} x \\ y \end{bmatrix}$ which yields the maximum profit.

The problem may be stated as follows:

We consider all vectors (x, y) in the positive quadrant of the plane $\mathbf{R}^2$—the entries x and y nonnegative—which under the linear mapping in Eq. (4),

$$\begin{bmatrix} u \\ v \end{bmatrix} = \begin{bmatrix} 3 & 2 \\ 1 & 4 \end{bmatrix} \begin{bmatrix} x \\ y \end{bmatrix},$$

are mapped into vectors (u, v) satisfying the constraint imposed by the machine limitations,

$$\begin{bmatrix} u \\ v \end{bmatrix} \leq \begin{bmatrix} 18 \\ 12 \end{bmatrix}.$$

From the class of such vectors (x, y)—the feasible schedules—we wish to find a particular vector (x_0, y_0) for which the value assigned by the "profit" functional (linear mapping into $\mathbf{R}$)

$$P = [18 \quad 32]$$

is maximized. (Recall the original profit data.)

The treatment generalizes to higher-dimensional problems. Analogous to the role of the 2×2 matrix in Eq. (5), an $n \times k$ matrix M defines a linear mapping of the vector space $\mathbf{R}^k$, consisting of k-tuples of numbers $(x_1, x_2, \ldots, x_k)$, into the vector space $\mathbf{R}^n$, consisting of n-tuples of numbers $(u_1, u_2, \ldots, u_n)$. Here, linearity is defined by the obvious generalizations of properties 1 and 2, spelled out above in the "$\mathbf{R}^2$ into $\mathbf{R}$" and "$\mathbf{R}^2$ into $\mathbf{R}^2$" cases. In particular, the linear mappings of $\mathbf{R}^k$ into $\mathbf{R}$ are precisely the functionals expressed by $1 \times k$ row matrices.

Composition of Linear Mappings

In a succession of steps, the statement of the production-scheduling problem above has been completed. But the factory has other financial concerns. Let's elaborate on the situation.

> The operating expenses associated with machines U and V include the rent on the machines, the cost of power (gas and electricity), and the labor cost. The expense in dollars per hour of operation on each machine in the respective categories is as follows:

	Machine	
	U	*V*
Rent	4	8
Power	2	3
Labor	6	8

Entry is the expense, in dollars, per hour of operation.

Let r, p, and l represent the expenses in dollars in the respective categories corresponding to the machine-usage vector (u, v)—u hours on U and v hours on V. It should be clear that the three functions, (u, v) into r, (u, v) into p, and (u, v) into l, are all linear mappings. In matrix terms they are given as follows:

$$[r] = [4 \quad 8]\begin{bmatrix} u \\ v \end{bmatrix},$$

$$[p] = [2 \quad 3]\begin{bmatrix} u \\ v \end{bmatrix},$$

$$[l] = [6 \quad 8]\begin{bmatrix} u \\ v \end{bmatrix}.$$

These three simultaneous functionals determine a single mapping given by the matrix equation

$$\begin{bmatrix} r \\ p \\ l \end{bmatrix} = \begin{bmatrix} 4 & 8 \\ 2 & 3 \\ 6 & 8 \end{bmatrix}\begin{bmatrix} u \\ v \end{bmatrix}. \tag{6}$$

Here we have the mapping of the machine-usage vector (u, v) in $\mathbf{R}^2$ into the operating-expenses vector (r, p, l) in the vector space $\mathbf{R}^3$ consisting of triples of numbers. In line with the earlier observation, this mapping, determined by a 3×2 matrix, is a linear mapping of $\mathbf{R}^2$ into $\mathbf{R}^3$.

But the attitude has been that the machine-usage vector (u, v) is determined by the production-schedule vector (x, y)—x units of product X and y units of product Y. The linear mapping from (x, y) to (u, v) was given in Eq. (4). We repeat

$$\begin{bmatrix} u \\ v \end{bmatrix} = \begin{bmatrix} 3 & 2 \\ 1 & 4 \end{bmatrix} \begin{bmatrix} x \\ y \end{bmatrix}. \tag{4}$$

As an example, suppose the factory produces 4 units of X and 2 units of Y. Thus $(x, y) = (4, 2)$. From Eq. (4), the machine usage required is

$$\begin{bmatrix} u \\ v \end{bmatrix} = \begin{bmatrix} 3 & 2 \\ 1 & 4 \end{bmatrix} \begin{bmatrix} 4 \\ 2 \end{bmatrix} = \begin{bmatrix} 16 \\ 12 \end{bmatrix},$$

that is, 16 hours on U and 12 hours on V. This machine-usage vector, $(u, v) = (16, 12)$, is substituted into Eq. (6):

$$\begin{bmatrix} r \\ p \\ l \end{bmatrix} = \begin{bmatrix} 4 & 8 \\ 2 & 3 \\ 6 & 8 \end{bmatrix} \begin{bmatrix} 16 \\ 12 \end{bmatrix} = \begin{bmatrix} 160 \\ 68 \\ 192 \end{bmatrix}.$$

We see that the production of $x = 4$ units of X and $y = 2$ units of Y gives rise to operating expenses of \$160 for rent, \$68 for power, and \$192 for labor. (The profit is over and above.)

In passing from $(x, y) = (4, 2)$ through $(u, v) = (16, 12)$ to $(r, p, l) = (160, 68, 192)$, two successive linear mappings are involved:

> The mapping of $\mathbf{R}^2$ into $\mathbf{R}^2$ determined by a matrix M (of the linear programming problem) and the mapping of $\mathbf{R}^2$ into $\mathbf{R}^3$ determined by a matrix E (the "operating expenses" matrix).

These matrices are

$$E = \begin{bmatrix} 4 & 8 \\ 2 & 3 \\ 6 & 8 \end{bmatrix} \quad \text{and} \quad M = \begin{bmatrix} 3 & 2 \\ 1 & 4 \end{bmatrix}.$$

And we have

$$\begin{bmatrix} 160 \\ 68 \\ 192 \end{bmatrix} = E \begin{bmatrix} 16 \\ 12 \end{bmatrix}, \qquad \text{where} \quad \begin{bmatrix} 16 \\ 12 \end{bmatrix} = M \begin{bmatrix} 4 \\ 2 \end{bmatrix}.$$

In general, we can start with any vector (x, y) in $\mathbf{R}^2$ and apply the linear mapping M, i.e., the mapping determined by the matrix M, to obtain an image vector (u, v), in $\mathbf{R}^2$, and then apply the linear mapping E to (u, v) to obtain an image (r, p, l) in $\mathbf{R}^3$. The succession of mappings—M followed by E—constitutes a single mapping, $E \circ M$ (E "of" M) of $\mathbf{R}^2$ into $\mathbf{R}^3$—(x, y) into (r, p, l). The formation of a function in this manner, from the succession of two functions, is called the *composition* of the functions; $E \circ M$ is called the *composite* function.

With the matrices E and M above, the composite mapping $E \circ M$ maps (x, y) in $\mathbf{R}^2$ into (r, p, l) in $\mathbf{R}^3$ as follows:

$$\begin{bmatrix} r \\ p \\ l \end{bmatrix} = E\begin{bmatrix} u \\ v \end{bmatrix}, \qquad \text{where } \begin{bmatrix} u \\ v \end{bmatrix} = M\begin{bmatrix} x \\ y \end{bmatrix},$$

or simply

$$\begin{bmatrix} r \\ p \\ l \end{bmatrix} = E\left\{M\begin{bmatrix} x \\ y \end{bmatrix}\right\}.$$

Since matrix multiplication is associative, we can regroup the factors:

$$\begin{bmatrix} r \\ p \\ l \end{bmatrix} = \{EM\}\begin{bmatrix} x \\ y \end{bmatrix} = \left\{\begin{bmatrix} 4 & 8 \\ 2 & 3 \\ 6 & 8 \end{bmatrix}\begin{bmatrix} 3 & 2 \\ 1 & 4 \end{bmatrix}\right\}\begin{bmatrix} x \\ y \end{bmatrix},$$

$$\begin{bmatrix} r \\ p \\ l \end{bmatrix} = \begin{bmatrix} 20 & 40 \\ 9 & 16 \\ 26 & 44 \end{bmatrix}\begin{bmatrix} x \\ y \end{bmatrix}.$$

We see that the composition of the two linear mappings—M followed by E—is the linear mapping of $\mathbf{R}^2$ into $\mathbf{R}^3$ determined by the matrix product EM.

To carry the idea one step further, we let T be the mapping that carries the operating-expenses vector (r, p, l) into the total operating expense

$$t = r + p + l.$$

That is, T is the (linear) functional $[1 \quad 1 \quad 1]$:

$$[t] = [1 \quad 1 \quad 1]\begin{bmatrix} r \\ p \\ l \end{bmatrix}.$$

Now the production schedule (x, y) maps into the total operating expense t in three steps:

$$[t] = T\left\{E\left\{M\begin{bmatrix} x \\ y \end{bmatrix}\right\}\right\} = [1 \quad 1 \quad 1]\left\{\begin{bmatrix} 4 & 8 \\ 2 & 3 \\ 6 & 8 \end{bmatrix}\left\{\begin{bmatrix} 3 & 2 \\ 1 & 4 \end{bmatrix}\begin{bmatrix} x \\ y \end{bmatrix}\right\}\right\}.$$

The passage from (x, y) to t in one step is given by the linear functional determined by the matrix product

$$TEM = T(EM) = [1 \quad 1 \quad 1]\begin{bmatrix} 20 & 40 \\ 9 & 16 \\ 26 & 44 \end{bmatrix} = [55 \quad 100].$$

In particular, for the production schedule $(x, y) = (4, 2)$, the total operating expense is given by

$$TEM\begin{bmatrix} 4 \\ 2 \end{bmatrix} = [55 \quad 100]\begin{bmatrix} 4 \\ 2 \end{bmatrix} = [420].$$

We see that the total operating expense associated with $x = 4$ units of X and $y = 2$ units of Y is \$420. This is consistent with the earlier calculation of \$160 for rent, \$68 for power, and \$192 for labor.

In the purely mathematical setting, matrices evolve as the devices for expressing linear mappings of vector spaces. Identifying vectors with column matrices, an $n \times k$ matrix determines a linear mapping of $\mathbf{R}^k$ into $\mathbf{R}^n$. It can be shown that the composition of linear mappings—one followed by another—always defines a linear mapping, and the matrix of the composite mapping is the matrix product obtained by multiplying the matrices of the constituent mappings. Indeed, from a mathematical point of view the rule for multiplying matrices is dictated by this requirement.

8-7 EXERCISES

1. Consider the mapping T of $\mathbf{R}^2$ into $\mathbf{R}^2$ defined by a 2×2 matrix as follows: For $\mathbf{c} = (x, y)$ in $\mathbf{R}^2$,

$$T(\mathbf{c}) = T(x, y) = (u, v),$$

where

$$\begin{bmatrix} u \\ v \end{bmatrix} = \begin{bmatrix} 1 & 3 \\ 5 & 2 \end{bmatrix}\begin{bmatrix} x \\ y \end{bmatrix}.$$

(a) For the following pairs of vectors $\mathbf{c}_1$ and $\mathbf{c}_2$, compute $T(\mathbf{c}_1)$, $T(\mathbf{c}_2)$, and $T(\mathbf{c}_1 + \mathbf{c}_2)$, and show that

$$T(\mathbf{c}_1 + \mathbf{c}_2) = T(\mathbf{c}_1) + T(\mathbf{c}_2).$$

(i) $\mathbf{c}_1 = (1, 0)$ and $\mathbf{c}_2 = (0, 1)$.
(ii) $\mathbf{c}_1 = (2, 1)$ and $\mathbf{c}_2 = (0, .5)$.
(iii) $\mathbf{c}_1 = (-1, 3)$ and $\mathbf{c}_2 = (1, -4)$.

(b) For the following vectors $\mathbf{c}$ and scalars s, compute $T(\mathbf{c})$ and $T(s\mathbf{c})$ and show that

$$T(s\mathbf{c}) = sT(\mathbf{c}).$$

(i) $\mathbf{c} = (2, 1)$ and $s = 5$.
(ii) $\mathbf{c} = (3, -2)$ and $s = -.5$.

2. Let F be a linear mapping of $\mathbf{R}^2$ into $\mathbf{R}^2$ with

$$F(1, 0) = (2, 1) \quad \text{and} \quad F(0, 1) = (5, -2).$$

Use the properties of linearity to find (a) $F(0, 5)$, (b) $F(1, -1)$, (c) $F(\frac{1}{3}, \frac{2}{3})$.

3. Let A be a functional on $\mathbf{R}^3$, i.e., a linear mapping of $\mathbf{R}^3$ into $\mathbf{R}$. Suppose

$$A(1, 0, 0) = 4, \qquad A(0, 1, 0) = 1, \qquad A(0, 0, 1) = 3.$$

(a) From the linearity properties of A, show that

$$A(x, y, z) = xA(1, 0, 0) + yA(0, 1, 0) + zA(0, 0, 1).$$

(b) Find
(i) $A(1, -1, -1)$. (ii) $A(\frac{1}{4}, \frac{1}{4}, \frac{1}{4})$. (iii) $A(.5, 0, 2)$.

(c) Express the functional A as a row matrix $[a_1 \quad a_2 \quad a_3]$.

4. Consider the general linear mapping T of $\mathbf{R}^3$ into $\mathbf{R}^2$, determined by a general 2×3 matrix. That is,

$$T(x, y, z) = (u, v),$$

where

$$\begin{bmatrix} u \\ v \end{bmatrix} = \begin{bmatrix} t_{11} & t_{12} & t_{13} \\ t_{21} & t_{22} & t_{23} \end{bmatrix} \begin{bmatrix} x \\ y \\ z \end{bmatrix}.$$

(a) In terms of the entries t_{ij}, find

$$T(1, 0, 0), \quad T(0, 1, 0), \quad \text{and} \quad T(0, 0, 1).$$

(b) What is the relation between the images of the vectors in part (a) and the entries in the matrix determining the mapping T?

5. Let F be a linear mapping of $\mathbf{R}^3$ into $\mathbf{R}^2$. Suppose

$$F(1, 0, 0) = (1, 2), \qquad F(1, 1, 0) = (3, 1), \qquad F(1, 1, 1) = (0, 0).$$

(a) Find $F(0, 1, 0)$. [*Hint:* $(0, 1, 0) = (1, 1, 0) - (1, 0, 0)$.]

(b) Find $F(0, 0, 1)$.

(c) Exhibit the 2×3 matrix that expresses the mapping F. (See Prob. 4.)

(d) Find $F(0, 1, 1)$, $F(-2, 3, 3)$, $F(.5, 0, .2)$.

6. Consider the composite mapping

$$\begin{bmatrix} s \\ t \end{bmatrix} = \begin{bmatrix} 2 & 1 \\ 3 & 1 \end{bmatrix} \begin{bmatrix} u \\ v \end{bmatrix}, \quad \text{where} \quad \begin{bmatrix} u \\ v \end{bmatrix} = \begin{bmatrix} -1 & 1 \\ 2 & 0 \end{bmatrix} \begin{bmatrix} x \\ y \end{bmatrix}.$$

(a) By computing the intermediate vector (u, v), find the vector (s, t) associated with the following vectors (x, y):
(i) $(x, y) = (0, 1)$. (ii) $(x, y) = (-1, 2)$. (iii) $(x, y) = (.5, .5)$.

(b) Compute the product

$$M = \begin{bmatrix} 2 & 1 \\ 3 & 1 \end{bmatrix} \begin{bmatrix} -1 & 1 \\ 2 & 0 \end{bmatrix}.$$

Then show for the vectors (x, y) in part (a) that the same images (s, t) are obtained in one step by the rule

$$\begin{bmatrix} s \\ t \end{bmatrix} = M \begin{bmatrix} x \\ y \end{bmatrix}.$$

7. Consider the following linear mappings:

$$S(x, y, z) = (u, v), \qquad \text{where } \begin{bmatrix} u \\ v \end{bmatrix} = \begin{bmatrix} -2 & 1 & 0 \\ 0 & 1 & -2 \end{bmatrix} \begin{bmatrix} x \\ y \\ z \end{bmatrix},$$

$$T(x, y) = (u, v, w), \qquad \text{where } \begin{bmatrix} u \\ v \\ w \end{bmatrix} = \begin{bmatrix} 1 & 0 \\ 3 & 2 \\ -2 & 0 \end{bmatrix} \begin{bmatrix} x \\ y \end{bmatrix}.$$

(a) For the following vectors (x, y) in $\mathbf{R}^2$, find $T(x, y)$; then apply the mapping S to $T(x, y)$ to find the image of (x, y) under the composite mapping $S \circ T$,

$$(S \circ T)(x, y) = S(T(x, y)).$$

(i) $(x, y) = (1, 0)$.
(ii) $(x, y) = (-2, 1)$.
(iii) $(x, y) = (10, -5)$.

(b) Find the matrix expressing $S \circ T$. Then employ this matrix to find $(S \circ T)(x, y)$ for the vectors (x, y) in part (a).

(c) For the following vectors (x, y, z) in $\mathbf{R}^3$, find $S(x, y, z)$; then apply the mapping T to $S(x, y, z)$ to find the image of (x, y, z) under the composite mapping $T \circ S$,

$$(T \circ S)(x, y, z) = T(S(x, y, z)).$$

(i) $(x, y, z) = (0, 1, 0)$.
(ii) $(x, y, z) = (1, 2, 1)$.
(iii) $(x, y, z) = (-3, 5, -2)$.

(d) Find the matrix expressing $T \circ S$. Then employ this matrix to find $(T \circ S)(x, y, z)$ for the vectors (x, y, z) in part (*c*).

8. Consider the following linear mapping M of $\mathbf{R}^2$ into $\mathbf{R}^2$:

$$M(x, y) = (u, v), \qquad \text{where } \begin{bmatrix} u \\ v \end{bmatrix} = \begin{bmatrix} 0 & -1 \\ 1 & 0 \end{bmatrix} \begin{bmatrix} x \\ y \end{bmatrix}.$$

The "powers" of the mapping M are defined by successive compositions:

$$M^2 = M \circ M, \quad M^3 = M \circ M^2, \quad M^4 = M \circ M^3, \quad \text{and so on.}$$

Find the matrices expressing (a) M^2, (b) M^4, (c) M^5, (d) M^{15}.

9. Let T_1 and T_2 be the following linear mappings of $\mathbf{R}^2$ into $\mathbf{R}^2$:

$$T_1(x, y) = (u, v), \qquad \text{where } \begin{bmatrix} u \\ v \end{bmatrix} = \begin{bmatrix} 1 & 0 \\ -2 & 3 \end{bmatrix} \begin{bmatrix} x \\ y \end{bmatrix},$$

$$T_2(x, y) = (u, v), \qquad \text{where } \begin{bmatrix} u \\ v \end{bmatrix} = \begin{bmatrix} -1 & 2 \\ 4 & 1 \end{bmatrix} \begin{bmatrix} x \\ y \end{bmatrix}.$$

The *sum* of the mappings, $T_1 + T_2$, is the mapping of $\mathbf{R}^2$ into $\mathbf{R}^2$, defined by the rule

$$(T_1 + T_2)(x, y) = T_1(x, y) + T_2(x, y).$$

(a) Find $(T_1 + T_2)(x, y)$ for the following vectors:
  (i) $(x, y) = (1, 0)$. (ii) $(x, y) = (0, 1)$.
  (iii) $(x, y) = (-1, 2)$. (iv) $(x, y) = (1.5, -.5)$.

(b) Show that $T_1 + T_2$ is the linear mapping whose matrix is the sum of the matrices expressing T_1 and T_2. That is, show that for all (x, y)

$$(T_1 + T_2)(x, y) = (u, v),$$

where

$$\begin{bmatrix} u \\ v \end{bmatrix} = \left\{\begin{bmatrix} 1 & 0 \\ -2 & 3 \end{bmatrix} + \begin{bmatrix} -1 & 2 \\ 4 & 1 \end{bmatrix}\right\}\begin{bmatrix} x \\ y \end{bmatrix}.$$

10. Let T be the linear mapping of $\mathbf{R}^2$ into $\mathbf{R}^2$:

$$T(x, y) = (u, v), \qquad \text{where } \begin{bmatrix} u \\ v \end{bmatrix} = \begin{bmatrix} -1 & 2 \\ 4 & 1 \end{bmatrix}\begin{bmatrix} x \\ y \end{bmatrix}.$$

For a real number s the *scalar multiple* sT is the mapping of $\mathbf{R}^2$ into $\mathbf{R}^2$, defined by the rule

$$(sT)(x, y) = sT(x, y).$$

(a) Find
  (i) $(3T)(1, 0)$. (ii) $(-2T)(-1, 2)$. (iii) $(sT)(0, 1)$.

(b) Show that sT is the linear mapping whose matrix is s times the matrix of T.

A FINAL APPLICATION

In large, we have studied two fundamental concepts of mathematics: probability and matrices. The important applications of these topics go far beyond the illustrations that we've considered. To conclude with a proper commencement, we'll take a brief look at an example of the two concepts at work in a fresh setting.

9-1. Markov Processes

In the "repeated-trials" processes of Chapters 4 and 5 the probabilities for the outcomes on a trial were always the same, irrespective of what may have occurred on preceding trials. There are many situations in which this assumption is not applicable. The following illustration is reminiscent of our earliest examples.

> Carol has three urns—red, yellow, and green, respectively. The red urn contains two marbles, one red and one yellow; the yellow urn contains three marbles, one yellow and two green; and the green urn contains three marbles, one red, one yellow, and one green. See Fig. 9-1.

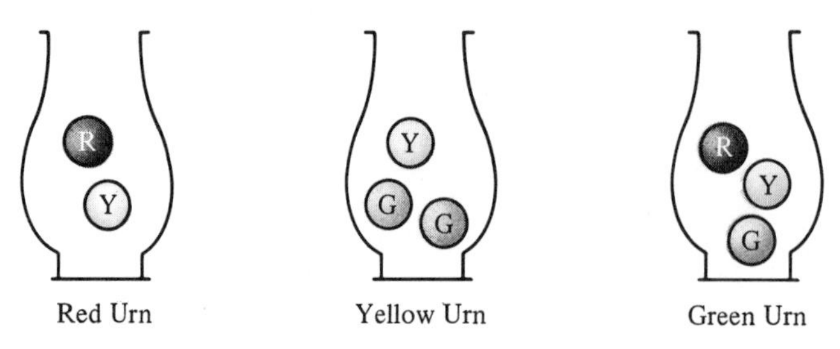

Figure 9-1 Carol's Urns.

On each of several trials, Carol withdraws a marble at random from one of the urns, inspects its color, and returns it to its urn. Then on the succeeding trial she draws from the urn of the same color as the marble just drawn. Thus if a red marble is drawn on the nth trial, the red urn is employed on the $(n + 1)$st trial.

We are interested in questions of the sort, Given that Carol drew from the red urn on the first trial, what is the probability that a yellow marble is drawn on the third trial?

Each trial in Carol's process is interpreted as a component process in which the basic outcomes are *transitions* from one color to another, or the same, color. For example, suppose a red marble is drawn on the nth trial, and then a yellow marble is drawn from the red urn on the $(n + 1)$st trial. In this case the outcome on the $(n + 1)$st trial is represented as (R, Y). But (R, Y) says more than "A yellow marble is drawn from the red urn"; (R, Y) also tells us that a red marble was drawn on the preceding trial and that the yellow urn is to be employed in the succeding trial. Thus (R, Y) represents the transition from color R to color Y.

We identify the three colors as the three *states* in the process. For a trial described by the (R, Y) combination, we say that the *input state* is R and the *output state* is Y. With each such combination there is an associated probability called the *transition probability*. The (R, Y) combination, for instance, has a transition probability of $\frac{1}{2}$, since the probability of passing to state Y, given that we enter in state R, is $\frac{1}{2}$. This is just the probability of drawing a yellow marble from the red urn, which contains one yellow and one red marble.

Considering all the possibilities, the transition probabilities may be tabulated as follows:

Transition Probabilities

		Output state		
		R	*Y*	*G*
Input state	*R*	$\frac{1}{2}$	$\frac{1}{2}$	0
	Y	0	$\frac{1}{3}$	$\frac{2}{3}$
	G	$\frac{1}{3}$	$\frac{1}{3}$	$\frac{1}{3}$

To pave the ground for generalization, we identify the states R, Y, and G as States 1, 2, and 3, respectively. With this convention, the legend in the display of transition probabilities may be left implicit. The square array of tabulated values becomes the *matrix of transition probabilities*. For Carol's process this is the matrix

$$M = \begin{bmatrix} \frac{1}{2} & \frac{1}{2} & 0 \\ 0 & \frac{1}{3} & \frac{2}{3} \\ \frac{1}{3} & \frac{1}{3} & \frac{1}{3} \end{bmatrix}.$$

The specific question posed was,

Given that Carol drew from the red urn on the first trial, what is the probability that a yellow marble is drawn on the third trial?

This question is rephrased as follows:

Given that the initial state is State 1, what is the probability that the terminal state after three trials is State 2?

The entries in the matrix M are the conditional probabilities corresponding to a single trial of the process. For instance, the "$\frac{1}{2}$" in Row I and Column II tells us the following:

Given that the initial state is State 1, the probability is $\frac{1}{2}$ that the terminal state after one trial is State 2.

Assuming that the intial state is State 1, the probabilities for the terminal state after two trials can be found from the tree of Fig. 9-2.

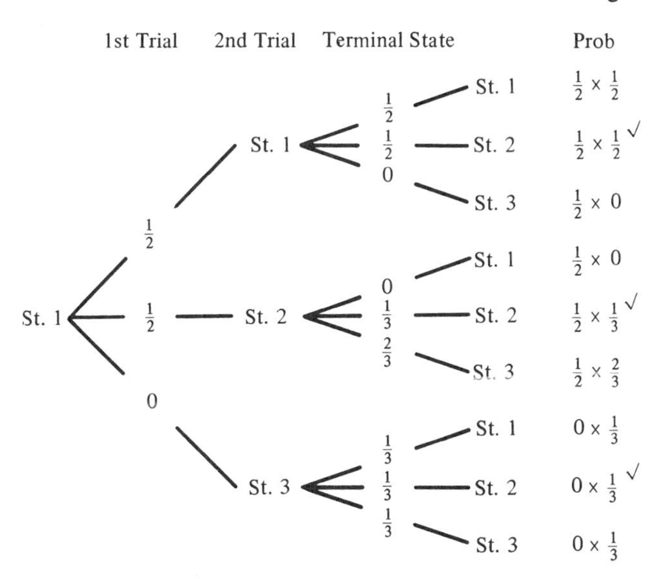

Figure 9-2.

In each component of the tree the branch probabilities have been lifted directly from the row of the matrix M corresponding to the state initiating the component.

In the figure we have checked the paths for which the terminal state is State 2. Adding the probabilities corresponding to these paths, we see that the probability of arriving at State 2 in two trials, given that the initial state is State 1, is

$$(\tfrac{1}{2} \times \tfrac{1}{2}) + (\tfrac{1}{2} \times \tfrac{1}{3}) + (0 \times \tfrac{1}{3}) = \tfrac{5}{12}.$$

Comparing this calculation with the tree of Fig. 9-2 and with the matrix M of transition probabilities, whose entries are the branch probabilities in the tree, we see that the value $\frac{5}{12}$ was obtained by multiplying, matrix fashion, the first row of M times the second column of M.

Applying the same reasoning to the other combinations, we conclude that the "two-trial" transition probabilities are given by the corresponding entries in the matrix M^2:

$$M^2 = M \times M = \begin{bmatrix} \frac{1}{2} & \frac{1}{2} & 0 \\ 0 & \frac{1}{3} & \frac{2}{3} \\ \frac{1}{3} & \frac{1}{3} & \frac{1}{3} \end{bmatrix} \begin{bmatrix} \frac{1}{2} & \frac{1}{2} & 0 \\ 0 & \frac{1}{3} & \frac{2}{3} \\ \frac{1}{3} & \frac{1}{3} & \frac{1}{3} \end{bmatrix} = \begin{bmatrix} \frac{1}{4} & \frac{5}{12} & \frac{1}{3} \\ \frac{2}{9} & \frac{1}{3} & \frac{4}{9} \\ \frac{5}{18} & \frac{7}{18} & \frac{1}{3} \end{bmatrix}.$$

The value $\frac{5}{12}$, which we computed as the probability of going from State 1 to State 2 in two trials, is the Row I and Column II entry in M^2. In like manner, the $\frac{4}{9}$ in Row II and Column III of M^2 tells us the following:

The probability of arriving at State 3 in two trials, given that we start in State 2, is $\frac{4}{9}$.

The argument extends to any number of trials. In particular, the transition probabilities for a sequence of three trials are the entries in the matrix M^3:

$$M^3 = M \times M^2 = \begin{bmatrix} \frac{1}{2} & \frac{1}{2} & 0 \\ 0 & \frac{1}{3} & \frac{2}{3} \\ \frac{1}{3} & \frac{1}{3} & \frac{1}{3} \end{bmatrix} \begin{bmatrix} \frac{1}{4} & \frac{5}{12} & \frac{1}{3} \\ \frac{2}{9} & \frac{1}{3} & \frac{4}{9} \\ \frac{5}{18} & \frac{7}{18} & \frac{1}{3} \end{bmatrix} = \begin{bmatrix} \frac{17}{72} & \frac{3}{8} & \frac{7}{18} \\ \frac{7}{27} & \frac{10}{27} & \frac{10}{27} \\ \frac{1}{4} & \frac{41}{108} & \frac{10}{27} \end{bmatrix}.$$

Inspecting the Row I and Column II entry in M^3, we can finally answer the question originally posed:

Given that Carol drew from the red urn (State 1) on the first trial, the probability that a yellow marble (State 2) is drawn on the third trial is $\frac{3}{8}$.

If we assume that Carol starts by drawing from the green urn (State 3), we can find the probability that she draws a red marble (State 1) on the 50th trial by simply computing M^{50} and inspecting the entry in Row III and Column I. (And a simple computation it is—for a computer.)

The General Case

To generalize Carol's process, we suppose that she has n urns of different colors, each containing a known assortment of marbles of these colors. In language less biased, we consider processes consisting of an indeterminant number of trials such that initially, and after each trial, the process is in one of n states. We label the states as

$$\text{State } 1, \text{State } 2, \ldots, \text{State } n.$$

For each combination of states (State i, State j), there is a transition probability, represented as p_{ij}, of passing from State i to State j on any given trial. That is, p_{ij} is the probability that the process is in State j after a trial, given that the process was in State i before the trial.

Processes fitting this description are called *Markov processes*, or *Markov chains*.

To develop the mathematical model of a Markov process we identify the $n \times n$ matrix M of transition probabilities. The entry in the ith row and jth column of M is p_{ij}, the probability of transition from State i to State j on a single trial.

$$M = \begin{bmatrix} p_{11} & p_{12} & \cdots & p_{1n} \\ p_{21} & p_{22} & \cdots & p_{2n} \\ \cdot & \cdot & & \cdot \\ \cdot & \cdot & & \cdot \\ \cdot & \cdot & & \cdot \\ p_{n1} & p_{n2} & \cdots & p_{nn} \end{bmatrix}.$$

The matrix M of transition probabilities is arithmetically characterized by the requirement that all entries are nonnegative, and the sum of the entries in each row is one.

Multiplying M by itself for k factors, where k is a positive integer, we identify the entry in the ith row and jth column of M^k as $p_{ij}^{(k)}$ (read "p sub i,j super k").

As in Carol's process, it can be argued that $p_{ij}^{(k)}$ is the probability that after k trials the process is in State j, given that initially, i.e., before the first trial, the process was in State i.

There are many important applications which can be interpreted as Markov processes.

Example: A psychologist is conditioning rats to master a maze. The success of a rat on any given trial is rated in one of three categories:

Poor, Fair, Good.

From many trials with many rats, it is found that the probabilities of achieving the respective grades on any given trial, depending on the performance in the preceding trial, are as follows:

		Probabilities for a Given Trial		
		Poor	*Fair*	*Good*
Performance on the Previous Trial	*Poor*	.3	.7	0
	Fair	0	.4	.6
	Good	0	0	1

Question: Given that the rat is a poor performer in advance of the first trial, find the probabilities associated with the respective grades on the fourth trial.

Solution: Identifying the levels of grade in appreciating order as States 1, 2, and 3, we interpret the sequence of trials as a Markov process with the following matrix of transition probabilities:

$$M = \begin{bmatrix} .3 & .7 & 0 \\ 0 & .4 & .6 \\ 0 & 0 & 1 \end{bmatrix}.$$

The transition probabilities for two, three, and four trials are found by computing M^2, M^3, and M^4:

$$M^2 = \begin{bmatrix} .09 & .49 & .42 \\ 0 & .16 & .84 \\ 0 & 0 & 1 \end{bmatrix},$$

$$M^3 = \begin{bmatrix} .027 & .259 & .714 \\ 0 & .064 & .936 \\ 0 & 0 & 1 \end{bmatrix},$$

$$M^4 = \begin{bmatrix} .0081 & .1225 & .8694 \\ 0 & .0256 & .9744 \\ 0 & 0 & 1 \end{bmatrix}.$$

From the nature of the situation it is natural to view the entries in the matrix M as rounded-off estimates of probabilities. This casts a cloud on the accuracy of the values in the matrix M^4. But let's assume that the entries in M are precise enough to justify accuracy to two decimal places in M^4.

The entries in the first row of M^4 provide the answers to the questions posed:

Given that the rat is a poor performer in advance of the first trial, the probabilities for his performance of the fourth trial are

.01 for a grade of Poor,
.12 for a grade of Fair, and
.87 for a grade of Good.

Comment: It is reasonable to protest that one should take into account more of the rat's recent accomplishments than just the one preceding trial in setting the probabilities. Perhaps so. Of course, there is a mathematical theory for handling the more complicated problem—but our present aims are modest.

9-1 EXERCISES

1. To avoid physical exertion in simulations of probabilistic phenomena, a student has contrived a marvelous coin-flipping machine. Unfortunately, many trials show the device to be biased. There is a probability of .6 that the coin will land in the orientation—Heads or Tails—that it had before the flip, and a probability of .4 that its orientation will change. In tabulated form, the transition probabilities are as follows:

		Output State	
		H	T
Input State	H	.6	.4
	T	.4	.6

(a) For the matrix M of transition probabilities, compute M^3.
(b) What is the value of $p_{12}^{(3)}$, the probability that the coin lands Tails on the third trial, given that it was lying Heads before the first trial?

2. Suppose the probabilities for the political affiliation of sons, given the affiliation of their fathers, are as shown in the following table:

		Son		
		Republican	*Democrat*	*Independent*
Father	*Rep*	·6	.1	.3
	Dem	.2	.6	.2
	Ind	.2	.3	.5

What is the probability that
(a) The grandson of a Republican is an Independent?
(b) The grandson of an Independent is a Democrat?
(c) The great-grandson of a Democrat is a Republican?

3. An urn contains three marbles, each of which is red or white. A trial of the process consists of adding a white marble to the urn, and then at random removing one of the four marbles.
(a) Construct the matrix M of transition probabilities, letting State 0 correspond to "no red marbles in the urn," State 1 to "one red," State 2 to "two reds," and State 3 to "three reds."

(b) For $k = 0, 1, 2, 3$, find the value of $p_{3k}^{(3)}$, the probability that after three trials there are k red marbles in the urn, given that before the first trial all three marbles were red.

4. A series of debates is being conducted on a proposed constitutional amendment. Suppose the probabilities for a listener's position on the amendment after hearing a debate, given his position before the debate, are as follows:

		Position Afterward		
		For	*Undecided*	*Against*
Position Before	*For*	.7	.2	.1
	Undecided	.3	.5	.2
	Against	.2	.2	.6

Find the probability that, after hearing three of these debates, a listener is *for* the amendment, given that before the first debate

(a) He was for the amendment.

(b) He was undecided.

(c) He was against the amendment.

5. A phenomenon has two stable states, States 1 and 4, and two unstable states, States 2 and 3. Given a stimulus—trial of the process—the phenomenon in State 2 has probability $\frac{2}{3}$ of falling into State 1 and probability $\frac{1}{3}$ of shifting to State 3. If it is in State 3, the probability is $\frac{2}{3}$ that it falls into State 4 and $\frac{1}{3}$ that it shifts to State 2. Once in States 1 or 4, the phenomenon remains there. Thus the matrix of transition probabilities is

$$M = \begin{bmatrix} 1 & 0 & 0 & 0 \\ \frac{2}{3} & 0 & \frac{1}{3} & 0 \\ 0 & \frac{1}{3} & 0 & \frac{2}{3} \\ 0 & 0 & 0 & 1 \end{bmatrix} = \frac{1}{3}\begin{bmatrix} 3 & 0 & 0 & 0 \\ 2 & 0 & 1 & 0 \\ 0 & 1 & 0 & 2 \\ 0 & 0 & 0 & 1 \end{bmatrix}.$$

Suppose the phenomenon starts in State 2.

(a) Find the probability that after four successive stimuli, the phenomenon is in (i) State 1, (ii) State 2, (iii) State 3, (iv) State 4.

(b) Find the probability, to three-decimal-place accuracy, that the phenomenon ultimately comes to rest in State 1.

9-2. Gambler's Ruin

In this, the last section of the book, we'll look at a relatively simple type of Markov process derived from a fascinating empirical problem. In common with the best applications of mathematics, the analysis calls on a variety of notions, giving us a chance to touch base with several of the topics we've studied.

We illustrate with a simple game of chance:

On each play of a certain game, the player either wins a dollar or he loses a dollar. The probability of winning is .4, and the probability of losing is .6.

Thus the player can expect to win about four-tenths of the time, in the long run. But, to be practical, this characterization of his prospects is not very illuminating. Indeed, considering the unfavorable odds, the realistic appraisal of the player's "long-run" prospect is simply that he's going to lose all of his money.

To put the problem in a more meaningful setting, we might imagine that the player starts with $90 and is firmly resolved to stop playing as soon as his bankroll reaches $100; and, needless to say, he is compelled to stop when he runs out of money. Until one or the other of these situations occurs, he continues playing. The interesting problem is to find the probabilities of success and failure in this venture. Respectful of what is usually the more likely outcome, problems of this type are classified under the heading "Gambler's Ruin."

In the interest of simplicity, let's switch to smaller numbers in the problem (an activity long ago identified as "moving to a spot where the light is better"). We'll suppose that the player starts with only one, two, or three dollars and that he stops when his bankroll reaches zero or four dollars.

The plays of the game are interpreted as trials in a Markov process in which the states are naturally identified as 0, 1, 2, 3, and 4, corresponding to the size of the player's bankroll. The transition probabilities are easy to compute. For instance, if the process begins in State 2 (the player starts with $2), then there is a probability of .4 of transition to State 3 (the player wins a dollar) and a probability of .6 of transition to State 1 (he loses a dollar). The transition probabilities to the other states are zero. Thus we set

$$p_{20} = 0, \qquad p_{21} = .6, \qquad p_{22} = 0, \qquad p_{23} = .4, \qquad p_{24} = 0.$$

The mathematical model is constructed as though the plays go on indefinitely, imagining that the player participates by just watching with empty pockets once his bankroll reaches zero, and the happier analogy when he finds himself with $4. Thus the transition probabilities from States 0 and 4 are

$$p_{00} = 1, \qquad p_{01} = p_{02} = p_{03} = p_{04} = 0,$$

$$p_{40} = p_{41} = p_{42} = p_{43} = 0, \qquad p_{44} = 1.$$

All possibilities considered, we construct the following 5×5 matrix M of transition probabilities:

$$M = \begin{bmatrix} 1 & 0 & 0 & 0 & 0 \\ .6 & 0 & .4 & 0 & 0 \\ 0 & .6 & 0 & .4 & 0 \\ 0 & 0 & .6 & 0 & .4 \\ 0 & 0 & 0 & 0 & 1 \end{bmatrix}.$$

Based on the discussion of the previous section, we can gauge the player's prospects for two or three plays of the game by computing M^2 and M^3:

$$M^2 = \begin{bmatrix} 1 & 0 & 0 & 0 & 0 \\ .6 & .24 & 0 & .16 & 0 \\ 0 & 0 & .48 & 0 & .16 \\ 0 & .36 & 0 & .24 & .4 \\ 0 & 0 & 0 & 0 & 1 \end{bmatrix},$$

$$M^3 = \begin{bmatrix} 1 & 0 & 0 & 0 & 0 \\ .744 & 0 & .192 & 0 & .064 \\ .360 & .288 & 0 & .192 & .160 \\ .216 & 0 & .288 & 0 & .496 \\ 0 & 0 & 0 & 0 & 1 \end{bmatrix}.$$

We see, for instance, that if the player starts with \$2 and participates in three plays, then the probability is .36 that he will be ruined, and the probability is .16 that he will reach his \$4 goal. The probability is .288 that he will find himself with \$1 and .192 that he'll have \$3.

Of course, the Gambler's Ruin question relates to the probability of ultimate ruin or ultimate success. Thus it is not M^3 that interests us, but rather M^n, where n is a very large number. A good indication of what to expect for a large number of plays is provided by M^{25}. This is an easy problem for a computer. Displayed below is M^{25}, with approximations rounded off to three decimal places:

$$M^{25} = \begin{bmatrix} 1 & 0 & 0 & 0 & 0 \\ .877 & 0 & .001 & 0 & .123 \\ .692 & .001 & 0 & .001 & .308 \\ .415 & 0 & .001 & 0 & .585 \\ 0 & 0 & 0 & 0 & 1 \end{bmatrix}.$$

As we see from the entries in M^{25}, the probability that the player will still be actively participating after 25 plays is negligible. To three-decimal-place accuracy, M^{25} presents the solution of our problem. For the player with the \$4 goal, there is a probability of .877 that he will ultimately be ruined if he starts with \$1; starting with \$2, the probability is .692; and starting with \$3, the probability is .415.

A Theoretical Attack

The problem above was effectively resolved by computing M^{25}, where M is the 5×5 matrix of transition probabilities. But this direct attack is not very useful for treating the Gambler's Ruin problem in a more general setting. We'll stay with the matrix M for purposes of illustration, but now we'll pursue an approach which easily extends to the general problem.

As the form of M^{25} suggests, and as should be clear from the interpretation of the process, the entries in the matrix M^n for sufficiently large n will be arbitrarily close to the entries in a fixed matrix, which we romantically name M^∞. In effect, the entries in M^∞ are the transition probabilities associated with such a long sequence of plays that the probability that the player is still actively participating has become zero, for all reasonable intents and purposes. Thus outside the first and last columns all of the entries are zero. We represent M^∞ as follows:

$$M^\infty = \begin{bmatrix} 1 & 0 & 0 & 0 & 0 \\ Q_1 & 0 & 0 & 0 & P_1 \\ Q_2 & 0 & 0 & 0 & P_2 \\ Q_3 & 0 & 0 & 0 & P_3 \\ 0 & 0 & 0 & 0 & 1 \end{bmatrix}.$$

The interpretation of the entries should be clear. For instance, Q_2 is the probability of ultimate ruin, given that the player starts with \$2, and P_3 is is the probability of ultimate success, given that the player starts with \$3.

Since

$$M \times M^n = M^{n+1}$$

for all n, and since M^n and M^{n+1} are both very close to M^∞ for large n, it follows that

$$M \times M^\infty = M^\infty.$$

Computing $M \times M^\infty$, we have

$$M \times M^\infty = \begin{bmatrix} 1 & 0 & 0 & 0 & 0 \\ .6 & 0 & .4 & 0 & 0 \\ 0 & .6 & 0 & .4 & 0 \\ 0 & 0 & .6 & 0 & .4 \\ 0 & 0 & 0 & 0 & 1 \end{bmatrix} \begin{bmatrix} 1 & 0 & 0 & 0 & 0 \\ Q_1 & 0 & 0 & 0 & P_1 \\ Q_2 & 0 & 0 & 0 & P_2 \\ Q_3 & 0 & 0 & 0 & P_3 \\ 0 & 0 & 0 & 0 & 1 \end{bmatrix}$$

$$= \begin{bmatrix} 1 & 0 & 0 & 0 & 0 \\ .6 + .4Q_2 & 0 & 0 & 0 & .4P_2 \\ .6Q_1 + .4Q_3 & 0 & 0 & 0 & .6P_1 + .4P_3 \\ .6Q_2 & 0 & 0 & 0 & .6P_2 + .4 \\ 0 & 0 & 0 & 0 & 1 \end{bmatrix}.$$

Thus the equality

$$M \times M^\infty = M^\infty$$

is expressed by the following set of equations involving the entries in M^∞:

$$\begin{aligned} Q_1 &= .6 + .4Q_2, & P_1 &= .4P_2, \\ Q_2 &= .6Q_1 + .4Q_3, & P_2 &= .6P_1 + .4P_3, \\ Q_3 &= .6Q_2, & P_3 &= .6P_2 + .4. \end{aligned} \tag{1}$$

It is worth observing that we could have obtained these equations by the arguments of Chapter 2. Suppose, for instance, that the player starts with \$2. His experience in the overall game to the point of ruin or success is represented by the two-stage probability tree in Fig. 9-3. Since the process begins in State 2, the probability of Ruin is Q_2. And since this probability is equal to the sum of the probabilities for the paths leading to Ruin in the tree, we have

$$Q_2 = .6Q_1 + .4Q_2.$$

By the analogous argument,

$$P_2 = .6P_1 + .4P_3.$$

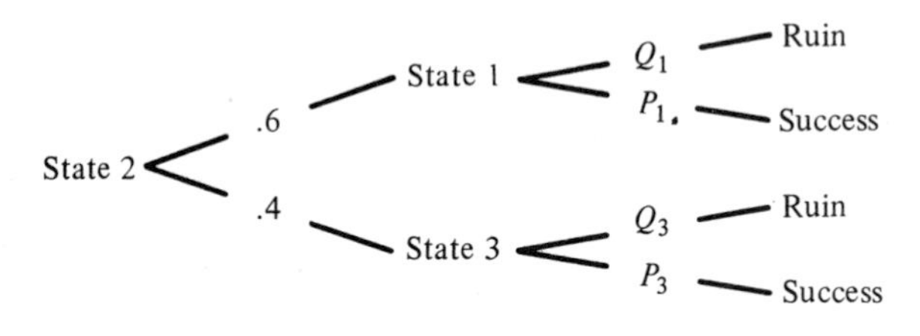

Figure 9-3.

In the interest of symbolic symmetry, we set $Q_0 = 1$ and $Q_4 = 0$. This makes sense since the probability of going from State 0 to Ruin is 1, and the probability of going from State 4 to Ruin is 0. The equations on the left in (1) are rewritten as follows:

$$\begin{aligned} Q_1 &= .6Q_0 + .4Q_2 \quad \text{or} \quad Q_2 = -\frac{.6}{.4}Q_0 + \frac{1}{.4}Q_1, \\ Q_2 &= .6Q_1 + .4Q_3 \quad \text{or} \quad Q_3 = -\frac{.6}{.4}Q_1 + \frac{1}{.4}Q_2. \\ Q_3 &= .6Q_2 + .4Q_4 \quad \text{or} \quad Q_4 = -\frac{.6}{.4}Q_2 + \frac{1}{.4}Q_3. \end{aligned} \tag{2}$$

The multiplier .6/.4 in these equations is the ratio of the probability $q = .6$ of failure on a single trial (losing a dollar) to the probability $p = .4$ of success (winning a dollar). We represent this ratio as r:

$$r = \frac{q}{p} = \frac{.6}{.4}.$$

The other multiplier in the equations on the right in Eq. (2) is expressed in terms of r:

$$\frac{1}{.4} = \frac{p+q}{p} = \frac{p}{p} + \frac{q}{q} = 1 + \frac{.6}{.4} = 1 + r.$$

The equations on the right in (2) are now written

$$\begin{aligned} Q_2 &= -rQ_0 + (1 + r)Q_1, \\ Q_3 &= -rQ_1 + (1 + r)Q_2, \\ Q_4 &= -rQ_2 + (1 + r)Q_3. \end{aligned} \tag{3}$$

We have reduced the Gambler's Ruin problem for the player with the \$4 goal to the problem of finding values Q_k which satisfy the system (3), with $Q_0 = 1$ and $Q_4 = 0$.

Having reached this point in the analysis of the particular problem, we are ready to extend the discussion to the general case. The particular solution can easily be extracted from the general solution.

The General Problem

Empirically, the general Gambler's Ruin problem may be described as follows:

> A player participates in a game of chance in which on each play there is a probability p of winning a dollar and a probability $q = 1 - p$ of losing a dollar. He continues playing until his bankroll reaches zero or n dollars, where n is some given integer. The problem is to compute the probabilities of ruin Q_k, where Q_k is the probability that the player's bankroll reaches zero before it reaches n dollars, given that he starts with k dollars, where $k = 0, 1, 2, \ldots, n$.

The plays of the game are interpreted as trials in a Markov process. The same argument that gave the 5×5 matrix M in the foregoing illustration, where $n = 4$, $p = .4$, and $q = .6$, produces an $(n + 1) \times (n + 1)$ matrix M of transition probabilities in the general case. The matrix is of the form

$$M = \begin{bmatrix} 1 & 0 & 0 & 0 & \ldots & 0 & 0 & 0 \\ q & 0 & p & 0 & \ldots & 0 & 0 & 0 \\ 0 & q & 0 & p & \ldots & 0 & 0 & 0 \\ \cdot & \cdot & \cdot & \cdot & \cdot & \cdot & \cdot & \cdot \\ \cdot & \cdot & \cdot & \cdot & \cdot & \cdot & \cdot & \cdot \\ \cdot & \cdot & \cdot & \cdot & \cdot & \cdot & \cdot & \cdot \\ 0 & 0 & 0 & 0 & \ldots & q & 0 & p \\ 0 & 0 & 0 & 0 & \ldots & 0 & 0 & 1 \end{bmatrix}.$$

For sufficiently large n the entries in M^n are arbitrarily close to the entries in a fixed matrix, which we identify as M^∞. As in the special case, the form of M^∞ is as follows:

$$M^\infty = \begin{bmatrix} 1 & 0 & 0 & 0 & \ldots & 0 & 0 & 0 \\ Q_1 & 0 & 0 & 0 & \ldots & 0 & 0 & P_1 \\ Q_2 & 0 & 0 & 0 & \ldots & 0 & 0 & P_2 \\ \cdot & \cdot & \cdot & \cdot & \cdot & \cdot & \cdot & \cdot \\ \cdot & \cdot & \cdot & \cdot & \cdot & \cdot & \cdot & \cdot \\ \cdot & \cdot & \cdot & \cdot & \cdot & \cdot & \cdot & \cdot \\ Q_{n-1} & 0 & 0 & 0 & \ldots & \cdot & \cdot & P_{n-1} \\ 0 & 0 & 0 & 0 & \ldots & 0 & 0 & 1 \end{bmatrix}.$$

The Q_k in the first column of M^∞ are the desired probabilities of ruin.

The argument leading to the system of equations (3) can be applied word for word to the general case. Of course, the dimension of the problem is generalized—there is an n-dollar goal rather than a four-dollar goal—and the specific probabilities, .6 and .4, are replaced by general values q and p. We are led to the following conclusion:

> **Conditions for Determining $\mathbf{Q}_k$:** Setting $r = q/p$, the values of Q_k, for $k = 0, 1, 2, \ldots, n$, are determined by the requirement that
>
> $$Q_0 = 1 \quad \text{and} \quad Q_n = 0,$$
>
> and, in general, Q_k satisfies the following system of equations:
>
> $$\begin{aligned} Q_2 &= -rQ_0 + (1 + r)Q_1 \\ Q_3 &= -rQ_1 + (1 + r)Q_2 \\ &\;\vdots \\ Q_n &= -rQ_{n-2} + (1 + r)Q_n. \end{aligned} \tag{4}$$

To find the Q_k satisfying these conditions we employ the notion of linear mappings of vector spaces discussed in Sec. 8-7. We view the desired Q_k in pairs as comprising the following vectors in the space $\mathbf{R}^2$:

$$(Q_0, Q_1), (Q_1, Q_2), (Q_2, Q_3), \ldots, (Q_{n-1}, Q_n).$$

Interpreting vectors as cclumn matrices, we consider the mapping of $\mathbf{R}^2$ into $\mathbf{R}^2$ determined by the 2×2 matrix

$$T = \begin{bmatrix} 0 & 1 \\ -r & 1 + r \end{bmatrix}.$$

Applying this mapping to the vector (Q_0, Q_1), we have

$$T\begin{bmatrix} Q_0 \\ Q_1 \end{bmatrix} = \begin{bmatrix} 0 & 1 \\ -r & 1 + r \end{bmatrix}\begin{bmatrix} Q_0 \\ Q_1 \end{bmatrix} = \begin{bmatrix} Q_1 \\ -rQ_0 + (1 + r)Q_1 \end{bmatrix}.$$

But from the first equation in system (4),

$$-rQ_0 + (1 + r)Q_1 = Q_2.$$

Therefore,

$$T\begin{bmatrix} Q_0 \\ Q_1 \end{bmatrix} = \begin{bmatrix} Q_1 \\ Q_2 \end{bmatrix}.$$

And by the same argument,

$$T\begin{bmatrix} Q_1 \\ Q_2 \end{bmatrix} = \begin{bmatrix} Q_2 \\ Q_3 \end{bmatrix}.$$

We note that $\begin{bmatrix} Q_2 \\ Q_3 \end{bmatrix}$ may be obtained from $\begin{bmatrix} Q_0 \\ Q_1 \end{bmatrix}$ in two steps:

$$\begin{bmatrix} Q_2 \\ Q_3 \end{bmatrix} = T\begin{bmatrix} Q_1 \\ Q_2 \end{bmatrix} = T\left(T\begin{bmatrix} Q_0 \\ Q_1 \end{bmatrix}\right) = TT\begin{bmatrix} Q_0 \\ Q_1 \end{bmatrix} = T^2\begin{bmatrix} Q_0 \\ Q_1 \end{bmatrix},$$

and, similarly,

$$\begin{bmatrix} Q_3 \\ Q_4 \end{bmatrix} = T\begin{bmatrix} Q_2 \\ Q_3 \end{bmatrix} = T\left(T^2\begin{bmatrix} Q_0 \\ Q_1 \end{bmatrix}\right) = T^3\begin{bmatrix} Q_0 \\ Q_1 \end{bmatrix},$$

and so on, leading finally to the conclusion that

$$\begin{bmatrix} Q_{n-1} \\ Q_n \end{bmatrix} = T^{n-1}\begin{bmatrix} Q_0 \\ Q_1 \end{bmatrix}. \tag{5}$$

We are not especially interested in the precise form of the 2×2 matrix T^{n-1}. The important fact is that this matrix, like all 2×2 matrices, defines a linear mapping of $\mathbf{R}^2$ into $\mathbf{R}^2$, the vectors being interpreted as column matrices. To make use of this fact, we observe a couple of vectors in $\mathbf{R}^2$ which are mapped in a simple fashion by the matrix T.

Let $\mathbf{v}_1 = (1, 1)$. Then under the mapping determined by T,

$$T\begin{bmatrix} 1 \\ 1 \end{bmatrix} = \begin{bmatrix} 0 & 1 \\ -r & 1+r \end{bmatrix}\begin{bmatrix} 1 \\ 1 \end{bmatrix} = \begin{bmatrix} 1 \\ -r + (1+r) \end{bmatrix} = \begin{bmatrix} 1 \\ 1 \end{bmatrix}.$$

Thus $\mathbf{v}_1 = (1, 1)$ is mapped by T into itself. Successive mappings by T, of course, continue to carry $(1, 1)$ into itself. In particular,

$$T^{n-1}\begin{bmatrix} 1 \\ 1 \end{bmatrix} = \begin{bmatrix} 1 \\ 1 \end{bmatrix}.$$

Let $\mathbf{v}_2 = (1, r)$. Then

$$T\begin{bmatrix} 1 \\ r \end{bmatrix} = \begin{bmatrix} 0 & 1 \\ -r & 1+r \end{bmatrix}\begin{bmatrix} 1 \\ r \end{bmatrix} = \begin{bmatrix} r \\ -r + (1+r)r \end{bmatrix} = \begin{bmatrix} r \\ r^2 \end{bmatrix},$$

$$T\begin{bmatrix} 1 \\ r \end{bmatrix} = r\begin{bmatrix} 1 \\ r \end{bmatrix}.$$

Mapping again,

$$T^2\begin{bmatrix} 1 \\ r \end{bmatrix} = T\left(T\begin{bmatrix} 1 \\ r \end{bmatrix}\right) = T\left(r\begin{bmatrix} 1 \\ r \end{bmatrix}\right) = rT\begin{bmatrix} 1 \\ r \end{bmatrix} = r\left(r\begin{bmatrix} 1 \\ r \end{bmatrix}\right) = r^2\begin{bmatrix} 1 \\ r \end{bmatrix}.$$

Continuing in this fashion, we reach

$$T^{n-1}\begin{bmatrix} 1 \\ r \end{bmatrix} = r^{n-1}\begin{bmatrix} 1 \\ r \end{bmatrix}.$$

Looking at T^{n-1} as a function mapping vectors into vectors, we see that

$$T^{n-1}(\mathbf{v}_1) = T^{n-1}(1, 1) = (1, 1) = \mathbf{v}_1$$

and $\qquad\qquad\qquad\qquad\qquad\qquad\qquad\qquad\qquad\qquad\qquad\qquad\qquad\qquad\qquad\qquad$ (6)

$$T^{n-1}(\mathbf{v}_2) = T^{n-1}(1, r) = r^{n-1}(1, r) = r^{n-1}\mathbf{v}_2.$$

Of course, if $r = 1$, then $\mathbf{v}_1 = (1, 1)$ and $\mathbf{v}_2 = (1, r)$ are one and the same vector. Let's defer the $r = 1$ case for a moment, and suppose that r is not equal to one. Then $\mathbf{v}_1$ and $\mathbf{v}_2$ are two distinct points in the coordinate plane $\mathbf{R}^2$ lying on the line of points of the form $(1, y)$. Since we require that $Q_0 = 1$, it follows that (Q_0, Q_1) is on this line. Therefore, from the development in Sec. 7-2, (Q_0, Q_1) is of the form

$$(Q_0, Q_1) = (1 - a)\mathbf{v}_1 + a\mathbf{v}_2$$
$$= (1 - a)(1, 1) + a(1, r) \tag{7}$$

for some scalar a. The key to finding Q_1 and the other Q_k is the determination of the correct value of a in Eq. (7).

From Eq. (5) we know that

$$(Q_{n-1}, Q_n) = T^{n-1}(Q_0, Q_1).$$

Substituting for (Q_0, Q_1) from Eq. (7), we have

$$(Q_{n-1}, Q_n) = T^{n-1}((1 - a)\mathbf{v}_1 + a\mathbf{v}_2). \tag{8}$$

At this point we use the fact that T^{n-1} is a linear mapping. The linearity permits us to combine vectors by addition and scalar multiplication before we map or afterward—the result will be the same. Therefore, from Eq. (8)

$$(Q_{n-1}, Q_n) = (1 - a)T^{n-1}(\mathbf{v}_1) + aT^{n-1}(\mathbf{v}_2).$$

But from Eq. (6)

$$T^{n-1}(\mathbf{v}_1) = \mathbf{v}_1 \quad \text{and} \quad T^{n-1}(\mathbf{v}_2) = r^{n-1}\mathbf{v}_2.$$

Thus we conclude that

$$(Q_{n-1}, Q_n) = (1 - a)(\mathbf{v}_1) + ar^{n-1}(\mathbf{v}_2)$$
$$= (1 - a)(1, 1) + ar^{n-1}(1, r). \tag{9}$$

Extracting the second coordinates of the vectors in Eq. (9), we have

$$Q_n = (1 - a) + ar^{n-1}r = 1 - a(1 - r^n). \tag{10}$$

Now we can easily find the correct value of the scalar a from the requirement

$$Q_n = 0.$$

Making this substitution in Eq. (10), we have

$$0 = 1 - a(1 - r^n)$$

or

$$a = \frac{1}{1 - r^n} = \frac{-1}{r^n - 1}. \tag{11}$$

For those problems in which $r = q/p$ is not equal to 1, i.e., q is not equal to p, the value of a in Eq. (11) generates the values of Q_k.

The first coordinate in Eq. (7) gives

$$Q_0 = (1 - a) + a = 1,$$

as desired.

Looking at the second coordinate in Eq. (7), we have

$$Q_1 = (1 - a) + ar = 1 + a(r - 1).$$

Substituting the value of a from Eq. (11),

$$Q_1 = 1 - \frac{r - 1}{r^n - 1} = \frac{(r^n - 1) - (r - 1)}{r^n - 1} = \frac{r^n - r}{r^n - 1}.$$

The remaining Q_k are computed from the equations in the system (4):

$$Q_2 = -rQ_0 + (1+r)Q_1 = -r + \frac{(1+r)(r^n - r)}{r^n - 1}$$

$$= \frac{-r(r^n - 1) + (r^n - r) + r(r^n - r)}{r^n - 1} = \frac{r^n - r^2}{r^n - 1},$$

and, similarly,

$$Q_3 = -rQ_1 + (1+r)Q_2 = \frac{r^n - r^3}{r^n - 1},$$

and so on, giving the formula

$$Q_k = \frac{r^n - r^k}{r^n - 1}, \qquad k = 0, 1, 2, \ldots, n, \tag{12}$$

where $r = \dfrac{q}{p} \neq 1$.

Equation (12) gives the general solution of the Gambler's Ruin problem for those cases in which $r = q/p$ is not equal to one.

For the case in which $r = 1$, that is, $p = q = \frac{1}{2}$, a variation on the argument is required. The only change is that we replace the vector $\mathbf{v}_2 = (1, r)$, which in the $r = 1$ case is the same as $\mathbf{v}_1 = (1, 1)$, by the vector

$$\mathbf{v}_2 = (1, 2).$$

We find $T(\mathbf{v}_2)$ by multiplying by the matrix T, setting $r = 1$:

$$T\begin{bmatrix} 1 \\ 2 \end{bmatrix} = \begin{bmatrix} 0 & 1 \\ -r & 1+r \end{bmatrix}\begin{bmatrix} 1 \\ 2 \end{bmatrix} = \begin{bmatrix} 0 & 1 \\ -1 & 2 \end{bmatrix}\begin{bmatrix} 0 \\ 1 \end{bmatrix} = \begin{bmatrix} 2 \\ 3 \end{bmatrix},$$

and

$$T^2\begin{bmatrix} 1 \\ 2 \end{bmatrix} = \begin{bmatrix} 0 & 1 \\ -1 & 2 \end{bmatrix}\begin{bmatrix} 2 \\ 3 \end{bmatrix} = \begin{bmatrix} 3 \\ 4 \end{bmatrix},$$

and so on, reaching

$$T^{n-1}\begin{bmatrix} 1 \\ 2 \end{bmatrix} = \begin{bmatrix} n \\ 1+1 \end{bmatrix}.$$

Using the new $\mathbf{v}_2 = (1, 2)$ exactly as we used the old $\mathbf{v}_2$, the analogue of Eq. (11) is

$$a = -\frac{1}{n}.$$

And the generated solutions Q_k are

$$Q_k = 1 - \frac{k}{n}, \qquad k = 0, 1, 2, \ldots, n, \tag{13}$$

for the case $p = q = \frac{1}{2}$.

Comment: The $p = q = \frac{1}{2}$ case represents a fair game. As one should suspect, the expected value of the player's winnings in this instance is zero. This is easy to show. Representing the player's winning as the random variable X, we find that the range of X is the set $\{-k, n - k\}$.

(The player is either ruined, in which instance he loses his initial bankroll of k dollars, or he reaches his goal, thereby winning $n - k$ dollars.) The respective probabilities are

$$Q_k = \Pr(X = -k) = 1 - \frac{k}{n},$$

$$P_k = \Pr(X = n - k) = 1 - Q_k = \frac{k}{n}.$$

Thus

$$E(X) = (-k)\left(1 - \frac{k}{n}\right) + (n - k)\frac{k}{n} = 0.$$

We conclude the discussion of Gambler's Ruin, and the book, with two examples.

Example: In the game of Roulette there are 38 equally likely numbers that may occur. A player betting a dollar on "red" wins a dollar if any one of a set of 18 of these numbers occur; otherwise, he loses his dollar.

Suppose a player starts with \$90 and bets \$1 on "red" on each play, stopping when he reaches a goal of \$100 or when he runs out of money. What is the probability of failure in this venture?

Solution: The desired value is Q_k for $k = 90$ in the Gambler's Ruin problem with $n = 100$ and

$$p = \tfrac{18}{38} \quad \text{and} \quad q = \tfrac{20}{38}.$$

Thus

$$\frac{q}{p} = r = \tfrac{20}{18} = \tfrac{1}{.9}.$$

Since $r \neq 1$, we apply Eq. (12),

$$Q_k = \frac{r^n - r^k}{r^n - 1}.$$

It is easier to work with $1/r = .9$, so we rewrite Q_k by dividing the numerator and denominator by r^n:

$$Q_k = \frac{1 - (1/r)^{n-k}}{1 - (1/r)^n}.$$

Substituting the given values of r, n, and k (fortunately a calculator is at hand), we have

$$Q_{90} = \frac{1 - (.9)^{10}}{1 - (.9)^{100}} = \frac{1 - .349}{1 - .000} = .651.$$

We see that the probability of ruin is .651. (Thus the probability of success—winning \$10 rather than losing \$90—is about 35%. It doesn't look like a good gamble.)

For the final example we return to our old friend Adam with whom we began the adventure. He's again at his task of drawing cards from a deck. But now he's induced Freshman Bob to join him in an unimaginative game. As things develop, Bob finds himself faced with a sophisticated problem.

Example: On each play of a game, Adam shuffles a standard 52-card deck and draws a single card. If the card is red, Adam gives Bob one cent; if it is black, Bob pays Adam one cent.

They each started with 10 pennies, informally agreeing to play until one or the other has all 20 pennies. But after a while we find Bob with 15 cents, thoroughly bored, and suggesting that they stop. Adam is still enjoying the game and offers to pay Bob one cent outright for firmly agreeing to persevere to the end. Or, if Bob prefers, Adam will keep his penny but remove one black card from the deck, thereby increasing Bob's chances on each play.

Which of these options is the better choice for Bob?

Solution: The first of Bob's options represents a Gambler's Ruin problem with

$$n = 20, \quad k = 16, \quad \text{and} \quad p = q = \frac{1}{2}.$$

Since $p = q$, we apply Eq. (13):

$$Q_k = 1 - \frac{k}{n}.$$

Substituting the given values, we have

$$Q_{16} = 1 - \tfrac{16}{20} = \tfrac{4}{20} = .2.$$

The second option is a Gambler's Ruin problem with

$$n = 20, \quad k = 15, \quad p = \frac{26}{51}, \quad \text{and} \quad q = \frac{25}{51}.$$

Thus

$$r = \frac{q}{p} = \frac{25}{26}.$$

From Eq. (12) the probability of ruin is

$$Q_{15} = \frac{r^{20} - r^{15}}{r^{20} - 1} = \frac{(\frac{25}{26})^{20} - (\frac{25}{26})^{15}}{(\frac{25}{26})^{20} - 1}.$$

Again calling on a calculator, we find

$$Q_{15} = .182.$$

We see that Bob is less likely to be ruined under the second option. Thus he is better advised to have Adam remove the one black card from the deck.

We didn't compute the expected length of the game under the respective options, which might be of more interest to Bob. But that analysis would call for a probability space with an infinite number of elements in the sample set. So we'll leave the "expected length" question to a treatise more concerned with "infinite mathematics."

9-2 EXERCISES

Some of these problems require a calculator.

1. (a) In the illustration developed in the text we considered the Gambler's Ruin problem with

$$n = 4, \quad p = .4, \quad \text{and} \quad q = .6.$$

Use the general formula (12) to find the precise values of

$$Q_1, \quad Q_2, \quad \text{and} \quad Q_3.$$

Compare these values with the corresponding entries given for the matrix M^{25}, where M is the 5×5 matrix of transition probabilities.

(b) The example at the beginning of the section concerned the probabilities of success and failure for a player starting with \$90 and having a \$100 goal in a game in which on each play the probability of winning a dollar is $p = .4$ and the probability of losing a dollar is $q = .6$. Use Eq. (12) in the form

$$Q_k = \frac{1 - (1/r)^{n-k}}{1 - (1/r)^n}$$

to find the probabilities,

$$Q_{90} \quad \text{and} \quad P_{90} = 1 - Q_{90},$$

for the Gambler's Ruin problem with $n = 100$, $p = .4$, and $q = .6$. (Three-decimal-place approximation is appropriate.)

2. In the final example of the book we found that in the Gambler's Ruin problem with

$$n = 20, \quad p = \tfrac{26}{51}, \quad \text{and} \quad q = \tfrac{25}{51}$$

the value of Q_{15} is .182. This case corresponded to Adam removing one black card from the deck. But suppose Adam removes two black cards. What, then, is the probability that Bob, starting with 15 cents, will be ruined before he reaches the 20-cent goal? In other words, find Q_{15} in the Gambler's Ruin problem with

$$n = 20, \quad p = \tfrac{26}{50}, \quad \text{and} \quad q = \tfrac{24}{50}.$$

3. Recall the "Roulette" example. Betting on "red" on each play, the probability of winning a dollar is $p = \frac{18}{38}$ and the probability of losing a dollar is $q = \frac{20}{38}$. Find the probability of ruin for a player with a \$25 goal who

(a) Starts with \$20.
(b) Starts with \$15.
(c) Starts with \$10.

4. Find the probabilities of ruin Q_k in the Gambler's Ruin problem for the following values of n, k, p, and q:
(a) $n = 20, k = 10, p = q = .5$.
(b) $n = 20, k = 10, p = .49, q = .51$.
(c) $n = 20, k = 18, p = q = .5$.
(d) $n = 20, k = 18, p = .4, q = .6$.

5. Consider the game with payoff matrix

$$G = \begin{bmatrix} 1 & -1 & -1 \\ -1 & 1 & -1 \\ -1 & -1 & 1 \end{bmatrix}.$$

Interpreting the payoff in dollars, one of the players wins a dollar from the other on each play. Suppose the row player R starts with one, two, or three dollars and is resolved to stop when his bankroll reaches four dollars; and, of course, he must stop when his bankroll reaches zero. If the column player C employs his optimal strategy, what is the probability that he will ruin the row player R if R starts with
(a) One dollar?
(b) Two dollars?
(c) Three dollars?

Solutions for Odd-Numbered Exercises

Chapter 1

Exercises 1-1, page 6

1. (a) Interpretation of elements: "10¢" represents the statement "The two coins drawn by Henry total ten cents," and analogously for the other elements. The set does qualify as a basic outcome set.

(b) "At least one nickel" is shorthand for "At least one of the two coins drawn by Henry is a nickel." Not a basic outcome set because both statements are true for the "15¢" occurrence.

(c) Interpretation analogous to parts (a) and (b). Does qualify as a basic outcome set.

3. Not a basic outcome set because none of the four statements is true when "One of the cookies has exactly three raisins and the other does not have exactly three raisins." The foregoing statement completes a basic outcome set. (Not the only solution.)

5. The basic outcome "An odd number shows" does not contain enough information to decide whether "The number 5 shows" is true or false.

7. (a) (i), (ii).

(b) The relevant statements (i) and (ii) are equivalent, both having the truth set {One head, Two heads, Three heads}.

9. (a) {HH, HT, TH, TT}.

(b) For example, {first flip H, first flip T}.

Exercises 1-2, page 12

1. (a) Both marbles drawn are red.

(b) {Both marbles are red, The first marble is red and the second is white, The first marble is white and the second is red, Both marbles are white}

3. (a)

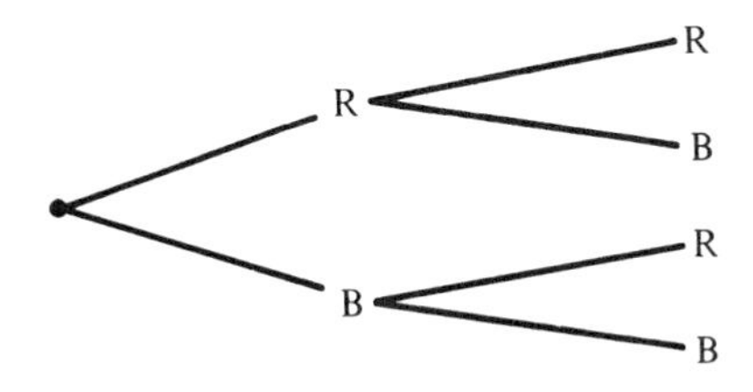

(b) Standing alone, the initial component does not represent a basic outcome analysis.

5.

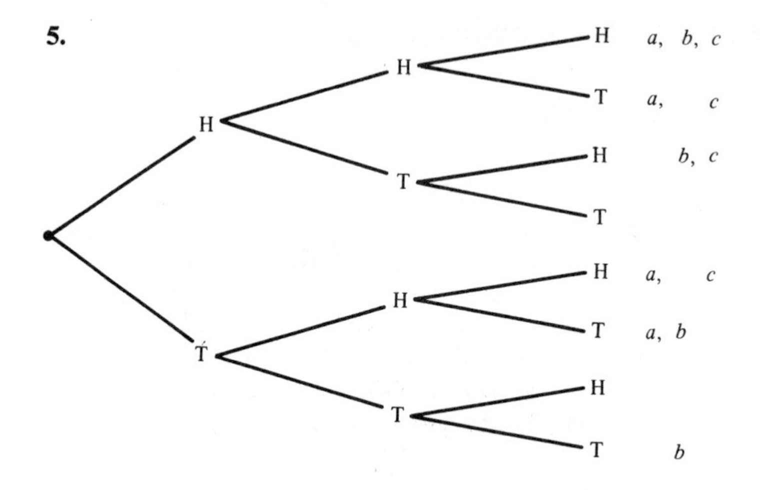

7.

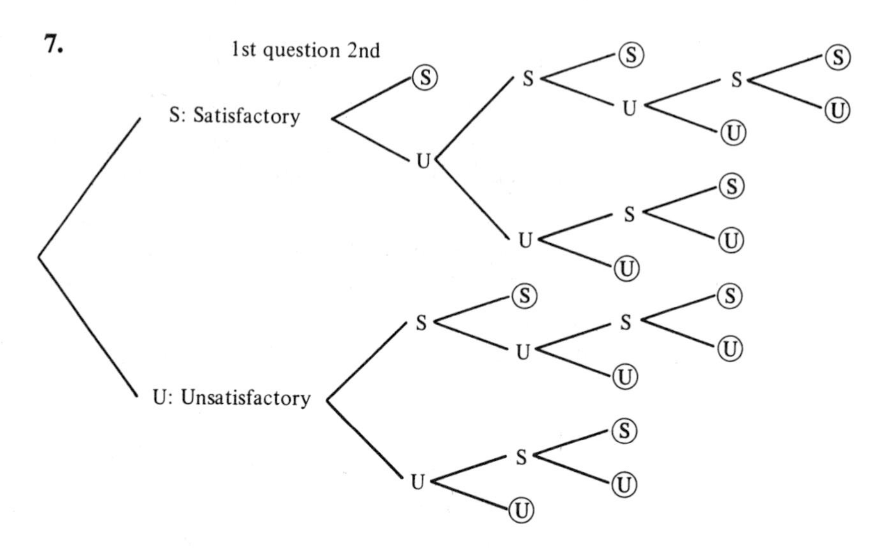

9. For example,

(a) We may let the first roll of the die correspond to Larry and the second to Matt, with the number showing on the die representing one more than the number of ducks bagged. With this association,

i) The die shows 1 on both rolls.
ii) The first roll is 3 and the second roll is 4.
iii) The sum for the two rolls is seven.
iv) The first roll shows a larger number than the second.

(b) Urns L and M each contain six marbles, numbered 0 through 5. The process is to draw one marble from each of the urns.

Exercises 1-3, page 19

1.

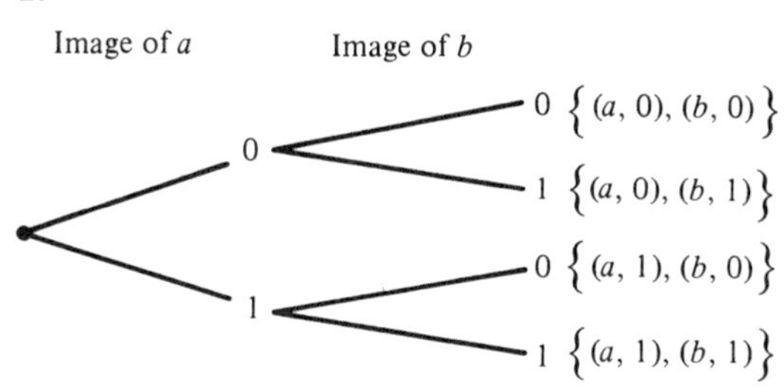

3. {(1, 1), (2, 1), (3, 1)}, {(1, 1), (2, 1), (3, 2)}, etc. In all, eight sets should be listed.

5. (a), (b).

7. 10^5, or 100,000.

9. (a) 9^3, or 729.
(b) 504.

Exercises 1-4, page 27

1. (a) 5^3, or 125. (b) $5 \times 4 \times 3$, or 60.

3.

Image of 0 — Image of 1

1 — 2, 3, 4
2 — 1, 3, 4
3 — 1, 2, 4
4 — 1, 2, 3

5. (a) 20. (b) 25. (c) 720. (d) 120. (e) 840.

7. (a) 20. (b) 300. (c) 1. (d) 210. (e) 495.

9. (a) {1, 2, 3}, {1, 2, 4}, {1, 2, 5}, {1, 3, 4}, {1, 3, 5}, {1, 4, 5}, {2, 3, 4}, {2, 3, 5}, {2, 4, 5}, {3, 4, 5}.
(b) (1, 3, 4), (1, 4, 3), (3, 1, 4), (3, 4, 1), (4, 1, 3), (4, 3, 1).

11. (a) 6840. (b) 1140.

13. (a) 2^{10}, or 1024. (b) $C(10, 7)$, or 120.

Exercises 1-5, page 34

1. (a) 7. (b) 23.

3.

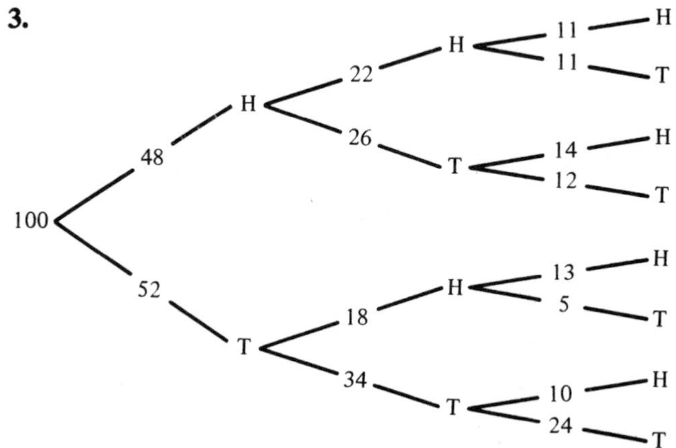

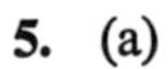
5. (a)

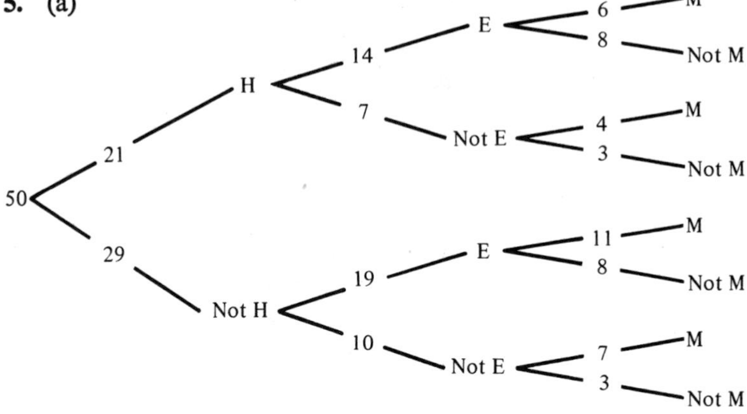

(b) 18.

7. (a)

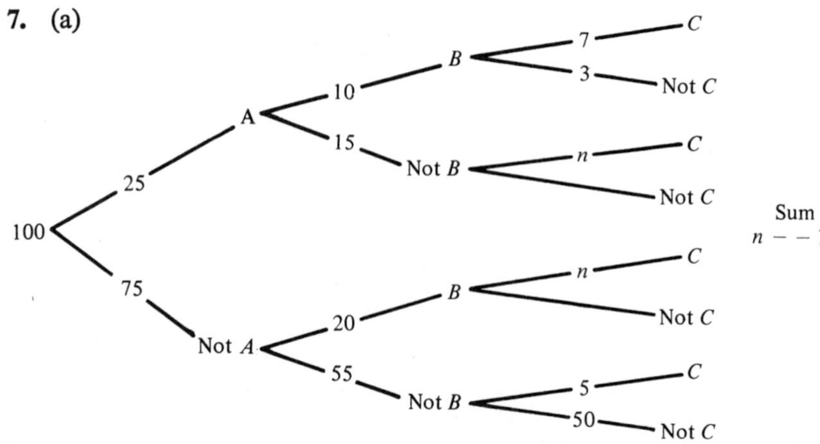

(b) 22. (c) 21. (d) 11.

Chapter 2

Exercises 2-1, page 43

1. (a)

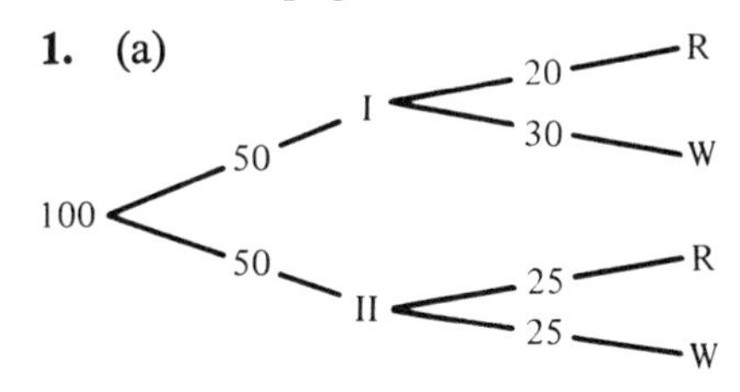

(b) $(20 + 25) \div 100 = .45$.

3.

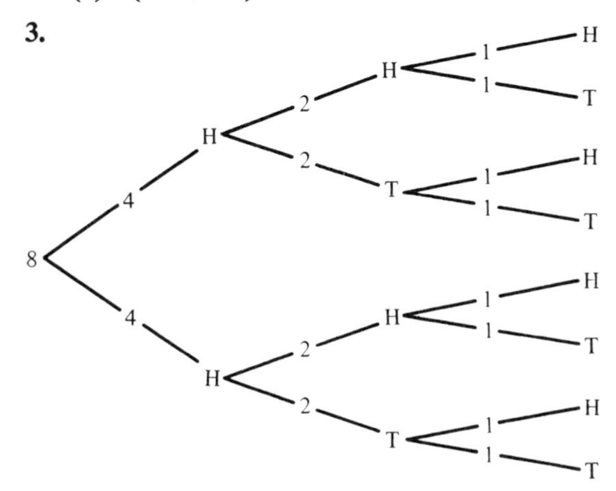

The probability that exactly two of the flips are Heads is $\frac{3}{8}$.

5. (a)

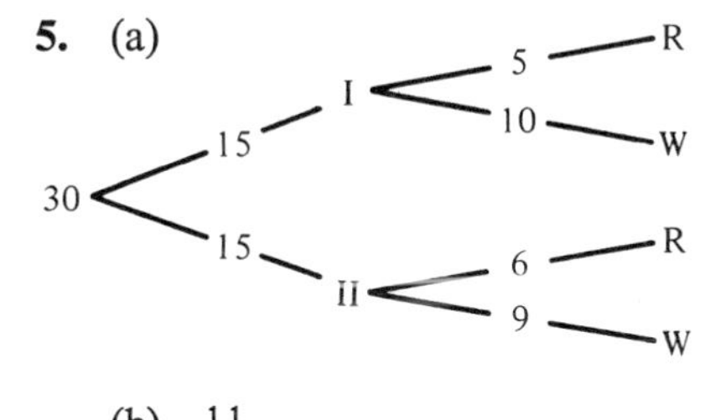

(b) $\frac{11}{30}$.

7.

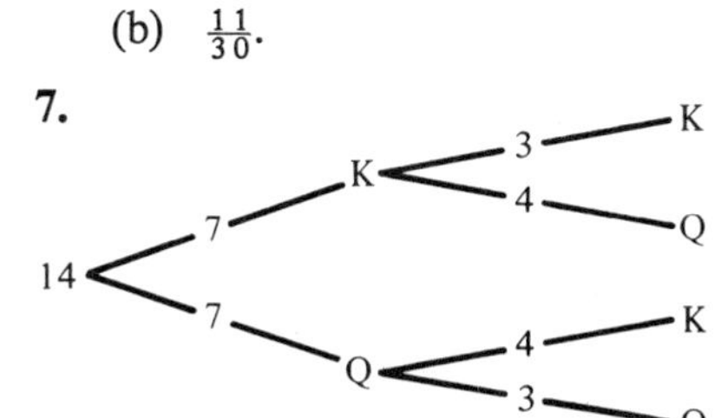

(a) $\frac{3}{14}$, (b) $\frac{8}{14}$.

9. Frequency tree:

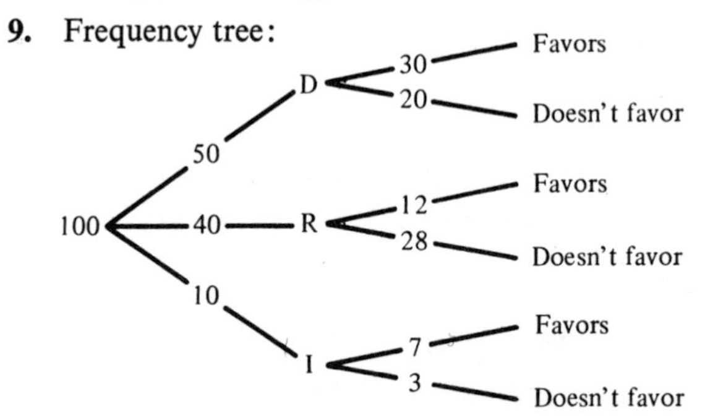

As matters stand, the probability that a randomly chosen voter favors the ordinance is $(30 + 12 + 7) \div 100 = .49$. It looks like the proponents of the ordinance had better intensify their efforts.

Exercises 2-2, page 51

1. (a)

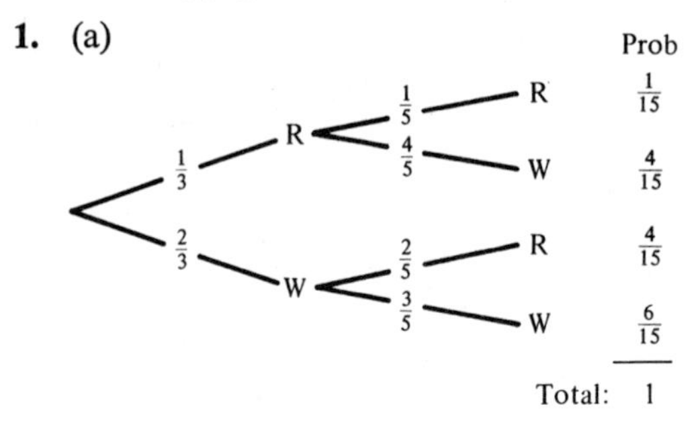

(b) $\frac{1}{15} + \frac{4}{15} = \frac{1}{3}$.

(c)

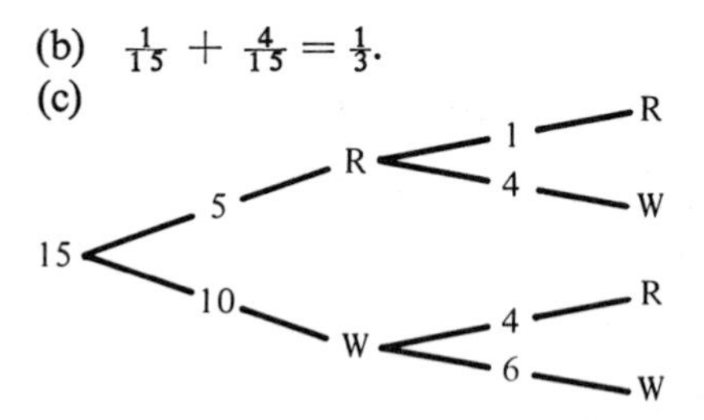

The probability that the second draw is red is $(1 + 4) \div 15$, or $\frac{1}{3}$.

3. (a)

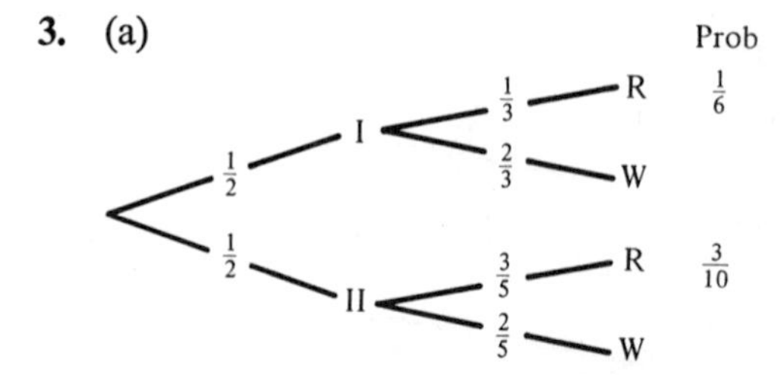

(b) $(\frac{1}{2} \times \frac{1}{3}) + (\frac{1}{2} \times \frac{3}{5}) = \frac{14}{30} = \frac{7}{15}$.

(c) For example,

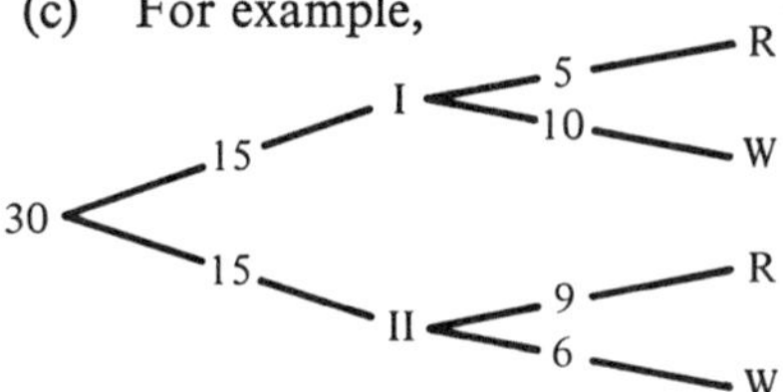

The desired probability is $(5 + 9) \div 30 = \frac{7}{15}$.

5.

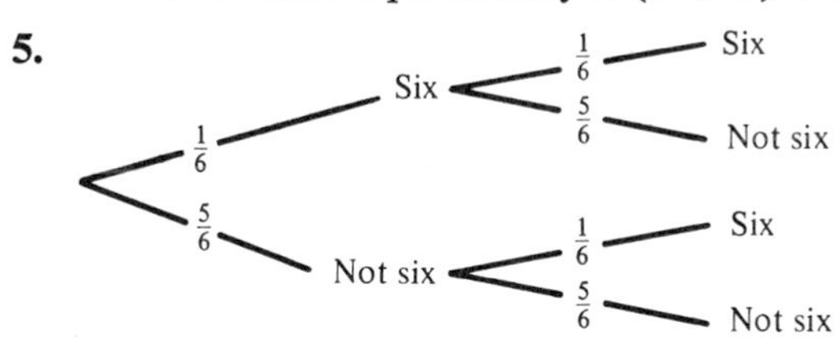

(a) $\frac{1}{36}$, (b) $\frac{25}{36}$, (c) $\frac{10}{36}$.

7. (a)

1/4 H 13/51 D 12/50 H

(b) $\frac{1}{4} \times \frac{13}{51} \times \frac{12}{50} = \frac{13}{850}$.

9. (a) $\frac{1}{4} \times \frac{12}{51} \times \frac{11}{50}$, or $\frac{11}{850}$.

(b) $\dfrac{P(13, 3)}{P(52, 3)} = \dfrac{13 \times 12 \times 11}{52 \times 51 \times 50} = \dfrac{11}{850}$.

(c) $C(13, 3) \div C(52, 3) = \left(\dfrac{13 \times 12 \times 11}{3 \times 2 \times 1}\right) \div \left(\dfrac{52 \times 51 \times 50}{3 \times 2 \times 1}\right) = \dfrac{11}{850}$.

11.

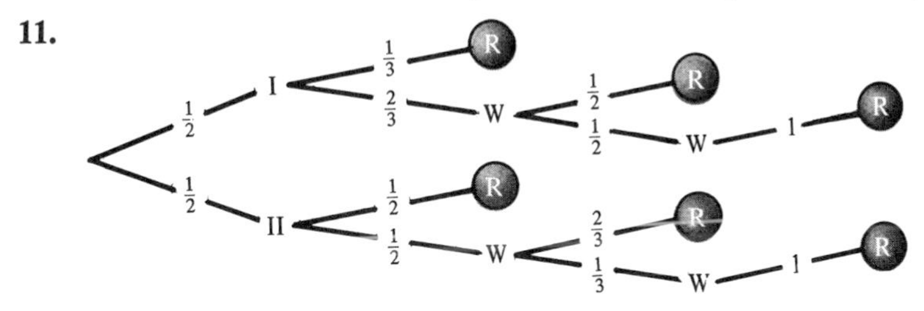

(a) $\frac{5}{12}$. (b) $\frac{1}{3}$. (c) $\frac{1}{4}$.

Exercises 2-3, page 58

1. (a) $\frac{12}{51}$. (b) $\frac{4}{51}$.

3. (a) $\frac{2}{5}$.

(b) Condition: A red marble is drawn. Outcome: Urn I is selected.

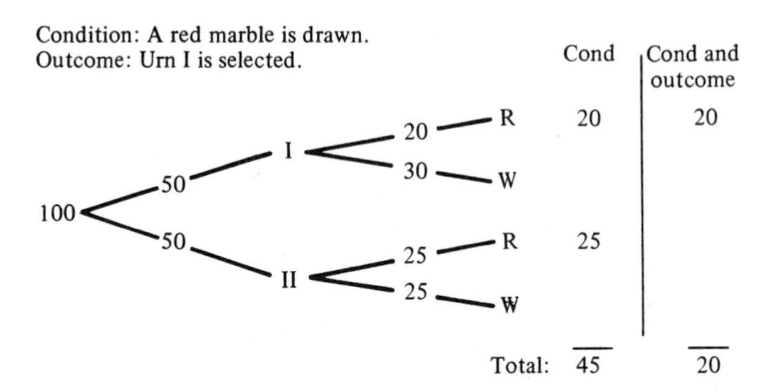

The desired conditional probability is $\frac{20}{45}$, or $\frac{4}{9}$.

(c) Condition: A red marble is drawn. Outcome: Urn I is selected.

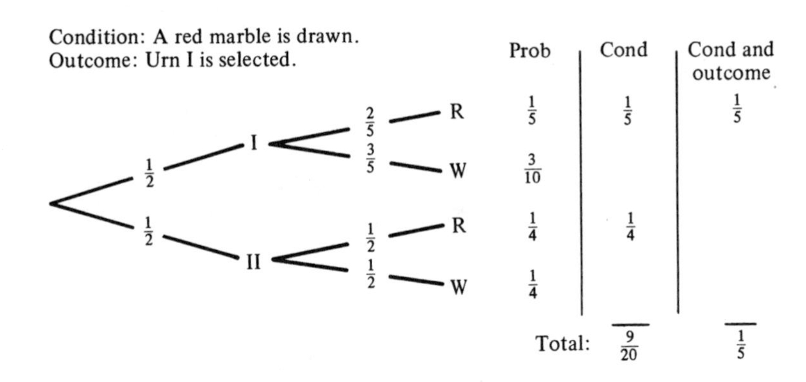

Answer: $\frac{1}{5} \div \frac{9}{20}$, or $\frac{4}{9}$.

5. (a) $\frac{5}{25}$, or $\frac{1}{5}$.

(b) Condition: The student fails. Outcome: The student takes test T_1.

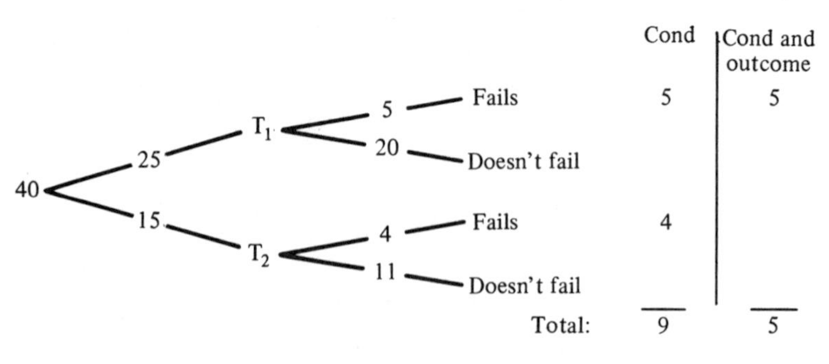

The answer is $\frac{5}{9}$.

7. (a)

First card	Branch prob	Second card	Branch prob	Prob × 33	7(b) Cond	7(b) Cond and outcome	7(c) Cond	7(c) Cond and outcome
J	$\frac{1}{3}$	J	$\frac{3}{11}$	3			3	
		Q	$\frac{4}{11}$	4			4	4
		K	$\frac{4}{11}$	4			4	4
Q	$\frac{1}{3}$	J	$\frac{4}{11}$	4	4			
		Q	$\frac{3}{11}$	3	3	3		
		K	$\frac{4}{11}$	4	4	4		
K	$\frac{1}{3}$	J	$\frac{4}{11}$	4	4		4	
		Q	$\frac{4}{11}$	4	4	4	4	4
		K	$\frac{3}{11}$	3	3	3	3	3
			Total:	33	22	14	22	15

(b) $\frac{14}{22}$. (c) $\frac{15}{22}$.

9. Condition: Favors the Ordinance. Outcome: Is an Independent.

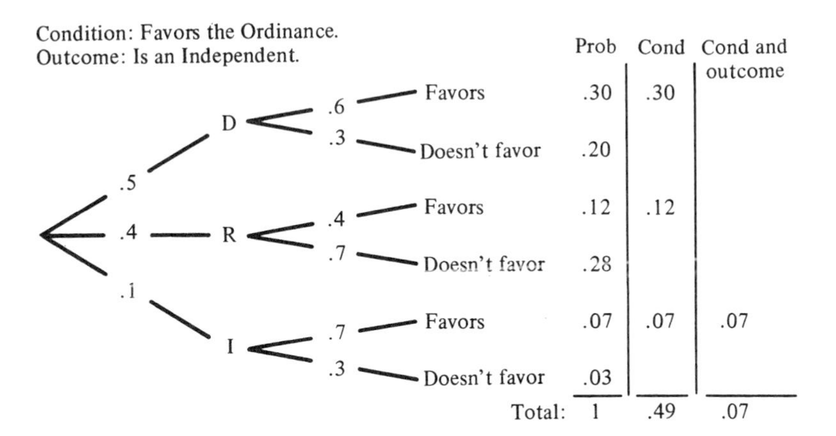

Answer: $\frac{.07}{.49} = \frac{1}{7}$.

11. (a) $\frac{10}{29}$. (b) $\frac{9}{29}$. (c) $\frac{10}{29}$.

Chapter 3

Exercises 3-1, page 67

1. $\bar{A} = \{s_3, s_4, s_7, s_8\}$, $\bar{B} = \{s_2, s_4, s_6, s_8\}$, $A \cup B = \{s_1, s_2, s_3, s_5, s_6, s_7\}$, $A \cap B = \{s_1, s_5\}$, $\bar{A} \cup C = \{s_2, s_3, s_4, s_6, s_7, s_8\}$, $A \cap \bar{C} = \{s_1, s_5\}$.

3. (a) $\Pr(A) = .6$, (b) $\Pr(\bar{A}) = .4$, (c) $\Pr(A \cap B) = .45$, (d) $\Pr(A \cup B) = .7$.

5. (a) $\Pr(A \cap B) = 0$, $\Pr(A \cup B) = 1$.
(b) $\Pr(A \cap B) = 0$, $\Pr(A \cup B) = .55$.
(c) $\Pr(A \cap B) = .6$, $\Pr(A \cup B) = 1$.

7. (a)

Probability Space $\mathfrak{M}$
Sample Set: $S = \{m_1, m_2, m_3\}$

Event	*Prob*	*Event*	*Prob*
$\varnothing$	0	$\{m_1, m_2\}$	.7
$\{m_1\}$	.2	$\{m_1, m_3\}$	.5
$\{m_2\}$	.5	$\{m_2, m_3\}$	.8
$\{m_3\}$	.3	S	1

(b) One marble is drawn at random from an urn containing 2 red, 5 white, and 3 black marbles.

9.

Probability Space $\mathfrak{A}$
Sample Set: $S = \{a_1, a_2, a_3, a_4\}$

Event	*Prob*	*Event*	*Prob*	*Event*	*Prob*
$\varnothing$	0	$\{a_1, a_2\}$	.3	$\{a_1, a_2, a_3\}$	.6
$\{a_1\}$	.2	$\{a_1, a_3\}$	.6	$\{a_1, a_2, a_4\}$	.5
$\{a_2\}$	.1	$\{a_1, a_4\}$	.5	$\{a_1, a_3, a_4\}$	.9
$\{a_3\}$	.4	$\{a_2, a_3\}$	.5	$\{a_2, a_3, a_4\}$	.8
$\{a_4\}$	.3	$\{a_2, a_4\}$	.4	S	1
		$\{a_3, a_4\}$	.7		

Exercises 3-2, page 75

1. (a) $A \cap B = \{s_2, s_4\}$,
$A \cap \bar{B} = \{s_1, s_3\}$,
$\bar{A} \cap B = \{s_6, s_8\}$,
$\bar{A} \cap \bar{B} = \{s_5, s_7\}$.

(b) $A \cap B = \{s_1, s_2\}$,
$A \cap \bar{B} = \{s_3, s_4\}$,
$\bar{A} \cap B = \varnothing$,
$\bar{A} \cap \bar{B} = \{s_5, s_6, s_7, s_8\}$.

(c) $A \cap B = \varnothing$,
$A \cap \bar{B} = \{s_1, s_2, s_3, s_4\}$,
$\bar{A} \cap B = \{s_7, s_8\}$,
$\bar{A} \cap \bar{B} = \{s_5, s_6\}$.

(d) $A \cap B = \{s_1, s_2, s_3, s_4\}$,
$A \cap \bar{B} = \varnothing$,
$\bar{A} \cap B = \{s_5, s_6, s_7, s_8\}$,
$\bar{A} \cap \bar{B} = \varnothing$.

3. Each element in A is in $A \cap B$ or in $A \cap \bar{B}$, depending on whether or not it belongs to B. Furthermore, any element in $A \cap B$ or in $A \cap \bar{B}$ is necessarily in A. Therefore,

$$A = (A \cap B) \cup (A \cap \bar{B}). \tag{1}$$

No element belongs to both B and $\bar{B}$; thus no element belongs to both $A \cap B$ and $A \cap \bar{B}$. Therefore,

$$(A \cap B) \cap (A \cap \bar{B}) = \varnothing. \tag{2}$$

Applying Axiom 3, with $X = A \cap B$ and $Y = A \cap \bar{B}$,

$$\Pr(A) = \Pr((A \cap B) \cup (A \cap \bar{B})) = \Pr(A \cap B) + \Pr(A \cap \bar{B}). \tag{3}$$

We are given $\Pr(A \cap B) = .25$ and $\Pr(A \cap \bar{B}) = .15$. Inserting these values into Eq. (3), we have $\Pr(A) = .25 + .15 = .4$, as desired.

5. Eq. (1): The elements in $A \cup B$ are precisely the elements in A together with the elements in B which do not belong to A. Eq. (2): Argument leading to line (1) in the solution of Prob. 3 above. Eq. (3): Since every element in $\bar{A} \cap B$ is in $\bar{A}$, it follows that no element in $\bar{A} \cap B$ is also in A. Eq. (4): Argument leading to line (2) in the solution of Prob. 3 above.

Proof of the Theorem: From (1) and (3) and Axiom 3:

$$\Pr(A \cup B) = \Pr(A \cup (\bar{A} \cap B)) = \Pr(A) + \Pr(\bar{A} \cap B). \tag{5}$$

From (2) and (4) and Axiom 3:

$$\Pr(B) = \Pr((A \cap B) \cup (\bar{A} \cap B)) = \Pr(A \cap B) + \Pr(\bar{A} \cap B). \tag{6}$$

From (6):

$$\Pr(\bar{A} \cap B) = \Pr(B) - \Pr(A \cap B). \tag{7}$$

Substituting (7) into (5), $\Pr(A \cup B) = \Pr(A) + \Pr(B) - \Pr(A \cap B)$, which is the desired result.

7. Since there are no elements in the empty set $\varnothing$,

$$\varnothing \cap S = \varnothing. \tag{1}$$

And clearly

$$\varnothing \cup S = S. \tag{2}$$

Applying Axiom 3:

$$\Pr(\varnothing \cup S) = \Pr(S) = \Pr(\varnothing) + \Pr(S). \tag{3}$$

Subtracting $\Pr(S)$ from both sides in (3), $0 = \Pr(\varnothing)$, as desired.

Exercises 3-3, page 83

1. (a) $\Pr(A \mid B) = \dfrac{\Pr(A \cap B)}{\Pr(B)} = \dfrac{\Pr\{h_1\}}{\Pr\{h_1, h_4\}} = \dfrac{.1}{.5} = \dfrac{1}{5}.$

(b) $\frac{2}{3}$. (c) $\frac{5}{8}$.

3. Probability tree:

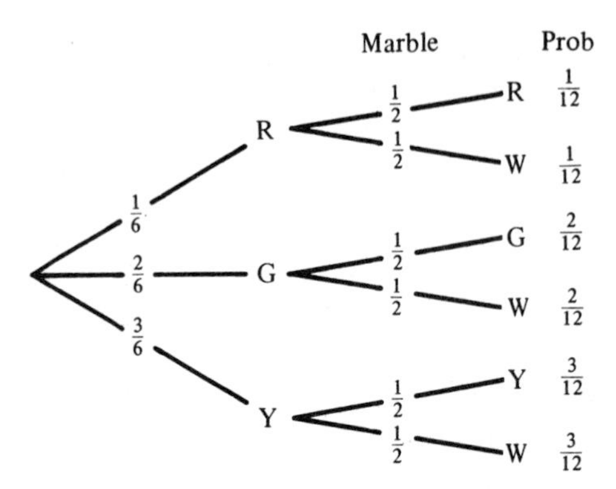

$\Pr(A) = \frac{11}{12}$, $\Pr(B) = \frac{4}{6}$, $\Pr(A \cap B) = \frac{7}{12}$. Since $\Pr(A \cap B) \neq \Pr(A) \times \Pr(B)$, it follows that events A and B are not independent. (See Prob. 11 below.)

5. Independent pairs of events are A and B, A and C, and B and C.

7. (a) $\Pr(B) = \frac{5}{12}$. (b) $\Pr(A \cup B) = \frac{8}{12}$. (c) $\Pr(A \cup \bar{B}) = \frac{2}{3}$. (d) $\Pr(A \mid B) = \frac{1}{5}$. (e) $\Pr(B \mid A) = \frac{1}{4}$.

9. (a) Let A_1 be the event "The first card is a Spade" and A_2 be the event "The second card is a Spade," and let A be the event "Both cards are Spades." Thus $A = A_1 \cap A_2$, $\Pr(A_1) = \frac{1}{4}$, and $\Pr(A_2 \mid A_1) = \frac{12}{51}$. By the reasoning of Prob. 8, $\Pr(A) = \Pr(A_1 \cap A_2) = \Pr(A_1) \times \Pr(A_2 \mid A_1) = \frac{1}{4} \times \frac{12}{51} = \frac{1}{17}$. The event of interest, "At least one of the cards is not a Spade," is the complementary event $\bar{A}$. We have $\Pr(\bar{A}) = 1 - \Pr(A) = 1 - \frac{1}{17} = \frac{16}{17}$.

(b) By the argument of part (a), with B being the event "Both cards are Kings," $\Pr(B) = \frac{1}{13} \times \frac{3}{51} = \frac{1}{221}$. The desired probability is $\Pr(\bar{B}) = 1 - \Pr(B) = \frac{220}{221}$.

11. (a) Weak probability tree, with values as given or deduced as noted:

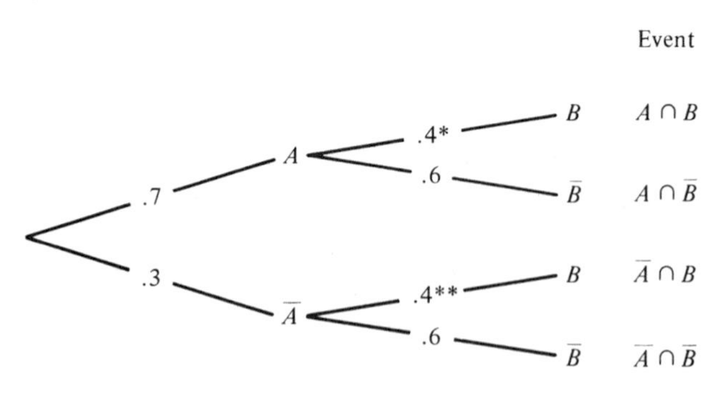

Notes: *This branch probability .4 is precisely $\Pr(B \mid A)$. **If we represent this branch probability as x, $\Pr(B) = \Pr(A \cap B) + \Pr(\bar{A} \cap B) = (.7 \times .4) + (.3 \times x) = .4$, yielding $x = .4$. But this branch probability is precisely $\Pr(B \mid \bar{A})$. Thus $\Pr(B \mid \bar{A}) = .4$. We conclude that B is also independent of $\bar{A}$.

(b) By the argument leading to line (3) in the solution of Prob. 3 of Sec. 3-2, $\Pr(B) = \Pr(A \cap B) + \Pr(\bar{A} \cap B)$. Since A and B are independent, $\Pr(B) = \Pr(A) \times \Pr(B) + \Pr(\bar{A} \cap B)$. Thus

$$\Pr(\bar{A} \cap B) = \Pr(B) - \Pr(A) \times \Pr(B)$$
$$= (1 - \Pr(A)) \times \Pr(B) = \Pr(\bar{A}) \times \Pr(B).$$

Therefore, $\bar{A}$ and B are independent. (It also follows that $\bar{A}$ and $\bar{B}$ are independent. Thus if $\bar{A}$ and $\bar{B}$ are not independent, then A and B are also not independent. In the case of Prob. 3, it is immediate that $\bar{A}$ and $\bar{B}$ are not independent.)

Exercises 3-4, page 90

1. (a) $\Pr(X = -1) = .1$, $\Pr(X = 1) = .1$, $\Pr(X = 2) = .4$, $\Pr(X = 3) = .4$.

   (b) $(-1 \times .1) + (1 \times .1) + (2 \times .4) + (3 \times .4) = 2$.

3. $(0 \times .1) + (1 \times .2) + (2 \times .3) + (3 \times .4) = 2$.

5.

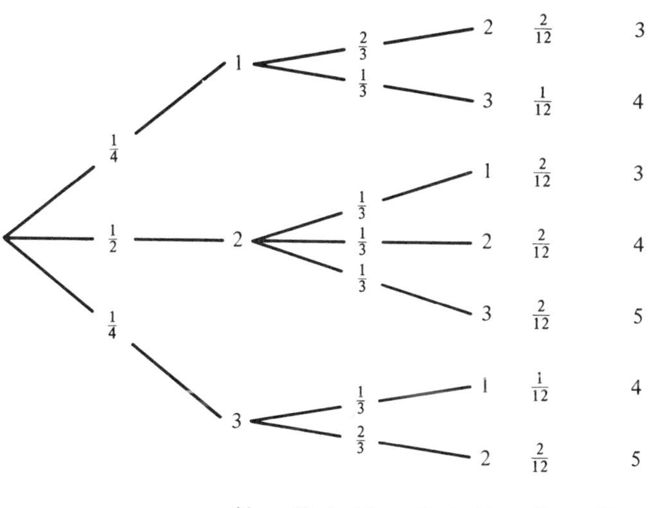

$$E(X) = \frac{(4 \times 3) + (4 \times 4) + (4 \times 5)}{12} = \frac{(4 \times 12)}{12} = 4.$$

(b) Following the procedure of part (a) with second-stage components appropriately adjusted, we find again that $E(X) = 4$.

7. (a)

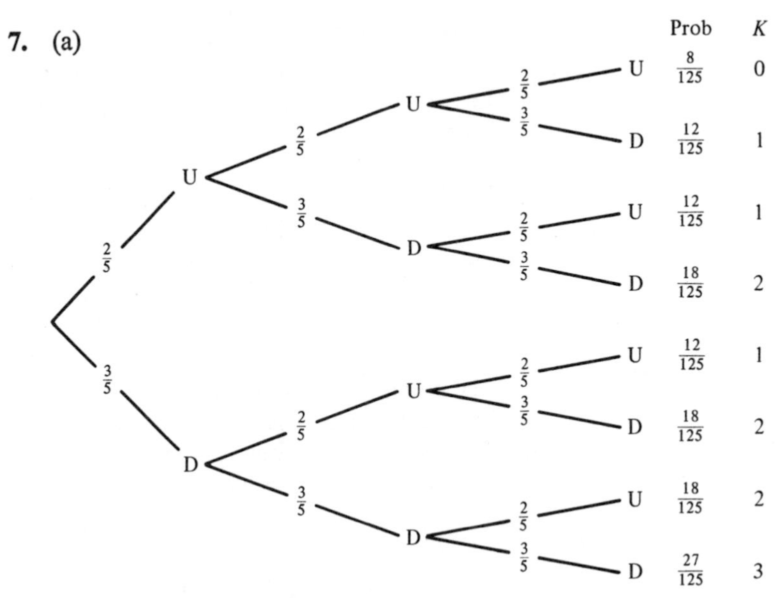

(b) $\Pr(K = 0) = \frac{8}{125}$, $\Pr(K = 1) = \frac{36}{125}$, $\Pr(K = 2) = \frac{54}{125}$, $\Pr(K = 3) = \frac{27}{125}$.
(c) $E(K) = \frac{9}{5}$.

9. Three and a half, if we assume the professor is alert enough not to return to a drawer already searched.

11. (a) 7, (b) $\dfrac{21 \times 21}{36} = \dfrac{49}{4}$.

13. (a) Zero.
(b) One-twelfth of a cent per play, regardless of whether Carl plays honestly or somehow biases his flip also.

Exercises 3-5, page 97

1. $E(X) = 2$, $E(Y) = .7$, $E(X + Y) = 2.7$.

3. (a)

Element of S	e_1	e_2	e_3	e_4	e_5	e_6
Weight	$\frac{2}{18}$	$\frac{1}{18}$	$\frac{2}{18}$	$\frac{4}{18}$	$\frac{7}{18}$	$\frac{2}{18}$
X	1	−1	2	2	−2	0
Y	−2	1	2	0	−1	−4
$2X - Y$	4	−3	2	4	−3	4
X^2	1	1	4	4	4	0
XY	−2	−1	4	0	2	0

(b) $E(2X - Y) = (4 \times \frac{8}{18}) + ((-3) \times \frac{8}{18}) + (2 \times \frac{2}{18}) = \frac{12}{18} = \frac{2}{3}$. Similarly, $E(X) = -\frac{1}{18}$ and $E(Y) = -\frac{14}{18}$. Thus $2E(X) - E(Y) = -\frac{2}{18} + \frac{14}{18} = \frac{12}{18} = \frac{2}{3}$.

(c) $E(X^2) = \frac{55}{18}$, $E(XY) = \frac{17}{18}$.

5. (a) $E(X_1) = E(X_2) = E(X_3) = 11$. (You should be able to give an argument that would lead one to anticipate that these three random variables have the same expected values.) $E(X_1 + X_2 + X_3) = 33$.

   (b) $E(X_1 + X_2 + X_3)$ is the number of cents that Henry can expect to find in his three-coin draw, on the average in the long run.

7. Let the random variables X_1, X_2, X_3, X_4 represent the shim thickness in mils for the four respective draws. It should be clear that the four random variables all have the same range and associated probabilities. We have $E(X_1) = E(X_2) = E(X_3) = E(X_4) = 12.5$. Thus $E(X_1 + X_2 + X_3 + X_4) = 4 \times 12.5 = 50$. The expected value of the thickness of the stack is 50 mils.

9. Let $\Pr(X = 0) = 1 - a$ and $\Pr(X = 1) = a$. Then $E(X) = a$. Let $\Pr(Y = 1) = b$ and $\Pr(Y = 2) = 1 - b$. Then $E(Y) = (1 \times b) + (2 \times (1 - b)) = (2 - b)$. Therefore,

$$E(X) \times E(Y) = a(2 - b). \tag{1}$$

The random variable XY has range $\{0, 1, 2\}$. Thus

$$E(XY) = (0 \times \Pr(XY = 0)) + (1 \times \Pr(XY = 1)) + (2 \times \Pr(XY = 2)). \tag{2}$$

The event $XY = 1$ is the event $(X = 1) \cap (Y = 1)$. From the assumed independence,

$$\begin{aligned}\Pr(XY = 1) = \Pr((X = 1) \cap (Y = 1)) &= \Pr(X = 1) \times \Pr(Y = 1) \\ &= ab.\end{aligned} \tag{3}$$

Also, $XY = 2$ is the event $(X = 1) \cap (Y = 2)$. Again from independence,

$$\begin{aligned}\Pr(XY = 2) = \Pr((X = 1) \cap (Y = 2)) &= \Pr(X = 1) \times \Pr(Y = 2) \\ &= a(1 - b).\end{aligned} \tag{4}$$

From Eqs. (2), (3), and (4),

$$\begin{aligned}E(XY) &= (1 \times (ab)) + (2 \times (a(1 - b))) \\ &= ab + 2a - 2ab = a(2 - b).\end{aligned} \tag{5}$$

Comparing Eqs. (1) and (5), we have the desired conclusion: $E(XY) = E(X) \times E(Y)$.

Exercises 3-6, page 105

1. Representing the random variable as H, $\mu = E(H) = 77$, $\sigma^2 = E((H - \mu)^2) = 8$, and $\sigma = \sqrt{8} = 2.83$. Thus the heights of the players have a mean of 77 inches and a standard deviation of 2.83 inches.

3. $\mu = 12$ and $\sigma = \sqrt{6} = 2.44$. The amount of money in Tom's choice has a mean of 12 cents and a standard deviation of 2.44 cents.

5. (a) $\mu = -.5$.
(b) $E(|X - \mu|) = 1.55$.
(c) $\sigma^2 = 3.025$.
(d) $\sigma = \sqrt{3.025} = 1.74$.

7. $\mu = -\frac{5}{12}$ and $\sigma^2 = \frac{4}{3} - (\frac{5}{12})^2 = \frac{167}{144}$.

9. $\mu = -.1$; i.e., on the average he loses 10 cents per play. $\sigma = \sqrt{4.99}$, or about 2.23 dollars.

11. (a) $\mu = 1$, and $\sigma = \sqrt{2}/2 = .707$.
(b) $\mu = 1$, and $\sigma = \sqrt{3}/3 = .577$.

13. $E(X^2) = 3$.

Chapter 4

Exercises 4-1, page 115

1.

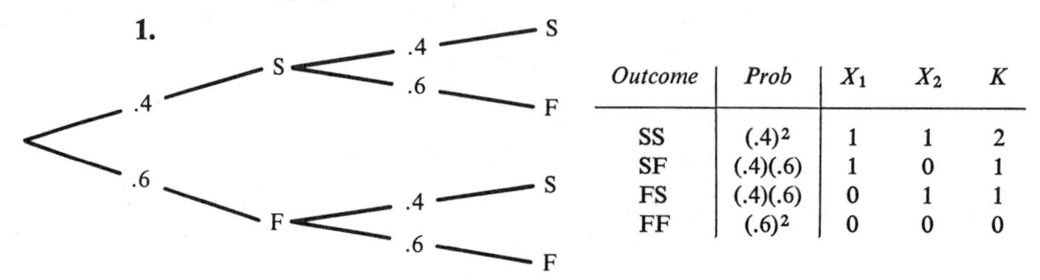

Outcome	*Prob*	X_1	X_2	K
SS	$(.4)^2$	1	1	2
SF	$(.4)(.6)$	1	0	1
FS	$(.4)(.6)$	0	1	1
FF	$(.6)^2$	0	0	0

3. (a)

Element	*Prob*	X_1	X_2	X_3	K
SSS	$(\frac{1}{4})^3$	1	1	1	3
SSF	$(\frac{1}{4})^2(\frac{3}{4})$	1	1	0	2
SFS	$(\frac{1}{4})^2(\frac{3}{4})$	1	0	1	2
SFF	$(\frac{1}{4})(\frac{3}{4})^2$	1	0	0	1
FSS	$(\frac{1}{4})^2(\frac{3}{4})$	0	1	1	2
FSF	$(\frac{1}{4})(\frac{3}{4})^2$	0	1	0	1
FFS	$(\frac{1}{4})(\frac{3}{4})^2$	0	0	1	1
FFF	$(\frac{3}{4})^3$	0	0	0	0

(b)

k	0	1	2	3
$\Pr(K = k)$	$\frac{27}{64}$	$\frac{27}{64}$	$\frac{9}{64}$	$\frac{1}{64}$

5. (a) SSSS, SSSF, SSFS, SSFF, SFSS, . . . , FFFF. (In all, 16 basic outcomes should be listed.)
(b) $\frac{1}{16}$.
(c)

k	0	1	2	3	4
$\Pr(K = k)$	$\frac{1}{16}$	$\frac{4}{16}$	$\frac{6}{16}$	$\frac{4}{16}$	$\frac{1}{16}$

7. (a) .3456. (b) .3456. (c) .5248.

9.

k	0	1	2	3	4	5	6
Prob of k Heads	$\frac{1}{64}$	$\frac{6}{64}$	$\frac{15}{64}$	$\frac{20}{64}$	$\frac{15}{64}$	$\frac{6}{64}$	$\frac{1}{64}$

Exercises 4-2, page 120

1. $\frac{40}{243}$.

3.

k	0	1	2	3	4	5
$b(k; 5, \frac{1}{3})$	$\frac{32}{243}$	$\frac{80}{243}$	$\frac{80}{243}$	$\frac{40}{243}$	$\frac{10}{243}$	$\frac{1}{243}$

5. (a) $\dfrac{20 \times 19 \times 18 \times 17}{1 \times 2 \times 3 \times 4} \times (.8)^{16} \times (.2)^4$.

(b) $\dfrac{12 \times 11 \times 10 \times 9}{1 \times 2 \times 3 \times 4} \times (.6)^8 \times (.4)^4$.

(c) $\dfrac{25 \times 24 \times 23 \times 22 \times 21}{1 \times 2 \times 3 \times 4 \times 5} \times \left(\dfrac{3}{4}\right)^{20} \times \left(\dfrac{1}{4}\right)^5$.

(d) $\dfrac{100 \times 99 \times 98 \times 97}{1 \times 2 \times 3 \times 4} \times (.95)^{96} \times (.05)^4$.

7. (a) $C(4, 2) \times (3/8)^2 \times (5/8)^2 = \frac{675}{2048}$.
(b) $C(12, 8) \times (1/2)^{12} = \frac{495}{4096}$.

9. (a) $(.99)^{100}$.
(b) $100 \times (.01) \times (.99)^{99} = (.99)^{99}$.
(c) $\dfrac{100 \times 99}{2} \times (.01)^2 \times (.99)^{98} = \dfrac{1}{2} \times (.99)^{99}$.
(d) $1 - ((.99)^{100} + (.99)^{99} + \frac{1}{2} \times (.99)^{99})$.

11. One.

13. $\Pr(K = 8)$ is larger. [Set up the ratio of $\Pr(K = 8)$ to $\Pr(K = 7)$, and divide out common factors.]

Exercises 4-3, page 127

1. (a)

k	0	1	2	3	4
$\Pr(K = k)$	$\frac{1}{16}$	$\frac{4}{16}$	$\frac{6}{16}$	$\frac{4}{16}$	$\frac{1}{16}$

(b) $\mu = E(K) = (0 \times \frac{1}{16}) + (1 \times \frac{4}{16}) + (2 \times \frac{6}{16}) + (3 \times \frac{4}{16}) + (4 \times \frac{1}{16})$ $= 2$; $\sigma^2 = E(K - \mu)^2 = (0 \times \frac{6}{16}) + (1 \times \frac{8}{16}) + (4 \times \frac{2}{16}) = 1$.
(c) $\mu = np = 4 \times \frac{1}{2} = 2$; $\sigma^2 = npq = 4 \times \frac{1}{2} \times \frac{1}{2} = 1$.

3. (a) $\mu = 60$, $\sigma^2 = 36$, $\sigma = 6$.
(b) $\mu = 90$, $\sigma^2 = 36$, $\sigma = 6$.
(c) $\mu = 320$, $\sigma^2 = 64$, $\sigma = 8$.
(d) $\mu = 100$, $\sigma^2 = 90$, $\sigma = \sqrt{90} = 9.487$.

5. $\mu = 300$ and $\sigma = 10$.

7. $\mu = 90$ and $\sigma = 3$.

9. $\mu = 36$ and $\sigma = 6.00$.

Exercises 4-4, page 135

1. (a) $\mu = 80$ and $\sigma = 4$.
(b) (i) $x = 1.5$, (ii) $x = -1$.

3. (a) $k = 60$. (b) $k = 66$. (c) $k = 48$.

5. (a) .10. (b) .02. (c) .06.

7. (a) .06. (b) .07. (c) .01.

9. (a) .13. (b) .13. (c) .13. (d) .11.

Exercises 4-5, page 142

1. The approximation technique is justified for parts (b), (c), and (d).

3. (a) .83. (b) .08. (c) .10. (d) .92.

5. .14.

7. .18 (approximately, of course).

9. .31.

11. .02

13. (a) .34. (b) .69.

15. .27.

Exercises 4-6, page 151

1. (a) $\mu_i = 2$, $\sigma_i^2 = \frac{2}{3}$.
(b) $\mu = 300$, $\sigma = 10$.
(c) .69.

3. (a) $\mu = 210$, $\sigma = 7$.
(b) .08.

5. .85.

7. .73.

9. (a) Exactly .1375.
   (b) Approximately .34.
   (c) Approximately .02.

11. k greater than 33.

Exercises 4-7, page 155

1. 46%.

3. (a) Less than 1%. (b) 22%. (c) 62%. (d) 15%.

5. .63.

7. (a) .49. (b) .49.

Chapter 5

Exercises 5-1, 164

1. (a) and (b) Straightforward normal curve approximations.
   (c) The indication is that there are two red marbles in the urn, because the experience would be so highly unlikely if there were only one red marble.

3. (a) $k_0 = 60$. It is a description of a hypothesis test consisting of a binomial process of $n = 100$ trials with unknown probability p of success on each trial. The null hypothesis is $p \leq .5$, which hypothesis is rejected if the critical outcome $K \geq k_0 = 60$ occurs, i.e., if 60 or more successes occur. The level of significance is .025. On one line. $n = 100$, $H_0: p \leq .5$, $C: K \geq 60$, Sig: .025. With the obvious adjustments, this same description applies to the remaining parts: (b) $k_0 = 334$. (c) $k_0 = 336$. (d) $k_0 = 261$. (e) $k_0 = 269$. (f) $k_0 = 919$.

5. (a) $C: K \geq 372$. (b) $C: K \geq 375$. (c) $C: K \geq 379$.

7. Swift says yes, Wilson says no.

9. At 9: 36 p.m.

Exercises 5-2, page 171

1. $.65 < p < .75$.

3. $.53 < p < .60$.

5. (a) $.28 < p < .32$.
   (b) $.48 < p < .52$.
   (c) $.69 < p < .73$.

7. (a) Yes. (b) Yes. (c) No. (d) No.

9. n between 1102 and 1238, inclusive.

Exercises 5-3, page 178

1. (a) No.
(b) No. But a Type II error is committed.
(c) Yes.

3. If $p \leq .5$, the x value for $k = 220$ is 2 or greater, so $\Pr(K \geq 220) < .05$. Similarly, if $p > .6$, the x value for $k = 220$ is -2.04 or less, so $\Pr(K < 220) < .05$.

5. Any value k_0 between 340 and 345, inclusive.

7. No. Statistical inferences are not to be drawn from curiosities observed after the fact.

9. (a) $\frac{1}{10}$. (b) $\frac{7}{8}$.

Chapter 6

Exercises 6-1, page 188

1. (a) $\begin{bmatrix} -2 & 3 & -4 \\ 3 & -4 & 5 \\ -4 & 5 & -6 \end{bmatrix}$. (b) $\begin{bmatrix} 2 & -3 & 4 \\ -3 & 4 & -5 \\ 4 & -5 & 6 \end{bmatrix}$.

3. (a) $[1.2 \quad -.2]$. (b) $\begin{bmatrix} .4 \\ .2 \end{bmatrix}$.

5. (a) (a_1) $[.5 \quad .5 \quad -.5]$.
(a_2) $[0 \quad .6 \quad .2]$.
(a_3) $[\frac{1}{12} \quad \frac{7}{12} \quad \frac{1}{12}]$.
(b) (b_1) $\begin{bmatrix} -\frac{1}{3} \\ \frac{2}{3} \end{bmatrix}$. ($b_2$) $\begin{bmatrix} -.1 \\ .3 \end{bmatrix}$. ($b_3$) $\begin{bmatrix} \frac{1}{12} \\ \frac{1}{12} \end{bmatrix}$.

7. (a) $\begin{bmatrix} 0 \\ -\frac{1}{4} \end{bmatrix}$. (b) For example, $\begin{bmatrix} p \\ q \end{bmatrix} = \begin{bmatrix} .74 \\ .26 \end{bmatrix}$.

9. (a) Payoff matrix with Mike as R and Matt as C:

$$\begin{bmatrix} -1 & 2 & -6 \\ 2 & -1 & 2 \\ -6 & 2 & -1 \end{bmatrix}.$$

(b) Row strategy: $[\frac{1}{6} \quad \frac{4}{6} \quad \frac{1}{6}]$. Expectation: $[\frac{1}{6} \quad 0 \quad \frac{1}{6}]$. (Strategy $[.17 \quad .66 \quad .17]$ shows that the game favors the row player, Mike.)

Exercises 6-2, page 196

1. (a) (a_1) -1.2, (a_2) $-.5$, (a_3) $-.6$.
(b) The strategy in (a_2) is the most attractive of the three.

3. (a) (a_1) $-.2$, (a_2) 0.
(b) (b_1) $\frac{1}{3}$, (b_2) $.1$.
(c) (c_1) No, because row strategy (a_2) shows that no column expectation can have a maximum less than 0. (*C* can't win more than *R* can keep him from winning.) (c_2) Yes; strategy [0 .59 .41] does the trick. (c_3) No, because column strategy (b_2) shows that no row expectation can have a minimum greater than .1.

5. For strategy [.3 0 .7], the expectation is [1.5 .5 .5], which has minimum .5. For strategy $\begin{bmatrix} 0 \\ .5 \\ .5 \end{bmatrix}$, the expectation is $\begin{bmatrix} .5 \\ -.5 \\ .5 \end{bmatrix}$, which has maximum .5. Since the respective minimum and maximum are equal, the strategies are optimal. The value of the game is .5.

7. Row expectation is $[-\frac{2}{11} \quad -\frac{2}{11} \quad \frac{10}{11}]$, and column expectation is $\begin{bmatrix} -\frac{3}{11} \\ -\frac{2}{11} \\ -\frac{2}{11} \end{bmatrix}$. The strategies are optimal since the minimum in the row expectation, $-\frac{2}{11}$, is equal to the maximum in the column expectation.

9. Strategy $[\frac{6}{14} \quad \frac{5}{14} \quad \frac{3}{14}]$ produces expectation [1 .5 .5 .5]. The minimum is .5. As shown in the text, .5 is also the maximum in the expectation produced by the column strategy given in the problem.

Exercises 6-3, page 203

1. (a) Strictly determined. Player *R* always plays II because the entries are better in both columns. Player *C* knows this, and therefore always plays I. Optimal strategies are [0 1] and $\begin{bmatrix} 1 \\ 0 \end{bmatrix}$ with expectations [−2 1] and $\begin{bmatrix} -3 \\ -2 \end{bmatrix}$. Respective minimum and maximum are both −2, the value of the game.
(b) Not strictly determined.
(c) Strictly determined. Player *C* always plays II because the entries are better in both rows. Knowing this, *R* always plays I. Optimal strategies are [1 0] and $\begin{bmatrix} 0 \\ 1 \end{bmatrix}$. Minimum and maximum in the respective expectations are both 0, the value of the game.
(d) Strictly determined. *R* always plays I because both entries are better, and *C* always plays I because both entries are better. Optimal strategies are [1 0] and $\begin{bmatrix} 1 \\ 0 \end{bmatrix}$. Minimum and maximum in the respective expectations are both 0, the value of the game.

3. (a) Strictly determined. Optimal strategies are [1 0] and $\begin{bmatrix} 0 \\ 0 \\ 1 \\ 0 \end{bmatrix}$.

Verification: Strategy [1 0] produces expectation [−1 2 −1 0], and strategy $\begin{bmatrix}0\\0\\1\\0\end{bmatrix}$ produces expectation $\begin{bmatrix}-1\\-2\end{bmatrix}$. The strategies are optimal since the minimum in the row expectation is −1, the same as the maximum in the column expectation. The value of the game is −1.

(b) Not strictly determined.

(c) Strictly determined. Optimal strategies are [0 1 0] and $\begin{bmatrix}0\\0\\0\\1\end{bmatrix}$.

Verification: The expectations produced by these strategies are [7 9 8 7] and $\begin{bmatrix}6\\7\\7\end{bmatrix}$. The respective minimum and maximum are both 7, the value of the game.

5. All column strategies of the form $\begin{bmatrix}0\\p\\0\\q\end{bmatrix}$, e.g., $\begin{bmatrix}0\\1\\0\\0\end{bmatrix}$, $\begin{bmatrix}0\\0\\0\\1\end{bmatrix}$, $\begin{bmatrix}0\\.5\\0\\.5\end{bmatrix}$, are optimal. The optimal row strategy is [0 1 0], producing expectation [1 1 2 1] with minimum 1. All column strategies of the above form produce expectations with maximum also equal to 1, the value of the game.

Exercises 6-4, page 212

1. (a) [8]. (b) [8 0]. (c) $\begin{bmatrix}8\\4\end{bmatrix}$.

3. (a)

$$AB = \begin{bmatrix}15 & -2 & 13\\4 & 0 & 4\\2 & 4 & 6\end{bmatrix},$$

$$BA = \begin{bmatrix}11 & 6\\9 & 10\end{bmatrix}.$$

(b)

$$AB = \begin{bmatrix}6 & -8\\3 & -1\\-12 & 7\end{bmatrix}.$$

BA is not defined.

5. (a) (a_1) $\mathbf{r}G = [-4 \quad 6 \quad 0]$.
(a_2) $\mathbf{r}G = [.5 \quad -.5 \quad -.5]$.
(a_3) $\mathbf{r}G = [\frac{1}{11} \quad \frac{1}{11} \quad -\frac{5}{11}]$.

(b) (b_1) $G\mathbf{c} = \begin{bmatrix}6\\-7\end{bmatrix}$, ($b_2$) $G\mathbf{c} = \begin{bmatrix}-.2\\.2\end{bmatrix}$, ($b_3$) $G\mathbf{c} = \begin{bmatrix}\frac{1}{11}\\\frac{1}{11}\end{bmatrix}$.

7. $$(AB)C = A(BC) = \begin{bmatrix} -6 & 11 \\ 4 & -12 \end{bmatrix}.$$

Exercises 6-5, page 221

1. (a) $$\mathbf{r}G = [-.1 \quad .9], \ G\mathbf{c} = \begin{bmatrix} .8 \\ -.2 \end{bmatrix}.$$

(b)

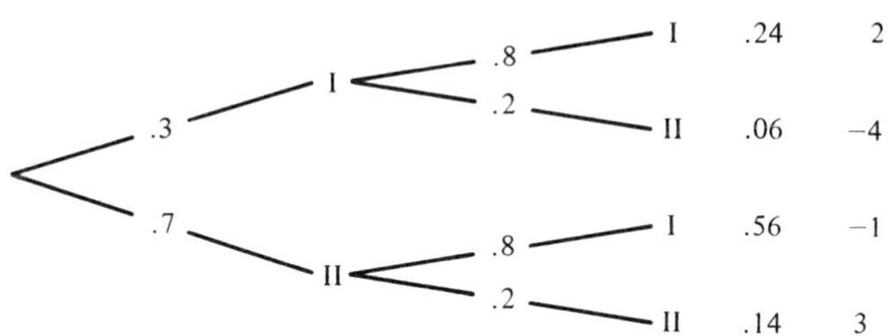

(c) $v = (2 \times .24) + ((-4) \times .06) + ((-1) \times .56) + (3 \times .14) = .1.$

(d) $\min \mathbf{r}G = -.1$ and $\max G\mathbf{c} = .8$.

3. (a) $\min \mathbf{r}_1 G = -.6$, $\min \mathbf{r}_2 G = -.5$, $\min \mathbf{r}_3 G = -1.4$. $\mathbf{r}_2$ is the best of the three.

(b) $\max G\mathbf{c}_1 = \frac{4}{3}$, $\max G\mathbf{c}_2 = .2$, $\max G\mathbf{c}_3 = .5$. $\mathbf{c}_2$ is the best of the three.

5. $\mathbf{r}_0 G = [-.1 \quad -.1]$. $G\mathbf{c}_0 = \begin{bmatrix} -.1 \\ -.1 \\ -.3 \end{bmatrix}$, $\min \mathbf{r}_0 G = \max G\mathbf{c}_0 = -.1$. The value of the game is $-.1$.

7. It is given that

$$\min \mathbf{r}_0 G = \max G\mathbf{c}_0. \tag{1}$$

Therefore, it follows from the duality theorem that for row strategy $\mathbf{r}_0'$,

$$\min \mathbf{r}_0' G \leq \min \mathbf{r}_0 G, \tag{2}$$

and for column strategy $\mathbf{c}_0'$,

$$\max G\mathbf{c}_0' \geq \max G\mathbf{c}_0. \tag{3}$$

Also, it is given that

$$\min \mathbf{r}_0' G = \max G\mathbf{c}_0'. \tag{4}$$

Therefore, it follows from the duality theorem that for row strategy $\mathbf{r}_0$,

$$\min \mathbf{r}_0 G \leq \min \mathbf{r}_0' G, \tag{5}$$

and for column strategy $\mathbf{c}_0$,

$$\max G\mathbf{c}_0 \geq \max G\mathbf{c}_0'. \tag{6}$$

But if x and y are numbers such that $x \le y$ and $y \le x$, then $x = y$. Therefore, from (2) and (5),

$$\min \mathbf{r}_0' G = \min \mathbf{r}_0 G, \tag{7}$$

and from (3) and (6),

$$\max G\mathbf{c}_0' = \max G\mathbf{c}_0. \tag{8}$$

Finally, from Eqs. (1), (4), (7), and (8), $\min \mathbf{r}_0 G = \min \mathbf{r}_0' G = \max G\mathbf{c}_0' = \max G\mathbf{c}_0$, which is the desired result.

Exercises 6-6, page 229

1. $\mathbf{r}_0 G = [0 \quad 0 \quad 0]$, $G\mathbf{c}_0 = \begin{bmatrix} 0 \\ 0 \\ 0 \end{bmatrix}$, $\min \mathbf{r}_0 G = \max G\mathbf{c}_0 = 0$. Therefore, $\mathbf{r}_0$ and $\mathbf{c}_0$ are dual optimal strategies; the value of the game is $v_G = 0$.

3. (a) Optimal strategies are $\mathbf{r}_0 = [.6 \quad .4]$, $\mathbf{c}_0 = \begin{bmatrix} .7 \\ .3 \end{bmatrix}$.

Verification: $\mathbf{r}_0 G = [-.2 \quad -.2]$, $G\mathbf{c}_0 = \begin{bmatrix} -.2 \\ -.2 \end{bmatrix}$, $\min \mathbf{r}_0 G = \max G\mathbf{c}_0 = -.2$. Therefore, $\mathbf{r}_0$ and $\mathbf{c}_0$ are dual optimal strategies; the value of the game is $v_G = -.2$. *Note:* The above format adapts to the remaining exercises.

(b) $\mathbf{r}_0 = [.7 \quad .3]$, $\mathbf{c}_0 = \begin{bmatrix} \frac{8}{15} \\ \frac{7}{15} \end{bmatrix}$, $\min \mathbf{r}_0 G = \max G\mathbf{c}_0 = -.2 = v_G$.

(c) $\mathbf{r}_0 = [\frac{1}{4} \quad \frac{3}{4}]$, $\mathbf{c}_0 = \begin{bmatrix} \frac{3}{4} \\ \frac{1}{4} \end{bmatrix}$, $\min \mathbf{r}_0 G = \max G\mathbf{c}_0 = .5 = v_G$.

(d) $\mathbf{r}_0 = [1 \quad 0]$, $\mathbf{c}_0 = \begin{bmatrix} 1 \\ 0 \end{bmatrix}$, $\min \mathbf{r}_0 G = \max G\mathbf{c}_0 = 0 = v_G$.

5. Subtracting $-.5$ from each entry in G_U, $G_F = \begin{bmatrix} 1.5 & -5.5 \\ -4.5 & 16.5 \end{bmatrix}$. Optimal strategies are $\mathbf{r}_0 = [\frac{3}{4} \quad \frac{1}{4}]$ and $\mathbf{c}_0 = \begin{bmatrix} \frac{11}{14} \\ \frac{3}{14} \end{bmatrix}$; $\min \mathbf{r}_0 G_F = \max G_F\mathbf{c}_0 = 0$. Therefore, G_F is a fair game.

7. (a) Optimal strategies: $\mathbf{r}_0 = [0 \quad 1 \quad 0]$, $\mathbf{c}_0 = \begin{bmatrix} 0 \\ 0 \\ 1 \end{bmatrix}$; $\min \mathbf{r}_0 G = \max G\mathbf{c}_0 = 0 = v_G$.

(b) (Player C never plays II because III is better in every entry, and ignoring Column II, player R never plays Row II because I is better in every entry.) Optimal strategies: $\mathbf{r}_0 = [\frac{5}{8} \quad 0 \quad \frac{3}{8}]$, $\mathbf{c}_0 = \begin{bmatrix} \frac{3}{8} \\ 0 \\ \frac{5}{8} \end{bmatrix}$; $\min \mathbf{r}_0 G = \max G\mathbf{c}_0 = \frac{1}{8} = v_G$.

9. Letting $T = a + b + c + d$, optimal strategies are $\mathbf{r}_0 = \left[\frac{c+d}{T}\right.$

$$\left.\frac{a+b}{T}\right], \quad \mathbf{c}_0 = \begin{bmatrix} \dfrac{b+d}{T} \\ \dfrac{a+c}{T} \end{bmatrix}.$$

Verification: $\mathbf{r}_0 G = \begin{bmatrix} \dfrac{a(c+d)-c(a+b)}{T} & \dfrac{-b(c+d)+d(a+b)}{T} \end{bmatrix}$

$$= \begin{bmatrix} \dfrac{ad-bc}{T} & \dfrac{ad-bc}{T} \end{bmatrix}. \quad G\mathbf{c}_0 = \begin{bmatrix} \dfrac{a(b+d)-b(a+c)}{T} \\ \dfrac{-c(b+d)+d(a+c)}{T} \end{bmatrix} = \begin{bmatrix} \dfrac{ad-bc}{T} \\ \dfrac{ad-bc}{T} \end{bmatrix}.$$

$\min \mathbf{r}_0 G = \max G\mathbf{c}_0 = \dfrac{ad-bc}{T}$. Thus the value of the game is $\dfrac{ad-bc}{T}$. This value is zero if and only if $ad - bc = 0$, or, equivalently, $a/c = b/d$.

Exercises 6-7, page 239

1. (a) $\begin{bmatrix} -1 & -3 & 10 \\ 5 & 3 & -5 \end{bmatrix}$.

(b) $\begin{bmatrix} -3 & 3 & -4 \\ 5 & -5 & -3 \end{bmatrix}$.

(c) $\begin{bmatrix} -1 & -9 & 27 \\ 10 & 10 & -11 \end{bmatrix}$.

3. $AB = \begin{bmatrix} -1 & 5 \\ 0 & 6 \end{bmatrix}$, $AC = \begin{bmatrix} 1 & 4 \\ 2 & 6 \end{bmatrix}$, $A(B+C) = \begin{bmatrix} 0 & 9 \\ 2 & 12 \end{bmatrix} = AB + AC$.

5. $(AB)C = A(BC) = \begin{bmatrix} 3 & 9 \\ 7 & 15 \end{bmatrix}$.

7. (a) $A_1BC_1 = [3]$, $A_1BC_2 = [4]$, $A_2BC_1 = [73]$, $A_2BC_2 = [25]$.

(b) $ABC = \begin{bmatrix} 3 & 4 \\ 73 & 25 \end{bmatrix}$.

9. (a) $AB = \begin{bmatrix} 5 & 2 \\ -6 & 3 \end{bmatrix}$.

(b) $[3 \quad -1] + [2 \quad 4] + [0 \quad -1] = [5 \quad 2]$.

(c) $-2[3 \quad -1] + 0[1 \quad 2] + 1[0 \quad 1] = [-6 \quad 2] + [0 \quad 1] = [-6 \quad 3]$.

(d) $\begin{bmatrix} 3 \\ -6 \end{bmatrix} + \begin{bmatrix} 2 \\ 0 \end{bmatrix} = \begin{bmatrix} 5 \\ -6 \end{bmatrix}$.

(e) $-1\begin{bmatrix} 1 \\ -2 \end{bmatrix} + 2\begin{bmatrix} 2 \\ 0 \end{bmatrix} + 1\begin{bmatrix} -1 \\ 1 \end{bmatrix} = \begin{bmatrix} -1 \\ 2 \end{bmatrix} + \begin{bmatrix} 4 \\ 0 \end{bmatrix} + \begin{bmatrix} -1 \\ 1 \end{bmatrix} = \begin{bmatrix} 2 \\ 3 \end{bmatrix}$.

11. (a) $\mathbf{r}G = [(r_1g_{11} + r_2g_{21}) \quad (r_1g_{12} + r_2g_{22}) \quad (r_1g_{13} + r_2g_{23})] = [r_1g_{11} \quad r_1g_{12} \quad r_1g_{13}] + [r_2g_{21} \quad r_2g_{22} \quad r_2g_{23}] = r_1[g_{11} \quad g_{12} \quad g_{13}] + r_2[g_{21} \quad g_{22} \quad g_{23}]$.

(b) $$G\mathbf{c} = \begin{bmatrix} g_{11}c_1 + g_{12}c_2 + g_{13}c_3 \\ g_{21}c_1 + g_{22}c_2 + g_{23}c_3 \end{bmatrix} = \begin{bmatrix} c_1 g_{11} \\ c_1 g_{21} \end{bmatrix} + \begin{bmatrix} c_2 g_{12} \\ c_2 g_{22} \end{bmatrix} + \begin{bmatrix} c_3 g_{13} \\ c_3 g_{23} \end{bmatrix}$$
$$= c_1\begin{bmatrix} g_{11} \\ g_{21} \end{bmatrix} + c_2\begin{bmatrix} g_{12} \\ g_{22} \end{bmatrix} + c_3\begin{bmatrix} g_{13} \\ g_{23} \end{bmatrix}.$$

Chapter 7

Exercises 7-1, page 247

1. (a) $(-1, 2)$. (b) $(-6, 8.8)$. (c) $(-1, -31)$.

3.

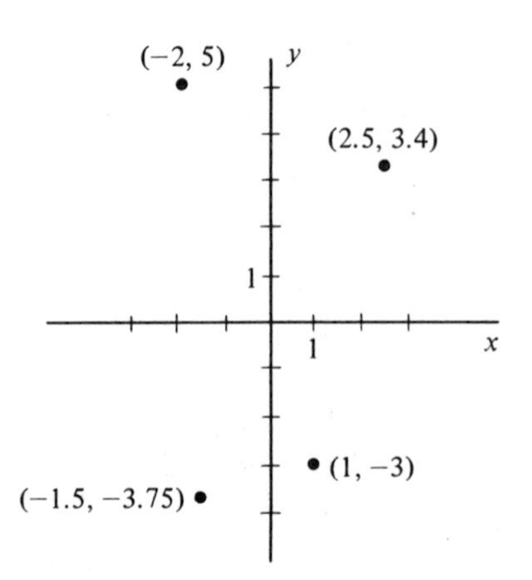

5. (a)–(c)

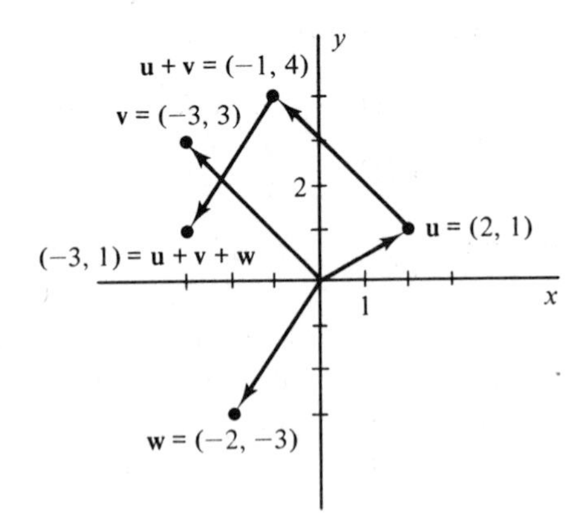

(d) $\mathbf{u} + \mathbf{v} + \mathbf{w}$ is the point reached by beginning at the origin and successively traveling along the rays representing $\mathbf{u}$, $\mathbf{v}$, and $\mathbf{w}$. (The language here is to be interpreted as presuming that the "$\mathbf{v}$ ray" and "$\mathbf{w}$ ray" have been displaced in the obvious manner so that the trip involves no jumping around.)

7.

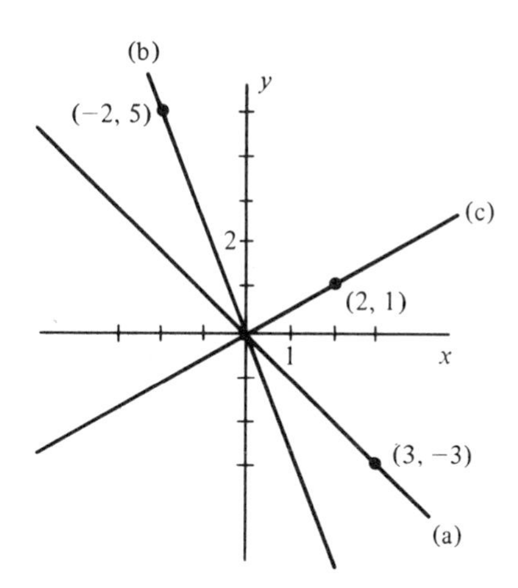

Exercises 7-2, page 256

1.

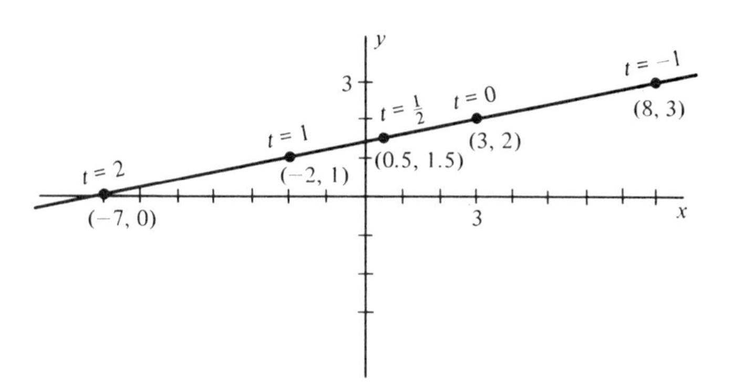

3. (a)

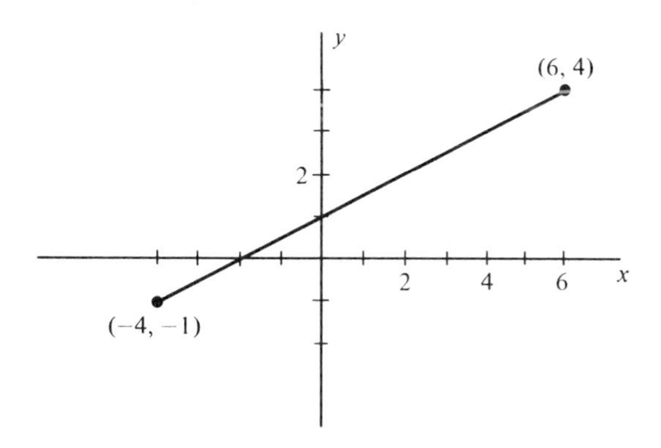

(b) $c_1 = \frac{2}{5}, c_2 = \frac{3}{5}$.

5.

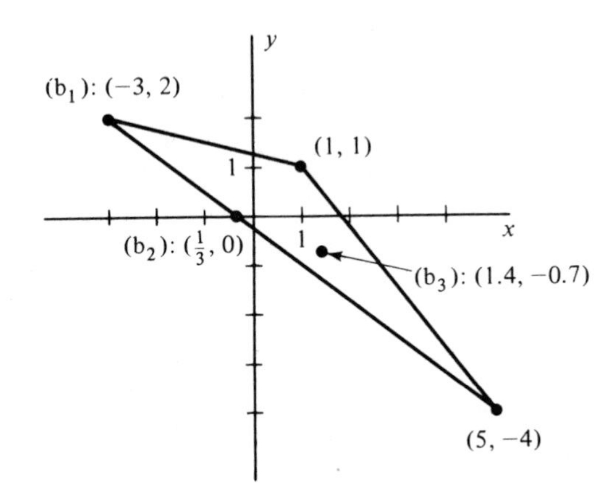
y
(b1): (−3, 2)
(1, 1)
1
x
1
(b2): (1/3, 0)
(b3): (1.4, −0.7)
(5, −4)

7.

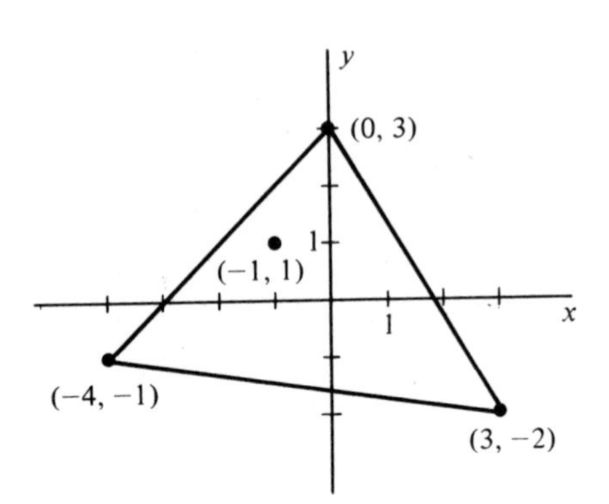
y
(0, 3)
1
(−1, 1)
x
1
(−4, −1)
(3, −2)

9.

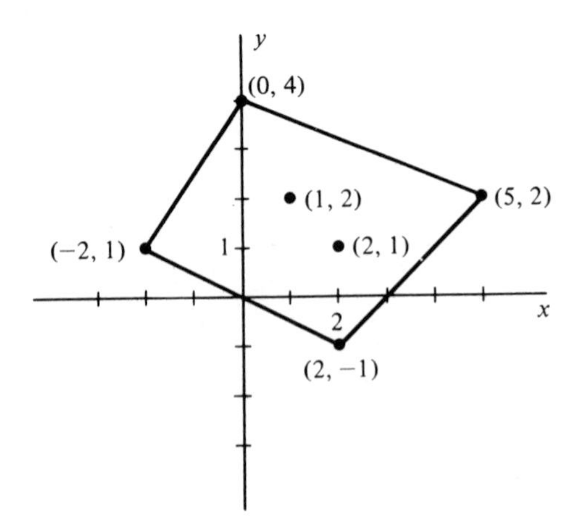
y
(0, 4)
(1, 2)
(5, 2)
(−2, 1)
1
(2, 1)
x
2
(2, −1)

11.

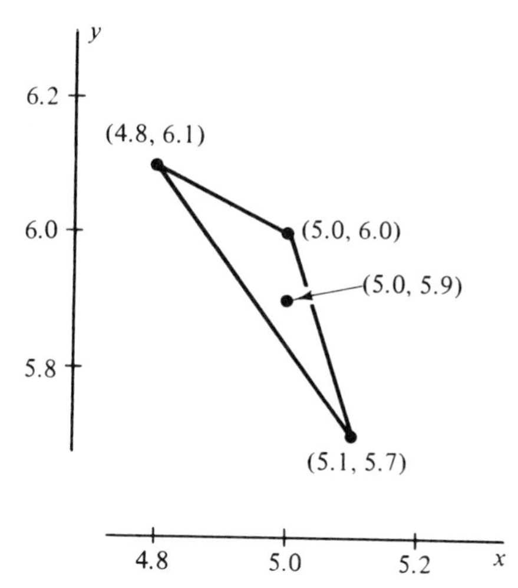

Exercises 7-3, page 269

1.

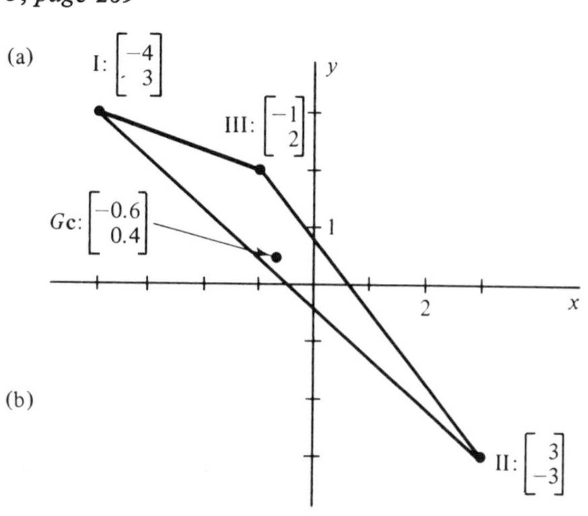

3. (a)

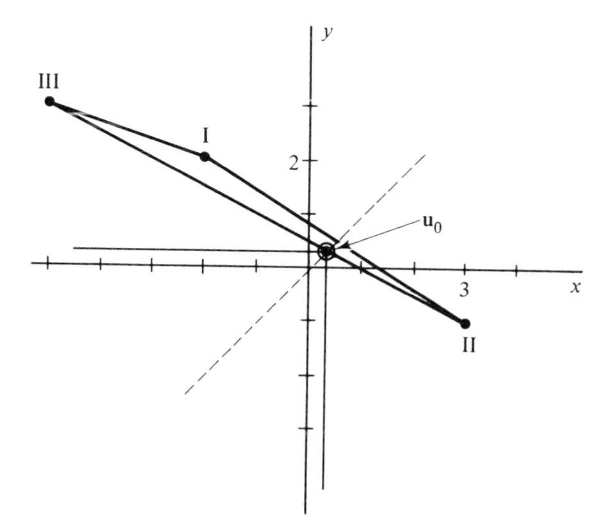

(b) Critical columns: II and III.

(c) $G' = \begin{bmatrix} 3 & -5 \\ -1 & 3 \end{bmatrix}$, $\mathbf{r}'_0 = [\frac{1}{3} \quad \frac{2}{3}]$ $\mathbf{c}'_0 = \begin{bmatrix} \frac{2}{3} \\ \frac{1}{3} \end{bmatrix}$.

(d) $\mathbf{r}_0 = [\frac{1}{3} \quad \frac{2}{3}]$, $\mathbf{c}_0 = \begin{bmatrix} 0 \\ \frac{2}{3} \\ \frac{1}{3} \end{bmatrix}$, $\mathbf{r}_0 G = [\frac{2}{3} \quad \frac{1}{3} \quad \frac{1}{3}]$, $G\mathbf{c}_0 = \begin{bmatrix} \frac{1}{3} \\ \frac{1}{3} \end{bmatrix}$, $\min \mathbf{r}_0 G = \max G\mathbf{c}_0 = \frac{1}{3} = v_G$.

5. (a) Optimal strategies: $\mathbf{r}_0 = [\frac{6}{13} \quad \frac{7}{13}]$, $\mathbf{c}_0 = \begin{bmatrix} \frac{11}{13} \\ \frac{2}{13} \\ 0 \end{bmatrix}$, $\mathbf{r}_0 G = [-\frac{1}{13} \quad -\frac{1}{13} \quad 0]$, $G\mathbf{c}_0 = \begin{bmatrix} -\frac{1}{13} \\ -\frac{1}{13} \end{bmatrix}$, $\min \mathbf{r}_0 G = \max G\mathbf{c}_0 = -\frac{1}{13} = v_G$.

(b) Optimal strategies: $\mathbf{r}_0 = [0 \quad 1]$, $\mathbf{c}_0 = \begin{bmatrix} 1 \\ 0 \\ 0 \end{bmatrix}$, $\mathbf{r}_0 G = [1 \quad 3 \quad 1]$, $G\mathbf{c}_0 = \begin{bmatrix} -1 \\ 1 \end{bmatrix}$, $\min \mathbf{r}_0 G = \max G\mathbf{c}_0 = 1 = v_G$.

(c) Optimal strategies: $\mathbf{r}_0 = [.4 \quad .6]$, $\mathbf{c}_0 = \begin{bmatrix} 0 \\ 0 \\ .4 \\ 0 \\ 0 \\ 0 \\ .6 \end{bmatrix}$, $\mathbf{r}_0 G = [.6 \quad .6 \quad .4 \quad .8 \quad .6 \quad .4]$, $G\mathbf{c}_0 = \begin{bmatrix} .4 \\ .4 \end{bmatrix}$, $\min \mathbf{r}_0 G = \max G\mathbf{c}_0 = .4 = v_G$.

7. (a) Optimal strategies: $\mathbf{r}_0 = [\frac{2}{5} \quad \frac{3}{5}]$, $\mathbf{c}_0 = \begin{bmatrix} \frac{2}{3} \\ \frac{1}{3} \\ 0 \end{bmatrix}$, $\mathbf{r}_0 G = [-1 \quad -1 \quad -1]$, $G\mathbf{c}_0 = \begin{bmatrix} -1 \\ -1 \end{bmatrix}$, $\min \mathbf{r}_0 G = \max G\mathbf{c}_0 = -1 = v_G$.

(b) For example, (1) $\mathbf{c}_0 = \begin{bmatrix} \frac{2}{3} \\ \frac{1}{3} \\ 0 \end{bmatrix}$ or (2) $\mathbf{c}_0 = \begin{bmatrix} 0 \\ \frac{3}{5} \\ \frac{2}{5} \end{bmatrix}$, or, mixing these two strategies, playing (1) with probability $\frac{3}{8}$ and (2) with probability $\frac{5}{8}$, we obtain (3) $\mathbf{c}_0 = \begin{bmatrix} .25 \\ .5 \\ .25 \end{bmatrix}$. In each case, $G\mathbf{c}_0 = \begin{bmatrix} -1 \\ -1 \end{bmatrix}$. Thus with $\mathbf{r}_0$ as in part (a), $\min \mathbf{r}_0 G = \max G\mathbf{c}_0 = -1 = v_G$.

Exercises 7-4, page 277

1. (a) 10. (b) 5. (c) 0. (d) 5. (e) 0. (f) 5.

3. (a) (a_1) $F(\mathbf{u}) = F(4, -1) = 14$ and $F(\mathbf{v}) = F(2, 3) = 0$. $F(\mathbf{u} + \mathbf{v}) = F(6, 2) = 14 = F(\mathbf{u}) + F(\mathbf{v})$. ($a_2$) $F(\mathbf{u}) = F(10, 25) = -20$ and $F(\mathbf{v}) = F(15, 5) = 35$. $F(\mathbf{u} + \mathbf{v}) = F(25, 30) = 15 = F(\mathbf{u}) + F(\mathbf{v})$.

(b) (b_1) $cF(\mathbf{v}) = 5 \times F(3, 1) = 5 \times 7 = 35$. $F(c\mathbf{v}) = F(15, 5) = 35 = cF(\mathbf{v})$. ($b_2$) $cF(\mathbf{v}) = \frac{1}{3} \times F(-3, 6) = \frac{1}{3} \times (-21) = -7$. $F(c\mathbf{v}) = F(-1, 2) = -7 = cF(\mathbf{v})$.

5.

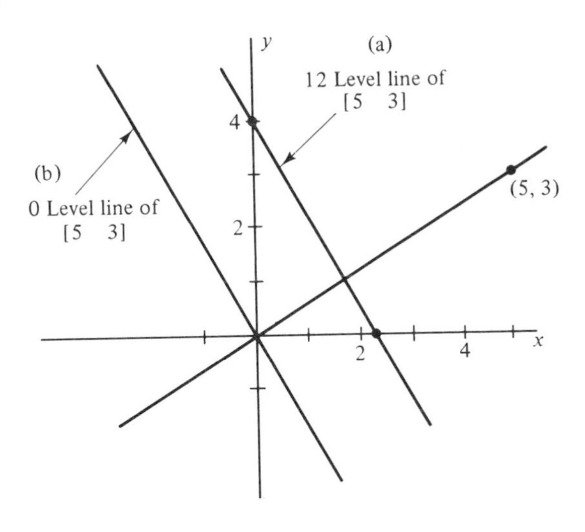

7. Some k level lines of $[-3 \quad 2]$:

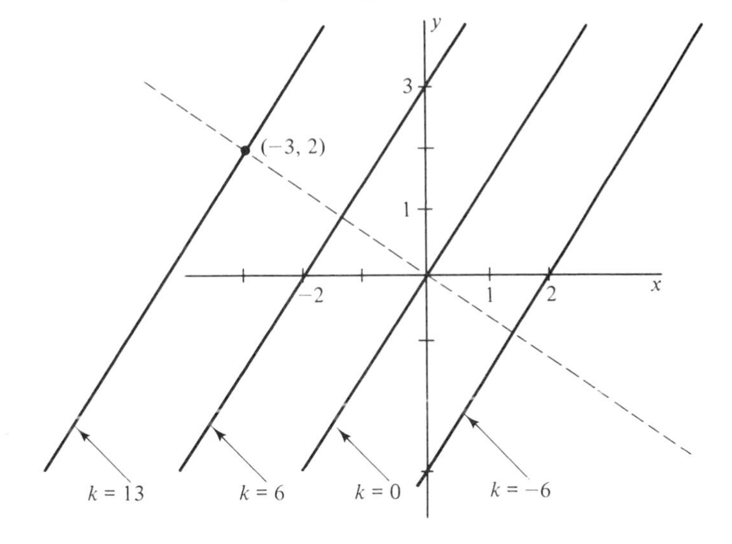

9. (a) $[1 \quad 1]$, or a functional of the form $[a \quad a]$.
(b) $[3 \quad -2]$, or a functional of the form $[3a \quad -2a]$.
(c) $[2 \quad -3]$, or a functional of the form $[2a \quad -3a]$.

Exercises 7-5, page 288

1. (a) 

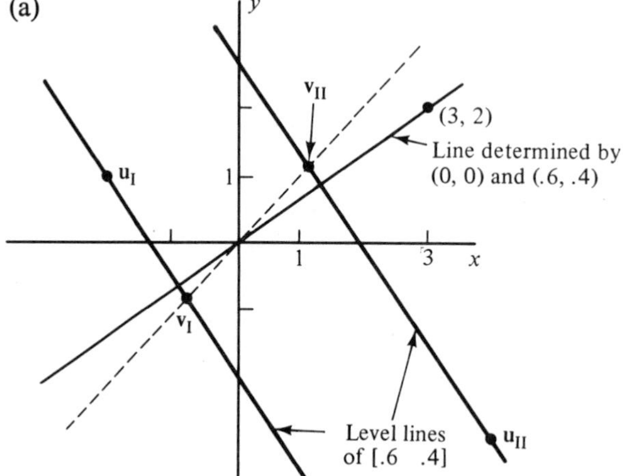

(b) The level line of [.6 .4] through $\mathbf{u}_I = (-2, 1)$ crosses the main diagonal at $\mathbf{v}_I = (-.8, -.8)$. The level line of [.6 .4] through $\mathbf{u}_{II} = (4, -3)$ crosses the main diagonal at $\mathbf{v}_{II} = (1.2, 1.2)$.

3. (a)

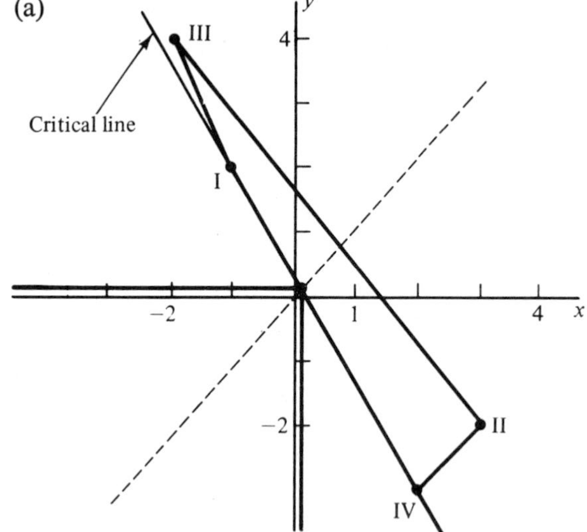

(b) Critical columns are I and IV. Optimal strategies: $\mathbf{r}_0 = [\frac{5}{8} \;\; \frac{3}{8}]$,

$$\mathbf{c}_0 = \begin{bmatrix} \frac{5}{8} \\ 0 \\ 0 \\ \frac{3}{8} \end{bmatrix}, \quad \mathbf{r}_0 G = [\tfrac{1}{8} \;\; \tfrac{9}{8} \;\; \tfrac{2}{8} \;\; \tfrac{1}{8}], \quad G\mathbf{c}_0 = \begin{bmatrix} \frac{1}{8} \\ \frac{1}{8} \end{bmatrix}, \quad \min \mathbf{r}_0 G = \max G\mathbf{c}_0$$

$= \frac{1}{8} = v_G$.

(c) The critical line shown in part (a) is determined by the points $(-1, 2)$ and $(2, -3)$, corresponding to Columns I and IV. This line is the $\frac{1}{8}$ level line of the functional $[\frac{5}{8} \;\; \frac{3}{8}] = \mathbf{r}_0$.

5. (a) Optimal strategies: $\mathbf{r}_0 = [1 \quad 0]$, $\mathbf{c}_0 = \begin{bmatrix} 1 \\ 0 \\ 0 \end{bmatrix}$, $\mathbf{r}_0 G = [1 \quad 3 \quad 1]$,

$G\mathbf{c}_0 = \begin{bmatrix} 1 \\ 1 \end{bmatrix}$, $\min \mathbf{r}_0 G = \max G\mathbf{c}_0 = 1$. The value of the game is 1.

(b) Optimal strategies: $\mathbf{r}_0 = [.5 \quad .5]$, $\mathbf{c}_0 = \begin{bmatrix} 0 \\ .55 \\ .45 \end{bmatrix}$, (Other strategies qualify as $\mathbf{c}_0$.) $\mathbf{r}_0 G = [-1 \quad -1 \quad -1]$, $G\mathbf{c}_0 = \begin{bmatrix} -1 \\ -1 \end{bmatrix}$, $\min \mathbf{r}_0 G = \max G\mathbf{c}_0 = -1$. The value of the game is -1.

7. (a)

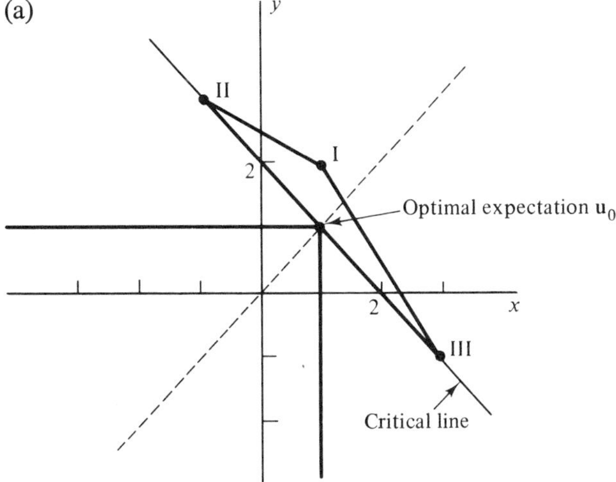

(b) Since the critical line appears to be perpendicular to the main diagonal, determined by (0, 0) and (.5, .5), it looks like $\mathbf{r}_0 = [.5 \quad .5]$. The optimal expectation $\mathbf{u}_0$ seems to be midway between the points corresponding to Columns II and III, which suggest $\mathbf{c}_0 = \begin{bmatrix} 0 \\ .5 \\ .5 \end{bmatrix}$. Finally, the optimal expectation $\mathbf{u}_0$ appears to be the point (1, 1), indicating that the value of the game is 1. Our guesses: $\mathbf{r}_0 = [.5 \quad .5]$, $\mathbf{c}_0 = \begin{bmatrix} 0 \\ .5 \\ .5 \end{bmatrix}$, $v_G = 1$.

Verification: $\mathbf{r}_0 G = [1.5 \quad 1 \quad 1]$, $G\mathbf{c}_0 = \begin{bmatrix} 1 \\ 1 \end{bmatrix}$, $\min \mathbf{r}_0 G = \max G\mathbf{c}_0 = 1$. Therefore $\mathbf{r}_0$ and $\mathbf{c}_0$ are optimal. The value of the game is $v_G = 1$.

Exercises 7-6, page 292

1. Optimal strategies for G_2: $\mathbf{r}_0 = [\frac{5}{8} \quad \frac{3}{8}]$, $\mathbf{c}_0 = \begin{bmatrix} 0 \\ \frac{3}{8} \\ \frac{5}{8} \\ 0 \end{bmatrix}$, which convert to optimal strategies for G_1: $\mathbf{r}_0 = [0 \quad \frac{3}{8} \quad \frac{5}{8} \quad 0]$, $\mathbf{c}_0 = \begin{bmatrix} \frac{5}{8} \\ \frac{3}{8} \end{bmatrix}$.

Verification: $\mathbf{r}_0 G_1 = [-\frac{1}{8} \quad -\frac{1}{8}]$, $G_1\mathbf{c}_0 = \begin{bmatrix} -\frac{4}{8} \\ -\frac{1}{8} \\ -\frac{1}{8} \\ -\frac{5}{8} \end{bmatrix}$, $\min \mathbf{r}_0 G_1 = \max G_1 \mathbf{c}_0$ $= -\frac{1}{8}$. Thus $\mathbf{r}_0$ and $\mathbf{c}_0$ are optimal; the value of the game is $-\frac{1}{8}$.

3. (a)

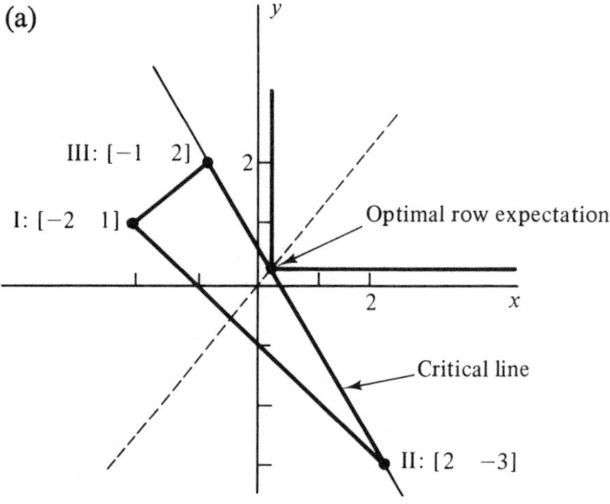

(b) Critical rows are II and III. Optimal strategies: $\mathbf{r}_0 = [0 \quad \frac{3}{8} \quad \frac{5}{8}]$, $\mathbf{c}_0 = \begin{bmatrix} \frac{5}{8} \\ \frac{3}{8} \end{bmatrix}$, $\min \mathbf{r}_0 G = \max G\mathbf{c}_0 = \frac{1}{8}$. The value of the game is $\frac{1}{8}$.

5. (a) Optimal strategies: $\mathbf{r}_0 = [\frac{2}{5} \quad \frac{3}{5} \quad 0 \quad 0]$, $\mathbf{c}_0 = \begin{bmatrix} \frac{2}{3} \\ \frac{1}{3} \end{bmatrix}$, $\min \mathbf{r}_0 G = \max G\mathbf{c}_0 = 0 = v_G$. (It is a fair game.)

(b) Optimal strategies: $\mathbf{r}_0 = [0 \quad 0 \quad \frac{1}{2} \quad \frac{1}{2} \quad 0]$, $\mathbf{c}_0 = \begin{bmatrix} \frac{5}{14} \\ \frac{9}{14} \end{bmatrix}$, $\min \mathbf{r}_0 G = \max G\mathbf{c}_0 = \frac{1}{2} = v_G$.

7. (a) Optimal strategies: $\mathbf{r}_0 = [\frac{4}{5} \quad \frac{1}{5}]$, $\mathbf{c}_0 = \begin{bmatrix} 0 \\ 0 \\ \frac{4}{5} \\ \frac{1}{5} \end{bmatrix}$, $\min \mathbf{r}_0 G = \max G\mathbf{c}_0$ $= \frac{14}{5} = v_G$.

(b) Optimal strategies: $\mathbf{r}_0 = [0 \quad 1 \quad 0 \quad 0]$, $\mathbf{c}_0 = \begin{bmatrix} \frac{1}{2} \\ \frac{1}{2} \end{bmatrix}$. (Other strategies qualify as $\mathbf{c}_0$). $\min \mathbf{r}_0 G = \max G\mathbf{c}_0 = 4 = v_G$. (It should be observed that the same polygon of expectations serves both problems.)

9. (a) Optimal strategies: $\mathbf{r}_0 = [\frac{11}{24} \quad \frac{13}{24} \quad 0]$, $\mathbf{c}_0 = \begin{bmatrix} 0 \\ \frac{5}{8} \\ \frac{3}{8} \\ 0 \end{bmatrix}$, $\min \mathbf{r}_0 G = \max G\mathbf{c}_0 = -\frac{1}{8} = v_G$.

(b) Optimal strategies: $\mathbf{r}_0 = [0 \quad 1 \quad 0]$, $\mathbf{c}_0 = \begin{bmatrix} 1 \\ 0 \\ 0 \end{bmatrix}$, $\min \mathbf{r}_0 G = \max G\mathbf{c}_0 = 2 = v_G$.

(c) Optimal strategies: $\mathbf{r}_0 = [\frac{3}{8} \quad \frac{5}{8} \quad 0]$, $\mathbf{c}_0 = \begin{bmatrix} 3/8 \\ 0 \\ 5/8 \end{bmatrix}$. (Other strategies qualify as $\mathbf{c}_0$.) $\min \mathbf{r}_0 G = \max G\mathbf{c}_0 = \frac{23}{8} = v_G$. (In part (c), Row III is eliminated from consideration because strategy [.5 .5 0] promises a better payoff in every column.)

Chapter 8

Exercises 8-1, page 312

1. (a) The X requirement is met for 5.8 days and the Y requirement for 6.2 days.

(b) (i) [5.2 6.8], (ii) [6.1 5.9], (iii) [7.4 5.6].

(c) The diet [.3 .7] in (ii).

(d) The desired optimal diet is the optimal row strategy $\mathbf{r}_0$ in the game with payoff matrix $M = \begin{bmatrix} 4 & 8 \\ 7 & 5 \end{bmatrix}$. Optimal strategies: $\mathbf{r}_0 = [\frac{1}{3} \quad \frac{2}{3}]$, $\mathbf{c}_0 = \begin{bmatrix} \frac{1}{2} \\ \frac{1}{2} \end{bmatrix}$

Verification: $\mathbf{r}_0 M = [6 \quad 6]$, $M\mathbf{c}_0 = \begin{bmatrix} 6 \\ 6 \end{bmatrix}$, $\min \mathbf{r}_0 M = \max M\mathbf{c}_0 = 6$. Therefore, the optimal diet is $[u \quad v] = [\frac{1}{3} \quad \frac{2}{3}]$, which meets the dietary requirements for 6 days.

3. (a) (i) $\begin{bmatrix} u \\ v \end{bmatrix} = \begin{bmatrix} 5.6 \\ 6.2 \end{bmatrix}$, (ii) $\begin{bmatrix} u \\ v \end{bmatrix} = \begin{bmatrix} 6.8 \\ 5.6 \end{bmatrix}$, (iii) $\begin{bmatrix} u \\ v \end{bmatrix} = \begin{bmatrix} 4 \\ 7 \end{bmatrix}$.

(b) For the schedule $\begin{bmatrix} .6 \\ .4 \end{bmatrix}$ in (i).

(c) Here the optimal production schedule is given by the optimal column strategy $\mathbf{c}_0$ in the game with payoff matrix M shown in the solution of Prob. 1(d). As there verified, $\mathbf{c}_0 = \begin{bmatrix} .5 \\ .5 \end{bmatrix}$. We conclude that the mixed lot of 1000 units that can be produced in least possible time consists of 500 units of X and 500 units of Y.

5. (a) The optimal production schedule is given by the optimal row strategy $\mathbf{c}_0$ in the game with payoff matrix $M = \begin{bmatrix} 3 & 8 & 5 \\ 7 & 2 & 4 \end{bmatrix}$.

Optimal strategies: $\mathbf{r}_0 = [.6 \quad .4]$, $\mathbf{c}_0 = \begin{bmatrix} .2 \\ 0 \\ .8 \end{bmatrix}$.

Verification: $\mathbf{r}_0 M = [4.6 \quad 5.6 \quad 4.6]$, $M\mathbf{c}_0 = \begin{bmatrix} 4.6 \\ 4.6 \end{bmatrix}$, $\min \mathbf{r}_0 M = \max M\mathbf{c}_0 = 4.6$. We conclude that for a total of 5000 units in the least possible time the factory should produce 1000 units of X, no units of Y, and 4000 units of Z. This schedule requires 4.6 days.

(b) A diet requires nutrients X, Y, and Z, to be obtained from foods U and V. A dollar's worth of U contains a 3-day supply of X, an 8-day supply of Y, and a 5-day supply of Z, while a dollar's worth of V contains a 7-day supply of X, a 2-day supply of Y, and a 4-day supply of Z. The problem is to determine how a one-dollar expenditure should be divided between foods U and V in order to satisfy all three dietary requirements for the longest possible time. From the solution $\mathbf{r}_0 = [.6 \quad .4]$ in part (a) it follows that the optimal diet consists of 60 cents worth of U and 40 cents worth of V. This diet meets the requirements for 4.6 days.

Exercises 8-2, page 323

1. (a) $[x \quad y] = [62 \quad 42]$, not feasible.
(b) $[x \quad y] = [62 \quad 54]$, feasible, cost is \$29.
(c) $[x \quad y] = [54 \quad 42]$, not feasible.
(d) $[x \quad y] = [60 \quad 60]$, feasible, cost is \$30.

3. (a)–(d)

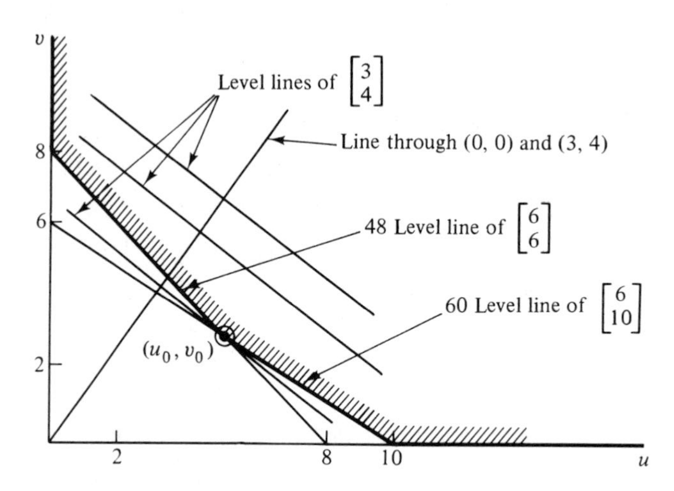

(e) The desired matrix $[u_0 \quad v_0]$ is determined by the intersection of the 48 level line of $\begin{bmatrix} 6 \\ 6 \end{bmatrix}$ and the 60 level line of $\begin{bmatrix} 6 \\ 10 \end{bmatrix}$. Solving the simultaneous equations, $[u_0 \quad v_0] = [5 \quad 3]$.

(f) The minimum cost diet consists of 5 packages of U and 3 packages of V. Its cost is \$27.

5. (a), (b)

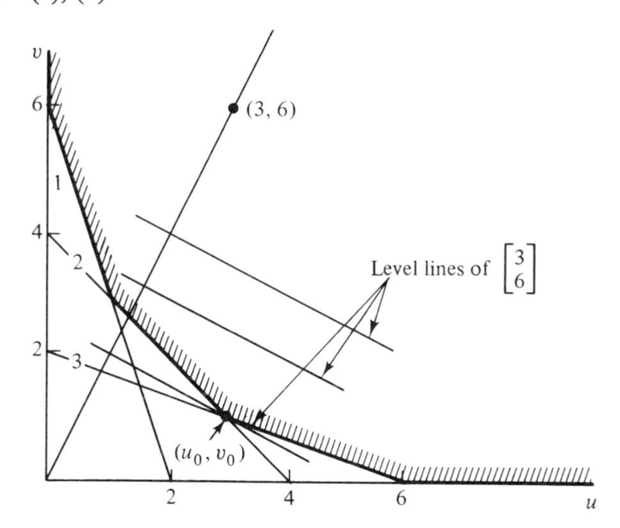

(b continued) The desired matrix $[u_0 \quad v_0]$ is at the intersection of the 16 level line of $\begin{bmatrix}4\\4\end{bmatrix}$, labeled 2, and the 12 level line of $\begin{bmatrix}2\\6\end{bmatrix}$, labeled 3. Thus $[u_0 \quad v_0] = [3 \quad 1]$.

7. Translation of the problem with insurance coverage in units of \$100: Given the matrices

$$M = \begin{bmatrix}10 & 3 & 3\\10 & 10 & 1\end{bmatrix}, \begin{bmatrix}3\\4\end{bmatrix} = B,$$

$$A = [240 \quad 100 \quad 50],$$

Find a matrix $[u \quad v]$ of nonnegative entries satisfying the constraint $[u \quad v]M \geq A$ which minimizes the value of $[u \quad v]B$. The graphical interpretation indicates that the solution of the problem is given by $[u_0 \quad v_0] = [20 \quad 4]$. We conclude that the man should buy 20 units of plan U and 4 units of plan V. The total monthly premium is \$76.

Exercises 8-3, page 333

1. (a) Requires 11 hours on U, 4 hours on V, and 25 hours on W. The schedule is *not* feasible.
(b) Requires 12 hours on U, 6 hours on V, and 24 hours on W. The schedule is feasible, and produces a profit of \$300.
(c) Requires 12 hours on U, 8 hours on V, and 20 hours on **W**. The schedule is feasible, and produces a profit of \$280.
(d) Requires 14 hours on U, 12 hours on V, and 18 hours on W. The schedule is feasible, and produces a profit of \$300.

3. (a) (i) $\mathbf{r}M = [29 \quad 23]$. $\mathbf{r}$ does *not* satisfy $\mathbf{r}M \geq A$. (ii) $\mathbf{r}M = [31 \quad 21]$. $\mathbf{r}$ satisfies $\mathbf{r}M \geq A$, $\mathbf{r}B = [160]$.

(b) (i) $M\mathbf{c} = \begin{bmatrix} 10 \\ 8 \\ 12 \end{bmatrix}$, (ii) $M\mathbf{c} = \begin{bmatrix} 11.5 \\ 5 \\ 11 \end{bmatrix}$. In both cases $M\mathbf{c} \leq B$. (i) $A\mathbf{c} = [140]$, (ii) $A\mathbf{c} = [140]$.

5. For the matrices M, A, and B of Prob. 2: *The Minimization Problem:* Find a 1×2 row matrix $\mathbf{r}$ of nonnegative entries satisfying $\mathbf{r}M \geq A$ such that $\mathbf{r}B$ is minimized. *The Maximization Problem:* Find a 3×1 column matrix $\mathbf{c}$ of nonnegative entries satisfying $M\mathbf{c} \leq B$ such that $A\mathbf{c}$ is maximized.

Verification of the given solutions: $\mathbf{r}_0 M = [12 \quad 16 \quad 20] \geq [12 \quad 16 \quad 12] = A.\checkmark$ $M\mathbf{c}_0 = \begin{bmatrix} 96 \\ 60 \end{bmatrix} \leq \begin{bmatrix} 96 \\ 60 \end{bmatrix} = B.\checkmark$ $\mathbf{r}_0 B = A\mathbf{c}_0 = [276].\checkmark\checkmark$

7. (a) From the graphical interpretation, $\mathbf{r}_0 = [0 \quad 12.5]$.

(b) Verification of $\mathbf{r}_0$ and $\mathbf{c}_0$: $\mathbf{r}_0 M = [100 \quad 175 \quad 125] \geq [80 \quad 98 \quad 125] = A.\checkmark$ $M\mathbf{c}_0 = \begin{bmatrix} 200 \\ 80 \end{bmatrix} \leq \begin{bmatrix} 250 \\ 80 \end{bmatrix} = B.\checkmark$ $\mathbf{r}_0 B = A\mathbf{c}_0 = [1000].\checkmark\checkmark$ Therefore, the solution $\mathbf{r}_0 = [0 \quad 12.5]$ is correct.

Exercises 8-4, page 345

1. (a)–(c)

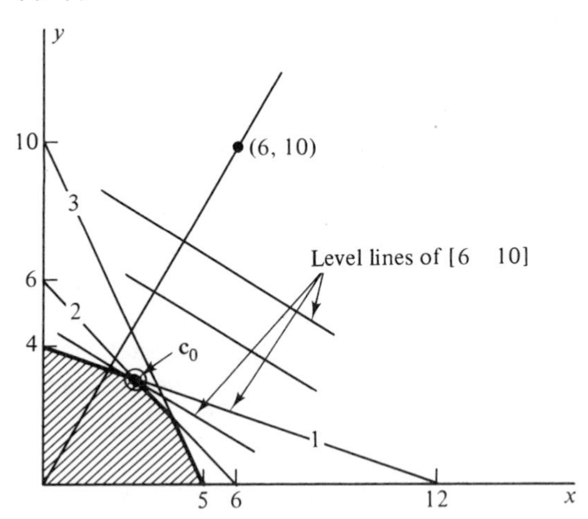

(c continued) $\mathbf{c}_0$ is at the intersection of the 24 level line of $[2 \quad 6]$ (labeled 1) and the 12 level line of $[2 \quad 2]$ (labeled 2). Solving, $\mathbf{c}_0 = \begin{bmatrix} 3 \\ 3 \end{bmatrix}$.

(d) $\mathbf{c}_0$ identifies both columns, and Rows I and II of M as critical. For $\mathbf{r}_0 = [u \quad v \quad 0]$, we set $\mathbf{r}_0 M = A$ (equality in both columns). Solving, $\mathbf{r}_0 = [1 \quad 2 \quad 0]$.

(e) $\mathbf{r}_0 M = [6 \quad 10] \geq [6 \quad 10] = A.\checkmark$ $\quad M\mathbf{c}_0 = \begin{bmatrix} 24 \\ 12 \\ 18 \end{bmatrix} \leq \begin{bmatrix} 24 \\ 12 \\ 20 \end{bmatrix} = B.\checkmark$

$\mathbf{r}_0 B = A\mathbf{c}_0 = [48].\checkmark\checkmark$

3. (a)

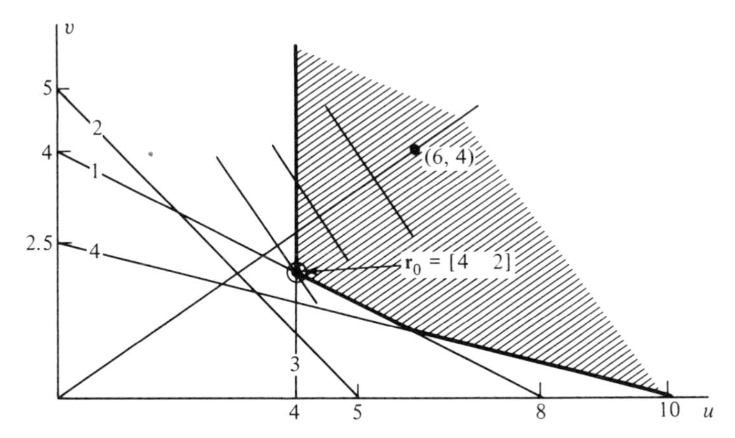

(b) Since $\mathbf{r}_0 = [4 \quad 2]$ identifies both rows and Columns I and III as critical, we set $\mathbf{c}_0 = \begin{bmatrix} x \\ 0 \\ z \\ 0 \end{bmatrix}$ and solve $M\mathbf{c}_0 = B$. Thus $\mathbf{c}_0 = \begin{bmatrix} 2 \\ 0 \\ 2 \\ 0 \end{bmatrix}$.

(c) $\mathbf{r}_0 M = [4 \quad 2]M = [8 \quad 12 \quad 8 \quad 12] \geq [8 \quad 10 \quad 8 \quad 10] = A.\checkmark$

$M\mathbf{c}_0 = M\begin{bmatrix} 2 \\ 0 \\ 2 \\ 0 \end{bmatrix} = \begin{bmatrix} 6 \\ 4 \end{bmatrix} \leq \begin{bmatrix} 6 \\ 4 \end{bmatrix} = B.\checkmark$ $\mathbf{r}_0 B = A\mathbf{c}_0 = [32].\checkmark\checkmark$

5.

$$M = \begin{bmatrix} 2 & 1 \\ \frac{1}{2} & \frac{1}{3} \\ 3 & 4 \end{bmatrix}, \begin{bmatrix} 150 \\ 40 \\ 320 \end{bmatrix} = B,$$

$$A = [12 \quad 7.5].$$

$$\mathbf{r}_0 = [1.5 \quad 18 \quad 0],\ \mathbf{c}_0 = \begin{bmatrix} 60 \\ 30 \end{bmatrix}.$$

Verification: $\mathbf{r}_0 M = [12 \quad 7.5] \geq [12 \quad 7.5] = A.\checkmark$ $M\mathbf{c}_0 = \begin{bmatrix} 150 \\ 40 \\ 300 \end{bmatrix} \leq \begin{bmatrix} 150 \\ 40 \\ 320 \end{bmatrix} = B.\checkmark$ $\mathbf{r}_0 B = A\mathbf{c}_0 = [945].\checkmark\checkmark$ Aleeta should produce 60 of the larger pots and 30 of the smaller, for a total income of $945.

7. She should buy 16 packs of the "Meat Loaf Special" and 8 packs of the "Combo Special," for a total cost of $24.40.

9. 200 tents and 1200 tarps, for a profit of \$7200.

11. 5 bags of assortment U, no V, and 5 bags of W, for a total cost of \$7.50.

Exercises 8-5, page 357

1. (a) Given:

$$M = \begin{bmatrix} 1 & 2 \\ 2 & 1 \end{bmatrix}, \begin{bmatrix} 20 \\ 22 \end{bmatrix} = B,$$
$$A = [16 \quad 14].$$

The Minimization Problem: Find a matrix $\mathbf{r} = [u \quad v]$ of nonnegative entries satisfying $\mathbf{r}M \geq A$ such that $\mathbf{r}B$ is minimized.

The Maximization Problem: Find $\mathbf{c} = \begin{bmatrix} x \\ y \end{bmatrix}$ of nonnegative entries satisfying $M\mathbf{c} \leq B$ such that $A\mathbf{c}$ is maximized. Verification of $\mathbf{r}_0$ and $\mathbf{c}_0$: $\mathbf{r}_0 M = [16 \quad 14] \geq [16 \quad 14] = A.\checkmark$ $M\mathbf{c}_0 = \begin{bmatrix} 20 \\ 22 \end{bmatrix} \leq \begin{bmatrix} 20 \\ 22 \end{bmatrix} = B.\checkmark$ $\mathbf{r}_0 B = A\mathbf{c}_0 = [212].\checkmark\checkmark$

(b) Shadow price of a unit of X is \$8, and of a unit of Y, \$6.

(c)
$$M = \begin{bmatrix} 1 & 2 \\ 2 & 1 \\ 1 & 1 \end{bmatrix}, \begin{bmatrix} 20 \\ 22 \\ 15 \end{bmatrix} = B$$
$$A = [16 \quad 14].$$

Solutions: $\mathbf{r}_0 = [4 \quad 6 \quad 0]$ (including no W) and $\mathbf{c}_0 = \begin{bmatrix} 8 \\ 6 \end{bmatrix}$.

Verification: $\mathbf{r}_0 M = [16 \quad 14] \geq [16 \quad 14] = A.\checkmark$ $M\mathbf{c}_0 = \begin{bmatrix} 20 \\ 22 \\ 14 \end{bmatrix} \leq \begin{bmatrix} 20 \\ 22 \\ 15 \end{bmatrix} = B.\checkmark$ $\mathbf{r}_0 B = A\mathbf{c}_0 = [212].\checkmark\checkmark$

(d) Maximum amount is \$14 per package of W.

(e) *Verification of* $\mathbf{r}_0$ *and* $\mathbf{c}_0$: $\mathbf{r}_0 M = [3.9 \quad 6.2]\begin{bmatrix} 1 & 2 \\ 2 & 1 \end{bmatrix} = [16.3 \quad 14] \geq [16.3 \quad 14] = A.\checkmark$ $M\mathbf{c}_0 = \begin{bmatrix} 1 & 2 \\ 2 & 1 \end{bmatrix}\begin{bmatrix} 8 \\ 6 \end{bmatrix} = \begin{bmatrix} 20 \\ 22 \end{bmatrix} \leq \begin{bmatrix} 20 \\ 22 \end{bmatrix} = B.\checkmark$ $\mathbf{r}_0 B = A\mathbf{c}_0 = [214.4].\checkmark\checkmark$ From parts (a) to (e), the minimal cost has increased by $214.4 - 212 = 2.4$, or \$2.40. The shadow price of a unit of X being \$8, this increase reflects the cost of the .3 unit increase in the X requirement.

3. (a) Solutions of dual problems, where $p > 21$: $\mathbf{r}_0 = [4 \quad 6 \quad 0]$, $\mathbf{c}_0 = \begin{bmatrix} 6 \\ 3 \end{bmatrix}$.

Verification: $\mathbf{r}_0M = [24 \quad 32] \geq [24 \quad 32] = A.\checkmark$ $M\mathbf{c}_0 = \begin{bmatrix} 24 \\ 24 \\ 21 \end{bmatrix}$ $\leq \begin{bmatrix} 24 \\ 24 \\ p \end{bmatrix} = B.\checkmark$ $\mathbf{r}_0B = A\mathbf{c}_0 = [240].\checkmark\checkmark$

(b) Each package of food U supplies 3 units of X and 2 units of Y and costs \$24; each package of V supplies 2 units of X and 4 units of Y and costs \$24; and each package of W supplies 2 units of X and 3 units of Y and costs p dollars. In part (a) we see that if $p < 21$, then the minimal cost diet includes no food W. As given by $\mathbf{c}_0 = \begin{bmatrix} 6 \\ 3 \end{bmatrix}$, the shadow prices for units of X and Y are \$6 and \$3, respectively. These shadow prices fix the value of a package of W at \$21.

5. (a) $\mathbf{r}_0 = [5 \quad 5]$, $\mathbf{c}_0 = \begin{bmatrix} 0 \\ 2 \\ 1 \end{bmatrix}$.

Verification: $\mathbf{r}_0M = [20 \quad 10 \quad 15] \geq [12 \quad 10 \quad 15] = A.\checkmark$ $M\mathbf{c}_0 = \begin{bmatrix} 4 \\ 3 \end{bmatrix} \leq \begin{bmatrix} 4 \\ 3 \end{bmatrix} = B.\checkmark$ $\mathbf{r}_0B = A\mathbf{c}_0 = [35].\checkmark\checkmark$

(b) $\mathbf{r}'_0 = [5.1 \quad 4.8]$, $\mathbf{c}_0 = \begin{bmatrix} 0 \\ 2 \\ 1 \end{bmatrix}$, $\mathbf{r}'_0M = [19.5 \quad 9.9 \quad 15] \geq [12 \quad 9.9 \quad 15]$ $= A'.\checkmark$ $M\mathbf{c}_0 = \begin{bmatrix} 4 \\ 3 \end{bmatrix} \leq \begin{bmatrix} 4 \\ 3 \end{bmatrix} = B.\checkmark$ $\mathbf{r}'_0B = A'\mathbf{c}_0 = [34.8].\checkmark\checkmark$ The minimal value of $\mathbf{r}B$ has decreased by .2, reflecting the value of the decrease of .1 in the "Y requirement" (conventional diet format) for which the "shadow price" per unit (second entry in $\mathbf{c}_0$) is 2.

(c) With $A'' = [13 \quad 10 \quad 15]$, the solutions $\mathbf{r}_0$ and $\mathbf{c}_0$ of part (a) still apply. Here, the increase in the "X requirement" (diet format) causes no increase in the minimal cost because the shadow price for X (first entry in $\mathbf{c}_0$) is zero.

Exercises 8-6, page 368

1. Optimal strategies in game G: $\mathbf{r}_0 = [\frac{1}{3} \quad \frac{2}{3}]$, $\mathbf{c}_0 = \begin{bmatrix} \frac{3}{4} \\ 0 \\ \frac{1}{4} \end{bmatrix}$, $\min \mathbf{r}_0G = \max G\mathbf{c}_0 = 6 = v_G$. Solution to the linear programming problems: $\mathbf{r}'_0 = (1/v_G)\mathbf{r}_0 = \frac{1}{6}\mathbf{r}_0 = [\frac{1}{18} \quad \frac{1}{9}]$, $\mathbf{c}'_0 = \frac{1}{6}\mathbf{c}_0 = \begin{bmatrix} \frac{1}{8} \\ 0 \\ \frac{1}{24} \end{bmatrix}$

Verification: $\mathbf{r}'_0G = [1 \quad \frac{10}{9} \quad 1] \geq [1 \quad 1 \quad 1] = A.\checkmark$ $G\mathbf{c}'_0 = \begin{bmatrix} 1 \\ 1 \end{bmatrix} \leq \begin{bmatrix} 1 \\ 1 \end{bmatrix} = B.\checkmark$ $\mathbf{r}'_0B = A\mathbf{c}'_0 = [\frac{1}{6}] = [1/v_G].\checkmark\checkmark$

3. Solutions to the linear programming problems: $\mathbf{r}_0' = [\frac{1}{4} \quad \frac{1}{4}]$, $\mathbf{c}_0' = \begin{bmatrix} \frac{2}{6} \\ 0 \\ \frac{1}{6} \end{bmatrix}$, $\mathbf{r}_0'B = A\mathbf{c}_0' = [\frac{1}{2}] = [d_0]$. Optimal strategies in the game G: $\mathbf{r}_0 = (1/d_0)\mathbf{r}_0' = [\frac{1}{2} \quad \frac{1}{2}]$, $\mathbf{c}_0 = (1/d_0)\mathbf{c}_0' = \begin{bmatrix} \frac{2}{3} \\ 0 \\ \frac{1}{3} \end{bmatrix}$.

Verification: $\mathbf{r}_0G = [2 \quad \frac{5}{2} \quad 2]$, $G\mathbf{c}_0 = \begin{bmatrix} 2 \\ 2 \end{bmatrix}$, $\min \mathbf{r}_0G = \max G\mathbf{c}_0 = 2 = v_G = (1/d_0)$.✓✓

5. (a) Adding 3 to all entries, $G^* = \begin{bmatrix} 10 & 0 \\ 0 & 5 \\ 5 & 2 \end{bmatrix}$. Solutions of linear programming problems for G^*, A, and B: $\mathbf{r}_0' = [.1 \quad .2 \quad 0]$, $\mathbf{c}_0' = \begin{bmatrix} .1 \\ .2 \end{bmatrix}$, $\mathbf{r}_0'B = A\mathbf{c}_0' = [.3]$.

(b) Value of the game: $v_G{}^* = 1/.3 = \frac{10}{3}$. Optimal strategies for G^* and G: $\mathbf{r}_0 = \frac{10}{3}\mathbf{r}_0' = [\frac{1}{3} \quad \frac{2}{3} \quad 0]$, $\mathbf{c}_0 = \frac{10}{3}\mathbf{c}_0' = \begin{bmatrix} \frac{1}{3} \\ \frac{2}{3} \end{bmatrix}$.

Verification (for game G): $\mathbf{r}_0G = [\frac{1}{3} \quad \frac{1}{3}]$, $G\mathbf{c}_0 = \begin{bmatrix} \frac{1}{3} \\ \frac{1}{3} \\ 0 \end{bmatrix}$, $\min \mathbf{r}_0G = \max G\mathbf{c}_0 = \frac{1}{3} = v_G$.

7. Linear programming format (Food U in "half-packages"): Given

$$M = \begin{bmatrix} \frac{1}{10} & \frac{1}{5} \\ \frac{1}{6} & \frac{1}{15} \end{bmatrix}, \quad \begin{bmatrix} 1 \\ 1 \end{bmatrix} = B,$$

$$A = [\ 1 \qquad 1\ ].$$

The Minimization Problem: Find a row matrix $\mathbf{r} = [u \quad v]$ of nonnegative entries satisfying $\mathbf{r}M \geq A$ such that $\mathbf{r}B$ is minimized. Optimal strategies in the game with payoff matrix M: $\mathbf{r}_0 = [\frac{1}{2} \quad \frac{1}{2}]$ $\mathbf{c}_0 = \begin{bmatrix} \frac{2}{3} \\ \frac{1}{3} \end{bmatrix}$; $\min \mathbf{r}_0M = \max M\mathbf{c}_0 = \frac{2}{15} = v_G$. Solution to the minimization problem: $\mathbf{r}_0' = (1/v_G)\mathbf{r}_0 = [\frac{15}{4} \quad \frac{15}{4}]$, from which we infer that the optimal diet consists of $\frac{15}{4}$ "half-packages"—or $7\frac{1}{2}$ packages—of U, and $\frac{15}{4}$, or $3\frac{3}{4}$, packages of V. The minimum cost is \$7.50, since $\mathbf{r}_0'B = 7.5$.

9. (a) Let $\mathbf{r} = [u \quad v]$, where u and v are nonnegative, represent the diet consisting of u packages of U and v packages of V. With M and A as given, $\mathbf{r}$ satisfies the requirements and respects the upper bound on Y precisely when $\mathbf{r}M \geq A$. The problem is to minimize the cost $\mathbf{r}B$, subject to these conditions on $\mathbf{r}$.

(b)

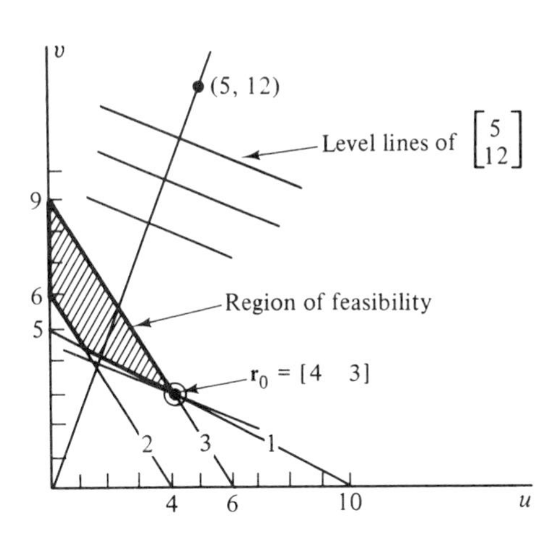

(c) With $\mathbf{r}_0 = [4 \quad 3]$, $\mathbf{r}_0 M = [20 \quad 36 \quad -36] \geq [20 \quad 24 \quad -36] = A$. Both rows and the first and third columns of M are critical. Thus we set $\mathbf{c}_0 = \begin{bmatrix} x \\ 0 \\ z \end{bmatrix}$ and solve $M\mathbf{c}_0 \leq B$ with equality in both rows. This gives $x = \frac{13}{4}$ and $z = \frac{1}{4}$.

(d) Solution: $\mathbf{r}_0 = [4 \quad 3]$, $\mathbf{c}_0 = \begin{bmatrix} \frac{13}{4} \\ 0 \\ \frac{1}{4} \end{bmatrix}$.

Verification: $\mathbf{r}_0 M \geq A$ verified in part (c).✓ $M\mathbf{c}_0 = \begin{bmatrix} 5 \\ 12 \end{bmatrix} \leq \begin{bmatrix} 5 \\ 12 \end{bmatrix} = B$.✓ $\mathbf{r}_0 B = A\mathbf{c}_0 = [56]$.✓✓

(e) The optimal diet consists of 4 packages of U and 3 packages of V, for a total cost of \$56.

Exercises 8-7, page 379

1. (a) (i) $T(\mathbf{c}_1) = (1, 5)$, $T(\mathbf{c}_2) = (3, 2)$, $T(\mathbf{c}_1 + \mathbf{c}_2) = T(1, 1) = (4, 7) = T(\mathbf{c}_1) + T(\mathbf{c}_2)$. (ii) $T(\mathbf{c}_1) = (5, 12)$, $T(\mathbf{c}_2) = (1.5, 1)$, $T(\mathbf{c}_1 + \mathbf{c}_2) = (6.5, 13)$. (iii) $T(\mathbf{c}_1) = (8, 1)$, $T(\mathbf{c}_2) = (-11, -3)$, $T(\mathbf{c}_1 + \mathbf{c}_2) = (-3, -2)$.

(b) (i) $T(\mathbf{c}) = (5, 12)$, $T(s\mathbf{c}) = T(10, 5) = (25, 60) = sT(\mathbf{c})$. (ii) $T(\mathbf{c}) = (-3, 11)$, $T(s\mathbf{c}) = T(-1.5, 1) = (1.5, -5.5) = sT(\mathbf{c})$.

3. (a) $A(x, y, z) = A(x(1, 0, 0) + y(0, 1, 0) + z(0, 0, 1))$. Thus from property 1, $A(x, y, z) = A(x(1, 0, 0)) + A(y(0, 1, 0)) + A(z(0, 0, 1))$, and from property 2, $A(x, y, z) = xA(1, 0, 0) + yA(0, 1, 0) + zA(0, 0, 1)$. It is given that $A(1, 0, 0) = 4$, $A(0, 1, 0) = 1$, and $A(0, 0, 1) = 3$. Therefore, $A(x, y, z) = 4x + y + 3z$.

(b) (i) $A(1, -1, -1) = 0$, (ii) $A(\frac{1}{4}, \frac{1}{4}, \frac{1}{4}) = 2$, (iii) $A(.5, 0, 2) = 8$.
(c) $[a_1 \quad a_2 \quad a_3] = [4 \quad 1 \quad 3]$.

5. (a) $F(0, 1, 0) = F(1, 1, 0) - F(1, 0, 0) = (2, -1)$.
(b) $F(0, 0, 1) = F(1, 1, 1) - F(1, 1, 0) = (-3, -1)$.
(c) $F(x, y, z) = (u, v)$, where $\begin{bmatrix} u \\ v \end{bmatrix} = \begin{bmatrix} 1 & 2 & -3 \\ 2 & -1 & -1 \end{bmatrix} \begin{bmatrix} x \\ y \\ z \end{bmatrix}$. Thus the matrix $\begin{bmatrix} 1 & 2 & -3 \\ 2 & -1 & -1 \end{bmatrix}$ expresses the mapping F.
(d) $F(0, 1, 1) = (-1, -2)$, $F(-2, 3, 3) = (-5, -10)$, $F(.5, 0, .2) = (-.1, .8)$.

7. (a) (i) $T(1, 0) = (1, 3, -2)$, $(S \circ T)(1, 0) = S(1, 3, -2) = (1, 7)$. (ii) $T(-2, 1) = (-2, -4, 4)$, $(S \circ T)(-2, 1) = S(-2, -4, 4) = (0, -12)$. (iii) $T(10, -5) = (10, 20, -20)$, $(S \circ T)(10, -5) = S(10, 20, -20) = (0, 60)$.
(b) The matrix of $S \circ T$, realized by multiplying the matrix of S times the matrix of T, is $\begin{bmatrix} 1 & 2 \\ 7 & 2 \end{bmatrix}$. The images of the vectors (x, y) are the same as given in part (a).
(c) (i) $S(0, 1, 0) = (1, 1)$, $(T \circ S)(0, 1, 0) = T(1, 1) = (1, 5, -2)$. (ii) $S(1, 2, 1) = (0, 0)$, $(T \circ S)(1, 2, 1) = T(0, 0) = (0, 0, 0)$. (iii) $S(-3, 5, 2) = (-1, 1)$, $(T \circ S)(-3, 5, 2) = T(-1, 1) = (-1, -1, 2)$.
(d) The matrix of $T \circ S$ is $\begin{bmatrix} -2 & 1 & 0 \\ -6 & 5 & -4 \\ 4 & -2 & 0 \end{bmatrix}$. The images of the vectors (x, y, z) are the same as given in part (c).

9. (a) (i) $(T_1 + T_2)(1, 0) = T_1(1, 0) + T_2(1, 0) = (1, -2) + (-1, 4) = (0, 2)$. (ii) $(T_1 + T_2)(0, 1) = (2, 4)$. (iii) $(T_1 + T_2)(-1, 2) = (4, 6)$. (iv) $(T_1 + T_2)(1.5, -.5) = (-1, 1)$.
(b) $(T_1 + T_2)(x, y) = (u, v)$, where $\begin{bmatrix} u \\ v \end{bmatrix} = \begin{bmatrix} 1 & 0 \\ -2 & 3 \end{bmatrix} \begin{bmatrix} x \\ y \end{bmatrix} + \begin{bmatrix} -1 & 2 \\ 4 & 1 \end{bmatrix} \begin{bmatrix} x \\ y \end{bmatrix} = \begin{bmatrix} x \\ -2x + 3y \end{bmatrix} + \begin{bmatrix} -x + 2y \\ 4x + y \end{bmatrix} = \begin{bmatrix} 2y \\ 2x + 4y \end{bmatrix} = \begin{bmatrix} 0 & 2 \\ 2 & 4 \end{bmatrix} \begin{bmatrix} x \\ y \end{bmatrix} = \left(\begin{bmatrix} -1 & 0 \\ -2 & 3 \end{bmatrix} + \begin{bmatrix} -1 & 2 \\ 4 & 1 \end{bmatrix} \right) \begin{bmatrix} x \\ y \end{bmatrix}$. (For a proof in one step, apply the distributive law for matrices.)

Chapter 9

Exercises 9-1, page 389

1. (a) $M = \begin{bmatrix} .6 & .4 \\ .4 & .6 \end{bmatrix}$, so $M^3 = \begin{bmatrix} .504 & .496 \\ .496 & .504 \end{bmatrix}$.
(b) $p_{12}^{(3)} = .496$.

3. (a)
$$M = \begin{bmatrix} 1 & 0 & 0 & 0 \\ \frac{1}{4} & \frac{3}{4} & 0 & 0 \\ 0 & \frac{1}{2} & \frac{1}{2} & 0 \\ 0 & 0 & \frac{3}{4} & \frac{1}{4} \end{bmatrix}.$$

(b) $p_{30}^{(3)} = \frac{3}{32}$, $p_{31}^{(3)} = \frac{9}{16}$, $p_{32}^{(3)} = \frac{21}{64}$, $p_{33}^{(3)} = \frac{1}{64}$.

5. (a) (i) $\frac{60}{81}$, (ii) $\frac{1}{81}$, (iii) 0, (iv) $\frac{20}{81}$.

(b) From (a), (i) above, $p_{21}^{(4)} = \frac{60}{81} = .74074$. Further computations show that $p_{21}^{(5)} = p_{21}^{(6)} = .74897$, $p_{21}^{(7)} = p_{21}^{(8)} = .74989$, which indicates that the desired probability is .750.

Exercises 9-2, page 402

1. (a) $$Q_1 = \frac{(\frac{3}{2})^4 - (\frac{3}{2})}{(\frac{3}{2})^4 - 1} = \frac{3^4 - 3 \times 2^3}{3^4 - 2^4} = \frac{57}{65} = .877.$$ Similarly, $Q_2 = \frac{9}{13} = .692$, $Q_3 = \frac{27}{65} = .415$. These three-decimal-place approximations are as shown in M^{25}.

(b) $$Q_{90} = \frac{1 - (1/r)^{10}}{1 - (1/r)^{100}},$$ where $r = \frac{q}{p} = \frac{3}{2}$. A calculator reveals $$Q_{90} = \frac{1 - .01734}{1 - .00000} = .983.$$ Thus $P_{90} = .017$.

3. We use the formula for Q_k given in Prob. 1(b) with $n = 25$ and $1/r = p/q = .9$.

(a) $$Q_{20} = \frac{1 - (.9)^5}{1 - (.9)^{25}} = \frac{1 - .59049}{1 - .07179} = .441.$$ Similarly, (b) $Q_{15} = .702$. (c) $Q_{10} = .856$.

5. As suggested by the symmetry in the payoff matrix, the optimal column strategy is to play each column with probability $\frac{1}{3}$. Thus the probability of R winning a dollar on a play is $p = \frac{1}{3}$, so $q = \frac{2}{3}$, and $r = q/p = 2$. The goal is $n = 4$.

(a) $$Q_1 = \frac{r^4 - r^1}{r^4 - 1} = \frac{2^4 - 2^1}{2^4 - 1} = \frac{14}{15} = .933.$$

(b) $$Q_2 = \frac{2^4 - 2^2}{2^4 - 1} = \frac{12}{15} = .8.$$

(c) $$Q_3 = \frac{2^4 - 2^3}{2^4 - 1} = \frac{8}{15} = .533.$$

INDEX

D

E

F

G

Q

R

S

T

U

V

W

Z